Regulatory microRNA

Regulatory microRNA

Special Issue Editors

Y-h. Taguchi
Hsiuying Wang

MDPI • Basel • Beijing • Wuhan • Barcelona • Belgrade

MDPI

Special Issue Editors
Y-h. Taguchi
Chuo University
Japan

Hsiuying Wang
National Chiao Tung University
Taiwan

Editorial Office
MDPI
St. Alban-Anlage 66
4052 Basel, Switzerland

This is a reprint of articles from the Special Issue published online in the open access journal *Cells* (ISSN 2073-4409) from 2018 to 2019 (available at: https://www.mdpi.com/journal/cells/special_issues/regulatory_microRNA)

For citation purposes, cite each article independently as indicated on the article page online and as indicated below:

LastName, A.A.; LastName, B.B.; LastName, C.C. Article Title. *Journal Name* **Year**, *Article Number*, Page Range.

ISBN 978-3-03897-768-1 (Pbk)
ISBN 978-3-03897-769-8 (PDF)

Contents

About the Special Issue Editors

Y-h. Taguchi received his Dr. Sci. in Physics from Tokyo Institute of Technology, 1988. He started his scientific career as a computational physicist before moving on to the field of bioinformatics. His current research interests are in tensor decomposition-based unsupervised feature extraction, on which he has published dozens of papers, including several on principal component analysis-based unsupervised feature extraction. He is an author of more than one hundred peer-reviewed papers, including those published in Scientific Reports, PLoS ONE and BMC Bioinformatics. He also serves as an academic editor of multiple journals, including PLoS ONE, and as a guest editor of special issues included in journals published by MDPI. In September 2018 he received recognition as a Top Reviewer for Assorted* by Publons.

Hsiuying Wang received her PhD degree from the Institute of Statistics, National Tsing Hua University, Taiwan, 1996. The research interests of Dr. Wang include theoretical statistics, industrial statistics, bioinformatics and medical data analysis. She has published more than fifty peer-reviewed papers in the fields of statistics, bioinformatics and data analysis.

Preface to "Regulatory microRNA"

MicroRNA is one of the oldest functional non-coding RNAs. In spite of a long history of investigation, there are many remaining questions. Generally, microRNAs are believed to downregulate target mRNAs. There are numerous target mRNAs for individual microRNAs which can regulate a wide range of biological processes, for example, in disease, development and differentiation. Thus, they are often used as biomarkers and therapeutic targets. New functions and target mRNAs are continuously being characterized for miRNAs. Thus, reviewing updated information on how miRNAs regulate target mRNAs, diseases and various other biological processes is important. We hope to bring readers up to date with the latest developments in the understanding of microRNA regulation as they read the excellent research covered by papers that are included in this book.

Y-h. Taguchi, Hsiuying Wang
Special Issue Editors

Article

Circular RNA circHIPK3 Promotes the Proliferation and Differentiation of Chicken Myoblast Cells by Sponging miR-30a-3p

Biao Chen [1,2,3], Jiao Yu [1,2,3], Lijin Guo [1,2,3], Mary Shannon Byers [4], Zhijun Wang [1,2,3], Xiaolan Chen [1,2,3], Haiping Xu [1,2,3] and Qinghua Nie [1,2,3,*]

[1]   Department of Animal Genetics, Breeding and Reproduction, College of Animal Science, South China Agricultural University, Guangzhou 510642, China; biaochen@stu.scau.edu.cn (B.C.); 13539763630@163.com (J.Y.); guolijin2016@163.com (L.G.); zhijunwang@stu.scau.edu.cn (Z.W.); xiaolanchen@stu.scau.edu.cn (X.C.); music-three@163.com (H.X.)
[2]   National-Local Joint Engineering Research Center for Livestock Breeding, Guangzhou 510642, China
[3]   Guangdong Provincial Key Lab of Agro-Animal Genomics and Molecular Breeding, and Key Laboratory of Chicken Genetics, Breeding and Reproduction, Ministry of Agriculture, Guangzhou 510642, China
[4]   Department of Biological Sciences, College of Life and Physical Sciences, Tennessee State University, Nashville, TN 37209, USA; maryshannonbyers@yahoo.com
*   Correspondence: nqinghua@scau.edu.cn; Tel.: +86-139-2219-5759

Received: 27 January 2019; Accepted: 17 February 2019; Published: 19 February 2019

Abstract: Circular RNAs and microRNAs widely exist in various species and play crucial roles in multiple biological processes. It is essential to study their roles in myogenesis. In our previous sequencing data, both miR-30a-3p and circular HIPK3 (circHIPK3) RNA, which are produced by the third exon of the *HIPK3* gene, were differentially expressed among chicken skeletal muscles at 11 embryo age (E11), 16 embryo age (E16), and 1-day post-hatch (P1). Here, we investigated their potential roles in myogenesis. Proliferation experiment showed that miR-30a-3p could inhibit the proliferation of myoblast. Through dual-luciferase assay and Myosin heavy chain (MYHC) immunofluorescence, we found that miR-30a-3p could inhibit the differentiation of myoblast by binding to Myocyte Enhancer Factor 2 C (*MEF2C*), which could promote the differentiation of myoblast. Then, we found that circHIPK3 could act as a sponge of miR-30a-3p and exerted a counteractive effect of miR-30a-3p by promoting the proliferation and differentiation of myoblasts. Taking together, our data suggested that circHIPK3 could promote the chicken embryonic skeletal muscle development by sponging miR-30a-3p.

Keywords: circular RNA; circHIPK3; microRNA; miR-30a-3p; skeletal muscle; proliferation; differentiation

1. Introduction

Skeletal muscles are important components in animals. Chicken skeletal muscle, which can provide high quality protein, is one of the most important meat source for humans. The development of skeletal muscle is regulated by multiple factors, including genetics, nutrition, disease, environment and so on [1,2]. Heritability estimates showed that chicken growth could be enhanced by genetic improvement [3]. The genetic factors which control skeletal muscle development include genes and non-coding RNAs.

MicroRNAs (miRNAs) have been shown to be involved in many biological processes, including muscle development [4]. Some myogenic miRNAs, including the miR-1 family, miR-206 and miR-133 family, regulate muscle development by targeting and inhibiting the expression of muscle-related gene [5,6]. Previous studies in our group showed that miR-203, miR-16, miR-29, and miR-1611 all played crucial roles in myoblast proliferation and differentiation [7–10]. Two other studies found that miR-30a-3p could suppress tumor growth [11,12]. However, the molecular function of chicken miR-30a-3p has not yet been reported.

Circular RNAs, which widely exist in the transcriptomes of different species and tissues, were previously considered as a kind of non-coding RNA, but they have now been demonstrated to have both coding and regulating functions [13–15]. Circular RNA, formed by the covalently joined 5′ and 3′ ends of linear RNA, possess a more stable structure than linear RNA. The functions of circular RNA include, acting as a miRNA sponge, participating in regulating the expression of its own linear RNA in different ways, coding protein, and deriving pseudogenes [16,17]. Previous studies found that circular RNAs were abundantly expressed in skeletal muscle tissue in many species [14,18,19]. Circular RNAs in chicken skeletal muscle could act as an miRNA sponge and regulate chicken muscle development [19]. A circular RNA, produced by *SVIL* could promote the proliferation and differentiation of myoblast cells by sponging miR-203 [20]. Circular RNA circFGFR2, generated by the *FGFR2* gene, could interact with miR-133a-5p and miR-29b-1-5p to regulate myoblast cells development [7].

A circular RNA produced by the third exon of the chicken *HIPK3* gene (circHIPK3—01, we named it as circHIPK3, hereinafter) has the highest expression level compare to other circular RNAs generated from *HIPK3* gene. It was also differentially expressed in different stages of skeletal development. We predicted it has three potential binding sites for miR-30a-3p. In this study, we aimed to examine the interaction of circHIPK3 and miR-30a-3p and their functions on myoblast proliferation and differentiation.

2. Materials and Methods

2.1. Ethics Statement

All animal experiments performed in this study met the requirements of the Institutional Animal Care and Use Committee at the South China Agricultural University (approval ID: SCAU#0014). All efforts were made to minimize the suffering of animals.

2.2. Primers

All primers that were used in this study were synthesized by Sangon (Sangon Biotech, Shanghai, China). The primers listed in Table 1 were designed by Premier Primer 5.0 software (Premier Bio-soft International, Palo Alto, CA, USA). Information of the qRT-PCR primers for *MYOD*, *MYOG*, *MYHC* and *GAPDH* were shown in our previous study [21].

2.3. Cell Culture and Transfection

Chicken primary myoblasts (CPMs) were isolated from the leg muscle of 10-day Yuhe chicken embryos (E10; Zhuhai Yuhe Company Ltd., Zhuhai, China), as described in our previous study [7]. Briefly, the legs of E10 chickens were collected, and the skin and bones were removed. Then, leg muscles were minced with scissors and trypsinized (Gibco, Grand Island, NY, USA) at 37 °C for 20 min). Digestion was done with complete 1640 medium-(RPMI), containing 20% fetal bovine serum (FBS), 1% nonessential amino acids, and 0.2% penicillin/streptomycin (Invitrogen, Carlsbad, CA, USA). The mixture was filtered and centrifuged at 500 g for 5 min. Following the serial plating, the cells were cultured in complete medium and incubated at 37 °C, in a 5% CO_2 humidified atmosphere. Chicken fibroblast DF-1 cells were cultured in Dulbecco's modified Eagle medium (DMEM) (Gibco, Grand Island, NY, USA) supplemented with 10% FBS and 0.2% penicillin/streptomycin (Invitrogen, Carlsbad, CA, USA), then incubated with 5% CO_2 at 37 °C humidity. DNA plasmids, miRNA mimic, mimic negative control (mimic NC), miRNA inhibitors, inhibitor negative control (inhibitor NC), small interfering RNA (siRNA), and siRNA negative control (siRNA NC) were transiently transfected into cells using Lipofectamine 3000 reagent (Invitrogen, Carlsbad, CA, USA).

2.4. RNA Exaction, cDNA Synthesis and Quantitative Real-Time PCR (qRT-PCR)

All RNAs were exacted using Trizol reagent (TaKaRa, Otsu, Japan) according to the manufacturer's instructions. The quality and concentrations of the RNA samples were detected by 1.5% agarose gel electrophoresis. Total RNA was employed to synthesize cDNA, using a Primescript RT Reagent Kit with gDNA Eraser (Perfect Real Time) (TaKaRa, Otsu, Japan). Synthesized cDNA libraries were diluted with RNase-free water at a ratio of 1:3 before real-time PCR. Relative mRNA expression levels were detected by qRT-PCR using SsoFast Eva Green Supermix (Bio-Rad, Hercules, CA, USA). *GAPDH* was used as an internal control. Reverse transcription for miRNA was conducted using ReverTra Ace qPCR RT Kit (Toyobo, Osaka, Japan). The specific bulge-loop miRNA qRT-PCR primer for miR-30a-3p and U6 were designed by RiboBio (RiboBio, Guangzhou, China). All qRT-PCR reactions were conducted with a CFX96 system (Bio-Rad, Hercules, CA, USA). All reactions were run in triplicates and fold expression changes were calculated using the comparative $2^{-\Delta\Delta Ct}$ method.

2.5. Validation of circHIPK3

Based on the NCBI reference sequences of *HIPK3* (NCBI accession number: NM_001199411.1), convergent and divergent primers were designed to validate the existence of circHIPK3. To confirm the cirHIPK3 junction, genomic DNA, and cDNA from CPMs were used for PCR reaction. All PCR products were sequenced by Sangon Biotech Co Ltd. Sequence analysis was conducted using DNASTAR software (DNASTAR 7.1, http://www.dnastar.com). For RNase R treatment, 2 mg of total RNA was incubated 20 min at 37 °C with RNase R (Epicentre Technologies, Madison, WI, USA), and employed to synthesize cDNA for qPCR. For the control group, the same amount of RNA was incubated 20 min at 37 °C and subsequently used to synthesize cDNA.

2.6. Plasmids Construction and RNA Oligonucleotides

For the construction of the circHIPK3 over-expression vector, exon 3 of *HIPK3* was amplified using cDNA, produced from CPMs and cloned into a pCD5ciR vector (Geneseed Biotech, Guangzhou, China) between *EcoRI* and *BamHI* restriction sites. The siRNAs to circHIPK3, which especially target the circHIPK3 rather than the linear HIPK3, were designed and synthesized by Geneseed using the sequence shown in Table 1. The gga-miR-30a-3p mimic, mimic NC, the gga-miR-30a-3p inhibitor and inhibitor NC were synthesized by RiboBio (Guangzhou, China). For the construction of pmirGLO Dual-Luciferase reporter vector, wild-type and mutated sequences in the 3'UTR region of *MEF2C* and the partial region of circHIPK3, which include the predicted binding sites of miR-30a-3p, were synthesized and inserted into pmirGLO vectors (Promega, Madison, WI, USA), according to instructions, using *NheI* and *XhoI* restriction sites. The gga-miR-30a sequence was also synthesized and inserted into pmirGLO vectors.

2.7. 5-Ethynyl-2′-Deoxyuridine (EdU) Assay

After 48 h of transfection, the treated CPMs and negative control groups in 24-well plates were incubated with 50 μM 5-ethynyl-20-deoxyuridine (RiboBio, Guangzhou, China) for 2 h at 37 °C. After washing twice, the cells were stained with C10310 EdU Apollo. EdU-stained cells were counted using a Leica DMi8 fluorescent microscope (400× magnification) (Leica, Wetzlar, Germany). The ratio of EDU-stained cells to Hoechst 33342-stained cells was calculated and represented the CPM proliferation rate. Detailed protocols were described in the manufacturer's instruction.

2.8. Flow Cytometry of the Cell Cycle

After 48 h of transient transfection with the over-expression plasmid (blank vector) and siRNA (siRNA NC), CPMs were collected from the 12-well plates and kept overnight in 70% ethanol at −20 °C. The cells were then incubated with 50 μg/mL PI (propidium iodide) (Sigma, Louis, MO, USA), 10 μg/mL RNase A (Takara, Otsu, Japan) and 0.2% (*v/v*) Triton X-100 (Sigma, Louis, MO, USA) at 4 °C for 30 min. Lastly, cells were detected with a BD AccuriC6 flow cytometer (BD Biosciences, San Jose, CA, USA), and the results were analyzed by FlowJo7.6 software.

2.9. Cell Counting Kit 8 (CCK-8) Assay

CPMs were seeded in a 48-well plate and cultured in complete medium. After transfection, cell proliferation was detected at 12, 24, 36, and 48 h using the TransDetect CCK Kit (TransGen Biotech, Beijing, China), following the manufacturer's protocol. Cells were added in 25 uL CCK solution to each well and incubated for 2 h at 37 °C in a 5% CO_2 cell incubator. Then absorbance of treated and control groups were measured with a Fluorescence/Multi-Detection Microplate Reader (BioTek, Winooski, VT, USA) by optical density at a wavelength of 450 nm.

2.10. Immunofluorescence

For immunofluorescence, after transfection, cells in 12-well plates were fixed for 30 min with 4% formaldehyde. Cells were then permeabilized by adding 0.1% Triton X-100 for 5 min and blocked for 30 min with goat serum. Following overnight incubation at 4 °C with anti-MYHC (B103; DHSB, Iowa City, IA, USA; 0.5 µg/mL), Fluorescein (FITC)-conjugated AffiniPure Goat Anti-Mouse IgG (H + L) (Bioworld, Minneapolis, MN, USA; 1:200) was added to the plate and incubated at room temperature for 1 h. Cell nuclei were stained with DAPI (1:50, Beyotime, Shanghai, China) for 5 min. The images were captured with fluorescence microscopy (Leica, Wetzlar, Germany). The area of cells labeled with anti-MYHC was measured using Photoshop software (Adobe Photoshop CC 2018, Adobe, San Jose, CA, USA), and the total myotube area was calculated as a percentage of the total image area covered by myotubes.

2.11. Binding Relationship Prediction and Dual-Luciferase Reporter Assay

To predict the relationship between target genes and miR-30a-3p, miRDB (http://mirdb.org/miRDB/) and RNAhybrid (http://bibiserv2.cebitec.uni-bielefeld.de/rnahybrid) were employed. After seeding DF-1 cells in the 96-well plate and culturing for 24 h, four groups (wild type pmirGLO plasmids and mimic as the treatment, mutated pmirGLO plasmids and mimic, wild type pmirGLO plasmids and mimic NC, mutated type pmirGLO plasmids and mimic NC) were set and co-transfected. For the confirmation of the target relationship between circHIPK3 and miR-30a-3p, another method of Dual-Luciferase reporter assay was employed. Three groups (circHIPK3 over-expression plasmid and miR-30a-3p mimic, pCD5ciR and miR-30a-3p mimic, pCD5ciR and mimic NC) were set and co-transfected with a pmirGLO vector containing the miR-30a sequence. After 48 h, Dual-GLO Luciferase Assay System kit (Promega, Madison, WI, USA) was employed to detect luminescent signals of firefly and Renilla Luciferase with a Fluorescence/Multi-Detection Microplate Reader (BioTek, Winooski, VT, USA). Firefly luciferase activities were normalized to Renilla luminescence in each well. Detailed protocols were described in the manufacturer's instruction.

2.12. Western Blotting

Briefly, cells were lysed in the radio immune precipitation assay (RIPA) buffer (Beyotime, Shanghai, China) containing phenylmethane sulfonyl fluoride (PMSF) protease inhibitor (Beyotime, Shanghai, China). After incubation on ice for 30 min, the samples were centrifuged at 10,000 g for 10 min at 4 °C, and the supernatant was collected. Proteins were separated by SDS-PAGE and blotted onto nitrocellulose membranes (Whatman, Maidstone, UK), then membranes were probed with primary and secondary antibodies. The primary antibodies used were anti-MYHC (1:1000, B103; DHSB, Iowa City, IA, USA), anti-GAPDH (1:1500, AB-P-R 001, Hangzhou Goodhere Biotech, Hangzhou, China), and anti-Tubulin (1:1000, Beyotime, Shanghai, China). The secondary antibodies used were goat anti-rabbit IgG-HRP (1:5000, BA1054, Boster, Wuhan, China) and peroxidase-goat anti-mouse IgG (1:2500, BA1051, Boster, Wuhan, China). Image J software (d1.47, National Institutes of Health, Bethesda, MD, USA) was used to quantify the band intensity.

2.13. Statistical Analysis

All results were presented as a mean ± SEM and were subjected to statistical analysis by two-tailed *t*-test. The level of significance was presented as * ($p < 0.05$), ** ($p < 0.01$) and *** ($p < 0.001$).

Table 1. Primers and RNA oligos used in this study.

Name	Nucleotide Sequences (5′→3′)	Tm. (°C)	Product Size (bp)	Application
QcircHIPK3	F: GTTTAATCCACGCTGACCTCA R: GACTTGTGAGGCCATACCTATA	61.3	130	qPCR for circHIPK3
QHIPK3	F: GGGGTATGTCCCGGAG R: CTTCGCTAATGGAACAACAC	61.3	261	qPCR for HIPK3
QMEF2C	F: AGGGTGTATGTGCAGGAACG R: AGCAATCTCGCAGTCACACA	60	288	qPCR for MEF2C
Convergent primers	F: TGGTACAAGCGGAGATGG R: TTGAGGTCAGCGTGGATTA	55	450	Amplification of partial sequence of exon 3 of HIPK3
Divergent primers	F: GCACGCCAAGGACAAATA R: TACGCTTCAATCCACATCG	58	782	Amplification of partial sequence of circHIPK3 which contain the joint site
β-actin	F: CTCCCCCATGCCATCCTCCGTCTG R: GCTGTGGCCATCTCCTGCTC	52–65	179	qPCR for β-actin
si-circHIPK3-001	CCCGGTATTATAGGTATGG	-	-	-
si-circHIPK3-002	GGTATTATAGGTATGGCCT	-	-	-
si-circHIPK3-003	ATTATAGGTATGGCCTCAC	-	-	-

Note: The nucleotide sequences of si-circHIPK3 represent the target sequences of each siRNA.

3. Results

3.1. circHIPK3 Differentially Expressed during Embryonic Leg Muscle Development

Previous circular RNA sequencing data from our lab revealed 11 circular RNAs were generated by the *HIPK3* gene (available in the Gene Expression Omnibus with accession number GSE89355). The genomic structure of chicken *HIPK3* and the regions, in which all the circular HIPK3 (circHIPK3) RNA were derived, are shown in Figure 1A. Interestingly, circHIPK3 (referred as circHIPK3—01 in Figure 1A), which was derived from exon3 of *HIPK3*, was the only exonic circular RNA. Compared with other circular RNAs derived from *HIPK3*, circHIPK3 had the highest expression level. Its expression level in E16 was significantly higher than in E11 and P1 (Figure 1B). The expression levels of circHIPK3 and *HIPK3* mRNA in E11, E12, E16, and E18 were detected by qRT-PCR (Figure 1C). The trend of the expression level of circHIPK3 was consistent with the result from the sequencing data. However, the expression patterns of circHIPK3 and *HIPK3* mRNA were not identical, which indicated that they might have different functions during leg muscle development. To confirm the sequence and the junction of circHIPK3, genomic DNA and cDNA were used for the PCR reaction, with convergent and divergent primers. The result of the PCR product electrophoresis showed the expectants of convergent primers were amplified with both templates. However, there was no PCR product of divergent primers with the genomic DNA template (Figure 1D). PCR products of divergent primers were analyzed by Sanger sequencing (Figure 1E). Sequencing results showed that circHIPK3 was generated from the third exon of *HIPK3*. The circHIPK3 was also validated by RNase R digestion. The result of qRT-PCR showed that RNase R had no impact on circHIPK3, whereas the levels of linear RNA, HIPK3 and β-actin, were significantly decreased (Figure 1F). These results validated the existence and differential expression of circHIPK3 during skeletal muscle development of chicken.

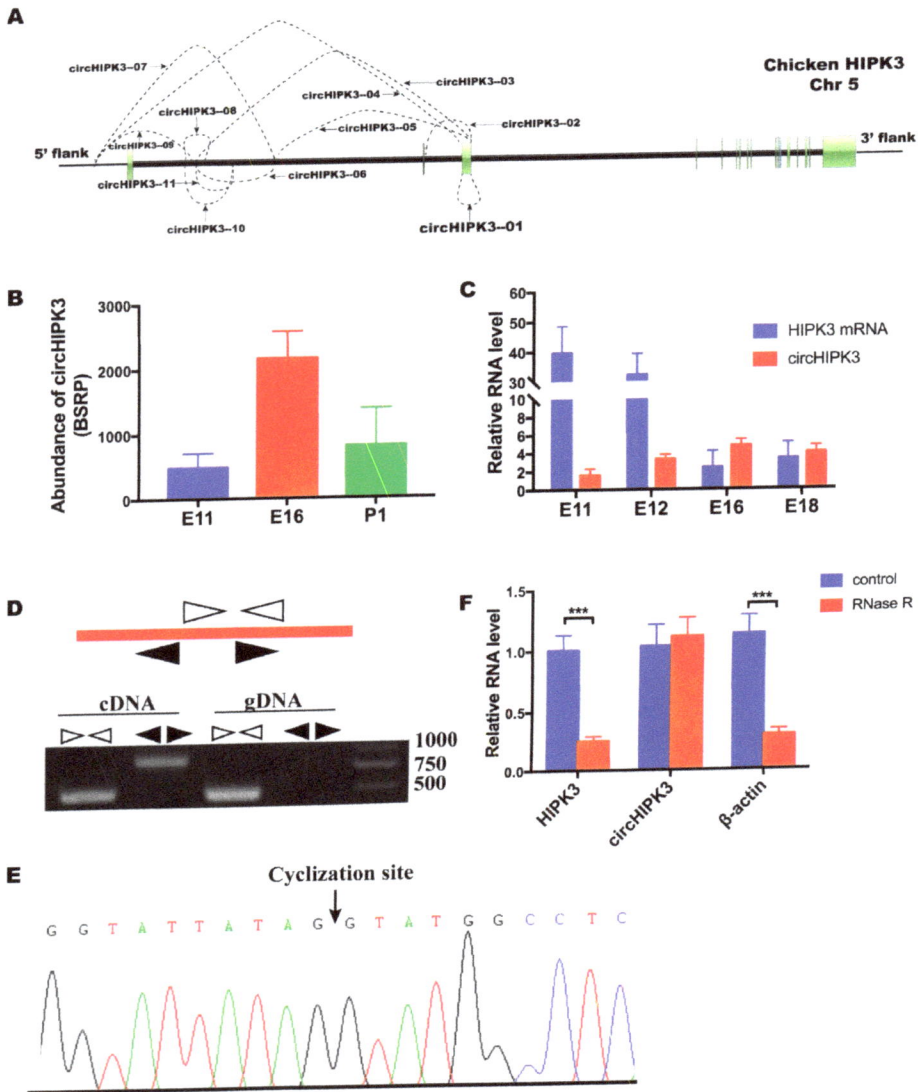

Figure 1. The differential expression and validation of circular HIPK3 (circHIPK3). (**A**) The schema of all circular RNA derived from *HIPK3*. The green rectangles represent the exons of *HIPK3*. (**B**) The RNA-Seq result showed that circHIPK3 was differentially expressed in E11, E16, and P1 of leg muscle. The expressed abundances were normalized as the number of back-spliced reads per million mapped reads (BSRP). (**C**) The expression profiles of circHIPK3 and *HIPK3* mRNA in E11, E12, E16 and E18. (**D**) Divergent primers amplified circHIPK3 in cDNA but not genomic DNA (gDNA). White triangles represent convergent primers and black triangles represent divergent primers. (**E**) Sanger sequencing confirmed the junction sequence of circHIPK3. (**F**) Quantitative real-time PCR (qRT-PCR) showed the resistance of circHIPK3 to RNase R digestion. In all panels, values represent mean ± SEM from three independent experiments. * $p < 0.05$; ** $p < 0.01$; *** $p < 0.001$.

3.2. circHIPK3 Interacts with miR-30a-3p

Many studies showed that circular RNAs exerted their functions by acting as the miRNA sponge. CircHIPK3 was predicted by miRDB and RNAhybrid to be a target of multiple miRNAs. Among these miRNAs, ggs-miR-30a-3p was chosen as a candidate because there were three potential binding sites in circHIPK3 (Figure 2A). The seed sequence of miR-30a-3p matched with three sites in circHIPK3 (Figure 2B). Besides, the prediction results from RNAhybrid indicated that the binding site 2 was the most stable format (Figure 2C). To identify the interactions between circHIPK3 and miR-30a-3p, the over-expression vector of circHIPK3 was constructed and transfected into DF-1 cells. The expression efficiency of the over-expression vector was detected by qPCR. Compared with the group transfected with pCD5ciR, the circHIPK3 over-expression vector expressed a higher level of circHIPK3 (Figure 2D). Then, circHIPK3 over-expression vector and miR-30a-3p mimic were co-transfected into DF-1 cells with a pmirGLO vector, containing the miR-30a sequence. Meanwhile, as the control group, the pCD5ciR and miR-30a-3p mimic (or mimic NC) were co-transfected with the pmirGLO vector, containing the miR-30a sequence. The results showed that the relative luminescence activity of the group with circHIPK3 over-expression vector and miR-30a-3p mimic was significantly higher than the group with pCD5ciR and miR-30a-3p mimic, but had no difference compared with the group pCD5ciR and mimic NC (Figure 2E). These results suggest that circHIPK3 could bind with miR-30a-3p mimic. In addition, sequences which contained binding site 2 or the mutated sequence were inserted into pmirGLO vector. Recombinant vectors with the wild type sequence was then co-transfected into DF-1 cells with miR-30a-3p mimic, meantime, three control groups were set (pmirGLO vector with mutated sequence and miR-30a-3p mimic, pmirGLO vector with wild type sequence and mimic NC, pmirGLO vector with mutated sequence and mimic NC). The results showed that the relative luminescence activity of the group with a wild-type plasmid and mimic was significantly decreased compared to the group transfected with mutated plasmid and mimic, and the group with wild type plasmids and mimic NC (Figure 2F). Moreover, the RNA level of circHIPK3 was significantly down-regulated after over-expression of miR-30a-3p mimic, compared to the group transfected with mimic NC (Figure 2G). Subsequently, the result of flow cytometry analysis showed that miR-30a-3p could reverse the effect of circHIPK3 on a cell cycle (Figure 2H). Altogether, these results indicated that miR-30a-3p could interact with circHIPK3.

3.3. miR-30a-3p Inhibits Myoblast Proliferation

To explore the function of miR-30a-3p on the proliferation of CPMs, miR-30a-3p mimic and inhibitor were transfected into CPMs with 100 nM to detect an over-expression effect and an inhibitory effect. The results showed that the two oligos of miR-30a-3p had the effect as expected compared with the mimic NC group, and inhibitor NC group, respectively, and could be used in the subsequent experiments (Figure 3A,B). After being transfected with miR-30a-3p mimic/mimic NC and miR-30a-3p inhibitor/inhibitor NC, flow cytometry analysis was performed in CPMs and the results showed that ectopic expression of miR-30a-3p suppressed the cell cycle markedly, while knock-down of miR-30a-3p significantly promoted the cell cycle (Figure 3C,D). Besides, CCK-8 assay was conducted to detect of proliferation vitality in CPMs. The results showed that the group which transfected with miR-30a-3p mimic had a lower proliferation vitality than mimic NC. In contrast, the group which transfected with miR-30a-3p inhibitor had a higher proliferation vitality than inhibitor NC group (Figure 3E,F). Furthermore, the EdU assay demonstrated that the rate of the cells, which were in the cell division in the ectopic expression group, was significantly less than in the mimic NC group, and the statistics of the cell proliferation rate of the miR-30a-3p over-expression group, were markedly lower than the control group (Figure 3G). Conversely, knock-down of miR-30a-3p dramatically increased the numbers of EdU strained cells compare with the inhibitor NC group (Figure 3H). Altogether, these results indicated that miR-30a-3p could suppress myoblast proliferation.

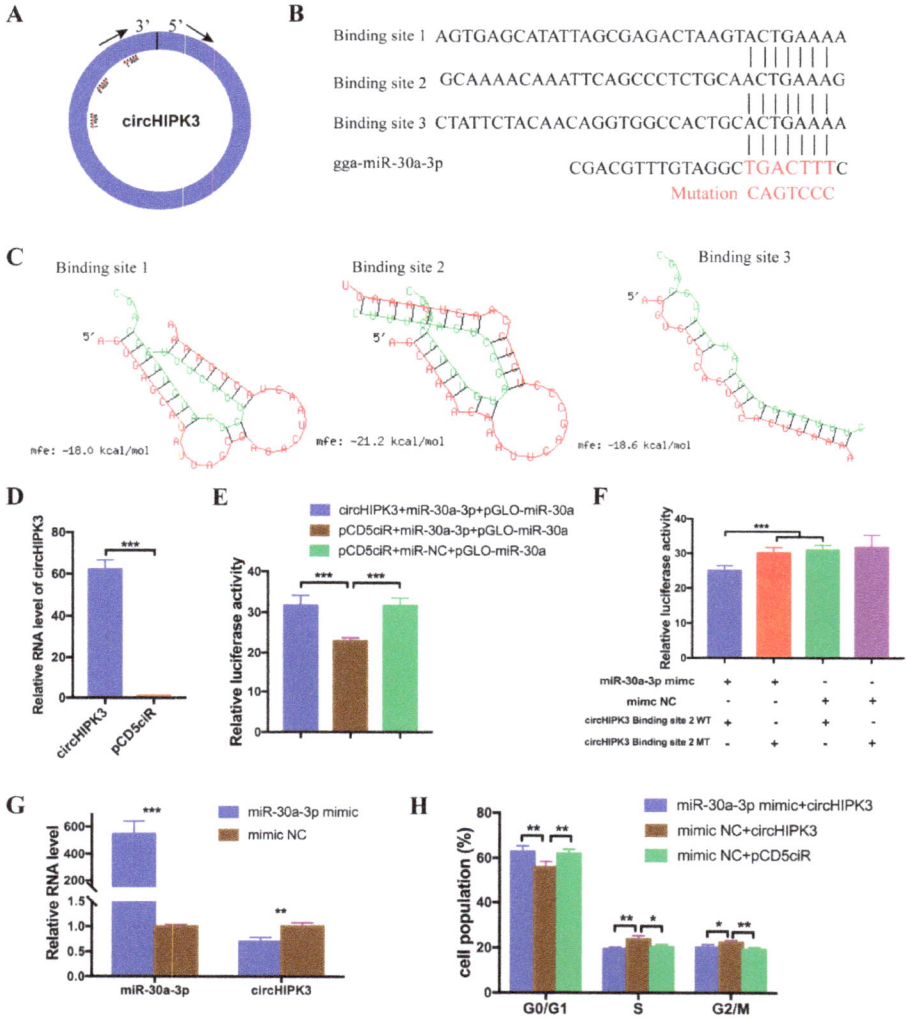

Figure 2. CircHIPK3 interacts with miR-30a-3p. (**A**) A schematic illustration showing the putative binding sites of miR-30a-3p on circHIPK3. (**B**) The potential binding site sequence of miR-30a-3p on circHIPK3. The seed sequences and mutant sequences were highlighted in red. (**C**) The potential interaction model between miR-30a-3p and circHIPK3 from RNAhybrid. (**D**) The expression efficiency of circHIPK3 over-expression vector in DF-1 cells. (**E**) Luminescence was measured after co-transfected with the luciferase reporter and miR-30a-3p mimic (or mimic NC) and circHIPK3 over-expression vector (or pCD5ciR). The relative levels of firefly luminescence normalized to Renilla luminescence are plotted. (n = 6). (**F**) Luminescence was measured after co-transfecting wild type or mutant linear sequence of circHIPK3 with miR-30a-3p mimic (or mimic NC) in DF-1 cells. (n = 6). (**G**) The RNA levels of miR-30a-3p and circHIPK3 from miR-30a-3p mimic transfected DF-1 cells. (**H**) The effect of co-transfected with miR-30a-3p mimic (or mimic NC) and circHIPK3 over-expression vector (or pCD5ciR) on cell-cycle progression of DF-1 cells. The plot of cell-cycle analysis in different cell-cycle phases was compared. In all panels, values represent mean ± SEM from three independent experiments. * $p < 0.05$; ** $p < 0.01$; *** $p < 0.001$.

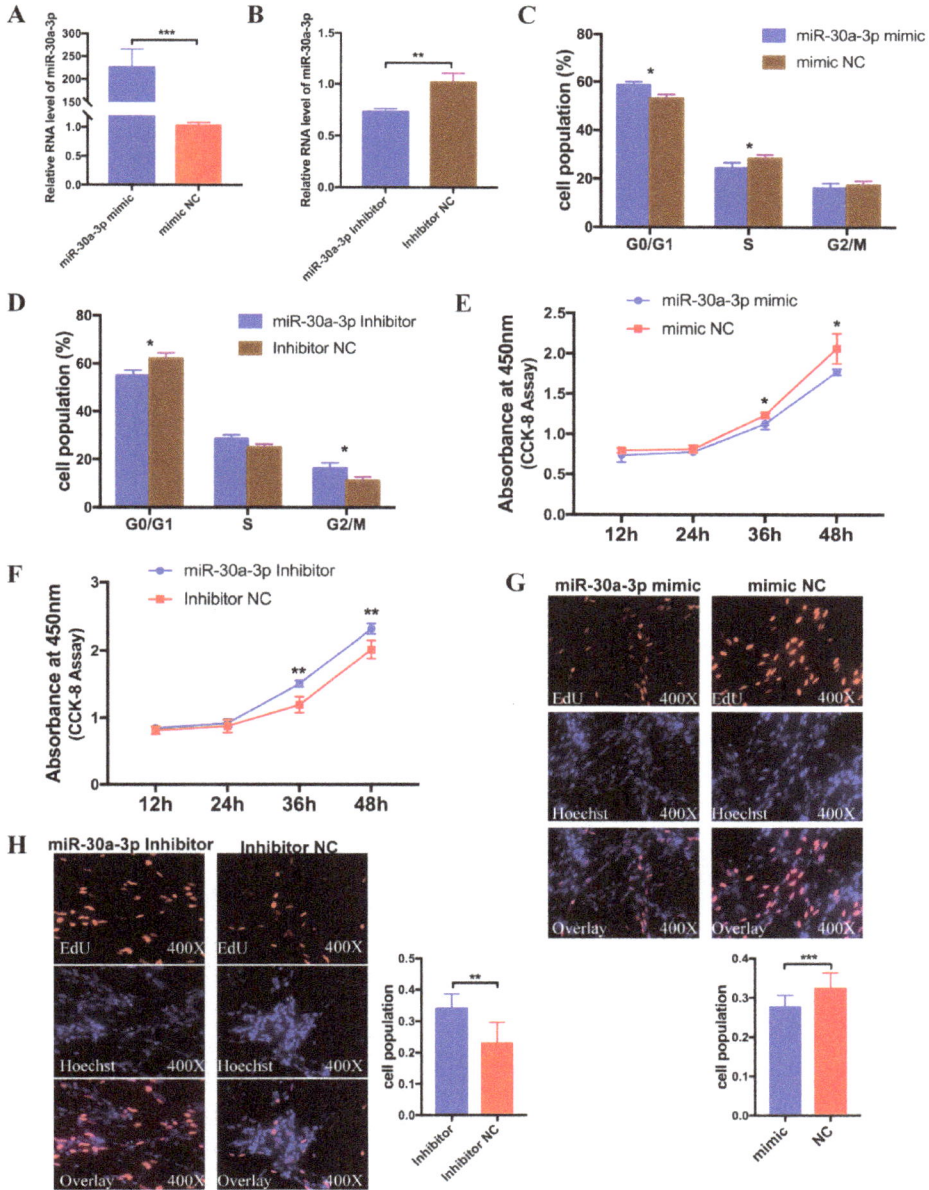

Figure 3. miR-30a-3p inhibits myoblast proliferation. (**A,B**) The over-expression and inhibitory effects of miR-30a-3p mimic and inhibitor in CPMs. (**C,D**) Effect of miR-30a-3p mimic and inhibitor on cell-cycle progression of chicken primary myoblasts (CPMs). The plot of cell-cycle analysis in different cell-cycle phases was compared. (**E,F**) The growth curves of CPMs were measured after the transfection of miR-30a-3p mimic and inhibitor. (**G,H**) 5-Ethynyl-2′-Deoxyuridine (EdU) assays for CPMs with over-expression and inhibition of miR-30a-3p. In all panels, values represent mean ± SEM from three independent experiments. * $p < 0.05$; ** $p < 0.01$; *** $p < 0.001$.

3.4. MEF2C Is a Target Gene of miR-30a-3p

To investigate the potential function of miR-30a-3p on CPM differentiation, we try to find the differentiation-related genes among the targets of miR-30a-3p. Interestingly, there are three potential binding sites in the 3′UTR region of *MEF2C* (Figure 4A). The potential interaction model from RNAhybrid showed that the binding site 3 was the most stable format (Figure 4B). To validate the target relationship between *MEF2C* and miR-30a-3p, wild-type and mutated-type sequences containing three binding sites, separately, were inserted into the pmirGLO vector for construction of pmirGLO dual-luciferase miRNA target expression vector. After co-transfection of vectors and mimics (or mimic NC), relative luminescence activities were detected by a Fluorescence/Multi-Detection Microplate Reader. For the Binding site 1, the relative luminescence activity of the group, with wild type reporter and mimic, was significantly lower compared with the control groups, which transfected with mutated reporters and mimic, wild type reporter and mimic NC, separately (Figure 4C). For the Binding site 2 and Binding site 3, the luminescence activity of the groups with wild type plasmids and mimic were all dramatically lower than the three control groups (mutated reporters and mimic, wild type reporter and mimic NC, mutated reporters and mimic NC) (Figure 4D,E). Particularly, the binding site 3 exerted the most significant interaction with miR-30a-3p compare with the control groups. The RNA expression level of *MEF2C* was significantly decreased after ectopic expression of miR-30a-3p mimic compared with the group transfected with mimic NC (Figure 4F).

3.5. miR-30a-3p Represses CPM Differentiation

After the confirmation of the target relationship between miR-30a-3p and *MEF2C*, we try to investigate the potential role of miR-30a-3p on CPM differentiation. First, the expression profile of miR-30a-3p was detected in the process of differentiation in CPMs. Interestingly, the expression level of miR-30a-3p was decreased in the first two days of differentiation medium (DM) compare to GM (growth medium), then increased in DM3 and DM4 (Figure 5A). Then, the expression of the myoblast differentiation marker genes, including *MYOD*, *MYOG*, and *MYHC* were evaluated by qPCR in myoblast transfected with miR-30a-3p mimic (or mimic NC) and inhibitor (or inhibitor NC). Over-expression of miR-30a-3p notably inhibited the expression of *MYOD*, *MYOG*, and *MYHC* compared with the groups transfected with mimic NC. Conversely, knock-down of miR-30a-3p promoted the expression of *MYOD*, *MYOG* and *MYHC* relative to the inhibitor NC group (Figure 5B,C). The protein level of *MYHC* was also detected by western blotting. Ectopic expression of miR-30a-3p inhibited the expression *MYHC*, conversely, knock-down of miR-30a-3p promoted the protein expression of *MYHC* (Figure 5D). MYHC immunofluorescence staining was employed on those transfected differentiated myoblasts at DM5. The results showed that the total area of myotubes of miR-30a-3p mimic transfected group was markedly less than that of the group transfected with mimic NC (Figure 5E). On the contrary, the areas of myotubes in miR-30a-3p inhibitor transfected group was more than that of control group (Figure 5F). To sum up, these results revealed that miR-30a-3p could repress CPM differentiation.

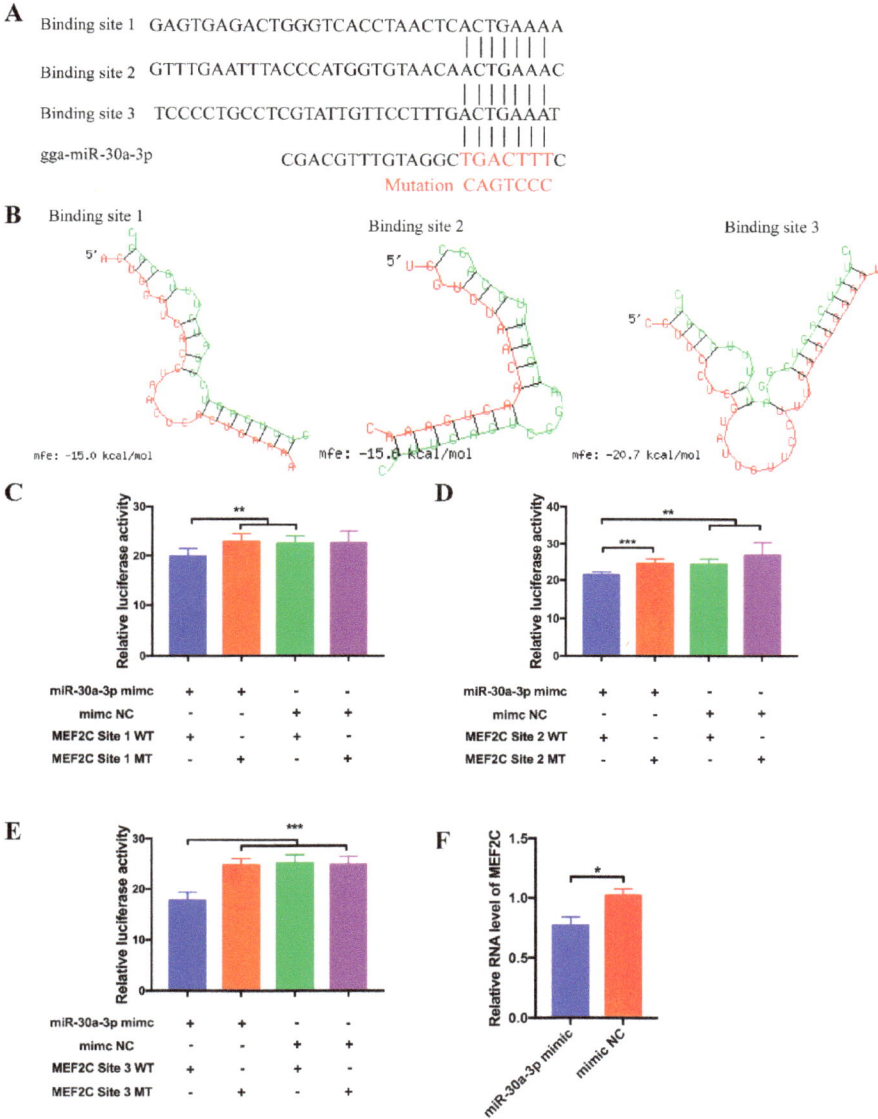

Figure 4. *MEF2C* is a target gene of miR-30a-3p. (**A**) The potential binding site sequence of miR-30a-3p on MEF2C. The seed sequences and mutant sequences were highlighted in red. (**B**) The potential interaction model between miR-30a-3p and MEF2C from RNAhybrid. (**C**–**E**) Luminescence was measured after co-transfecting wild type or mutant sequence of *MEF2C* with miR-30a-3p mimic (or mimic NC) in DF-1 cells. (n = 6). (**F**) The RNA level of *MEF2C* from miR-30a-3p mimic transfected CPMs. In all panels, values represent mean ± SEM from three independent experiments. * $p < 0.05$; ** $p < 0.01$; *** $p < 0.001$.

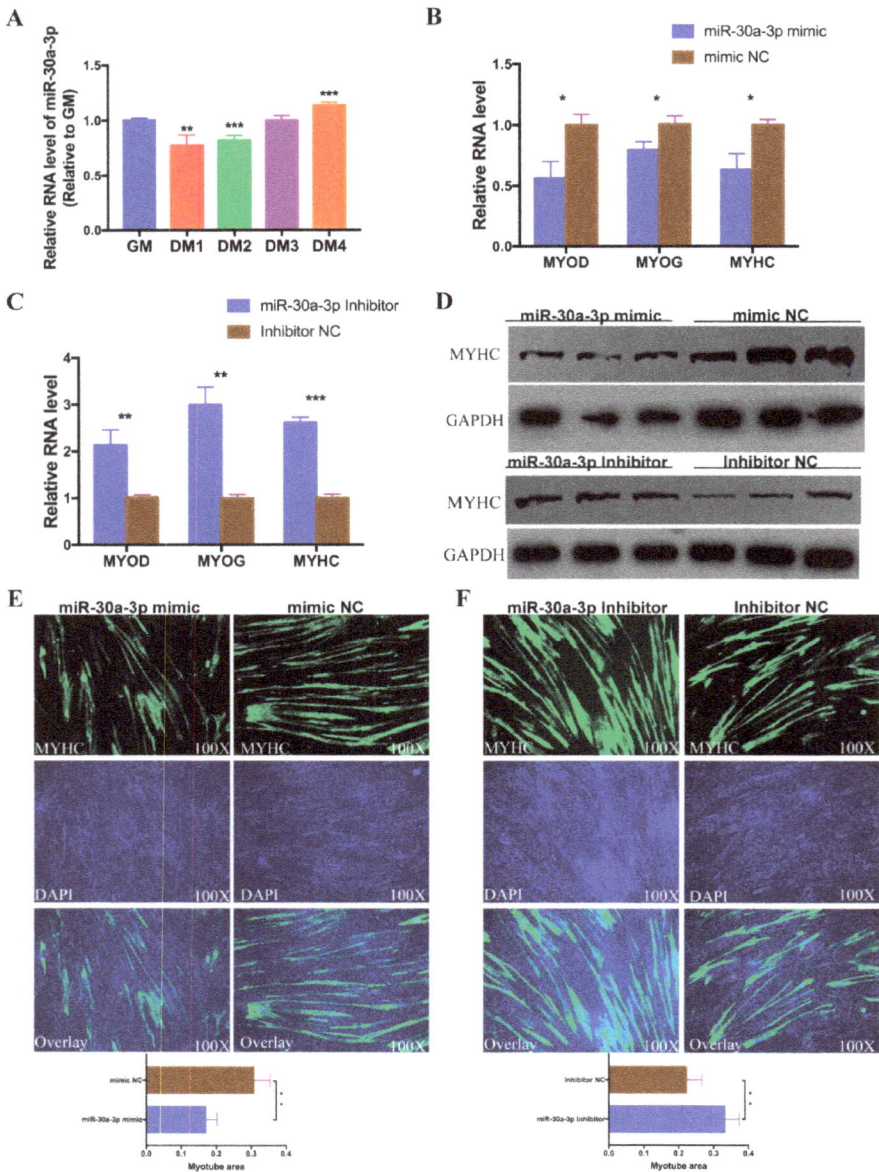

Figure 5. miR-30a-3p represses CPM differentiation. (**A**) The expression profile of miR-30a-3p in the process of CPMs induced differentiation. (**B**) The expression of *MYOD*, *MYOG* and *MYHC* in CPMs after over-expression of miR-30a-3p. (**C**) The expression of *MYOD*, *MYOG* and *MYHC* in CPMs after knock-down of miR-30a-3p. (**D**) The protein level of *MYHC* in CPMs after over-expression and knock-down of miR-30a-3p. (**E**) Immunofluorescence analysis of MYHC-staining cells after over-expression miR-30a-3p in CPMs. (**F**) Immunofluorescence analysis of MYHC-staining cells after knock-down of miR-30a-3p in CPMs. * $p < 0.05$; ** $p < 0.01$; *** $p < 0.001$.

3.6. circHIPK3 Promotes the Proliferation of CPMs

Given that circHIPK3 could act as a sponge of miR-30a-3p, and miR-30a-3p could repress the proliferation of CPMs, we hypothesized that circHIPK3 played an opposite role on the proliferation of CPMs. To investigate the role of circHIPK3 on skeletal muscle cell proliferation, the over-expression vector and siRNAs were transfected into CPMs. The over-expression vectors expressed high levels of circHIPK3 compared with the group transfected with pCD5ciR (Figure 6A), and all three siRNAs could significantly knock down the level of circHIPK3 relative to siRNA NC group (Figure 6B). Among the three siRNAs, si-circHIPK3-003 had the highest efficiency of interference effect and was chosen for the following experiments. After being transfected with circHIPK3 over-expression vector/pCD5ciR and si-circHIPK3-003/siRNA NC, flow cytometry analysis in CPMs was performed and the results showed that ectopic expression of circHIPK3 notably promoted the cell cycle, conversely, knock-down of circHIPK3 significantly retarded the cell cycle (Figure 6C,D). Moreover, CCK-8 assay was used to detect the proliferation vitality in CPMs. The results showed that the groups which transfected with circHIPK3 over-expression vectors, had higher proliferation vitality than the negative control group. In contrast, the groups which transfected with si-circHIPK3-003, had lower proliferation vitality than siRNA NC group (Figure 6E,F). The EdU assay indicated that the rate of the cells which were in the cell division in circHIPK3 over-expression group was significantly more than pCD5ciR group, the statistics of the cell proliferation rate of the circHIPK3 over-expression group were markedly higher (Figure 6G). Conversely, knock-down of circHIPK3 decreased the numbers of EdU strained cells dramatically (Figure 6H). In summary, these results demonstrated that circHIPK3 promoted the proliferation of CPMs.

3.7. circHIPK3 Promotes the Differentiation of CPMs

As the miR-30a-3p had the effect on CPM differentiation, and miR-30a-3p could interact with circHIPK3, we speculated that circHIPK3 might have the opposite effect on CPM differentiation. To confirm our hypothesis, the expression profile of circHIPK3 was detected in the process of differentiation in CPMs. Interestingly, the expression trend of circHIPK3 coincided with miR-30a-3p, and it dramatically decreased in the DM phase compare to GM but raised gradually in DM4 (Figure 7A), indicating that circHIPK3 may be related to the differentiation of CPMs. Then, the expression level of the myoblast differentiation marker genes including *MYOD*, *MYOG* and *MYHC* were evaluated by qPCR in CPMs transfected with circHIPK3 over-expression vector/pCD5ciR and siRNA/siRNA NC. The expression levels of *MYOG* and *MYHC* but not *MYOD* were significantly increased in CPMs transfected with the over-expression vector compare with the group transfected with pCD5ciR (Figure 7B). On the contrary, knock-down of circHIPK3 significantly inhibited the expression of *MYOD*, *MYOG* and *MYHC* compare with the siRNA NC group (Figure 7C). The relative protein level of *MYHC* was also decreased after knock-down of circHIPK3 (Figure 7D). Moreover, MYHC immunofluorescence staining was conducted on transfected differentiated CPMs at DM5. The results showed that the total area of myotubes of si-circHIPK3-003 transfected group was notably less than that of siRNA NC group, and the statistics of the myotube area rate of the si-circHIPK3-003 group were markedly lower than that of siRNA NC group (Figure 7E). These results suggested that circHIPK3 had counteractive effect of miR-30a-3p on CPM differentiation and knock-down of circHIPK3 suppressed the differentiation of CPMs.

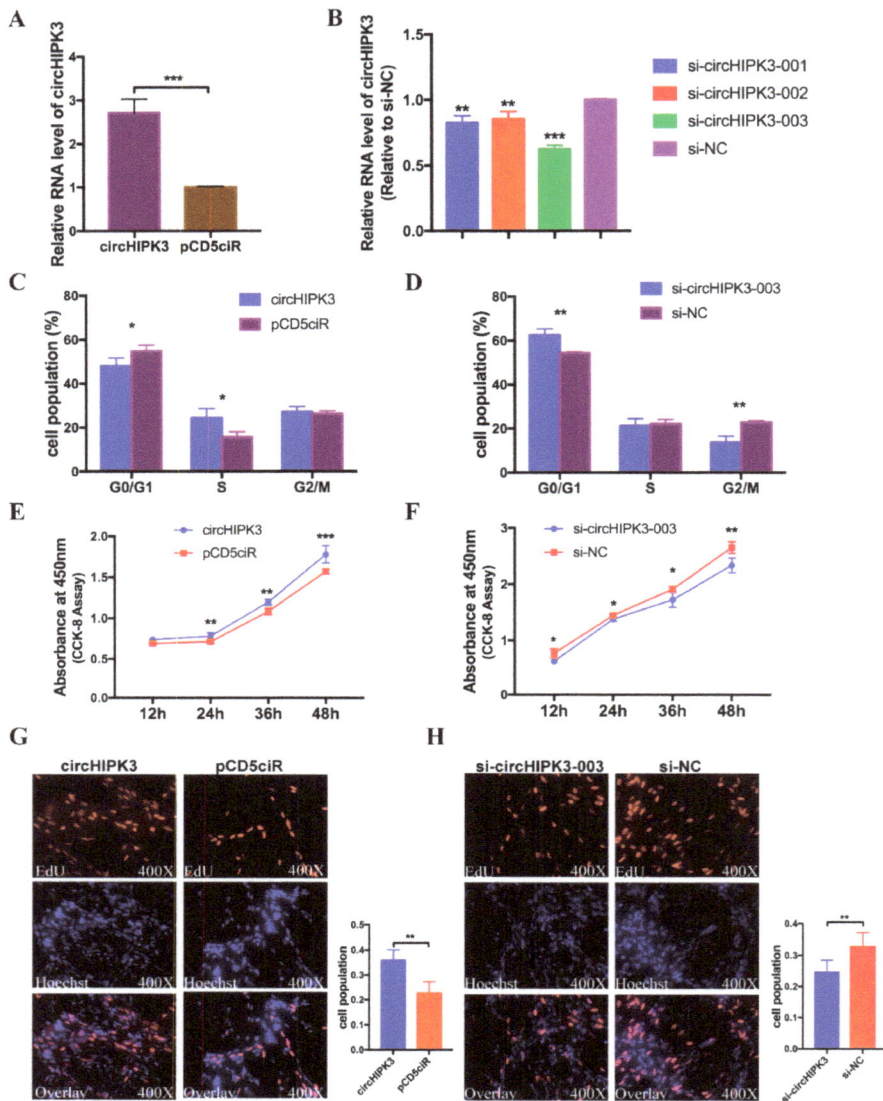

Figure 6. circHIPK3 promotes the proliferation of CPMs. (**A**) The over-expression effect of circHIPK3 over-expression vector in CPMs. (**B**) The interference effects of three siRNAs of circHIPK3 in CPMs. (**C**,**D**) Effect of circHIPK3 on cell-cycle progression of CPMs. The plot of cell-cycle analysis in different cell-cycle phases was compared. (**E**,**F**) The growth curves of CPMs were measured after the transfection of over-expression vector and siRNA of circHIPK3. (**G**,**H**) EdU assays for CPMs with over-expression and inhibition of circHIPK3. In all panels, the values represent mean ± SEM from three independent experiments. * $p < 0.05$; ** $p < 0.01$; *** $p < 0.001$.

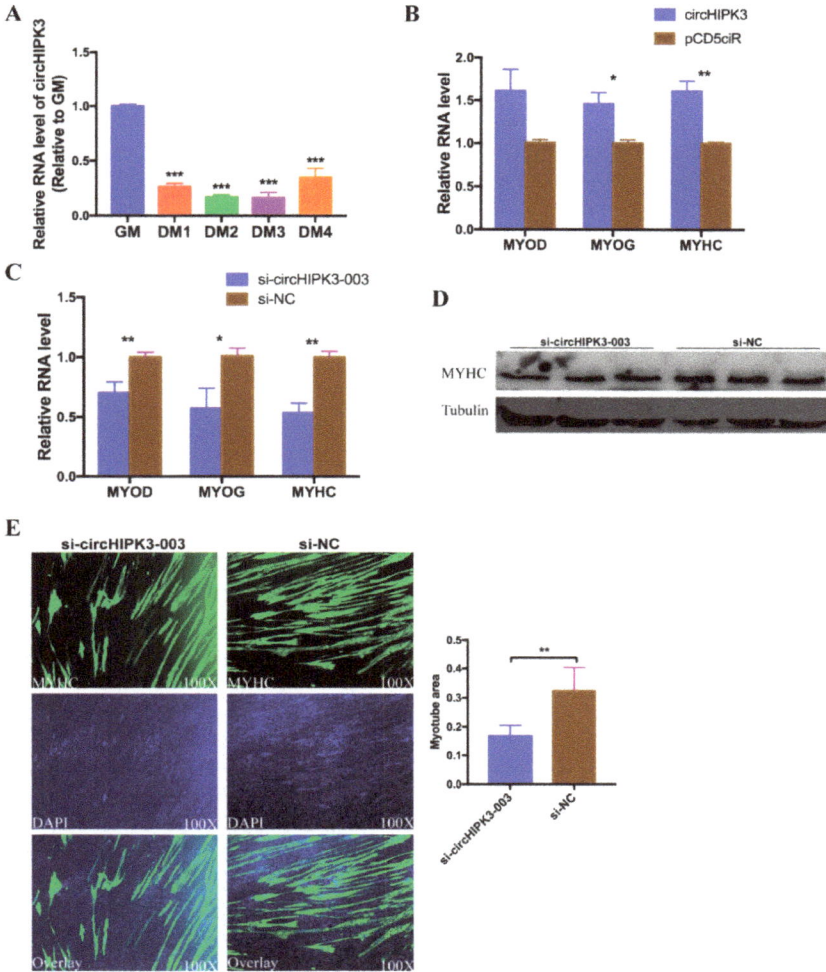

Figure 7. CircHIPK3 promotes the differentiation of CPMs. (**A**) The expression profile of circHIPK3 in the process of CPMs induced differentiation. (**B**) The expression level of *MYOD*, *MYOG* and *MYHC* in CPMs after over-expression of circHIPK3. (**C**) The expression of *MYOD*, *MYOG* and *MYHC* in CPMs after knock-down of circHIPK3. (**D**) The protein level of *MYHC* in CPMs after knock-down of circHIPK3. (**E**) Immunofluorescence analysis of MYHC-staining cells after knock-down of circHIPK3 in CPMs, * $p < 0.05$; ** $p < 0.01$; *** $p < 0.001$.

4. Discussion

Recently, circular RNAs have been identified as a new regulatory factor in multiple biological processes in different kinds of species [18,22–24]. Generally, the expression pattern of circular RNAs were corresponding to their linear parental transcript, the expression levels of circular RNAs were lower and had less functions compare to the corresponding mRNA [25,26]. However, some circular RNAs expressed independently high levels and exerted crucial roles in some special cell lines or tissues [14,27]. Many studies proved that circular RNAs were differentially expressed in different muscle developmental stages and played crucial roles in muscle development [28]. Circular RNA sequencing analysis, in different kinds of muscle cells or tissues, among many species, showed that most of the circular RNAs were exonic circular RNAs and differentially expressed during aging [29,30]. Functional circular RNAs in myoblast differentiation play a role by acting as an miRNA sponge or translating micropeptide [15,29,31,32]. Previous circular RNA sequencing data (GSE89355) showed that circHIPK3 was differentially expressed in three stages of chicken muscle development and expressed the highest level, while compared with the other ten circular RNAs, which was produced by chicken *HIPK3*. In this study, we found that circHIPK3 maintained a high expression level in myoblast and decreased sharply from DM1 to DM4, then the expression trend of circHIPK3 increased. This unique expression pattern suggested that circHIPK3 may have an important impact on CPM proliferation and differentiation.

MicroRNAs have been found to be widespread in various cell types or tissues and exerted a regulatory role in biological development [33–35]. It exerted important functions in muscle development and was associated with phenotypic changes in skeletal muscle [36,37]. miR-30a-3p is an inhibitor of proliferation in cancer cells [11,12]. In chicken skeletal muscle development, miR-30a-3p differentially expressed in three stages [38]. In this study, we demonstrated that miR-30a-3p inhibited the proliferation of CPMs and suppressed the differentiation by inhibiting the expression of *MEF2C*. According to the resources of microRNA viewer (http://people.csail.mit.edu/akiezun/microRNAviewer/index.html) (Supplementary Figure), miR-30a-3p was conserved in multiple species. The current study provided information to similar studies in other species.

Circular RNA that possesses miRNA response elements (MREs) is known to be a sponge for miRNAs [39]. Previous studies confirmed that the miR-30a family could bind to circHIPK3 in mice [40]. Given the homology of circHIPK3 and miR-30a family, we found that circHIPK3 possessed three potential binding sites for miR-30a-3p, through bioinformatic analyses by miRDB and RNAhybrid. Then two methods were conducted and we confirmed there are interactions between circHIPK3 and miR-30a-3p. Therefore, we hypothesized that circHIPK3 might play an opposite role of miR-30a-3p on skeletal muscle development.

CircHIPK3 was found to be involved in cell proliferation by acting as a sponge of multiple miRNAs, and *HIPK3* gene could produce several circular RNAs with different types in previous study [40,41]. miR-30a inhibited the proliferation by involving in different pathways in different kinds of cells [11,42,43]. In this study, we found that circHIPK3 promoted the proliferation and differentiation of CPMs by combing with miR-30a-3p. However, the underlying mechanism of miR-30a-3p retarding the myoblast proliferation needs further investigation.

In conclusion, our results suggested that miR-30a-3p could inhibit the proliferation of CPMs and repress the differentiation of CPMs by decreasing the expression of *MEF2C*, while circHIPK3 could promote the proliferation and differentiation of CPMs by sponging miR-30a-3p.

Supplementary Materials: The following are available online at http://www.mdpi.com/2073-4409/8/2/177/s1.

Author Contributions: B.C. carried out all the experiments, analyzed the data and wrote the paper. J.Y., L.G., Z.W. and X.C. participated in partial experiments. M.S.B. and H.X. joined the revision of the manuscript. Q.N. designed the experiments, analyzed data, and participated in its revision.

Funding: This research was funded by the grants from Science and Technology Planning Project of Guangzhou City (201604020007; 201504010017) and Guangdong Province (2018B020203001), and the Ten-Thousand Talents Program of China (W03020593).

Acknowledgments: We would like to thank Jing Liu and Zhuo Chen for their support.

Conflicts of Interest: The authors declare no conflict of interest.

References

1. Scanes, C.G.; Harvey, S.; Marsh, J.A.; King, D.B. Hormones and Growth in Poultry. *Poult. Sci.* **1984**, *63*, 2062–2074. [CrossRef] [PubMed]

2. Güller, I.; Russell, A.P. MicroRNAs in skeletal muscle: Their role and regulation in development, disease and function. *J. Physiol.* **2010**, *588*, 4075–4087. [CrossRef] [PubMed]

3. Flisar, T.; Malovrh, Š.; Terčič, D.; Holcman, A.; Kovač, M. Thirty-four generations of divergent selection for 8-week body weight in chickens. *Poult. Sci.* **2014**, *93*, 16–23. [CrossRef] [PubMed]

4. Bartel, D.P. MicroRNAs: Genomics, Biogenesis, Mechanism, and Function. *Cell* **2004**, *116*, 281–297. [CrossRef]

5. Chen, J.-F.; Mandel, E.M.; Thomson, J.M.; Wu, Q.; Callis, T.E.; Hammond, S.M.; Conlon, F.L.; Wang, D.-Z. The role of microRNA-1 and microRNA-133 in skeletal muscle proliferation and differentiation. *Nat. Genet.* **2006**, *38*, 228–233. [CrossRef] [PubMed]

6. Van Rooij, E.; Liu, N.; Olson, E.N. MicroRNAs flex their muscles. *Trends Genet.* **2008**, *24*, 159–166. [CrossRef] [PubMed]

7. Chen, X.; Ouyang, H.; Wang, Z.; Chen, B.; Nie, Q. A Novel Circular RNA Generated by FGFR2 Gene Promotes Myoblast Proliferation and Differentiation by Sponging miR-133a-5p and miR-29b-1-5p. *Cells* **2018**, *7*, 199. [CrossRef] [PubMed]

8. Luo, W.; Wu, H.; Ye, Y.; Li, Z.; Hao, S.; Kong, L.; Zheng, X.; Lin, S.; Nie, Q.; Zhang, X. The transient expression of miR-203 and its inhibiting effects on skeletal muscle cell proliferation and differentiation. *Cell Death Dis.* **2014**, *5*, e1347. [CrossRef] [PubMed]

9. Jia, X.; Ouyang, H.; Abdalla, B.A.; Xu, H.; Nie, Q.; Zhang, X. miR-16 controls myoblast proliferation and apoptosis through directly suppressing Bcl2 and FOXO1 activities. *Biochim. Biophys. Acta* **2017**, *1860*, 674–684. [CrossRef] [PubMed]

10. Ma, M.; Cai, B.; Jiang, L.; Abdalla, B.A.; Li, Z.; Nie, Q.; Zhang, X. lncRNA-Six1 Is a Target of miR-1611 that Functions as a ceRNA to Regulate Six1 Protein Expression and Fiber Type Switching in Chicken Myogenesis. *Cells* **2018**, *7*, 243. [CrossRef]

11. Qi, B.; Wang, Y.; Chen, Z.-J.; Li, X.-N.; Qi, Y.; Yang, Y.; Cui, G.-H.; Guo, H.-Z.; Li, W.-H.; Zhao, S. Down-regulation of miR-30a-3p/5p promotes esophageal squamous cell carcinoma cell proliferation by activating the Wnt signaling pathway. *World J. Gastroenterol.* **2017**, *23*, 7965–7977. [CrossRef]

12. Wang, W.; Lin, H.; Zhou, L.; Zhu, Q.; Gao, S.; Xie, H.; Liu, Z.; Xu, Z.; Wei, J.; Huang, X.; et al. MicroRNA-30a-3p inhibits tumor proliferation, invasiveness and metastasis and is downregulated in hepatocellular carcinoma. *Eur. J. Surg. Oncol.* **2014**, *40*, 1586–1594. [CrossRef] [PubMed]

13. Memczak, S.; Jens, M.; Elefsinioti, A.; Torti, F.; Krueger, J.; Rybak, A.; Maier, L.; Mackowiak, S.D.; Gregersen, L.H.; Munschauer, M.; et al. Circular RNAs are a large class of animal RNAs with regulatory potency. *Nature* **2013**, *495*, 333–338. [CrossRef]

14. Rybak-Wolf, A.; Stottmeister, C.; Glažar, P.; Jens, M.; Pino, N.; Giusti, S.; Hanan, M.; Behm, M.; Bartok, O.; Ashwal-Fluss, R.; et al. Circular RNAs in the Mammalian Brain Are Highly Abundant, Conserved, and Dynamically Expressed. *Mol. Cell* **2015**, *58*, 870–885. [CrossRef] [PubMed]

15. Legnini, I.; Di Timoteo, G.; Rossi, F.; Morlando, M.; Briganti, F.; Sthandier, O.; Fatica, A.; Santini, T.; Andronache, A.; Wade, M.; et al. Circ-ZNF609 Is a Circular RNA that Can Be Translated and Functions in Myogenesis. *Mole. Cell* **2017**, *66*, 22–37. [CrossRef] [PubMed]

16. Chen, L.-L. The biogenesis and emerging roles of circular RNAs. *Nat. Rev. Mol. Cell Biol.* **2016**, *17*, 205–211. [CrossRef] [PubMed]

17. Dong, R.; Zhang, X.-O.; Zhang, Y.; Ma, X.-K.; Chen, L.-L.; Yang, L. CircRNA-derived pseudogenes. *Cell Res.* **2016**, *26*, 747–750. [CrossRef]

18. Abdelmohsen, K.; Panda, A.C.; De, S.; Grammatikakis, I.; Kim, J.; Ding, J.; Noh, J.H.; Kim, K.M.; Mattison, J.A.; de Cabo, R.; et al. Circular RNAs in monkey muscle: Age-dependent changes. *Aging* **2015**, *7*, 903–910. [CrossRef] [PubMed]

19. Ouyang, H.; Chen, X.; Wang, Z.; Yu, J.; Jia, X.; Li, Z.; Luo, W.; Abdalla, B.A.; Jebessa, E.; Nie, Q.; et al. Circular RNAs are abundant and dynamically expressed during embryonic muscle development in chickens. *DNA Res.* **2017**, *3*. [CrossRef] [PubMed]

20. Ouyang, H.; Chen, X.; Li, W.; Li, Z.; Nie, Q.; Zhang, X. Circular RNA circSVIL Promotes Myoblast Proliferation and Differentiation by Sponging miR-203 in Chicken. *Front. Genet.* **2018**, *9*. [CrossRef]

21. Guo, L.; Huang, W.; Chen, B.; Jebessa Bekele, E.; Chen, X.; Cai, B.; Nie, Q. gga-mir-133a-3p Regulates Myoblasts Proliferation and Differentiation by Targeting PRRX1. *Front. Genet.* **2018**, *9*. [CrossRef] [PubMed]

22. Li, C.; Li, X.; Ma, Q.; Zhang, X.; Cao, Y.; Yao, Y.; You, S.; Wang, D.; Quan, R.; Hou, X.; et al. Genome-wide analysis of circular RNAs in prenatal and postnatal pituitary glands of sheep. *Sci. Rep.* **2017**, *7*, 97165–97177. [CrossRef] [PubMed]

23. Liang, G.; Yang, Y.; Niu, G.; Tang, Z.; Li, K. Genome-wide profiling of Sus scrofa circular RNAs across nine organs and three developmental stages. *DNA Res.* **2017**, *24*, 523–535. [CrossRef] [PubMed]

24. Piwecka, M.; Glažar, P.; Hernandez-Miranda, L.R.; Memczak, S.; Wolf, S.A.; Rybak-Wolf, A.; Filipchyk, A.; Klironomos, F.; Cerda Jara, C.A.; Fenske, P.; et al. Loss of a mammalian circular RNA locus causes miRNA deregulation and affects brain function. *Science* **2017**, *22*, eaam8526. [CrossRef] [PubMed]

25. Salzman, J.; Chen, R.E.; Olsen, M.N.; Wang, P.L.; Brown, P.O. Cell-Type Specific Features of Circular RNA Expression. *PLoS Genet.* **2013**, *9*. [CrossRef]

26. Guo, J.U.; Agarwal, V.; Guo, H.; Bartel, D.P. Expanded identification and characterization of mammalian circular RNAs. *Genome Biol.* **2014**, *15*. [CrossRef] [PubMed]

27. Conn, S.J.; Pillman, K.A.; Toubia, J.; Conn, V.M.; Salmanidis, M.; Phillips, C.A.; Roslan, S.; Schreiber, A.W.; Gregory, P.A.; Goodall, G.J. The RNA Binding Protein Quaking Regulates Formation of circRNAs. *Cell* **2015**, *160*, 1125–1134. [CrossRef]

28. Greco, S.; Cardinali, B.; Falcone, G.; Martelli, F. Circular RNAs in Muscle Function and Disease. *Int. J. Mol. Sci.* **2018**, *19*. [CrossRef]

29. Wei, X.; Li, H.; Yang, J.; Hao, D.; Dong, D.; Huang, Y.; Lan, X.; Plath, M.; Lei, C.; Lin, F.; et al. Circular RNA profiling reveals an abundant circLMO7 that regulates myoblasts differentiation and survival by sponging miR-378a-3p. *Cell Death Dis.* **2017**, *8*, e3153. [CrossRef]

30. Chen, J.; Zou, Q.; Lv, D.; Wei, Y.; Raza, M.A.; Chen, Y.; Li, P.; Xi, X.; Xu, H.; Wen, A.; et al. Comprehensive transcriptional landscape of porcine cardiac and skeletal muscles reveals differences of aging. *Oncotarget* **2017**, *9*, 1524–1541. [CrossRef]

31. Li, H.; Yang, J.; Wei, X.; Song, C.; Dong, D.; Huang, Y.; Lan, X.; Plath, M.; Lei, C.; Ma, Y.; et al. CircFUT10 reduces proliferation and facilitates differentiation of myoblasts by sponging miR-133a. *J. Cell. Physiol.* **2018**, *233*, 4643–4651. [CrossRef] [PubMed]

32. Li, H.; Wei, X.; Yang, J.; Dong, D.; Hao, D.; Huang, Y.; Lan, X.; Plath, M.; Lei, C.; Ma, Y.; et al. circFGFR4 Promotes Differentiation of Myoblasts via Binding miR-107 to Relieve Its Inhibition of Wnt3a. *Mol. Ther. Nucleic Acids* **2018**, *11*, 272–283. [CrossRef] [PubMed]

33. Rogg, E.M.; Abplanalp, W.T.; Bischof, C.; John, D.; Schulz, M.H.; Krishnan, J.; Fischer, A.; Poluzzi, C.; Schaefer, L.; Bonauer, A.; et al. Analysis of Cell Type-Specific Effects of MicroRNA-92a Provides Novel Insights Into Target Regulation and Mechanism of Action. *Circulation* **2018**, *138*, 2545–2558. [CrossRef]

34. Yasmeen, S.; Kaur, S.; Mirza, A.H.; Brodin, B.; Pociot, F.; Kruuse, C. miRNA-27a-3p and miRNA-222-3p as Novel Modulators of Phosphodiesterase 3a (PDE3A) in Cerebral Microvascular Endothelial Cells. *Mol. Neurobiol.* **2019**, *2*. [CrossRef] [PubMed]

35. Treiber, T.; Treiber, N.; Meister, G. Regulation of microRNA biogenesis and its crosstalk with other cellular pathways. *Nat. Rev. Mol. Cell Biol.* **2019**, *20*, 5–20. [CrossRef]

36. Wei, W.; He, H.-B.; Zhang, W.-Y.; Zhang, H.-X.; Bai, J.-B.; Liu, H.-Z.; Cao, J.-H.; Chang, K.-C.; Li, X.-Y.; Zhao, S.-H. miR-29 targets Akt3 to reduce proliferation and facilitate differentiation of myoblasts in skeletal muscle development. *Cell Death Dis.* **2013**, *4*, e668. [CrossRef]

37. Li, G.; Luo, W.; Abdalla, B.A.; Ouyang, H.; Yu, J.; Hu, F.; Nie, Q.; Zhang, X. miRNA-223 upregulated by MYOD inhibits myoblast proliferation by repressing IGF2 and facilitates myoblast differentiation by inhibiting ZEB1. *Cell Death Dis.* **2017**, *8*, e3094. [CrossRef]

38. Jebessa, E.; Ouyang, H.; Abdalla, B.A.; Li, Z.; Abdullahi, A.Y.; Liu, Q.; Nie, Q.; Zhang, X. Characterization of miRNA and their target gene during chicken embryo skeletal muscle development. *Oncotarget* **2018**, *9*. [CrossRef]

39. Hansen, T.B.; Jensen, T.I.; Clausen, B.H.; Bramsen, J.B.; Finsen, B.; Damgaard, C.K.; Kjems, J. Natural RNA circles function as efficient microRNA sponges. *Nature* **2013**, *495*, 384–388. [CrossRef]

40. Shan, K.; Liu, C.; Liu, B.H.; Chen, X.; Dong, R.; Liu, X.; Zhang, Y.Y.; Liu, B.; Zhang, S.J.; Wang, J.J.; et al. Circular Noncoding RNA HIPK3 Mediates Retinal Vascular Dysfunction in Diabetes Mellitus. *Circulation* **2017**, *136*, 1629–1642.

41. Zheng, Q.; Bao, C.; Guo, W.; Li, S.; Chen, J.; Chen, B.; Luo, Y.; Lyu, D.; Li, Y.; Shi, G.; et al. Circular RNA profiling reveals an abundant circHIPK3 that regulates cell growth by sponging multiple miRNAs.

Nat. Commun. **2016**, *7*, 11215. [CrossRef]

42. Cheng, C.-W.; Wang, H.-W.; Chang, C.-W.; Chu, H.-W.; Chen, C.-Y.; Yu, J.-C.; Chao, J.-I.; Liu, H.-F.; Ding, S.; Shen, C.-Y. MicroRNA-30a inhibits cell migration and invasion by downregulating vimentin expression and is a potential prognostic marker in breast cancer. *Breast Cancer Res. Treat.* **2012**, *134*, 1081–1093. [CrossRef] [PubMed]
43. Peng, R.; Zhou, L.; Zhou, Y.; Zhao, Y.; Li, Q.; Ni, D.; Hu, Y.; Long, Y.; Liu, J.; Lyu, Z.; et al. MiR-30a Inhibits the Epithelial—Mesenchymal Transition of Podocytes through Downregulation of NFATc3. *Int. J. Mol. Sci.* **2015**, *16*, 24032–24047. [CrossRef] [PubMed]

Article

A Novel Circular RNA Generated by FGFR2 Gene Promotes Myoblast Proliferation and Differentiation by Sponging miR-133a-5p and miR-29b-1-5p

Xiaolan Chen [1,2], Hongjia Ouyang [1,3], Zhijun Wang [1,2], Biao Chen [1,2] and Qinghua Nie [1,2,*]

[1] Department of Animal Genetics, Breeding and Reproduction, College of Animal Science, South China Agricultural University, Guangzhou 510642, China; xiaolanchen@stu.scau.edu.cn (X.C.); oyolive@stu.scau.edu.cn (H.O.); zhijunwang@stu.scau.edu.cn (Z.W.); biaochen@stu.scau.edu.cn (B.C.)
[2] National-Local Joint Engineering Research Center for Livestock Breeding, Guangdong Provincial Key Lab of Agro-Animal Genomics and Molecular Breeding, and the Key Lab of Chicken Genetics, Breeding and Reproduction, Ministry of Agriculture, Guangzhou 510642, China
[3] College of Animal Science & Technology, Zhongkai University of Agriculture and Engineering, Guangzhou 510225, China
*   Correspondence: nqinghua@scau.edu.cn; Tel.: +86-20-8528-5759; Fax: +86-20-8528-0740

Received: 30 September 2018; Accepted: 2 November 2018; Published: 6 November 2018

Abstract: It is well known that fibroblast growth factor receptor 2 (*FGFR2*) interacts with its ligand of fibroblast growth factor (*FGF*) therefore exerting biological functions on cell proliferation and differentiation. In this study, we first reported that the *FGFR2* gene could generate a circular RNA of circFGFR2, which regulates skeletal muscle development by sponging miRNA. In our previous study of circular RNA sequencing, we found that circFGFR2, generated by exon 3–6 of *FGFR2* gene, differentially expressed during chicken embryo skeletal muscle development. The purpose of this study was to reveal the real mechanism of how circFGFR2 affects skeletal muscle development in chicken. In this study, cell proliferation was analyzed by both flow cytometry analysis of the cell cycle and 5-ethynyl-2′-deoxyuridine (EdU) assays. Cell differentiation was determined by analysis of the expression of the differentiation marker gene and Myosin heavy chain (MyHC) immunofluorescence. The results of flow cytometry analysis of the cell cycle and EdU assays showed that, overexpression of circFGFR2 accelerated the proliferation of myoblast and QM-7 cells, whereas knockdown of circFGFR2 with siRNA reduced the proliferation of both cells. Meanwhile, overexpression of circFGFR2 accelerated the expression of myogenic differentiation 1 (*MYOD*), myogenin (*MYOG*) and the formation of myotubes, and knockdown of circFGFR2 showed contrary effects in myoblasts. Results of luciferase reporter assay and biotin-coupled miRNA pull down assay further showed that circFGFR2 could directly target two binding sites of miR-133a-5p and one binding site of miR-29b-1-5p, and further inhibited the expression and activity of these two miRNAs. In addition, we demonstrated that both miR-133a-5p and miR-29b-1-5p inhibited myoblast proliferation and differentiation, while circFGFR2 could eliminate the inhibition effects of the two miRNAs as indicated by rescue experiments. Altogether, our data revealed that a novel circular RNA of circFGFR2 could promote skeletal muscle proliferation and differentiation by sponging miR-133a-5p and miR-29b-1-5p.

Keywords: circular RNA; circFGFR2; *FGFR2*; miR-133a-5p; miR-29b-1-5p; skeletal muscle; proliferation; differentiation

---

1. Introduction

Circular RNA is a large class of endogenous RNA with a covalently closed loop. It was actually discovered in plants, mouse, and yeast twenty years ago [1–3]. However, it has been regarded as unvalued mis-splicing product of mRNA in the last decades as a few kinds and a small quantity of circular RNAs have been found [4]. In addition, circular RNA has no 5′ caps and 3′ tails, and it could be easily abandoned by traditional sequencing technology [5]. Fortunately, with the rapid development of high throughput sequencing technology, the mysterious veil of circular RNA was revealed step by step [6]. Large amounts of circular RNAs were discovered in many species, including human [7], monkey [8], and pig [9].

Nowadays, circular RNA is considered as an up-rising star in the small RNAs interaction network with regulatory potency [10]. The diverse functions of circular RNA act as miRNA sponge, participating in regulating the expression of its own linear RNA in different ways [10,11], sequestering proteins [12,13], coding protein in vitro [14–16], and deriving pseudogenes [17]. Acting as miRNA sponge is a well-studied function of circular RNA, also known as a competing endogenous RNA mechanism (ceRNA). The CeRNA mechanism is that messenger RNAs, transcribed pseudogenes, and long noncoding RNAs competitively combine with the same miRNA response elements (MREs), and then eliminate the inhibition of miRNA on their target genes. Circular RNA interacted with miRNA are ubiquitous in a variety of tissues. A well-known example is that ciRS-7 has more than 70 highly conserved target sites of miR-7 and can extremely repress the activity of miR-7 [18]. This is the strongest evidence for a circRNA function as the miRNA sponge has thrust circRNAs into the spotlight and spurred a multitude of studies searching for functional circRNA sponges [19–21].

In previous work [22], we used leg muscle tissues of two female XingHua chickens from each at days E11, E16, and P1 for circRNA sequencing to comprehensively identify stably expressed circRNAs during skeletal muscle development at the embryonic stage. As a result, 13,377 potential circRNAs were identified and abundantly expressed among different development stages. Furthermore, the differentially expressed genes (DEGs) analysis showed 462 of them were differentially expressed at different development stages. CircFGFR2 was one of the DEcircRNAs with high expression during skeletal muscle development. Through divergent reverse-transcription PCR and RNase R treatment, in previous work [22], we confirmed that circFGFR2 was a stable exonic circular RNA formed by 3–6 exons of fibroblast growth factor receptor 2 (*FGFR2*), with a length of 636 bp. As a member of *FGFRs* family, *FGFR2* interacts with fibroblast growth factor (*FGF*) to exert biological effects on cell proliferation and differentiation as well as skeletal development [23]. The different expression level of circRNAs implied that they could potentially regulate skeletal muscle development. We previously revealed that circRBFOX2 could interact with miR-206 to regulate skeletal muscle cell proliferation and differentiation [22]. Considering all of that, we assumed that circFGFR2 was another candidate circRNA that probably affects skeletal muscle development.

In comparison to circular RNA, miRNAs are extremely well studied non-coding RNAs that suppress protein expression by targeting the 3′-UTR (Untranslated Region) of their mRNA with Argonaute effector protein [24,25]. The MiR-133 family has two members of miR-133a and miR-133b, which are found to specifically express in skeletal muscle and cardiac [26]. MiR-133a-5p belongs to the miR-133a cluster. Many studies have shown that miR-133 families are involved in regulating the proliferation and differentiation of various kinds of skeletal muscle cells [27,28]. However, the role of miR-133a-5p on skeletal muscle development has not been reported in poultry. MiR-29b-1-5p is a mature miRNA and belongs to the miR-29b cluster of the miR-29 family. This family has other clusters of miR-29a and miR-29c [29]. In chicken, the gga-miR-29b cluster contains gga-miR-29b-1-5p, gga-miR-29b-2-5p, and gga-miR-1701. MiR-29s are efficient regulators in the process of cell proliferation [30], differentiation [31], apoptosis [32–34] as well as DNA methylation [35,36] in different cell types. In skeletal muscle, miR-29s could participate in regulating skeletal myogenesis through different pathways. In mouse C2C12 cells, they could down-regulate *Rybp* (Ring1 and YY1-binding Protein) [37], AKT serine/threonine kinase 3 (*AKT3*) [38], and histone deacetylase 4 (*HDAC4*) [39] to regulate the differentiation of skeletal muscle cell. In addition, miR-29s were also related to some muscle diseases, including muscle atrophy [40], dystrophic muscle pathogenesis [41], and Duchenne muscular dystrophy [42]. Obviously, miR-29s play important roles in muscle development.

In this study, we aim to investigate the effects of circFGFR2 on skeletal muscle cell development, and to reveal its regulatory mechanism by interacting miR-133a-5p and miR-29b-1-5p.

2. Materials and Methods

2.1. Ethics Statement

This study was carried out in accordance with the principles of the Basel Declaration and recommendations of the Statute on the Administration of Laboratory Animals, the South China Agriculture University Institutional Animal Care and Use Committee. The protocol was approved by the South China Agriculture University Institutional Animal Care and Use Committee (approval, 19 November 2017, ID: 2017046).

2.2. Primers

All primers used in this study were designed by Premier Primer 5.0 software (Premier Bio-soft International, Palo Alto, CA, USA) and synthesized by Sangon (Sangon Biotech, Shanghai, China). The detailed information of all primers is listed in Table 1.

Table 1. Primers used in this study.

Name	Nucleotide Sequences (5′→3′)	Annealing Temperature (°C)	Size	Application
circFGFR2	F: ACATCGTATTGGCGGCTAT R: ACCCCATCCTTAGTCCAAC	60	267	qRT-PCR for circFGFR2
FGFR2-1	F: GTCCGCTGTATGTGATTGTAG R: TGAATGTCATCTGCTCCTCT	56	129	qRT-PCR for FGFR2 gene
FGFR2-2	F: AGCCGCCAACCAAATACCAAATR: CGACAACATCGAGATGGTAAGT	56	636	Amplification of the whole linear sequence of circFGFR2
MYOD	F: GCTACTACACGGAATCACCAAAT R: CTGGGCTCCACTGTCACTCA	58	200	qRT-PCR
MYOG	F: CGGAGGCTGAAGAAGGTGAA R: CGGTCCTCTGCCTGGTCAT	60	320	qRT-PCR
β-actin	F: ACCACAGGACTCCATACCCAAGAAG R: GCCGAGAGAGAAATTGTGCGTGAC	52–60	146	qRT-PCR

2.3. RNA Extraction, cDNA Synthesis and Quantitative Real-Time PCR

The total RNA was extracted from cells by using RNAiso reagent (TaKaRa, Otsu, Japan). The quality and concentration of all obtained RNA samples were determined by 1.5% agarose gel electrophoresis and evaluated for optical density 260/280 ratio by Nanodrop 2000 spectrophotometer (Thermo, Waltham, MA, USA). For mRNA and circFGFR2 expression analysis, cDNA synthesis for mRNA was performed using a PrimeScript RT Reagent Kit (Perfect Real Time) (TaKaRa, Otsu, Japan). The β-actin gene was used as an internal control for quantitative real-time PCR (qRT-PCR) analysis. The reverse transcription reaction for miRNA was performed using ReverTra Ace qPCR RT Kit (Toyobo, Osaka, Japan). The specific Bulge-loop miRNA qRT-PCR Primer for miR-133a-5p, miR-29b-1-5p and U6 were designed by RiboBio (RiboBio, Guangzhou, China). qRT-PCR was performed on a Bio-Rad CFX96 Real-Time Detection System (Bio-Rad, Hercules, CA, USA) using iTaq™ Universal SYBR® Green Supermix Kit (Bio-Rad, Hercules, CA, USA). Each sample was assayed in triplicate, following the manufacturer's instructions. The specificity of the product was evaluated by the melting curve, and the quantitative values were obtained from the threshold PCR cycle number (Ct) at which the increase in signal is associated with an exponential growth at which the PCR product starts to be detected. The relative mRNA level in each sample was indicated by $2^{-\Delta\Delta Ct}$.

2.4. RNA Oligonucleotides and Plasmids Construction

The gga-miR-133a-5p mimic, gga-miR-29b-1-5p mimic and mimic control duplexes, the 3' end biotinylated gga-miR-133a-5p, gga-miR-29b-1-5p and mimic control duplexes, siRNA target against circFGFR2 (si-circFGFR2, 5'-CGATGTTGTCGAGCCGCCA-3') and non-specific siRNA negative control were synthesized by RiboBio (Guangzhou, China). For circFGFR2 overexpression plasmids construction, the linear sequences of circFGFR2 was amplified by PCR with primer *FGFR2-2*, and the cDNA template was synthesized from the RNA of chicken primary myoblast by RT-PCR. Then, the obtained linear sequences were cloned into *KpnI* and *BamHI* restriction sites of a circular expression vector-the pCD2.1-ciR vector (Geneseed Biotech, Guangzhou, China) according to the manufacturer's protocol, so as to generate the pCD2.1-circFGFR2 overexpression vector. For pmirGLO dual-luciferase reporter construction: the whole linear sequences of circFGFR2 were cloned into *XhoI and SalI* restriction sites of pmirGLO vector to generate the wild reporter vector (PGLO-WT reporter vector), which includes the predicted binding sites of miR-133a-5p and miR-29b-1-5p. PGLO-MT1 and PGLO-MT2 were two mutational reporter vectors of miR-133a-5p which were cloned into *XhoI and SalI* restriction sites of pmirGLO vector by PCR mutagenesis. We changed one of miR-133a-5p binding seed sequences from "CCAG" to "TTGA" in PGLO-MT1, while in PGLO-MT2 we changed another miR-133a-5p binding seed sequence (which included the binding site of miR-29b-1-5p) from "CCAG" to "GTTG". All luciferase reporters were constructed by Hongxun Biotech (Suzhou, China).

2.5. Cell Culture

Chicken embryo fibroblast cell line (DF-1) cells were cultured in high-glucose Dulbecco's modified Eagle's medium (Gibico, Grand Island, NY, USA) with 10% (*v/v*) fetal bovine serum (FBS) (Gibco, Grand Island, NY, USA) and 0.2% penicillin/streptomycin (Invitrogen, Carlsbad, CA, USA). Quail muscle cell line (QM-7) cells were cultured in high-glucose M199 medium (Gibco, USA) with 10% (*v/v*) FBS, 10% tryptose phosphate broth solution (Sigma, Louis, MO, USA) and 0.2% penicillin/streptomycin (Invitrogen, Carlsbad, CA, USA). Chicken primary myoblasts were isolated from the leg muscles of 11-day embryo age (E11) chickens. Leg tissues were collected from E11 chickens by completely removing skin and bones. Leg muscle was minced into sections of approximately 1 mm with scissors and then digested with 0.25% trypsin (Gibco, Grand Island, NY, USA) at 37 °C in a shaking water bath (90 oscillations/min) for 20 min. Digestions were terminated by adding equal values of complete medium-(RPMI)1640 medium with 20% FBS, 1% nonessential amino acids and 0.2% penicillin/streptomycin (Invitrogen, Carlsbad, CA, USA). The mixture was filtered through a nylon mesh with 70 mm pores (BD Falcon, Greiner, Germany). The filtered cells were centrifuged at $500 \times g$ for 5 min, and maintained in complete medium at 37 °C in a 5% CO_2, humidified atmosphere. Serial plating was performed to enrich myoblasts and to remove fibroblasts.

2.6. Transfections

Transfections were performed with Lipofectamine 3000 reagent (Invitrogen, Carlsbad, CA, USA) according to the manufacturer's instruction. Nucleic acids were diluted in OPTI-MEM Medium (Gibco, Grand Island, NY, USA).

2.7. 5-Ethynyl-2'-Deoxyuridine (EdU) Assays

After cells were transfected for 48 h, myoblasts were exposed to 50 μM 5-ethynyl-2'-deoxyuridine (EdU) (RiboBio, Guangzhou, China) for 2 h at 37 °C. Next, the cells were fixed in 4% paraformaldehyde (PFA) for 30 min and 2 mg/mL glycine solution was used to neutralize the 4% PFA. Cells were, then, permeabilized with 0.5% Triton X-100. Subsequently, 1× Apollo reaction cocktail (RiboBio, Guangzhou, China) was added to the cells and incubated for 30 min. The cells were stained with Hoechst 33342 for 30 min for DNA content analysis. Finally, the EdU-stained cells were visualized under a fluorescence microscope (Nikon, Tokyo, Japan or Leica, Wetzlar, Germany). The analysis of myoblast proliferation (ratio of EdU+ to all myoblasts) was performed using images of randomly selected fields obtained on the fluorescence microscope.

2.8. Flow Cytometry Analysis of the Cell Cycle

Myoblast cultures in growth medium (GM) were collected after a 48 h or 36 h-transfection and then fixed in 70% ethanol overnight at −20 °C. After incubation in 50 μg/mL propidium iodide (PI) (Sigma, Louis, MO, USA) containing 10 μg/mL RNase A (TaKaRa, Otsu, Japan) and 0.2% (v/v) Triton X-100 (Sigma, Louis, MO, USA) for 30 min at 4 °C, the cells were analyzed by using a BD AccuriC6 flow cytometer (BD Biosciences, San Jose, CA, USA) and FlowJo7.6 software (Treestar Incorporated, Ashland, OR, USA).

2.9. Immunofluorescence

For immunofluorescence, cells were seeded in 24-well plates. After transfection for 48 h, cells were fixed in 4% formaldehyde for 20 min then washed three times with PBS for 5 min. Subsequently, the cells were permeabilized by adding 0.1% Triton X-100 for 5 min and blocked with goat serum for 30 min. After incubation with MyHC (B103; DSHB, Iowa City, IA, USA; 0.5 μg/mL) at 37 °C for 2 h, the Fluorescein (FITC)-conjugated AffiniPure Goat Anti-Mouse IgG (H + L) (Bioworld, Minneapolis, MN, USA; 1:200) or FITC (Bioworld, Minneapolis, MN, USA; 1:50) was added and the cells were incubated at room temperature for 1 h. The cell nuclei were stained with 4′,6-diamidino-2-phenylindole (DAPI, Beyotime, Shanghai, China; 1:50) for 5 min. Images were obtained with a fluorescence microscope (Leica, Wetzlar, Germany). The area of cells labeled with anti-MyHC was measured by using ImageJ software (National Institutes of Health, Bethesda, MD, USA), and the total myotube area was calculated as a percentage of the total image area covered by myotubes.

2.10. Luciferase Reporter Assay

To investigate the binding sites of circFGFR2 with miR-133a-5p/miR-29b-1-5p, DF-1 cells were seeded in 96-well plates and then co-transfected with 100 ng of PGLO-WT reporter vector or mutant vectors PGLO-MT1 or PGLO-MT2, and 50 nM of miR-133a-5p/miR-29b-1-5p mimics or mimic control duplexes by using Lipofectamine 3000 reagent (Invitrogen, Carlsbad, CA, USA). To investigate whether circFGFR2 could inhibit the activity of miR-133a-5p/miR-29b-1-5p, DF-1 cells were seeded in 96-well plates and then co-transfected with 100 ng of PGLO-WT reporter vector or circFGFR2 overexpression vector, and 50 nM of miR-133a-5p/miR-29b-1-5p mimics or mimic control duplexes by using Lipofectamine 3000 reagent (Invitrogen, Carlsbad, CA, USA). After 48 h post transfection, luciferase activity analysis was performed using a Fluorescence/Multi-Detection Microplate Reader (BioTek, Winooski, VT, USA) and a Dual-GLO® Luciferase Assay System Kit (Promega, Madison, WI, USA). Firefly luciferase activities were normalized to Renilla luminescence in each well.

2.11. Biotin-Coupled miRNA Pull Down Assay

Transfection procedure: the 100 nM of 3′ end biotinylated miR-133a-5p, miR-29b-1-5p or mimic NC (RiboBio, Guangzhou, China) were transfected into QM-7 cells along with 30 μg circFGFR2 expression vector in T75 cell culture bottle. At 24 h after transfection, the cells were harvested and washed in PBS, then lysed in lysis buffer. A total of 100 μL washed streptavidin magnetic beads were blocked for 2 h and then added to each reaction tube to pull down the biotin-coupled RNA complex. All the tubes were incubated for 4 h on a rotator at a low speed (10 r/min). The beads were washed with lysis buffer five times and RNAiso reagent (TaKaRa, Otsu, Japan) was used to recover RNAs specifically interacting with miRNA. The abundance of circFGFR2 in bound fractions was evaluated by qRT-PCR analysis.

2.12. Statistical Analysis

In all panels, results are expressed as the mean $\pm$ S.E.M. of three independent experiments. For two group comparison analysis, statistical significance of differences between means was analyzed by unpaired Student's t-test. For multiple comparison analysis, data were analyzed by one-way ANOVA followed by both least significant difference (LSD) and Duncan test through Statistical Package for the Social Sciences software (SPSS 17.0, Chicago, IL, USA). We considered $p < 0.05$ to be statistically significant. * $p < 0.05$; ** $p < 0.01$. NC, negative control.

3. Results

3.1. CircFGFR2 Promotes Myoblast Proliferation

To investigate the role of circFGFR2 in skeletal muscle cell proliferation, we conducted overexpression and knocked down experiments by transfecting circFGFR2 overexpression vector and siRNAs (pCD2.1-circFGFR2 and si-circFGFR2) into chicken primary myoblast and QM-7 cell. The relative expression of circFGFR2 was detected after 48 h post transfection by qRT-PCR. Result showed that both the effect of overexpression and knockdown had reached a significant level in both myoblast and QM-7 cell (Figure 1A–D), and si-circFGFR2 specifically downregulated the expression of circFGFR2 but not linear mRNA of *FGFR2* (Figure 2B). Furthermore, we detected the proliferation process of both chicken primary myoblast and QM-7 cell by flow cytometry for cell cycle analysis and 5-ethynyl-2'-deoxyuridine (EdU) incorporation assays after transfecting with pCD2.1-circFGFR2/pCD2.1-cir, or si-circFGFR2/control. Cell cycle analysis showed that overexpression of circFGFR2 increased the cell population in S phase and decreased the cell population in G1/0 and G2/M phases (Figure 1E) while knockdown of circFGFR2 decreased the cell population in S phase and increased the cell population in G1/0 phase, as observed in chicken primary myoblast (Figure 1F). Meanwhile, the result of EdU strain assay showed that there were significantly more cells in the pCD2.1-circFGFR2 transfected group than in the control group (Figure 1G,H), whereas knockdown of circFGFR2 significantly decreased the numbers of EdU strained cells (Figure 1G,I). These results indicated that circFGFR2 could promote the proliferation rate of chicken primary myoblast. As expected, we obtained similar results in QM-7 cell (Figure 1J–N). These results suggested that circFGFR2 could significantly promote the proliferation of myoblast and QM-7 cell.

3.2. CircFGFR2 Promotes Myoblast Differentiation

Myogenesis is a complex process including myoblast proliferation, differentiation and myotube formation and is controlled by myogenic regulatory factors (MRFs), *MYOD*, *MYOG*, myogenic factor 5 (*Myf5*), and myogenic factor 6 (*Myf6*, also known as myogenic regulatory factor 4, *MRF4*). These factors activate muscle-specific genes to coordinate myoblasts to terminally withdraw from the cell cycle and subsequently fuse into multinucleated myotubes [43]. Following proliferation, the initiation of terminal differentiation and fusion begins with the expression of myogenin, which together with *MYOD*, activates the muscle specific structural and contractile genes to stimulate myoblast differentiation [44]. To address the potential role of circFGFR2 in primary myoblast differentiation, the expression of differentiation marker genes, including *MYOG* and *MYOD* were analyzed by qRT-PCR after transfecting with pCD2.1-circFGFR2/pCD2.1-cir, or si-circFGFR2/control. Result showed that overexpression of circFGFR2 significantly promoted the expression of *MYOD* and *MYOG* while knockdown of circFGFR2 significantly inhibited the expression of *MYOD* and *MYOG* (Figure 2A,B). It indicated that circFGFR2 may promote chicken primary myoblast differentiation. Subsequently, we induced chicken primary myoblast differentiation in vitro, as soon as the muscle cells started to differentiate into myotubes (the first day of differentiation, DM1), we transfected them with pCD2.1-circFGFR2/pCD2.1-cir. MyHC immunofluorescence staining was carried out on the differentiated myoblasts after 36 h post transfection (DM3). According to the immunofluorescence staining, we found that the areas of myotubes of pCD2.1-circFGFR2 transfected group were prominently greater than that of the control group (Figure 2C,D). Conversely, the areas of myotubes in the si-circFGFR2 transfected group were lower than that of the control group (Figure 2E,F). The result showed that circFGFR2 could promote the formation of myotubes and promote the early differentiation of chicken primary myoblast.

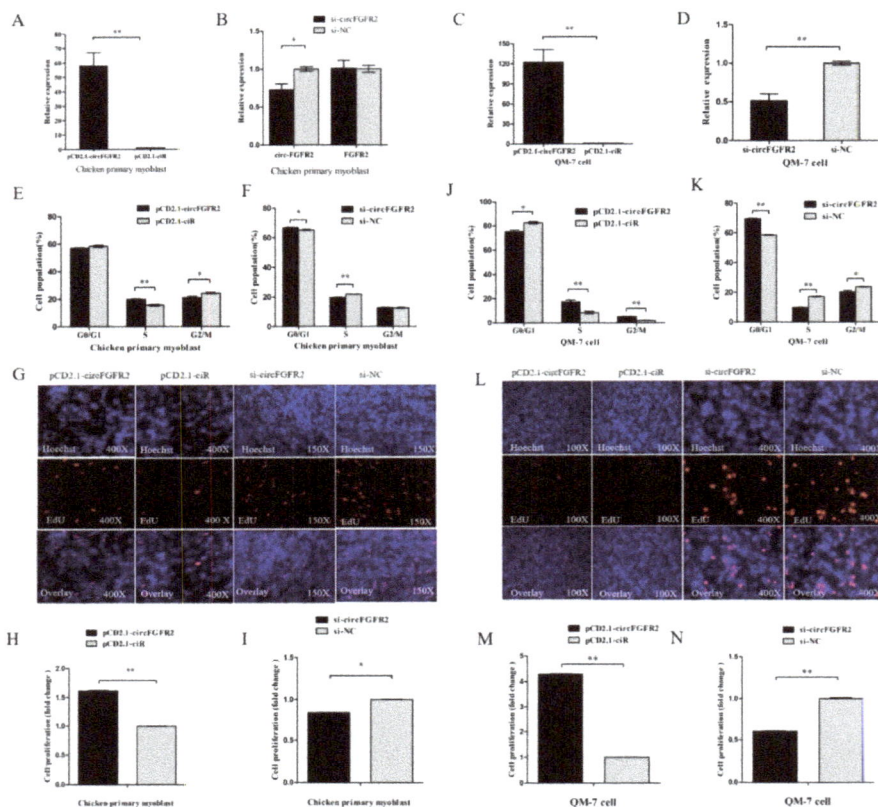

Figure 1. CircFGFR2 promotes myoblast proliferation. (**A,B**) The relative expression of circFGFR2 after transfected chicken primary myoblasts with 1 μg pCD2.1-circFGFR2 or 50 nM si-circFGFR2 for 48 h. (**C,D**) The relative expression of circFGFR2 after transfected QM-7 cells with 1 μg pCD2.1-circFGFR2 or 50nM si-circFGFR2 for 48 h. (**E,F**) Cell cycle analysis of chicken primary myoblasts transfected with 1 μg circFGFR2 pCD2.1-circFGFR2 or 50 nM si-circFGFR2 for 36 h. (**G**) 5-ethynyl-2′-deoxyuridine (EdU) assays for chicken primary myoblasts transfected with 1 μg circFGFR2 pCD2.1-circFGFR2 or 50 nM si-circFGFR2 for 36 h. (**H,I**) The percentage of EdU-stained chicken primary myoblasts after overexpression or knockdown of circFGFR2 for 36 h. (**J,K**) Cell cycle analysis of QM-7 cells transfected with 1 μg circFGFR2 pCD2.1-circFGFR2 or 50 nM si-circFGFR2 for 48 h. (**L**) EdU assays for QM-7 cells transfected with 1 μg circFGFR2 pCD2.1-circFGFR2 or 50 nM si-circFGFR2 for 48 h. (**M,N**) The percentage of EdU-stained chicken primary myoblasts after overexpression or knockdown of circFGFR2 for 48 h. In all panels, the results are shown as mean ± S.E.M., and the data are represented by three independent assays. Statistical significance of differences between means was assessed using an unpaired Student's *t*-test (* $p < 0.05$; ** $p < 0.01$).

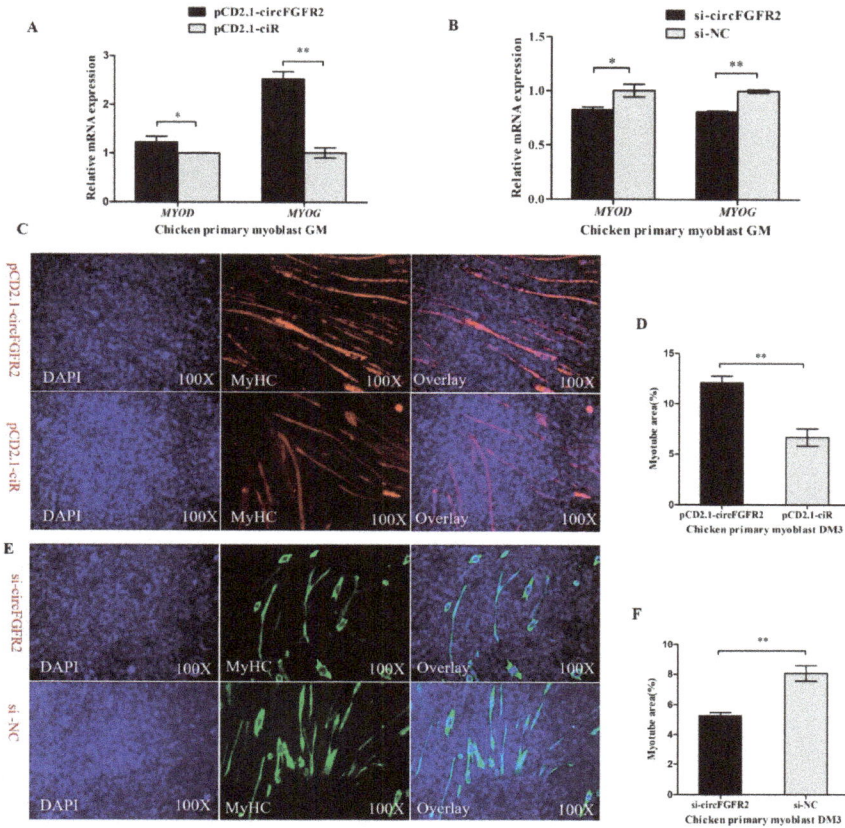

Figure 2. CircFGFR2 promotes myoblast differentiation. (**A**) Overexpression of circFGFR2 promotes mRNA expression of MYOD and MYOG. (**B**) Knockdown of circFGFR2 inhibits the mRNA expression of MYOD and MYOG. (**C,D**) Overexpression of circFGFR2 facilitates the formation of myotubes. (**E,F**) Down-regulation of circFGFR2 suppresses the formation of myotubes. In all panels, data are presented as mean ± S.E.M. of three biological replicates. Statistical significance of differences between means was assessed using an unpaired Student's *t*-test (* $p < 0.05$; ** $p < 0.01$).

3.3. CircFGFR2 Interacts with miR-133a-5p and miR-29b-1-5p, and Inhibits the Expression of miR-133a-5p and miR-29b-1-5p in Myoblast

Circular RNA has been shown to act as miRNA sponge and circFGFR2 could promote myoblast proliferation and differentiation. We hypothesized that circFGFR2 exerts function by acting as miRNA sponge as well as regulating the expression of miRNA. To screen potential miRNAs that bind to circFGFR2, we used RNAhybrid to conduct the putative combination site between circFGFR2 and miR-133a-5p/miR-29b-1-5p. Interestingly, we found that circFGFR2 has two potential miR-133a-5p binding sites (binding site 1and binding site 2) and one potential miR-29b-1-5p binding site. The potential miR-29b-1-5p binding site shares six of seven nucleotides with the binding site 2 of miR-133a-5p. The mature sequence of miR-133a-5p/miR-29b-1-5p and the predicted binding sites of these two miRNAs are shown in Figure 3A–D.

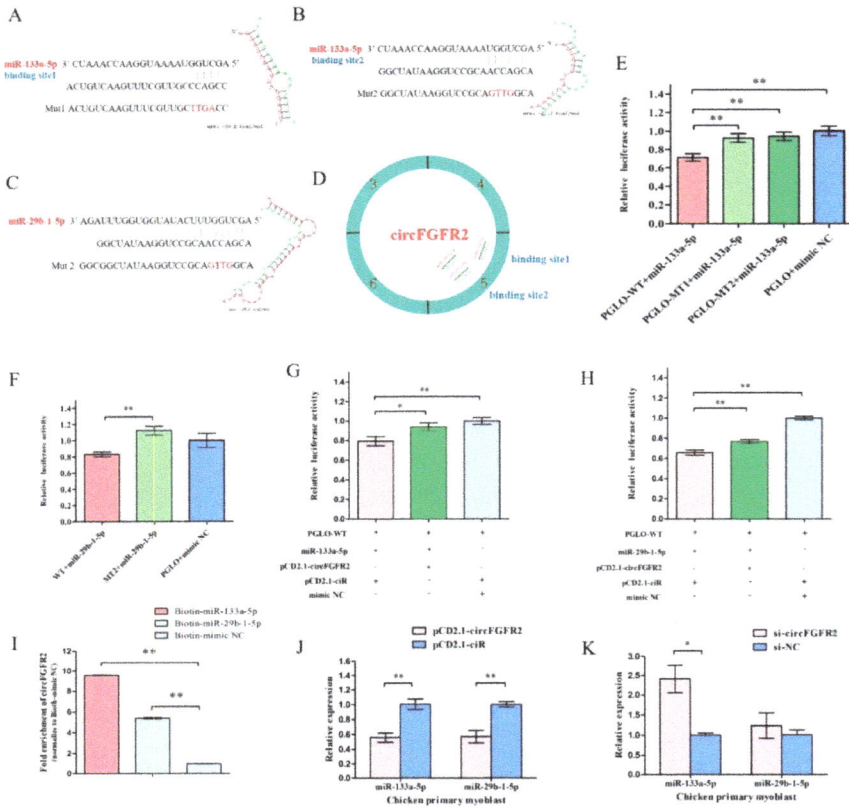

Figure 3. CircFGFR2 sponges with miR-133a-5p and miR-29b-1-5p, and inhibits the expression of miR-133a-5p and miR-29b-1-5p in myoblast. (**A–C**) The potential binding sites of miR-133a-5p and miR-29b-1-5p in circFGFR2. The mutant sequences in binding sites are highlighted in red. (**D**) A schematic drawing showing the putative binding sites of miR-133a-5p/miR-29b-1-5p associated with circFGFR2. (**E,F**) Luciferase assay was conducted by co-transfecting wild type or mutant linear sequence of circFGFR2 with miR-133a-5p/miR-29b-1-5p mimic or mimic-NC in DF-1 cells. (**G,H**) Luciferase assay was conducted by co-transfecting wild type circFGFR2 linear sequence and miR-133a-5p/miR-29b-1-5p mimic or mimic-NC and with circFGFR2 overexpression vector (pCD2.1-circFGFR2) or empty vector (pCD2.1-ciR). (**I**) Biotin-coupled miRNA pull down assay from the myoblast lysates after transfection with 3′ end biotinylated miR-133a-5p, miR-29b-1-5p or mimic NC. The expression level of circFGFR2 was quantified by qRT–PCR, and fold enrichment in the streptavidin captured fractions are plotted. (**J,K**) qRT–PCR analysis of the relative expression of miR-133a-5p and miR-29b-1-5p after overexpression or inhibition of circFGFR2. In all panels, results are expressed as the mean ± S.E.M. of three independent experiments. For two group comparison analysis, statistical significance of differences between means was analyzed by unpaired Student's *t*-test. For multiple comparison analysis, data were analyzed by one-way ANOVA followed by both least significant difference (LSD) and Duncan test through SPSS software. We considered $p < 0.05$ to be statistically significant. * $p < 0.05$; ** $p < 0.01$. NC, negative control.

To investigate the binding site of circFGFR2 with miR-133a-5p/miR-29b-1-5p, we constructed a dual-luciferase reporter by inserting the wild type (WT) or mutant (MT) linear sequence of circFGFR2 into the 3′ end of *firefly* luciferase of pmirGLO (PGLO) luciferase vector to generate a wild type reporter (PGLO-WT) and two mutant reporters (PGLO-MT1 and PGLO-MT2). PGLO-MT1 vector contains the mutated seed sequences for the binding site 1 of mir-133a-5p, and PGLO-MT2 contains the mutated seed sequence for miR-133a-5p binding site 2 and miR-29b-1-5p binding site. The mutant sequences are shown in Figure 3A–C. Then DF-1 cells were co-transfected with PGLO-WT, PGLO-MT1/PGLO-MT2/PGLO luciferase vector and co-transfected with miR-133a-5p/miR-29b-1-5p mimic/control duplexes, respectively. The relative luciferase activity in DF-1 cell line was significantly decreased when miR-133a-5p/miR-29b-1-5p mimic were co-transfected with PGLO-WT reporter (Figure 3E,F) compared with the miR-133a-5p/miR-29b-1-5p mimic and their correspondent mutant reporter co-transfected group. This result demonstrated that miR-133a-5p and miR-29b-1-5p could really combine with the predicted binding sites and miR-133a-5p could combine with both binding site 1 and site 2.

To study the effect of circFGFR2 on the activity of miR-133a-5p/miR-29b-1-5p, we conducted another luciferase reporter assay by co-transfected pCD2.1-circFGFR2 (circFGFR2 overexpression vector)/pCD2.1-ciR (the empty overexpression vector), miR-133a-5p/miR-29b-1-5p/mimic NC with PGLO-WT reporter vector. Luciferase reporter assay showed that the relative luciferase activity was significantly decreased when cells were co-transfected miR-133a-5p/miR-29b-1-5p mimic with PGLO-WT reporter, while the relative luciferase activity was significantly increased when cells were co-transfected the miR-133a-5p/miR-29b-1-5p mimic with pCD2.1-circFGFR2 (Figure 3G,H). It suggested that circFGFR2 could combine with exogenetic miR-133a-5p and miR-29b-1-5p and eliminate the activity of both miRNAs.

Subsequently, we also conducted biotin-coupled miRNA pull down assay to further confirm the interaction between circFGFR2 and miR-133a-5p/miR-29b-1-5p by using biotin-coupled miR-133a-5p and miR-29b-1-5p mimics. Compared with the negative control, we observed more than 8-fold enrichment of circFGFR2 in miR-133a-5p-captured fraction and more than 5-fold enrichment of circFGFR2 in miR-29b-1-5p-captured fraction (Figure 3I), which demonstrated that circFGFR2 could directly sponge miR-133a-5p and miR-29b-1-5p. The greater enrichment observed in miR-133a-5p-captured fraction is probably due to the fact that circFGFR2 contained two binding sites for miR-133a-5p but only one for miR-29b-1-5p.

In addition, the qRT-PCR result showed that overexpression of circFGFR2 could significantly decrease the expression of miR-133a-5p and miR-29b-1-5p (Figure 3J), while knockdown of circFGFR2 could up-regulate the expression of miR-133a-5p and miR-29b-1-5p in chicken primary myoblast (Figure 3K).

3.4. MiR-133a-5p and miR-29b-1-5p Inhibit Myoblast Proliferation

As circFGFR2 had an effect on myoblast proliferation, we also confirmed that circFGFR2 could inhibit the expression and activity of miR-133a-5p and miR-29b-1-5p. We speculated that miR-133a-5p and miR-29b-1-5p had a potential effect on myoblast proliferation. To confirm our hypothesis, we synthesized miR-133a-5p and miR-29b-1-5p mimic. In chicken primary myoblast, we detected the expression of miR-133a-5p and miR-29b-1-5p after transfected chicken primary myoblast with 50 nM miR-133a-5p or miR-29b-1-5p mimic for 48 h. The expression of miR-133a-5p or miR-29b-1-5p was significantly increased by mimic (Figure 4A,B). Subsequently, in chicken primary myoblast, flow cytometry analysis showed that overexpression of miR-133a-5p or miR-29b-1-5p could prominently increase the numbers of cells that progressed to G0/G1 and reduced the numbers of S phase cells (Figure 4C,D). Meanwhile, we found similar results in QM-7 cells as indicated by cycle analysis. MiR-133a-5p or miR-29b-1-5p overexpression significantly increased the number of QM-7 cells that progressed to G0/G1 and reduced the number of S phase cells (Figure 4E,F). Furthermore, the EdU assay demonstrated that overexpression of miR-133a-5p and miR-29b-1-5p dramatically decreased the numbers of EdU strained cells (Figure 4G–J) in both chicken primary myoblast and QM-7 cell, which indicated that miR-133a-5p and miR-29b-1-5p could inhibit the proliferation rate of skeletal muscle cells. These results revealed that miR-133a-5p and miR-29b-1-5p could suppress myoblast proliferation.

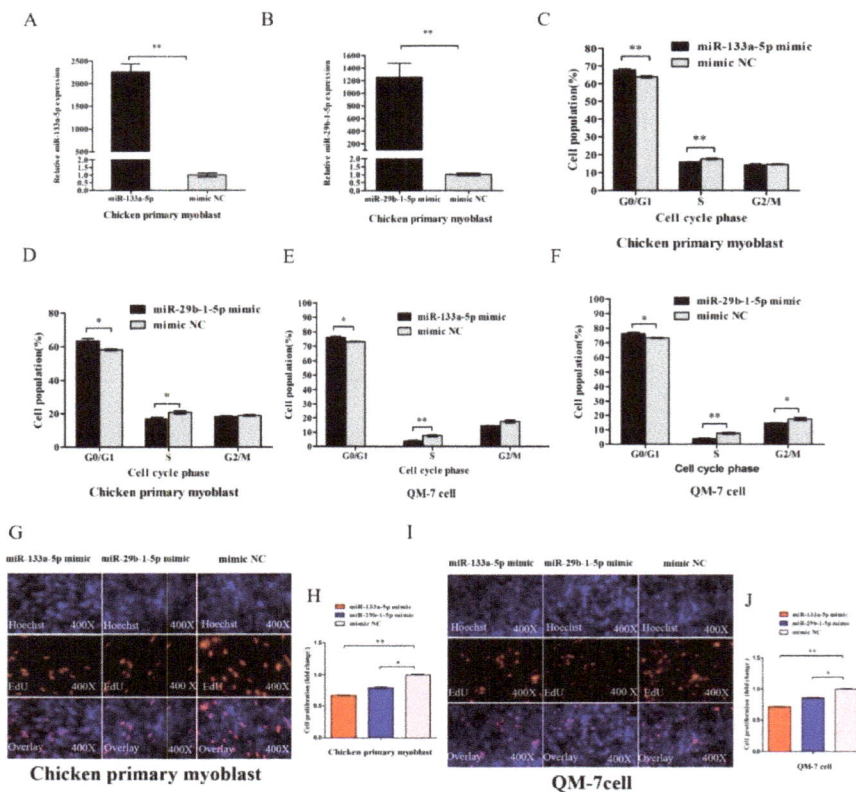

Figure 4. miR-133a-5p and miR-29b-1-5p inhibit myoblast proliferation. (**A,B**) The relative expression of miR-133a-5p and miR-29b-1-5p after transfected chicken primary myoblast with 50 nM miR-133a-5p and miR-29b-1-5p mimic for 48 h. (**C,D**) Cell cycle analysis of chicken primary myoblasts transfected with 50 nM miR-133a-5p and miR-29b-1-5p mimic for 36 h. (**E,F**) Cell cycle analysis of QM-7 cell transfected with 50 nM miR-133a-5p and miR-29b-1-5p mimic for 48 h. (**G,H**) EdU assay of chicken primary myoblasts transfected with 50 nM miR-133a-5p or miR-29b-1-5p mimic for 36 h. (**I,J**) EdU assay of QM-7 cell transfected with 50 nM miR-133a-5p or miR-29b-1-5p mimic for 48 h. In all panels, results are expressed as the mean ± S.E.M. of three independent experiments, and statistical significance of differences between means was assessed using an unpaired Student's *t*-test (* $p < 0.05$; ** $p < 0.01$). NC, negative control.

3.5. CircFGFR2 Eliminates the Inhibition Effect of miR-133a-5p and miR-29b-1-5p on Myoblast Proliferation

Considering the interaction between circFGFR2 and miR-133a-5p/miR-29b-1-5p, rescue experiments were conducted by co-transfecting circFGFR2 with miR-133a-5p/miR-29b-1-5p mimics to assess whether the inhibition on proliferation of two miRNAs could be blocked by circFGFR2 overexpression. As expected, flow cytometry analysis and EdU assay confirmed that circFGFR2 could eliminate the inhibition from overexpressed miR-133a-5p (Figure 5A–F) or miR-29b-1-5p on the proliferation of both chicken primary myoblast and QM-7 cell (Figure 5G–L).

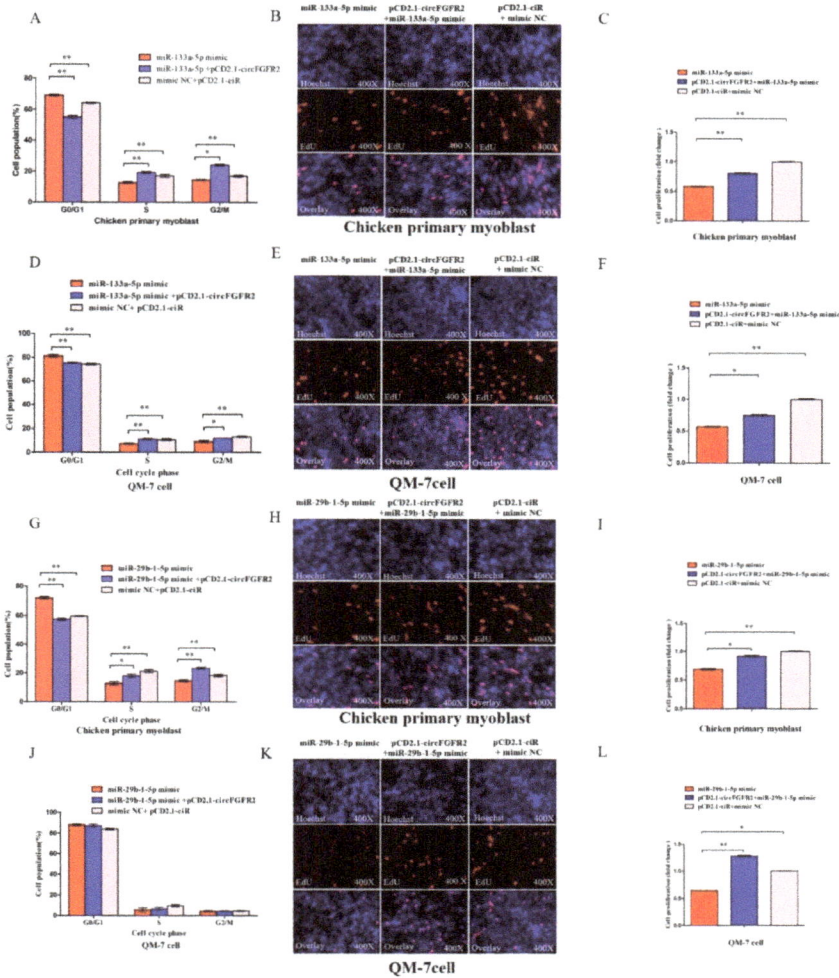

Figure 5. CircFGFR2 eliminates the inhibition effect of miR-133a-5p and miR-29b-1-5p on myoblast proliferation. (**A**) Cell cycle analysis of chicken primary myoblasts after co-transfection with the listed nucleic acids (miR-133a-5p, circFGFR2 overexpression vector and miR-133a-5p, empty overexpression vector and mimic NC, respectively) for 36 h. (**B,C**) EdU assays of chicken primary myoblasts after co-transfection with the listed nucleic acids (miR-133a-5p, circFGFR2 overexpression vector and miR-133a-5p, empty overexpression vector and mimic NC, respectively) for 36 h. (**D**) Cell cycle analysis of QM-7 cells after co-transfection with the listed nucleic acids (miR-133a-5p, circFGFR2 overexpression vector and miR-133a-5p, empty overexpression vector and mimic NC, respectively) for 48 h. (**E,F**) EdU assays of QM-7 cells after co-transfection with the listed nucleic acids (miR-133a-5p, circFGFR2 overexpression vector and miR-133a-5p, empty overexpression vector and mimic NC, respectively) for 48 h. (**G**) Cell cycle analysis of chicken primary myoblasts after co-transfection with the listed nucleic acids (miR-29b-1-5p, circFGFR2 overexpression vector and miR-29b-1-5p, empty overexpression vector and mimic NC, respectively) for 36 h. (**H,I**) EdU assays of chicken primary myoblasts after co-transfection with the listed nucleic acids (miR-29b-1-5p, circFGFR2 overexpression vector and miR-29b-1-5p, empty overexpression vector and mimic NC, respectively) for 36 h. (**J**) Cell cycle analysis

of QM-7 cells after co-transfection with the listed nucleic acids (miR-29b-1-5p, circFGFR2 overexpression vector and miR-29b-1-5p, empty overexpression vector and mimic NC, respectively) for 48 h. (**K,L**) EdU assays of QM-7 cells after co-transfection with the listed nucleic acids (miR-29b-1-5p, circFGFR2 overexpression vector and miR-29b-1-5p, empty overexpression vector and mimic NC, respectively) for 48 h. In all panels, results are expressed as the mean ± S.E.M. of three independent experiments. For two group comparison analysis, statistical significance of differences between means was analyzed by unpaired Student's *t*-test. For multiple comparison analysis, data were analyzed by one-way ANOVA followed by both least significant difference (LSD) and Duncan test through SPSS software. We considered $p < 0.05$ to be statistically significant. * $p < 0.05$; ** $p < 0.01$. NC, negative control.

3.6. miR-133a-5p and miR-29b-1-5p Repress Myoblast Differentiation

To unveil the potential roles of miR-133a-5p and miR-29b-1-5p in chicken primary myoblast differentiation, the expression of the myoblast differentiation marker genes including *MYOG* and *MYOD* were evaluated by qRT-PCR in myoblast transfected with miR-133a-5p or miR-29b-1-5p. Overexpression of miR-133a-5p notably inhibited the expression of *MYOD* and *MYOG*, and overexpression of miR-29b-1-5p could also inhibit the expression of *MYOD* and *MYOG* (Figure 6A,B). Furthermore, we synthesized miR-133a-5p and miR-29b-1-5p inhibitor to down-regulate the expression of miR-133a-5p or miR-29b-1-5p, and we found that down-regulation of miR-133a-5p or miR-29b-1-5p accelerated the expression of *MYOD* and *MYOG* (Figure 6C,D). Subsequently, we induced chicken primary myoblast differentiation in vitro, and we transfected them with miR-133a-5p or miR-29b-1-5p mimic/inhibitor at DM1. MyHC immunofluorescence staining was carried out on the transfected differentiated myoblasts at DM3. According to immunofluorescence staining, we found that the total areas of myotubes of miR-133a-5p or miR-29b-1-5p mimic transfected group were prominently less than that of the control group (Figure 6E,F). On the contrary, the areas of myotubes in miR-133a-5p or miR-29b-1-5p (Figure 6G,H) inhibitor transfected group were more than that of the control group. The results demonstrated that miR-133a-5p and miR-29b-1-5p could repress chicken primary myoblast differentiation.

3.7. CircFGFR2 Eliminates the Inhibition Effect of miR-133a-5p and miR-29b-1-5p on Myoblast Differentiation

We further performed a rescue experiment to investigate whether the suppressing effects of miR-133a-5p and miR-29b-1-5p on myoblast differentiation could be eliminated by circFGFR2 overexpression. As shown in Figure 7A, the expressions of *MYOD* and *MYOG* in miR-133a-5p and circFGFR2 co-transfected group were dramatically elevated compared with the miR-133a-5p transfected group. For miR-29b-1-5p, circFGFR2 also eliminated its repression effect on *MYOD* and *MYOG* (Figure 7B). Further MyHC immunofluorescence showed that overexpression of circFGFR2 eliminated the inhibition on myotube formation at DM3 caused by either miR-133a-5p or miR-29b-1-5p (Figure 7C–F). Taken together, these results demonstrated that circFGFR2 could eliminate the inhibition effect of miR-133a-5p and miR-29b-1-5p on myoblast differentiation.

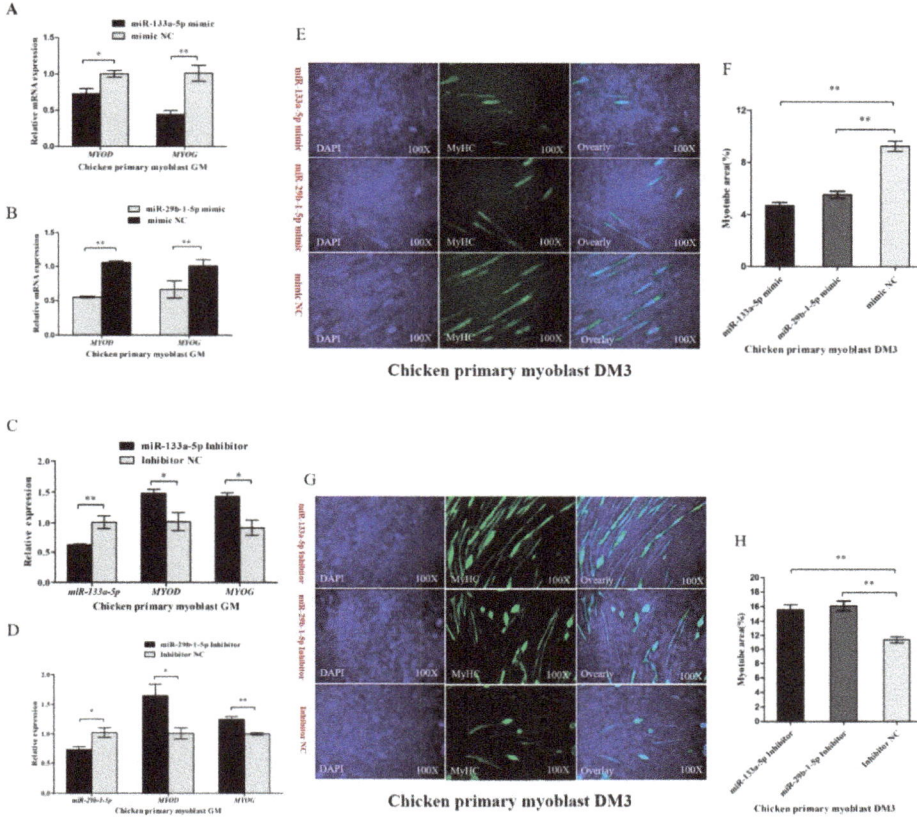

Figure 6. miR-133a-5p and miR-29b-1-5p repress myoblast differentiation. (**A,B**) Overexpression of miR-133a-5p and miR-29b-1-5p reduced the expression of *MYOD* and *MYOG*. (**C,D**) Inhibition of miR-133a-5p and miR-29b-1-5p accelerated the expression of *MYOD* and *MYOG*. (**E,F**) Immunofluorescence analysis of MyHC-staining cells after overexpression miR-133a-5p or miR-29b-1-5p. (**G,H**) Immunofluorescence analysis of MyHC-staining cells after down-regulation of miR-133a-5p or miR-29b-1-5p. In all panels, results are expressed as the mean ± S.E.M. of three independent experiments, and statistical significance of differences between means was assessed using an unpaired Student's *t*-test (* $p < 0.05$; ** $p < 0.01$). NC, negative control.

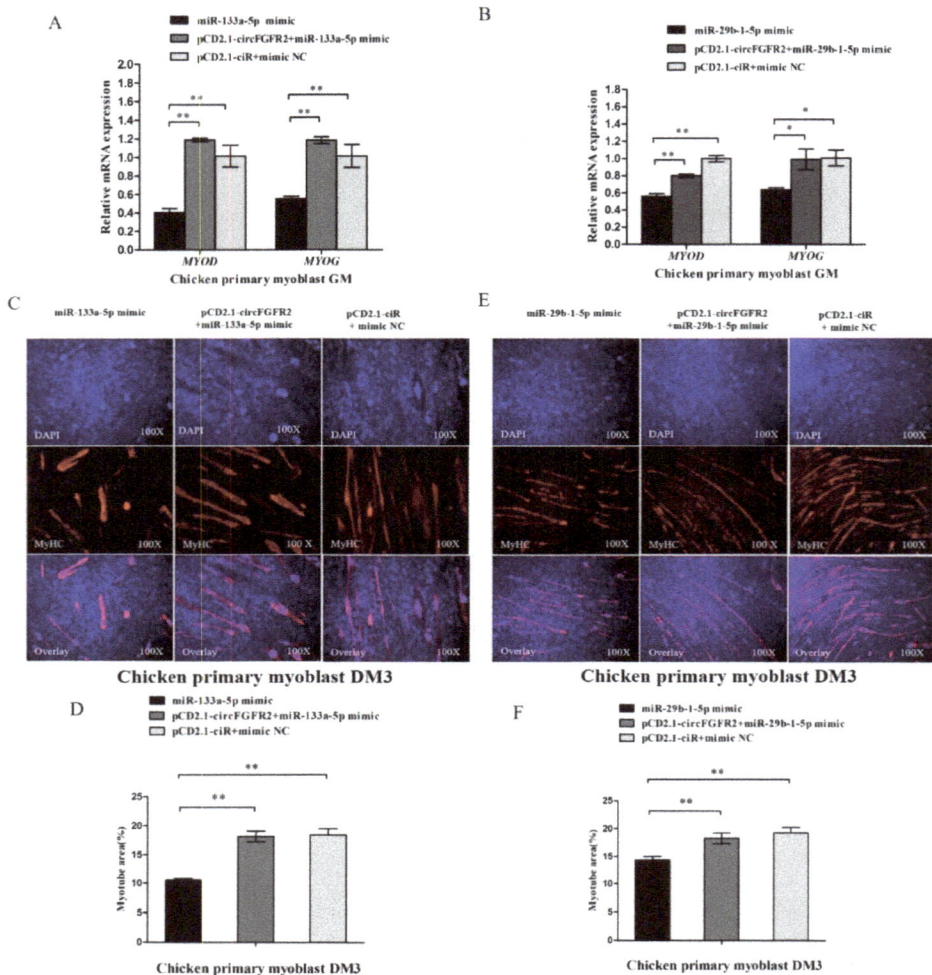

Figure 7. CircFGFR2 eliminates the inhibition effect of miR-133a-5p and miR-29b-1-5p on myoblast differentiation. (**A**) The mRNA expression of *MYOD* and *MYOG* of chicken primary myoblasts after co-transfection with the listed nucleic acids (miR-133a-5p, circFGFR2 overexpression vector and miR-133a-5p, empty overexpression vector and mimic NC, respectively). (**B**) The mRNA expression of *MYOD* and *MYOG* of chicken primary myoblasts after co-transfection with the listed nucleic acids (miR-29b-1-5p, circFGFR2 overexpression vector and miR-29b-1-5p, empty overexpression vector and mimic NC, respectively). (**C,D**) The myotube area of chicken primary myoblasts after co-transfection with the listed nucleic acids (miR-133a-5p, circFGFR2 overexpression vector and miR-133a-5p, empty overexpression vector and mimic NC, respectively). (**E,F**) The myotubes area of chicken primary myoblasts after co-transfection with the listed nucleic acids (miR-29b-1-5p, circFGFR2 overexpression vector, and miR-29b-1-5p, empty overexpression vector and mimic NC, respectively). In all panels, results are expressed as the mean ± S.E.M. of three independent experiments, and statistical significance of differences between means were analyzed by one-way ANOVA followed by both least significant difference (LSD) and Duncan test through SPSS software. We considered $p < 0.05$ to be statistically significant. * $p < 0.05$; * $p < 0.01$. NC, negative control.

4. Discussion

In recent years, circular RNAs have been successfully identified in various cell types across different species [7,9]. They have shown features of dynamic and tissue-specific expression, which indicate a distinct function in diverse tissues [45,46]. CircFGFR2 is a highly expressed DGcircRNA among millions of circRNAs during embryonic muscle development according to our previous circRNA sequencing results [22], which indicates that it has a potential effect in regulating skeletal muscle development. Here we primarily confirmed that circFGFR2 has a crucial function on skeletal muscle development. In both chicken primary myoblast and QM-7 cell, cell cycle analysis demonstrated that overexpression of circFGFR2 could significantly increase the cell numbers in S phase and reduce the cell numbers in G0/G1 phase, while downregulation of circFGFR2 showed the opposite effects. In addition, EdU incorporation assay confirmed that circFGFR2 elevated the cell proliferation rate as shown by overexpression and knockdown of circFGFR2. The results strongly supported that circFGFR2 could promote skeletal muscle cell proliferation. Skeletal myogenesis comes after cell cycle termination, which is coordinated by various regulatory transcription factors, including *MYOD*, *MYOG*, myogenic factor 5 (*Mrf5*), the muscle regulatory factor 4 (*Mrf4*), and myocyte enhancer factor-2 (*Mef2*) families [47,48]. *MYOD* and *MYOG* can regulate most myogenesis-related genes thus facilitating myoblast differentiation into myotubes [49,50]. *MyHC* is a differentiation marker gene of muscle and forms the backbone of the sarcomere thick filaments [51]. The circFGFR2 exerts a function in skeletal muscle cell proliferation, we detected whether circFGFR2 was also involved in skeletal muscle cell differentiation by monitoring the impact of circFGFR2 on the expression of *MYOD* and *MYOG*. As expected, circFGFR2 could promote the expression of *MYOD* and *MYOG*. MyHC immunofluorescence suggested that circFGFR2 accelerated the formation of myotubes, which confirmed another important role of circFGFR2 in skeletal muscle cell, i.e., it can facilitate myoblast differentiation.

Circular RNA is known to be a functional molecule transcribed from protein-encoding genes which contain MREs like other mRNAs or lncRNAs [52]. However, circular RNA was capable of escaping from degradation as it has no poly A tail could not be recognized by exonuclease compared with mRNAs or lncRNAs [5]. In addition, the expression level of some circular RNAs were not lower than their linear mRNAs [53]. Based on that advantage, they are efficient to act as ceRNA, which are enriched for stable miRNA binding sites and regulate the activity of miRNA. Bioinformatics technology is universally applicable for the analysis of the binding relationship of ceRNA and miRNA [54]. In this study, using the bioinformatics program RNAhybrid, we found that circFGFR2 had two possible binding sites for miR-133a-5p and one site for miR-29b-1-5p. Subsequently, we confirmed that miR-133a-5p and miR-29b-1-5p were actually combined with the predicated sites of circFGFR2 but not with *FGFR2* mRNA as indicated by two dual-luciferase reporter assays. Biotin-coupled miRNA pull down is an efficient method to verify the combined relationship between circular RNA and miRNA [18,19,55]. In this study, biotin-miR-133a-5p and biotin-miR-29b-1-5p were efficient in enriching circFGFR2, and overexpression of circFGFR2 significantly inhibits the expression of miR-133a-5p and miR-29b-1-5p which confirm the interacted relationship between circFGFR2 and miR-133a-5p/miR-29b-1-5p.

miR-133a-5p and miR-29b-1-5p belong to two miRNA families, miR-133 and miR-29, respectively. These two families have been well-studied miRNAs, and found to be involved in skeletal muscle cell proliferation and differentiation [27,38,56]. In mouse C2C12 cell line, miR-133 which contain a seed sequence of "UUGGUCC" could promote myoblast differentiation and inhibit cell proliferation, and miR-29 which contains a seed sequence of "AGCACCA" could reduce proliferation and facilitate differentiation [28,56]. The roles of miR-133a-5p and miR-29b-1-5p in avian skeletal muscle development still remain unclear. Here we first reported that miR-133a-5p and miR-29b-1-5p could repress the proliferation and differentiation of skeletal muscle cell. The roles of these two miRNAs were different from the studied miR-133 and miR-29 in mouse. We compared the sequence of miR-133a-5p and miR-29b-1-5p with other miR-133s and miR-29s in both chicken and mouse, and found that the mature sequences of gga-miR-133a-5p and gga-miR-29b-1-5p were different from the studied miR-133 and miR-29. Since the seed sequence was different, and miRNA exerts function by targeting the 3′-UTR of their target genes, it is possibly that the function of gga-miR-133a-5p and gga-miR-29b-1-5p was different from the miR-133 and miR-29 which have been studied in mouse. On the other hand, the roles of gga-miR-133a-5p or gga-miR-29b-1-5p were opposite to the effect of circFGFR2 in myoblast. It is therefore reasonable that circFGFR2 could act as a molecular sponge for miR-133a-5p and miR-29b-1-5p. To confirm this, we further performed rescue experiments and found that circFGFR2 eliminated the inhibition effect of miR-133a-5p and miR-29b-1-5p on myoblast proliferation and differentiation. Considering all of this, we declared that circFGFR2 regulates skeletal muscle cell proliferation and differentiation by inhibiting the expression and activity of miR-133a-5p and miR-29b-1-5p in poultry.

5. Conclusions

In conclusion, we found that a novel circular RNA of circFGFR2, generated by the FGFR2 gene, could regulate myoblast proliferation and differentiation by acting as a sponge of miR-133a-5p and miR-29b-1-5p in poultry.

Author Contributions: X.C. conceived the study, carried out all experiments, analyzed data, and wrote the paper. H.O. provided essential logistical help. Z.W. and B.C. participated in partial experiments. Q.N. conceived the study, and participated in its design and coordination.

Funding: This research was funded by the Natural Scientific Foundation of China (31472090), the China Agriculture Research System (CARS-41-G03) and the Graduate Student Overseas Study Program from the South China Agricultural University (2017LHPY025).

Acknowledgments: We thank the Chicken Breeding Farm of South China Agricultural University for providing the eggs for hatching chickens. We thank Endashaw Jebessa for his edit of the manuscript.

Conflicts of Interest: The authors declare that they have no conflict of interest.

References

1. Sanger, H.L.; Klotz, G.; Riesner, D.; Gross, H.J. Kleinschmidt, A.K. Viroids are single-stranded covalently closed circular RNA molecules existing as highly base-paired rod-like structures. *Proc. Natl. Acad. Sci. USA* **1976**, *73*, 3852–3856. [CrossRef] [PubMed]
2. Capel, B.; Swain, A.; Nicolis, S.; Hacker, A.; Walter, M.; Koopman, P.; Goodfellow, P.; Lovell-Badge, R. Circular transcripts of the testis-determining gene Sry in adult mouse testis. *Cell* **1993**, *73*, 1019–1030. [CrossRef]
3. Arnberg, A.C.; Van Ommen, G.J.; Grivell, L.A.; Van Bruggen, E.F.; Borst, P. Some yeast mitochondrial RNAs are circular. *Cell* **1980**, *19*, 313–319. [CrossRef]
4. Cocquerelle, C.; Mascrez, B.; Hetuin, D.; Bailleul, B. Mis-splicing yields circular RNA molecules. *FASEB J.* **1993**, *7*, 155–160. [CrossRef] [PubMed]
5. Jeck, W.R.; Sharpless, N.E. Detecting and characterizing circular RNAs. *Nat. Biotechnol.* **2014**, *32*, 453–461. [CrossRef] [PubMed]
6. Zhang, Z.; Qi, S.; Tang, N.; Zhang, X.; Chen, S.; Zhu, P.; Ma, L.; Cheng, J.; Xu, Y.; Lu, M.; et al. Discovery of replicating circular RNAs by RNA-seq and computational algorithms. *PLoS Pathog.* **2014**, *10*, e1004553. [CrossRef] [PubMed]
7. Salzman, J.; Gawad, C.; Wang, P.L.; Lacayo, N.; Brown, P.O. Circular RNAs are the predominant transcript isoform from hundreds of human genes in diverse cell types. *PLoS ONE* **2012**, *7*, e30733. [CrossRef] [PubMed]
8. Abdelmohsen, K.; Panda, A.C.; De, S.; Grammatikakis, I.; Kim, J.; Ding, J.; Noh, J.H.; Kim, K.M.; Mattison, J.A.; de Cabo, R.; et al. Circular RNAs in monkey muscle: Age-dependent changes. *Aging* **2015**, *7*, 903–910. [CrossRef] [PubMed]
9. Veno, M.T.; Hansen, T.B.; Veno, S.T.; Clausen, B.H.; Grebing, M.; Finsen, B.; Holm, I.E.; Kjems, J. Spatio-temporal regulation of circular RNA expression during porcine embryonic brain development. *Genome Biol.* **2015**, *16*, 245. [CrossRef] [PubMed]
10. Qu, S.; Yang, X.; Li, X.; Wang, J.; Gao, Y.; Shang, R.; Sun, W.; Dou, K.; Li, H. Circular RNA: A new star of noncoding RNAs. *Cancer Lett.* **2015**, *365*, 141–148. [CrossRef] [PubMed]

11. Chen, L.L. The biogenesis and emerging roles of circular RNAs. *Nat. Rev. Mol. Cell Biol.* **2016**, *17*, 205–211. [CrossRef] [PubMed]

12. Du, W.W.; Yang, W.; Chen, Y.; Wu, Z.K.; Foster, F.S.; Yang, Z.; Li, X.; Yang, B.B. Foxo3 circular RNA promotes cardiac senescence by modulating multiple factors associated with stress and senescence responses. *Eur. Heart J.* **2017**, *38*, 1402–1412. [CrossRef] [PubMed]

13. Du, W.W.; Yang, W.; Liu, E.; Yang, Z.; Dhaliwal, P.; Yang, B.B. Foxo3 circular RNA retards cell cycle progression via forming ternary complexes with p21 and CDK2. *Nucleic Acids Res.* **2016**, *44*, 2846–2858. [CrossRef] [PubMed]

14. Chen, C.Y.; Sarnow, P. Initiation of protein synthesis by the eukaryotic translational apparatus on circular RNAs. *Science* **1995**, *268*, 415–417. [CrossRef] [PubMed]

15. Wang, Y.; Wang, Z. Efficient backsplicing produces translatable circular mRNAs. *RNA* **2015**, *21*, 172–179. [CrossRef] [PubMed]

16. Abe, N.; Matsumoto, K.; Nishihara, M.; Nakano, Y.; Shibata, A.; Maruyama, H.; Shuto, S.; Matsuda, A.; Yoshida, M.; Ito, Y.; et al. Rolling Circle Translation of Circular RNA in Living Human Cells. *Sci. Rep.* **2015**, *5*, 16435. [CrossRef] [PubMed]

17. Dong, R.; Zhang, X.O.; Zhang, Y.; Ma, X.K.; Chen, L.L.; Yang, L. CircRNA-derived pseudogenes. *Cell Res.* **2016**, *26*, 747–750. [CrossRef] [PubMed]

18. Hansen, T.B.; Jensen, T.I.; Clausen, B.H.; Bramsen, J.B.; Finsen, B.; Damgaard, C.K.; Kjems, J. Natural RNA circles function as efficient microRNA sponges. *Nature* **2013**, *495*, 384–388. [CrossRef] [PubMed]

19. Wang, K.; Long, B.; Liu, F.; Wang, J.X.; Liu, C.Y.; Zhao, B.; Zhou, L.Y.; Sun, T.; Wang, M.; Yu, T.; et al. A circular RNA protects the heart from pathological hypertrophy and heart failure by targeting miR-223. *Eur. Heart J.* **2016**, *37*, 2602–2611. [CrossRef] [PubMed]

20. Yang, C.; Yuan, W.; Yang, X.; Li, P.; Wang, J.; Han, J.; Tao, J.; Li, P.; Yang, H.; Lv, Q.; et al. Circular RNA circ-ITCH inhibits bladder cancer progression by sponging miR-17/miR-224 and regulating p21, PTEN expression. *Mol. Cancer* **2018**, *17*, 19. [CrossRef] [PubMed]

21. Zheng, Q.; Bao, C.; Guo, W.; Li, S.; Chen, J.; Chen, B.; Luo, Y.; Lyu, D.; Li, Y.; Shi, G.; et al. Circular RNA profiling reveals an abundant circHIPK3 that regulates cell growth by sponging multiple miRNAs. *Nat. Commun.* **2016**, *7*, 11215. [CrossRef] [PubMed]

22. Ouyang, H.; Chen, X.; Wang, Z.; Yu, J.; Jia, X.; Li, Z.; Luo, W.; Abdalla, B.A.; Jebessa, E.; Nie, Q.; et al. Circular RNAs are abundant and dynamically expressed during embryonic muscle development in chickens. *DNA Res.* **2017**. [CrossRef] [PubMed]

23. Ornitz, D.M.; Marie, P.J. Fibroblast growth factor signaling in skeletal development and disease. *Genes Dev.* **2015**, *29*, 1463–1486. [CrossRef] [PubMed]

24. Ambros, V. The functions of animal microRNAs. *Nature* **2004**, *431*, 350–355. [CrossRef] [PubMed]

25. Baek, D.; Villen, J.; Shin, C.; Camargo, F.D.; Gygi, S.P.; Bartel, D.P. The impact of microRNAs on protein output. *Nature* **2008**, *455*, 64–71. [CrossRef] [PubMed]

26. Chen, J.F.; Mandel, E.M.; Thomson, J.M.; Wu, Q.; Callis, T.E.; Hammond, S.M.; Conlon, F.L.; Wang, D.Z. The role of microRNA-1 and microRNA-133 in skeletal muscle proliferation and differentiation. *Nat. Genet.* **2006**, *38*, 228–233. [CrossRef] [PubMed]

27. Luo, Y.; Wu, X.; Ling, Z.; Yuan, L.; Cheng, Y.; Chen, J.; Xiang, C. microRNA133a targets Foxl2 and promotes differentiation of C2C12 into myogenic progenitor cells. *DNA Cell Biol.* **2015**, *34*, 29–36. [CrossRef] [PubMed]

28. Mishima, Y.; Abreu-Goodger, C.; Staton, A.A.; Stahlhut, C.; Shou, C.; Cheng, C.; Gerstein, M.; Enright, A.J.; Giraldez, A.J. Zebrafish miR-1 and miR-133 shape muscle gene expression and regulate sarcomeric actin organization. *Genes Dev.* **2009**, *23*, 619–632. [CrossRef] [PubMed]

29. Kriegel, A.J.; Liu, Y.; Fang, Y.; Ding, X.; Liang, M. The miR-29 family: Genomics, cell biology, and relevance to renal and cardiovascular injury. *Physiol. Genom.* **2012**, *44*, 237–244. [CrossRef] [PubMed]

30. Lee, J.; Lim, S.; Song, B.W.; Cha, M.J.; Ham, O.; Lee, S.Y.; Lee, C.; Park, J.H.; Bae, Y.; Seo, H.H.; et al. MicroRNA-29b inhibits migration and proliferation of vascular smooth muscle cells in neointimal formation. *J. Cell Biochem.* **2015**, *116*, 598–608. [CrossRef] [PubMed]

31. Li, Z.; Hassan, M.Q.; Jafferji, M.; Aqeilan, R.I.; Garzon, R.; Croce, C.M.; van Wijnen, A.J.; Stein, J.L.; Stein, G.S.; Lian, J.B. Biological functions of miR-29b contribute to positive regulation of osteoblast differentiation. *J. Biol. Chem.* **2009**, *284*, 15676–15684. [CrossRef] [PubMed]

32. Fu, Q.; Shi, H.; Shi, M.; Meng, L.; Zhang, H.; Ren, Y.; Guo, F.; Jia, B.; Wang, P.; Ni, W.; et al. bta-miR-29b attenuates apoptosis by directly targeting caspase-7 and NAIF1 and suppresses bovine viral diarrhea virus replication in MDBK cells. *Can. J. Microbiol.* **2014**, *60*, 455–460. [CrossRef] [PubMed]

33. Shen, L.; Song, Y.; Fu, Y.; Li, P. MiR-29b mimics promotes cell apoptosis of smooth muscle cells via targeting on MMP-2. *Cytotechnology* **2018**, *70*, 351–359. [CrossRef] [PubMed]

34. Mott, J.L.; Kobayashi, S.; Bronk, S.F.; Gores, G.J. mir-29 regulates Mcl-1 protein expression and apoptosis. *Oncogene* **2007**, *26*, 6133–6140. [CrossRef] [PubMed]

35. Fabbri, M.; Garzon, R.; Cimmino, A.; Liu, Z.; Zanesi, N.; Callegari, E.; Liu, S.; Alder, H.; Costinean, S.; Fernandez-Cymering, C.; et al. MicroRNA-29 family reverts aberrant methylation in lung cancer by targeting DNA methyltransferases 3A and 3B. *Proc. Natl. Acad. Sci. USA* **2007**, *104*, 15805–15810. [CrossRef] [PubMed]

36. Fu, Q.; Shi, H.; Chen, C. Roles of bta-miR-29b promoter regions DNA methylation in regulating miR-29b expression and bovine viral diarrhea virus NADL replication in MDBK cells. *Arch. Virol.* **2017**, *162*, 401–408. [CrossRef] [PubMed]

37. Zhou, L.; Wang, L.; Lu, L.; Jiang, P.; Sun, H.; Wang, H. A novel target of microRNA-29, Ring1 and YY1-binding protein (Rybp), negatively regulates skeletal myogenesis. *J. Biol. Chem.* **2012**, *287*, 25255–25265. [CrossRef] [PubMed]

38. Wei, W.; He, H.B.; Zhang, W.Y.; Zhang, H.X.; Bai, J.B.; Liu, H.Z.; Cao, J.H.; Chang, K.C.; Li, X.Y.; Zhao, S.H. miR-29 targets Akt3 to reduce proliferation and facilitate differentiation of myoblasts in skeletal muscle development. *Cell Death Dis.* **2013**, *4*, e668. [CrossRef] [PubMed]

39. Winbanks, C.E.; Wang, B.; Beyer, C.; Koh, P.; White, L.; Kantharidis, P.; Gregorevic, P. TGF-beta regulates miR-206 and miR-29 to control myogenic differentiation through regulation of HDAC4. *J. Biol. Chem.* **2011**, *286*, 13805–13814. [CrossRef] [PubMed]

40. Li, J.; Chan, M.C.; Yu, Y.; Bei, Y.; Chen, P.; Zhou, Q.; Cheng, L.; Chen, L.; Ziegler, O.; Rowe, G.C.; et al. miR-29b contributes to multiple types of muscle atrophy. *Nat. Commun.* **2017**, *8*, 15201. [CrossRef] [PubMed]

41. Wang, L.; Zhou, L.; Jiang, P.; Lu, L.; Chen, X.; Lan, H.; Guttridge, D.C.; Sun, H.; Wang, H. Loss of miR-29 in myoblasts contributes to dystrophic muscle pathogenesis. *Mol. Ther.* **2012**, *20*, 1222–1233. [CrossRef] [PubMed]

42. Zanotti, S.; Gibertini, S.; Curcio, M.; Savadori, P.; Pasanisi, B.; Morandi, L.; Cornelio, F.; Mantegazza, R.; Mora, M. Opposing roles of miR-21 and miR-29 in the progression of fibrosis in Duchenne muscular dystrophy. *Biochim. Biophys. Acta* **2015**, *1852*, 1451–1464. [CrossRef] [PubMed]

43. Abmayr, S.M.; Pavlath, G.K. Myoblast fusion: Lessons from flies and mice. *Development* **2012**, *139*, 641–656. [CrossRef] [PubMed]

44. Sassoon, D.A. Myogenic regulatory factors: Dissecting their role and regulation during vertebrate embryogenesis. *Dev. Biol.* **1993**, *156*, 11–23. [CrossRef] [PubMed]

45. Salzman, J.; Chen, R.E.; Olsen, M.N.; Wang, P.L.; Brown, P.O. Cell-type specific features of circular RNA expression. *PLoS Genet.* **2013**, *9*, e1003777. [CrossRef]

46. Westholm, J.O.; Miura, P.; Olson, S.; Shenker, S.; Joseph, B.; Sanfilippo, P.; Celniker, S.E.; Graveley, B.R.; Lai, E.C. Genome-wide analysis of drosophila circular RNAs reveals their structural and sequence properties and age-dependent neural accumulation. *Cell Rep.* **2014**, *9*, 1966–1980. [CrossRef] [PubMed]

47. Dodou, E.; Xu, S.M.; Black, B.L. mef2c is activated directly by myogenic basic helix-loop-helix proteins during skeletal muscle development in vivo. *Mech. Dev.* **2003**, *120*, 1021–1032. [CrossRef]

48. Blum, R.; Vethantham, V.; Bowman, C.; Rudnicki, M.; Dynlacht, B.D. Genome-wide identification of enhancers in skeletal muscle: The role of MYOD1. *Genes Dev.* **2012**, *26*, 2763–2779. [CrossRef] [PubMed]

49. Berkes, C.A.; Tapscott, S.J. MYOD and the transcriptional control of myogenesis. *Semin. Cell Dev. Biol.* **2005**, *16*, 585–595. [CrossRef] [PubMed]

50. Cao, Y.; Kumar, R.M.; Penn, B.H.; Berkes, C.A.; Kooperberg, C.; Boyer, L.A.; Young, R.A.; Tapscott, S.J. Global and gene-specific analyses show distinct roles for MYOD and Myog at a common set of promoters. *EMBO J.* **2006**, *25*, 502–511. [CrossRef] [PubMed]

51. Tajsharghi, H.; Oldfors, A. Myosinopathies: Pathology and mechanisms. *Acta Neuropathol.* **2013**, *125*, 3–18. [CrossRef] [PubMed]

52. Ashwal-Fluss, R.; Meyer, M.; Pamudurti, N.R.; Ivanov, A.; Bartok, O.; Hanan, M.; Evantal, N.; Memczak, S.; Rajewsky, N.; Kadener, S. circRNA biogenesis competes with pre-mRNA splicing. *Mol. Cell* **2014**, *56*, 55–66. [CrossRef] [PubMed]

53. Jeck, W.R.; Sorrentino, J.A.; Wang, K.; Slevin, M.K.; Burd, C.E.; Liu, J.; Marzluff, W.F.; Sharpless, N.E. Circular RNAs are abundant, conserved, and associated with ALU repeats. *RNA* **2013**, *19*, 141–157. [CrossRef] [PubMed]

54. Dudekula, D.B.; Panda, A.C.; Grammatikakis, I.; De, S.; Abdelmohsen, K.; Gorospe, M. CircInteractome: A web tool for exploring circular RNAs and their interacting proteins and microRNAs. *RNA Biol.* **2016**, *13*, 34–42. [CrossRef] [PubMed]

55. Lal, A.; Thomas, M.P.; Altschuler, G.; Navarro, F.; O'Day, E.; Li, X.L.; Concepcion, C.; Han, Y.C.; Thiery, J.; Rajani, D.K.; et al. Capture of microRNA-bound mRNAs identifies the tumor suppressor miR-34a as a regulator of growth factor signaling. *PLoS Genet.* **2011**, *7*, e1002363. [CrossRef] [PubMed]

56. Zhou, L.; Wang, L.; Lu, L.; Jiang, P.; Sun, H.; Wang, H. Inhibition of miR-29 by TGF-beta-Smad3 signaling through dual mechanisms promotes transdifferentiation of mouse myoblasts into myofibroblasts. *PLoS ONE* **2012**, *7*, e33766. [CrossRef] [PubMed]

cells

Article

Relationship between Altered miRNA Expression and DNA Methylation of the DLK1-DIO3 Region in Azacitidine-Treated Patients with Myelodysplastic Syndromes and Acute Myeloid Leukemia with Myelodysplasia-Related Changes

Michaela Dostalova Merkerova [1,*], Hana Remesova [1], Zdenek Krejcik [1], Nikoleta Loudova [1], Andrea Hrustincova [1], Katarina Szikszai [1], Jaroslav Cermak [1], Anna Jonasova [2] and Monika Belickova [1]

[1] Institute of Hematology and Blood Transfusion, U Nemocnice 1, 128 20 Prague 2, Czech Republic; hana.remesova@uhkt.cz (H.R.); zdenek.krejcik@uhkt.cz (Z.K.); nikoleta.loudova@uhkt.cz (N.L.); andrea.mrhalkova@uhkt.cz (A.H.); katarina.szikszai@uhkt.cz (K.S.); jaroslav.cermak@uhkt.cz (J.C.); monika.belickova@uhkt.cz (M.B.)
[2] General University Hospital, 128 08 Prague, Czech Republic; atjonas@hotmail.com
*   Correspondence: Michaela.Merkerova@uhkt.cz; Tel.: +42-02-2197-7231

Received: 14 August 2018; Accepted: 11 September 2018; Published: 14 September 2018

Abstract: The *DLK1–DIO3* region contains a large miRNA cluster, the overexpression of which has previously been associated with myelodysplastic syndromes (MDS). To reveal whether this overexpression is epigenetically regulated, we performed an integrative analysis of miRNA/mRNA expression and DNA methylation of the regulatory sequences in the region (promoter of the *MEG3* gene) in CD34+ bone marrow cells from the patients with higher-risk MDS and acute myeloid leukemia with myelodysplasia-related changes (AML-MRC), before and during hypomethylating therapy with azacytidine (AZA). Before treatment, 50% of patients showed significant miRNA/mRNA overexpression in conjunction with a diagnosis of AML-MRC. Importantly, increased level of *MEG3* was associated with poor outcome. After AZA treatment, the expression levels were reduced and were closer to those seen in the healthy controls. In half of the patients, we observed significant hypermethylation in a region preceding the *MEG3* gene that negatively correlated with expression. Interestingly, this hypermethylation (when found before treatment) was associated with longer progression-free survival after therapy initiation. However, neither expression nor methylation status were associated with future responsiveness to AZA treatment. In conclusion, we correlated expression and methylation changes in the *DLK1–DIO3* region, and we propose a complex model for regulation of this region in myelodysplasia.

Keywords: microRNA; myelodysplastic syndromes; acute myeloid leukemia; azacitidine; 14q32; MEG3

1. Introduction

Myelodysplastic syndromes (MDS) represent a heterogeneous spectrum of hematopoietic stem cell (HSC) disorders characterized by inefficient hematopoiesis, peripheral blood cytopenia, and dysplasia in one or more myeloid cell lineages; they also have a tendency to evolve into acute myeloid leukemia with myelodysplasia-related changes (AML-MRC), which develops in approximately 30–40% of patients [1,2]. In recent years, the hypomethylating agent azacitidine (AZA) has become the standard therapy for higher-risk MDS and AML-MRC. AZA prolongs patient survival, improves clinical outcomes and life quality, and delays progression to AML. However, response to AZA therapy occurs in only approximately 40–50% of MDS patients [3–5]. AZA is a cytidine analog, which at low doses functions as a DNA methyltransferase inhibitor, causing DNA hypomethylation; at high doses, AZA shows direct cytotoxicity to abnormal bone marrow (BM) hematopoietic cells through its incorporation into DNA and RNA, resulting in cell death [6].

The *DLK1–DIO3* genomic region is located on chromosome 14q32 and contains three paternally expressed protein-coding genes (*DLK1, RTL1*, and *DIO3*), three maternally expressed noncoding genes (*MEG3, MEG8*, and antisense *RTL1*), many small nucleolar RNAs (snoRNAs), and one of the largest miRNA clusters (54 miRNAs) in the human genome [7] (Figure 1). Expression within the *DLK1–DIO3* region is under the control of three differentially methylated regions (DMRs): Intergenic DMR (IG-DMR), *DLK1*-DMR, and *MEG3*-DMR [8]. The *MEG3*-DMR overlaps with the promoter and the 5′-end of the *MEG3* gene and includes several binding sites for CCCTC-binding factor (CTCF). CTCF blocks the interaction between enhancers and promoters, causing transcriptional repression. It exerts its regulatory function by binding to unmethylated DNA, thus preventing the expression of target genes [9].

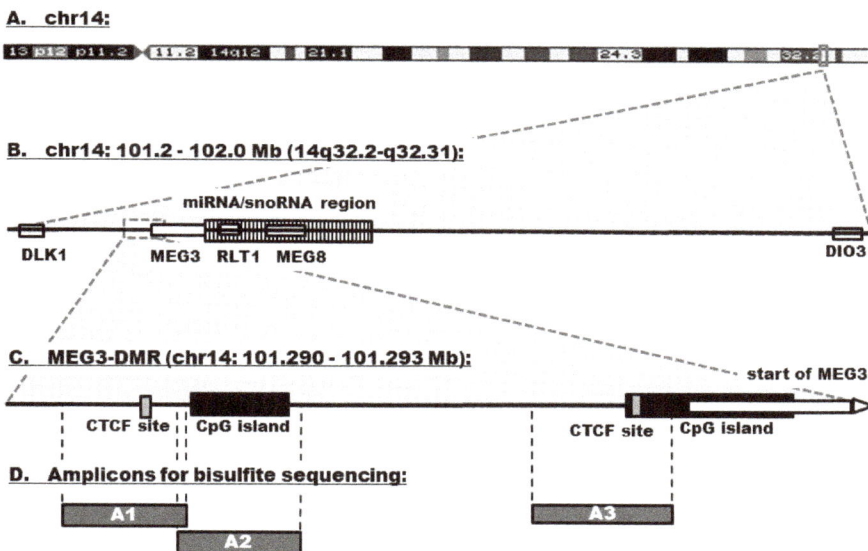

Figure 1. Genomic organization of the *DLK1–DIO3* domain. (**A**) The region is located on chromosome 14q32.2. (**B**) It contains three paternally expressed protein-coding genes (*DLK1, RTL1*, and *DIO3*), three maternally expressed noncoding genes (*MEG3, MEG8*, and antisense *RTL1*), many snoRNAs, and a large cluster of 54 miRNAs (striped bar). (**C**) The expression within the region is under the control of *MEG3*-DMR that overlaps with the promoter and the 5′-end of the *MEG3* gene. *MEG3*-DMR contains several CTFC binding sites (gray bars) and CpG islands (black bars). (**D**) The schema shows localization of amplicons (A1–A3) in the region that were used for methylation analyses.

The miRNAs in the *DLK1–DIO3* region possess oncogenic and tumor suppressor properties and are frequently deregulated in various cancers [7]. Their deregulation has been linked to abnormal induction of apoptosis and suppression of proliferation [10–12]. They are also involved in hematopoietic stem/progenitor cell (HSPC) differentiation [13–15]. The upregulation of miRNAs in the *DLK1–DIO3* region has been reported in acute promyelocytic leukemia (APL, a subtype of AML with t(15;17)) [16–19] and MDS [20–23]. In APL, a comparison of diagnostic and remission patient samples showed that strong upregulation of these miRNAs was correlated with hypermethylation at *MEG3*-DMR, including the CTCF binding site motifs [24]. Only one publication has studied the methylation of this region in MDS patients; Benetatos et al. (2010) showed that hypermethylation of the *MEG3* promoter occurred in 35% of MDS patients [25].

Using SNP array-based karyotyping in cytogenetically normal MDS patients, we previously demonstrated that upregulation of the miRNAs located in the *DLK1–DIO3* domain seen in MDS is probably not caused by any chromosomal aberration or uniparental disomy [22,26]. Due to the similarity between MDS and AML, it may be expected that overexpression of these miRNAs is associated with aberrant hypermethylation of the locus. In this study, we performed a thorough, integrative analysis investigating the expression of miRNAs and mRNAs encoded in the *DLK1–DIO3* region and compared these data with the methylation status of *MEG3*-DMR in CD34+ BM cells (i.e., undifferentiated hematopoietic stem/progenitor cells (HSPCs), including blasts) from higher-risk MDS/AML-MRC patients, particularly focusing on the effects of hypomethylating AZA therapy.

2. Materials and Methods

2.1. Patients

BM samples from 12 patients with higher-risk MDS (N = 7) or AML-MRC (N = 5) without previous malignancy, chemotherapy, or transplantation were obtained between the years 2010–2013 from the First Department of Medicine, Department of Hematology, General University Hospital, and from the Institute of Hematology and Blood Transfusion in Prague, Czech Rep. Paired samples (N = 24) before and during AZA therapy (collected at the time of the best response, between cycles 4 and 11) were collected from each patient. Patient age ranged from 63–82 years (average 69), and the female/male distribution was 6/6. AZA was administered at 75 mg/m^2/day for 7 consecutive days every 28 days. The hematological evaluation of the response to treatment was performed according to the International Working Group (IWG) criteria for MDS and AML [27,28]. Seven patients were considered as responders, and five patients were nonresponders. The detailed clinical characteristics of all patients are summarized in Table S1. All samples from this cohort were used for parallel investigation of miRNA/mRNA expression and DNA methylation. BM samples obtained from age-matched healthy donors with no adverse medical histories were used as controls. Due to limited amounts of sample, we had to use different controls for each type of analysis. Written informed consent was obtained from all test subjects in accordance with the approval of the Institutional Review Board (No. NS9634-4/2008).

2.2. Cell Separation and Nucleic Acid Extraction

Mononuclear cells (MNCs) were separated from BM aspirates by Ficoll-Paque density gradient centrifugation (GE Healthcare, Munich, Germany). CD34+ cells were isolated from MNCs by the Direct CD34 Progenitor Cell Isolation MACS Kit (Miltenyi Biotec, Bergisch Gladbach, Germany). Total RNA was extracted from separated CD34+ cells using the acid guanidinium-thiocyanate-phenol-chloroform method, and the RNA samples were incubated with DNase I (Qiagen, Hilden, Germany) to prevent genomic DNA contamination. The salting-out method was used to isolate DNA from CD34+ cells. The concentrations of DNA and RNA were assessed using a NanoDrop and a Qubit 2.0 Fluorometer (both from Thermo Fisher Scientific, Waltham, MA, USA). The integrity of total RNA was evaluated by an Agilent Bioanalyzer 2100 (Agilent Technologies, Santa Clara, CA, USA).

2.3. miRNA Expression

The miRNA expression data were retrieved from our previous project, which studied miRNA expression changes on a genome-wide level before and during AZA therapy using Agilent Human miRNA Microarrays, Release 19.0, 8 × 60K (Agilent Technologies, Santa Clara, CA, USA) [29]. That study included data on miRNA expression from a complete cohort of 12 patients and four controls. For this publication, we have focused on only miRNAs located in the *DLK1–DIO3* region.

2.4. Gene Expression

Data (NCBI GEO accession number GSE77750) from our previous genome-wide analysis performed using Illumina microarrays (HumanHT-12 v4 Expression, Illumina, San Diego, CA, USA) were used [30]. For this project, the expression of only five genes in the *DLK1–DIO3* region (*DLK1*, *RTL1*, *DIO3*, *MEG3*, and *MEG8*) and of the *CTCF* was further considered within our patient cohort (the same 12 patients with MDS/AML-MRC) and 10 controls.

For validation, the expression of *MEG3* was measured in an independent cohort of samples using RT-qPCR. The cohort included 79 patients with primary MDS (12 MDS with del(5q), 8 MDS with single lineage dysplasia (SLD), 8 MDS with ring sideroblasts and SLD (RS-SLD), 12 MDS with multilineage dysplasia (MLD), 8 MDS-RS-MLD, 10 MDS with excess blasts 1 (EB1), 21 MDS-EB2, 12 AML-MRC, and 13 age-matched healthy controls. SuperScript IV VILO Master Mix and the TaqMan Gene Expression Assay with TaqMan Universal Master Mix (all from Thermo Fisher Scientific, Waltham, MA, USA) were used. The data were normalized to *HTFR* as the reference gene and processed with the ddCt method.

2.5. Bisulfite Sequencing

Genomic DNA isolated from CD34+ BM cells from the 24 samples (the same set of paired samples from 12 patients as in the expression analyses) and three healthy controls was used for analysis of the methylation status of *MEG3*-DMR. Bisulfite conversion was performed using EpiTect Bisulfite Kit (Qiagen, Hilden, Germany). Based on the publication by Manodoro et al. [24], three regions with significant methylation changes in APL were chosen, and previously published primers for these amplicons (A1, A2, and A3) were used (Table S2). Amplicons were generated using FastStart High Fidelity PCR System (Roche Applied Science, Mannheim, Germany) and purified with Agencourt AMPure beads (Beckman Coulter, Krefeld, Germany). The purified PCR products were sequenced using the Roche 454 GS Junior System. Image processing was performed the GS RunBrowser software (Roche Applied Science). Methylation status of CpG sites was quantified using QUMA web-based tool [31].

2.6. Statistical Analyses

Statistical analyses were performed using the GraphPad Prism v7.03 (GraphPad Prism Software, La Jolla, CA, USA). Unpaired and paired t-tests were used to compare miRNA/mRNA expression levels and clinical parameters between different groups of samples. The chi-square test was applied for comparison of categorical clinical variables. Hierarchical clustering analysis was done using average linkage and Euclidean distance. Correlation analyses were done by computing Pearson correlation coefficients (r). Overall survival (OS) and progression-free survival (PFS) curves were generated by the Kaplan-Meier method, and the differences between groups were assessed by the log-rank test. OS and PFS were defined as the time from the beginning of treatment until death from any cause (OS) or disease progression (PFS). Differences were considered statistically significant at $p < 0.05$.

3. Results

3.1. Expression of miRNAs Encoded within the DLK1–DIO3 Region

Within the microarray miRNA expression data, signals of 25 miRNAs located in the *DLK1–DIO3* region were detected. Their levels varied by three orders of magnitude in microarray signal intensity. Before treatment, increased expression of these 25 miRNAs was detected in 6 of 12 samples (50% of patients), and in the remaining 6 samples, the miRNA expression was similar to that detected in controls (Figure 2A).

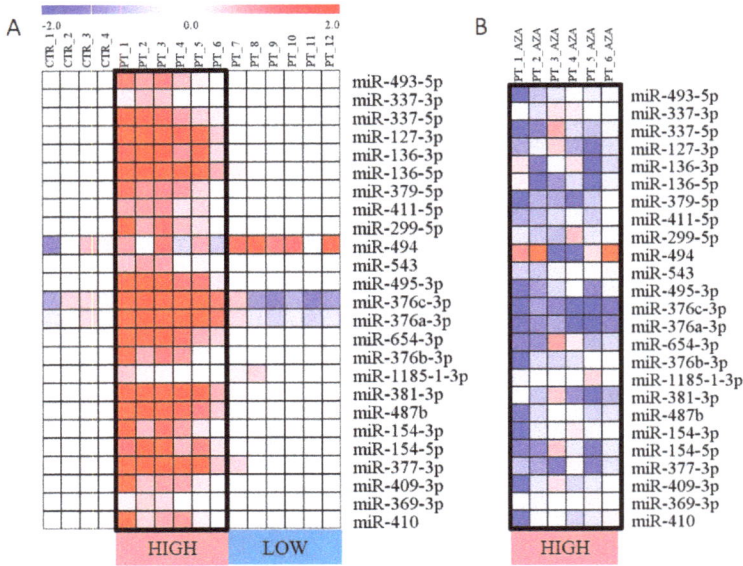

Figure 2. Expression of miRNAs encoded in the *DLK1–DIO3* region. (**A**) miRNA expression in untreated patients (PT) and healthy controls (CTR). The expression level is calculated as the binary logarithm of fold change (logFC) compared to the mean expression of controls. The miRNAs are aligned in the heatmap according to their location on chr14. (**B**) Reduction of miRNA expression following AZA treatment. The changes in miRNA levels are calculated as the logFC between paired samples (AZA treatment vs. pretreatment). Only those patients with miRNA overexpression before treatment are shown. Both heatmaps use a color gradient intensity scale to express visually the logFC values in a range of colors (blue—downregulation, red—upregulation, white—unchanged expression).

Furthermore, we compared miRNA expression as paired samples from pretreatment and during AZA therapy. miRNA expression was reduced following therapy specifically in the patients with overexpressed miRNAs before treatment, and their levels decreased to approximately half the pretreatment expression levels. In contrast, the levels of these miRNAs remained unchanged in the patients with baseline miRNA levels before AZA treatment (Figure 2B).

3.2. Expression of Genes Related to the DLK1–DIO3 Region

Expression of five genes encoded in the *DLK1–DIO3* region (*DLK1, RTL1, DIO3, MEG3,* and *MEG8*) and of the *CTCF* gene were measured using microarrays in the same cohort of patients as for the miRNA analysis. Signals of probes for the *RTL1* and *DIO3* genes were below the detection limit. Expression of *DLK1, MEG3,* and *MEG8* followed the same trend as that of the clustered miRNAs; their levels were increased in the untreated patients with elevated miRNA expression and subsequently reduced during AZA therapy. In the patients with baseline miRNA expression, the transcript levels remained unchanged compared with the levels seen in controls. Interestingly, expression of the *CTCF* gene was also moderately higher ($p = 0.185$) in the patients with increased miRNA expression in the *DLK1–DIO3* region and then reversed to baseline after therapy ($p = 0.026$) (Figure S1). Quantification of *MEG3* expression in a larger, independent cohort of patients showed significant upregulation ($p = 0.0049$) of *MEG3* in MDS/AML-MRC patients compared to healthy donors (Figure 3A). Interestingly, this upregulation occurred more frequently in patients with higher-risk diagnoses than in those with early subtypes of MDS (Figure 3B) and was associated with poor OS (hazard ratio [HR] = 2.6013, 95%

confidence interval [CI] = 1.209 to 5.648, $p = 0.0008$) and PFS (HR = 3.127, 95% CI = 1.411 to 6.933, $p < 0.0001$) in untreated patients (Figure 3C,D).

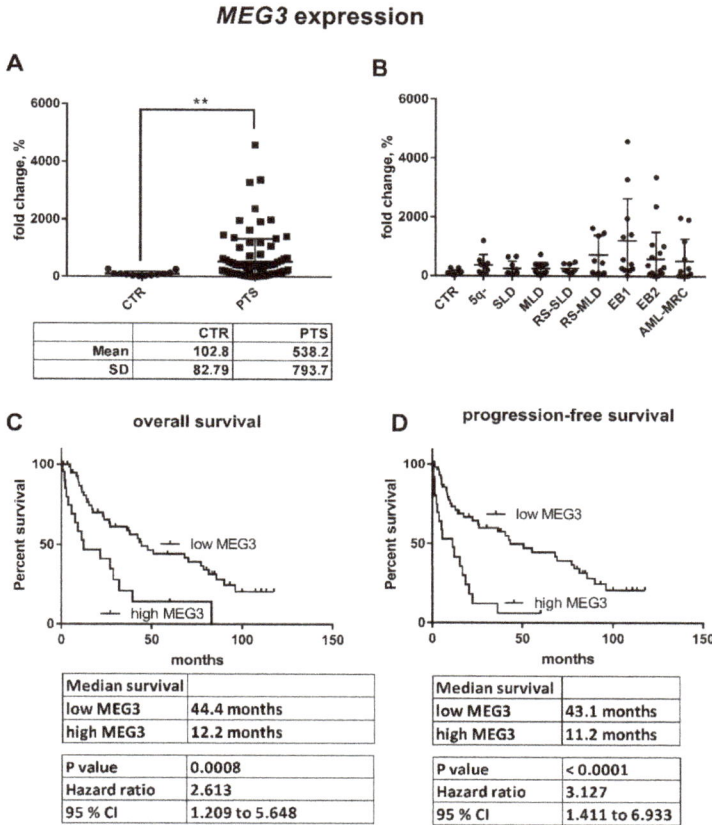

Figure 3. Relative *MEG3* expression in the validation cohort of MDS/AML-MRC patients. (**A**) The expression of *MEG3* was significantly different between healthy controls (CTR) and all patients (PTS) (** $p < 0.01$). (**B**) Further distribution of the patients according to their diagnosis revealed increased expression of *MEG3* particularly in the patients with advanced disease. (**C**) Overall survival and (**D**) progression-free survival of patients with low vs. high *MEG3* expression (the cut-off was set up to mean level of *MEG3*).

3.3. DNA Methylation in the MEG3-DMR

For DNA methylation analysis, amplicon libraries from three distinct regions overlapping the *MEG3*-DMR were prepared, and a total of 53,849 read sequences were generated using amplicon bisulfite sequencing. After a quality control check, 471 sequence reads on average were analyzed per amplicon per sample. To validate the reproducibility of the sequencing, we compared the methylation level of two overlapping CpGs independently sequenced in amplicons A1 and A2, which proved to have a high level of correlation ($r = 0.901$, $p < 0.001$).

Average methylation profiles examined throughout all three amplicons revealed several differentially methylated CpGs among various groups of samples (Figure 4A). The comparison between patients and controls showed significant hypermethylation of almost all CpGs in the A3 region in patients (overall hypermethylation >30% compared with the mean of the controls, which was observed in 50% of untreated patients) (Figure 4A,B). Following AZA therapy, a majority of CpGs from the A2 region (Figure 4C) and one CpG (#236) from the A1 region (Figure 4B) were significantly hypomethylated compared with the pretreatment status, particularly in the patients with increased miRNA levels. Finally, we examined methylation of CpGs in two CTCF binding sites located in our amplicons (CTCF_B: Amplicon A1 CpGs #311-319, and CTCF_D: Amplicon A3 CpGs #308-314) and detected significant hypermethylation of A3 CpG #314 in patient samples (Figure 4B).

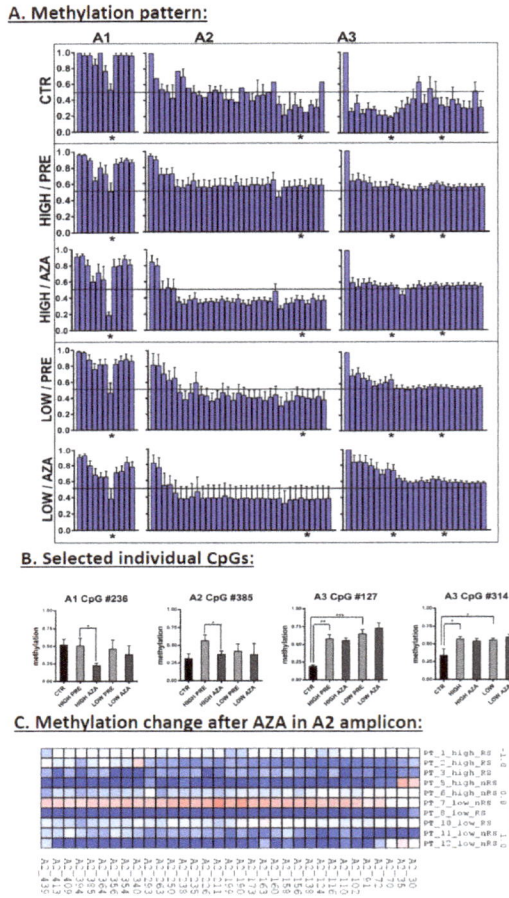

Figure 4. DNA methylation of the *MEG3*-DMR. (**A**) Average methylation profiles examined in the A1–A3 amplicons. Samples were separated into one control group (CTR) and four patient groups that were then divided according to the miRNA expression (HIGH or LOW groups) and AZA therapy (PRE—pretreatment, AZA—AZA-treatment). The horizontal line spanning the plots indicates 50% of methylation. (**B**) Methylation of individual CpGs that are marked in the first section by stars below the graphs. * $p < 0.05$, ** $p < 0.01$, *** $p < 0.001$. (**C**) Methylation change in the A2 amplicon after AZA treatment. Methylation change is expressed as the logFC in AZA-treated samples compared with paired pretreatment samples using a color gradient intensity scale (blue—hypomethylation, red—hypermethylation, white—unchanged methylation). RS—responders to AZA, nRS—nonresponders to AZA.

3.4. Correlation of Expression and Methylation Data

We performed a series of pairwise correlation analyses of the expression of selected miRNAs/mRNAs across all patient data (Table S3). Based on this analysis, we identified strong correlation of expression levels of majority of clustered miRNAs. Moreover, the level of *MEG3* positively correlated with the levels of *MEG8, DLK1,* and the miRNAs. However, *CTCF* expression was not significantly correlated with the levels of any of the tested genes/miRNAs.

Pairwise correlation analyses were performed also for DNA methylation data (Table S3). A significant positive correlation of the methylation levels of individual CpGs located nearby was detected. Four separate clusters regulated in common were identified: CpGs from the A1 amplicon, those from the A2 amplicon, and two separate CpG clusters within the A3 amplicon (A3_1—CpGs #54-171 and A3_2—CpGs #227-388).

Importantly, the correlation of methylation status of individual CpGs with miRNA/mRNA expression levels identified that methylation in a part of the A3 region (CpGs #54–231) is negatively correlated with the expression of the majority of the studied miRNAs/mRNAs, with the exception of miR-136-3p.

3.5. Clinical Characteristics with Relation to the Expression and Methylation Data

Because only a half of the patients showed overexpression of miRNAs/mRNAs from *DLK1–DIO3* region, we divided the cohort according to the expression activity in the locus and analyzed the two resulting groups of patients according to clinical features (age, sex, karyotype, BM blasts, diagnosis, risk category using criteria of the International Prognostic Scoring System (IPSS), and response to hypomethylating therapy) (Table 1). Of these characteristics, only the percentage of BM blasts (patients with low expression: $10.4 \pm 2.8\%$, patients with high expression: $31.6 \pm 9.0\%$; $p = 0.048$) and diagnosis (patients with low expression: 6 MDS patients, patients with high expression: 1 MDS and 5 AML-MRC patients; $p = 0.015$) were significantly associated with expression activity. Concerning cytogenetics, neither particular aberrations nor their classification based on IPSS showed a correlation with expression levels. The rest of the variables, including overall response rate (ORR) to AZA ($p = 0.558$), were not statistically significant.

Table 1. Clinical characteristics of patients and their stratification according to expression activity in the *DLK1–DIO3* locus.

Variable		Patients with Baseline miRNA Expression	Patients with Increased miRNA Expression	*p*-Value
Age (years)	Mean (range)	73 (65–88)	69 (63–81)	0.294
Sex, *n*	Male	4	2	0.248
	Female	2	4	
Karyotype, *n*	Normal	2	1	0.505
	Abnormal	4	5	
Marrow blasts	Mean (range)	10.4 (5.1–20.0)	31.6 (18.0–75.0)	0.048
Diagnosis, *n*	MDS	6	1	0.015
	AML-MRC	0	5	
IPSS, *n*	Intermediate-1/2	3	0	0.182
	High	3	6	
Response to AZA therapy	ORR, *n* (%)	3 (50%)	4 (67%)	0.558

IPSS—International Prognostic Scoring System, ORR—overall response rate.

Further, we investigated the survival of patients after the initiation of AZA therapy in relation to miRNA expression and DNA methylation in pretreatment samples. Concerning miRNA levels, we observed no differences in either OS ($p = 0.901$) or PFS ($p = 0.611$) between the patient with baseline vs. increased expression activity (Figure S2). For analysis of DNA methylation, we compared survival of the patients divided according to their overall methylation status in the three examined amplicons. Importantly, we found that PFS (but not OS) significantly correlated with methylation status of the A3 locus. Log-rank test showed that patients with higher methylation of this locus had a significantly longer PFS than those with lower methylation (HR = 0.220, 95% CI = 0.056 – 0.865, p = 0.030) (Figure 5).

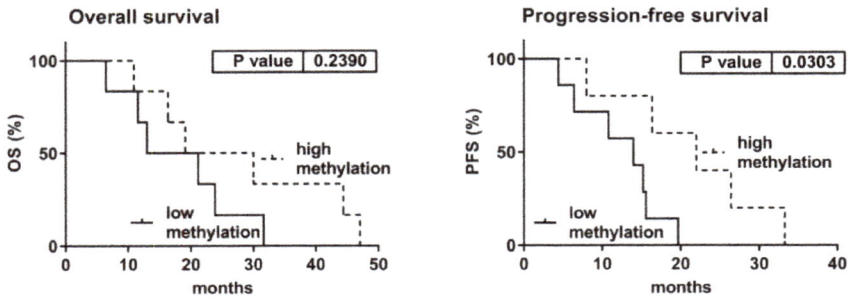

Figure 5. Patient survival after the initiation of AZA therapy according to DNA methylation of the A3 amplicon before treatment. DNA methylation in each sample was considered low or high according to the median value of overall methylation in the locus.

Finally, we investigated the potential application of the altered DNA methylation of the *MEG3*-DMR for prognostic purposes regarding the responsiveness to AZA treatment. However, we did not find any association between AZA response and methylation level in the analyzed amplicons. Although we observed a substantial decrease in DNA methylation in the A2 region following AZA therapy (Figure 4C), this change occurred in the majority of the patients regardless of their response status.

4. Discussion

In several recent publications, apparent overexpression of miRNAs located within the *DLK1–DIO3* region has been observed in MDS patients [20–23]. Following lenalidomide treatment, higher expression levels of these miRNAs diminished in the patients with low and intermediate-1 risk with a del(5q) aberration [22]. Overexpression of these miRNAs has been linked to hypermethylation of the *MEG3*-DMR in APL [24]. Therefore, we investigated the expression activity within the *DLK1–DIO3* region in the context of the methylation status of the *MEG3*-DMR in myelodysplasia. We focused on patients with higher-risk MDS/AML-MRC diagnoses treated with the hypomethylating agent AZA and examined gene expression levels and DNA methylation status in this chromosomal region following treatment. Although performed on a limited number of patients, our study provides a unique dataset focused particularly on the *DLK1–DIO3* region and includes information on miRNA and mRNA expression and methylation data. Furthermore, all the data were obtained from the same cohort of patients before and during hypomethylation therapy.

Initially, we studied expression within the *DLK1–DIO3* region. Our data confirmed that expression activity throughout the region was regulated uniformly because levels of the miRNAs and mRNAs significantly correlated within our patient cohort. In the untreated MDS/AML-MRC patients, 50% of samples showed significant overexpression of miRNAs and mRNAs encoded within the region. Other publications described increased expression of *DLK1–DIO3*-related miRNAs, even in somewhat higher proportions (approximately 90%) of MDS/AML-MRC patients [20–23].

Evaluation of expression data with clinical features showed that expression in the pretreatment samples was related to BM blast count, patient diagnosis, and outcome (OS and PFS). High expression was significantly associated with the diagnosis of AML-MRC and poor outcome, whereas low expression was related to MDS and favorable outcome. Because overexpression of these miRNAs has been linked to a specific cytogenetic aberration [t(15;17)] in primary AML (i.e., APL) [19], we inspected the chromosomes of our patients. However, no chromosomal abnormality or cytogenetic risk category was linked to the expression changes. Therefore, we conclude that the increased expression activity could be related to the process of leukemic transformation. Interestingly, *MEG3* overexpression has also been detected in CD34+ cells from patients with primary myelofibrosis (found in 65% of patients), and this upregulation was related to increased blast counts and higher percentages of circulating CD34+ cells [32].

As half of the patients in our cohort had increased expression activity in the *DLK1–DIO3* region, we examined whether this increase was affected by hypomethylating therapy. Indeed, we observed a reduction in expression to near normal levels after treatment with AZA. To distinguish whether this effect was directly caused by hypomethylation of DNA in the regulatory sequences of the region or if it was an indirect consequence of other changes caused by AZA exposure, we thoroughly examined the methylation pattern in the *MEG3*-DMR.

We investigated DNA methylation in the same regions (located in the *MEG3*-DMR) that were hypermethylated in APL so that we could compare methylation status in APL and myelodysplasia. In APL, the hypermethylation spread over all the three tested amplicons and diminished in remission, whereas we identified a distinct methylation pattern in our cohort of MDS/AML-MRC patients. In untreated patients, we revealed significant hypermethylation solely in the A3 amplicon (that closely precedes the *MEG3* start), whereas the remaining two amplicons had similar levels of methylation as the healthy controls. The hypermethylation in the A3 amplicon was seen in 50% of the MDS/AML-MRC untreated patients. For comparison, Benetatos et al. identified hypermethylation of the *MEG3* promoter in 35% of MDS and 48% of AML patients using conventional methylation-specific PCR [25]. In our patients, methylation of CpGs from the 5' upstream part of the A3 locus negatively correlated with the expression of genes encoded in the *DLK1–DIO3* region. Interestingly, hypermethylation of the A3 locus before treatment had a positive impact on PFS after AZA therapy initiation.

Various methylation patterns were observed in the remaining amplicons. Of them, the hypomethylation seen in the whole A2 region following AZA therapy was remarkable. Although significant in only the patients with increased miRNA levels, this methylation change was apparent in a majority of patients regardless of expression activity in the *DLK1–DIO3* region. Because of its widespread detection, this hypomethylation might be attributed to the hypomethylating effect of AZA and could therefore influence the deregulated expression of the miRNA cluster seen in a portion of our patients.

In APL, Manodoro et al. documented an atypical pattern of hypermethylation in the *DLK1–DIO3* locus associated with higher expression of the clustered miRNAs that they attributed to the presence of CTCF binding sites [24]. A CTCF insulator binds to several target sites (to their unmethylated forms) in the *MEG3*-DMR, inhibiting transcriptional activity in the region [9]. Here, we detected significant hypermethylation at one of the CTCF binding sites in the A3 amplicon. This hypermethylation occurred in a majority of patients and was not affected by AZA treatment. Interestingly, expression of the *CTCF* gene was higher in the untreated patients with increased miRNA expression, which might interfere with the potential impact of hypermethylation of its binding site in the *MEG3*-DMR. Unfortunately, expression of the CTCF insulator was not considered in the publication studying APL [24]. In summary, aberrant hypermethylation at one of the CTCF binding sites in the region might affect the activity of CTCF during transcriptional regulation of the *DLK1–DIO3* region, but this effect is probably not as large as in the APL study.

Finally, we investigated the potential application of the deregulation observed in the *DLK1–DIO3* locus for prognostic purposes regarding AZA treatment. Although we identified changes in expression and methylation levels before AZA therapy, these changes were not associated with future responsiveness to treatment. Concerning survival analyses, only methylation status of the A3 amplicon could be useful for the prediction of PFS during treatment. Kaplan-Meier curves suggested that A3 hypermethylation before treatment might serve as a potential positive prognostic marker for PFS after AZA initiation. However, since we evaluated only a small number of patients and no mutation experiments in animal models were done, strict interpretation of these data must be done with caution.

Epigenetic changes have frequently been demonstrated in MDS, especially in the subtypes with higher blast counts. Aberrant methylation has been found to correlate with poor prognosis and survival even in early stage patients [7,33]. Expression in the *DLK1–DIO3* region is known to be epigenetically regulated, and both expression and methylation of this locus have been recognized as altered in myelodysplasia. Here, we correlated changes in expression activity with specific methylation changes in its regulatory sequences in higher-risk MDS and AML-MRC patients treated with AZA and thoroughly investigated a complex regulatory mechanism within this region. Based on our data, we assumed that epigenetic alterations affect expression activity in the locus. Some of these changes have also been documented in APL [24]; however, other observations have had distinct features. Although further investigation in larger datasets and mutation experiments in animal models are required, we conclude that expression of miRNAs located in the *DLK1–DIO3* region is regulated, at least in part, in a different manner than in APL. It seems unlikely that the increased expression in this region in myelodysplasia can be simply attributed to an individual alteration in DNA methylation because we observed several changes that may affect the resulting state. Other mechanisms besides hypermethylation of the *MEG3*-DMR, such as the influence of transcription factors, may also play yet unrecognized roles in the regulation of the *DLK1–DIO3* region in myelodysplasia.

Supplementary Materials: The following are available at: http://www.mdpi.com/2073-4409/7/9/138/s1. Figure S1: Expression of genes within the DLK1–DIO3 region, Figure S2: Overall survival (A) and progression-free survival (B) of MDS/AML-MRC patients after the initiation of AZA therapy according to pretreatment levels of 14q32 miRNAs, Table S1: Patient characteristics, Table S2: List of primers used for the generation of amplicons for bisulfite sequencing, Table S3: Matrix showing pairwise correlations between expression of mRNAs/miRNAs and methylation of CpG sites.

Author Contributions: Conceptualization, M.D.M.; Formal analysis, M.D.M.; Funding acquisition, M.D.M.; Investigation, H.R., Z.K. and M.B.; Methodology, H.R., Z.K., N.L., A.H. and K.S.; Project administration, M.D.M. and M.B.; Supervision, J.C., A.J. and M.B.; Writing—original draft, M.D.M.; Writing—review & editing, H.R., Z.K., J.C., A.J. and M.B.

Funding: This work was supported by grant Nos. 16-33617A and 17-31398A and by the Project for Conceptual Development of Research Organization No. 00023736 from the Ministry of Health of the Czech Republic.

Acknowledgments: The authors thank Kyra Michalova and Zuzana Zemanova (Center of Oncocytogenetics, Faculty Hospital and First Faculty of Medicine, Charles University, Prague) for the cytogenetic data.

Conflicts of Interest: The authors declare that they have no competing interests.

References

1. Aul, C.; Bowen, D.T.; Yoshida, Y. Pathogenesis, etiology and epidemiology of myelodysplastic syndromes. *Haematologica* **1998**, *83*, 71–86. [PubMed]

2. Garcia-Manero, G. Myelodysplastic syndromes: 2014 update on diagnosis, risk-stratification, and management. *Am. J. Hematol.* **2014**, *89*, 97–108. [CrossRef] [PubMed]

3. Fenaux, P.; Mufti, G.J.; Hellstrom-Lindberg, E.; Santini, V.; Finelli, C.; Giagounidis, A.; Schoch, R.; Gattermann, N.; Sanzet, G.; List, A.; et al. Efficacy of azacitidine compared with that of conventional care regimens in the treatment of higher-risk myelodysplastic syndromes: A randomised, open-label, phase III study. *Lancet Oncol.* **2009**, *10*, 223–232. [CrossRef]

4. Kornblith, A.B.; Herndon, J.E.; Silverman, L.R.; Demakos, E.P.; Odchimar-Reissig, R.; Holland, J.F.; Powell, B.L.; De Castro, C.; Ellerton, J.; Larson, R.A.; et al. Impact of azacytidine on the quality of life of patients with myelodysplastic syndrome treated in a randomized phase III trial: A Cancer and Leukemia Group B study. *J. Clin. Oncol.* **2002**, *20*, 2441–2452. [CrossRef] [PubMed]

5. Vardiman, J.; Reichard, K. Acute Myeloid Leukemia with Myelodysplasia-Related Changes. *Am. J. Clin. Pathol.* **2015**, *144*, 29–43. [CrossRef] [PubMed]

6. Loiseau, C.; Ali, A.; Itzykson, R. New therapeutic approaches in myelodysplastic syndromes: Hypomethylating agents and lenalidomide. *Exp. Hematol.* **2015**, *43*, 661–672. [CrossRef] [PubMed]

7. Benetatos, L.; Hatzimichael, E.; Londin, E.; Vartholomatos, G.; Loher, P.; Rigoutsos, I.; Briasoulis, E. The microRNAs within the DLK1-DIO3 genomic region: Involvement in disease pathogenesis. *Cell. Mol. Life Sci.* **2013**, *70*, 795–814. [CrossRef] [PubMed]

8. Kagami, M.; O'Sullivan, M.J.; Green, A.J.; Watabe, Y.; Arisaka, O.; Masawa, N.; Matsuoka, K.; Fukami, M.; Matsubara, K.; Kato, F.; et al. The IG-DMR and the MEG3-DMR at human chromosome 14q32.2: Hierarchical interaction and distinct functional properties as imprinting control centers. *PLoS Genet.* **2010**, *6*, e1000992. [CrossRef] [PubMed]

9. Rosa, A.L.; Wu, Y.; Kwabi-Addo, B.; Coveler, K.J.; Reid Sutton, V.; Shaffer, L.G. Allele-specific methylation of a functional CTCF binding site upstream of MEG3 in the human imprinted domain of 14q32. *Chromosome Res.* **2005**, *13*, 809–818. [CrossRef] [PubMed]

10. Formosa, A.; Markert, E.K.; Lena, A.M.; Italiano, D.; Finazzi-Agro', E.; Levine, A.J.; Bernardiniet, S.; Garabadgiual, A.V.; Melino, G.; Candi, E. MicroRNAs, miR-154, miR-299-5p, miR-376a, miR-376c, miR-377, miR-381, miR-487b, miR-485-3p, miR-495 and miR-654-3p, mapped to the 14q32.31 locus, regulate proliferation, apoptosis, migration and invasion in metastatic prostate cancer cells. *Oncogene* **2014**, *33*, 5173–5182. [CrossRef] [PubMed]

11. Niu, C.S.; Yang, Y.; Cheng, C. MiR-134 regulates the proliferation and invasion of glioblastoma cells by reducing Nanog expression. *Int. J. Oncol.* **2013**, *42*, 1533–1540. [CrossRef] [PubMed]

12. Jin, Y.; Peng, D.; Shen, Y.; Xu, M.; Liang, Y.; Xiao, B.; Lu, J. MicroRNA-376c inhibits cell proliferation and invasion in osteosarcoma by targeting to transforming growth factor-alpha. *DNA Cell Biol.* **2013**, *32*, 302–309. [CrossRef] [PubMed]

13. Choong, M.L.; Yang, H.H.; McNiece, I. MicroRNA expression profiling during human cord blood-derived CD34 cell erythropoiesis. *Exp. Hematol.* **2007**, *35*, 551–564. [CrossRef] [PubMed]

14. Tenedini, E.; Roncaglia, E.; Ferrari, F.; Orlandi, C.; Bianchi, E.; Bicciato, S.; Tagliafico, E.; Ferrari, S. Integrated analysis of microRNA and mRNA expression profiles in physiological myelopoiesis: Role of hsa-mir-299-5p in CD34+ progenitor cells commitment. *Cell Death Dis.* **2010**, *1*. [CrossRef] [PubMed]

15. Wang, F.; Yu, J.; Yang, G.; Wang, X.; Zhang, J. Regulation of erythroid differentiation by miR-376a and its targets. *Cell Res.* **2011**, *21*, 1196–1209. [CrossRef] [PubMed]

16. Jongen-Lavrencic, M.; Sun, S.M.; Dijkstra, M.K.; Valk, P.J.M.; Löwenberg, B. MicroRNA expression profiling in relation to the genetic heterogeneity of acute myeloid leukemia. *Blood* **2008**, *111*, 5078–5085. [CrossRef] [PubMed]

17. Valleron, W.; Laprevotte, E.; Gautier, E.; Quelen, C.; Demur, C.; Delabesse, E.; Agirre, X.; Prósper, F.; Kiss, T.; Brousset, P. Specific small nucleolar RNA expression profiles in acute leukemia. *Leukemia* **2012**, *26*, 2052–2060. [CrossRef] [PubMed]

18. Ley, T.J.; Miller, C.; Ding, L.; Raphael, B.J.; Mungall, A.J.; Robertson, A.G.; Hoadley, K.; Triche, T.J.; Laird, P.W.; Baty, J.D.; et al. Genomic and epigenomic landscapes of adult de novo acute myeloid leukemia. *N. Engl. J. Med.* **2013**, *368*, 2059–2074. [PubMed]

19. Dixon-McIver, A.; East, P.; Mein, C.A.; Cazier, J.; Molloy, G.; Chaplin, T.; Lister, T.A.; Young, B.D.; Debernardi, S. Distinctive patterns of microRNA expression associated with karyotype in acute myeloid leukaemia. *PLoS ONE* **2008**, *3*, e2141. [CrossRef] [PubMed]

20. Votavova, H.; Grmanova, M.; Merkerova, M.D.; Belickova, M.; Vasikova, A.; Neuwirtova, R.; Cermak, J. Differential expression of microRNAs in CD34+ cells of 5q- syndrome. *J. Hematol. Oncol.* **2011**. [CrossRef] [PubMed]

21. Merkerova, M.D.; Krejcik, Z.; Votavova, H.; Belickova, M.; Vasikova, A.; Cermak, J. Distinctive microRNA expression profiles in CD34+ bone marrow cells from patients with myelodysplastic syndrome. *Eur. J. Hum. Genet.* **2011**, *19*, 313–319. [CrossRef] [PubMed]

22. Krejčík, Z.; Beličková, M.; Hruštincová, A.; Kléma, J.; Zemanová, Z.; Michalová, K.; Čermák, J.; Jonášová, A.; Merkerová, M.D. Aberrant expression of the microRNA cluster in 14q32 is associated with del(5q) myelodysplastic syndrome and lenalidomide treatment. *Cancer Genet.* **2015**, *208*, 156–161. [CrossRef] [PubMed]

23. Merkerova, M.D.; Krejcik, Z.; Belickova, M.; Hrustincova, A.; Klema, J.; Stara, E.; Zemanova, Z.; Michalova, K.; Cermak, J.; Jonasova, A. Genome-wide miRNA profiling in myelodysplastic syndrome with del(5q) treated with lenalidomide. *Eur. J. Haematol.* **2015**, *95*, 35–43. [CrossRef] [PubMed]

24. Manodoro, F.; Marzec, J.; Chaplin, T.; Miraki-Moud, F.; Moravcsik, E.; Jovanovic, J.V.; Wang, J.; Iqbal, S.; Taussig, D.; Grimwade, D.; et al. Loss of imprinting at the 14q32 domain is associated with microRNA overexpression in acute promyelocytic leukemia. *Blood* **2014**, *123*, 2066–2074. [CrossRef] [PubMed]

25. Benetatos, L.; Hatzimichael, E.; Dasoula, A.; Dranitsaris, G.; Tsiara, S.; Syrrou, M.; Georgiou, I.; Bourantas, K.L. CpG methylation analysis of the MEG3 and SNRPN imprinted genes in acute myeloid leukemia and myelodysplastic syndromes. *Leuk. Res.* **2010**, *34*, 148–153. [CrossRef] [PubMed]

26. Merkerova, M.D.; Bystricka, D.; Belickova, M.; Krejcik, Z.; Zemanova, Z.; Polak, J.; Hajkova, H.; Brezinova, J.; Michalova, K.; Cermak, J. From cryptic chromosomal lesions to pathologically relevant genes: Integration of SNP-array with gene expression profiling in myelodysplastic syndrome with normal karyotype. *Genes Chromosomes Cancer* **2012**, *51*, 419–428. [CrossRef] [PubMed]

27. Cheson, B.D.; Greenberg, P.L.; Bennett, J.M.; Lowenberg, B.; Wijermans, P.W.; Nimer, S.D.; Pinto, A.; Beran, M.; de Witte, T.M.; Stone, R.M.; et al. Clinical application and proposal for modification of the International Working Group (IWG) response criteria in myelodysplasia. *Blood* **2006**, *108*, 419–425. [CrossRef] [PubMed]

28. Cheson, B.D.; Bennett, J.M.; Kopecky, K.J.; Büchner, T.; Willman, C.L.; Estey, E.H.; Schiffer, C.A.; Doehner, H.; Tallman, M.S.; Lister, T.A.; et al. Revised recommendations of the International Working Group for Diagnosis, Standardization of Response Criteria, Treatment Outcomes, and Reporting Standards for Therapeutic Trials in Acute Myeloid Leukemia. *J. Clin. Oncol.* **2003**, *21*, 4642–4649. [CrossRef] [PubMed]

29. Krejcik, Z.; Belickova, M.; Hrustincova, A.; Votavova, H.; Jonasova, A.; Cermak, J.; Dyr, J.E.; Merkerova, M.D. MicroRNA profiles as predictive markers of response to azacitidine therapy in myelodysplastic syndromes and acute myeloid leukemia. *Cancer Biomark.* **2018**, *22*, 101–110. [CrossRef] [PubMed]

30. Belickova, M.; Merkerova, M.D.; Votavova, H.; Valka, J.; Vesela, J.; Pejsova, B.; Hajkova, H.; Klemaet, J.; Cermak, J.; Jonasova, A. Up-regulation of ribosomal genes is associated with a poor response to azacitidine in myelodysplasia and related neoplasms. *Int. J. Hematol.* **2016**, *104*, 566–573. [CrossRef] [PubMed]

31. Kumaki, Y.; Oda, M.; Okano, M. QUMA: Quantification tool for methylation analysis. *Nucleic Acids Res.* **2008**, *36*, W170–W175. [CrossRef] [PubMed]

32. Pennucci, V.; Zini, R.; Norfo, R.; Guglielmelli, P.; Bianchi, E.; Salati, S.; Sacchi, G.; Prudente, Z.; Tenedini, E.; Ruberti, S.; et al. Abnormal expression patterns of WT1-as, MEG3 and ANRIL long non-coding RNAs in CD34+ cells from patients with primary myelofibrosis and their clinical correlations. *Leuk. Lymphoma* **2015**, *56*, 492–496. [CrossRef] [PubMed]

33. Boultwood, J.; Wainscoat, J.S. Gene silencing by DNA methylation in haematological malignancies. *Br. J. Haematol.* **2007**, *138*, 3–11. [CrossRef] [PubMed]

cells

MDPI

Article

Integration of miRNA and mRNA Co-Expression Reveals Potential Regulatory Roles of miRNAs in Developmental and Immunological Processes in Calf Ileum during Early Growth

Duy N. Do [1,2], Pier-Luc Dudemaine [1], Bridget E. Fomenky [1,3] and Eveline M. Ibeagha-Awemu [1,*]

[1]   Agriculture and Agri-Food Canada, Sherbrooke Research and Development Centre,
      Sherbrooke, QC J1M 0C8, Canada; DuyNgoc.Do@AGR.GC.CA (D.N.D.);
      Pier-Luc.Dudemaine@AGR.GC.CA (P.-L.D.); bridget.fomenky.1@ulaval.ca (B.E.F.)
[2]   Department of Animal Science, McGill University, Ste-Anne-de-Bellevue, QC H9X 3V9, Canada
[3]   Département de Sciences Animale, Université Laval, Quebec, QC G1V 0A6, Canada
[*] Correspondence: eveline.ibeagha-awemu@agr.gc.ca; Tel.: +1-819-780-7249

Received: 13 August 2018; Accepted: 5 September 2018; Published: 11 September 2018

Abstract: This study aimed to investigate the potential regulatory roles of miRNAs in calf ileum developmental transition from the pre- to the post-weaning period. For this purpose, ileum tissues were collected from eight calves at the pre-weaning period and another eight calves at the post-weaning period and miRNA expression characterized by miRNA sequencing, followed by functional analyses. A total of 388 miRNAs, including 81 novel miRNAs, were identified. A total of 220 miRNAs were differentially expressed (DE) between the two periods. The potential functions of DE miRNAs in ileum development were supported by significant enrichment of their target genes in gene ontology terms related to metabolic processes and transcription factor activities or pathways related to metabolism (peroxisomes), vitamin digestion and absorption, lipid and protein metabolism, as well as intracellular signaling. Integration of DE miRNAs and DE mRNAs revealed several DE miRNA-mRNA pairs with crucial roles in ileum development (bta-miR-374a—*FBXO18*, bta-miR-374a—*GTPBP3*, bta-miR-374a—*GNB2*) and immune function (bta-miR-15b—*IKBKB*). This is the first integrated miRNA-mRNA analysis exploring the potential roles of miRNAs in calf ileum growth and development during early life.

Keywords: calf; Ileum; miRNA-mRNA integration; miRNA sequencing; growth; development

1. Introduction

MicroRNAs (miRNAs) are small (~22 nucleotides) endogenous RNA molecules that regulate gene expression post-transcriptionally by targeting principally the 3′ untranslated region (3′UTR) of genes and to some extend the 5′UTR, introns and coding region of mRNAs [1]. They play key roles in a wide range of biological processes [2]. In bovine, miRNAs have been shown to play important roles in embryonic development [3–5], mammary gland [6,7] and adipose tissue [8] functions and also in the regulation of production traits such as milk yield [6,9,10], milk quality [11] and diseases like mastitis [12,13] and Johne's disease [14]. The importance of miRNAs in gut development and disease (mostly inflammatory bowel disease) has been extensively studied in humans [15–18]. For instance, several miRNAs have been reported to be relevant for different aspects of inflammatory bowel disease (IBD), including miRNAs important for intestinal fibrosis (miR-29, miR-200b, miR-21, miR-192), epithelial barrier and immune function in IBD pathogenesis (miR-192, miR-21, miR-126, miR-155, miR-106a) [19].

Data from a few studies suggest important roles for miRNAs in calf's early life [20,21]. For instance, Liang et al. (2016a) [21] identified several dominantly expressed miRNAs (miR-143 (30% of read counts), miR-192 (15%), miR-10a (12%) and miR-10b (8%)) in ileum tissues of dairy calves collected at 30 min after birth and at 7, 21 and 42 days old. Furthermore, several temporally expressed miRNAs (miR-146, miR-191, miR-33, miR-7, miR-99/100, miR-486, miR-145, miR-196 and miR-211), regional specific miRNAs (miR-192/215, miR-194, miR-196, miR-205 and miR-31) and miRNAs (miR-15/16, miR-29 and miR-196) linked to bacterial abundance in the jejunum and ileum were also reported [21]. Moreover, several ileum miRNAs are reported to play important roles in host responses to *Mycobacterium avium* subspecies paratuberculosis infection, such as the role of bta-miR-196b in the proliferation of endothelial cells and bta-miR-146b in bacteria recognition and regulation of the inflammatory response [22].

During early life, calves undergo major physiological and digestive changes, including adaptation to diet changes from pre- to post-weaning. The interactions between transcriptional and post-transcriptional mechanisms are known to coordinate these developmental transitions via regulation of gene expression. Recently, we characterized the long non-coding RNA (lncRNA) expression in ileum tissues of calves and functional inference of identified lncRNA (623 known and 1505 novel) *cis* target genes revealed potential roles in growth and development as well as in posttranscriptional gene silencing by RNA or miRNA processing processes and in disease resistance mechanisms [23]. Moreover, we also observed that 122 miRNAs were significantly differentially expressed between the pre- and post-weaning periods in the rumen, suggesting important roles of miRNAs in calf gut during early life [24]. Therefore, we hypothesize that miRNAs might play important roles in ileum development during the pre- and post-weaning periods. Thus, in the current study, we performed integrated miRNA-mRNA co-expression analyses to uncover the potential roles of miRNAs in ileum development at the pre- and post-weaning periods.

2. Materials and Methods

2.1. Animals and Management

Procedures for animal management were conducted according to the Canadian national codes of practice for the care and handling of farm animals (http://www.nfacc.ca/codes-of-practice) and approved by the animal care and ethics committee of Agriculture and Agri-Food Canada (CIPA #442).

Experimental details have been described in our previous studies [23,24]. Briefly, sixteen 2–7 days-old Holstein calves were raised for a period of 96 days in individual pens. In the first week of the experiment, calves were fed milk replacer (6 L/day for the first four days and 9 L/day thereafter, Goliath XLR 27-16, La Coop, Montreal, QC, Canada), and then starter feed (Munerie Sawyerville Inc., Cookshire-Eaton, QC, Canada) was introduced (ad libitum) from the second week. After weaning, calves were fed with starter feed and hay ad libitum. The calves were weighed weekly until euthanization. At experiment D33 (pre-weaning), eight calves were humanely euthanized and another eight calves on D96 (post-weaning), for the collection of ileum tissue samples. Tissues were aseptically collected, snap frozen in liquid nitrogen, and stored at −80 °C until used.

2.2. Total RNA Purification

Ileum tissue (30 mg/sample) was used for total RNA isolation using miRNeasy Kit (Qiagen Inc., Toronto, ON, Canada). Potentially contaminating genomic DNA was removed by treating 10 μg of purified RNA (10 μg) with DNase (Turbo DNA-free™ Kit, Ambion Inc., Foster City, CA, USA). The RNA concentration (before and after digestion) and its quality (integrity) after DNase treatment were assessed with Nanodrop ND-1000 (NanoDrop Technologies, Wilmington, DE, USA) and Agilent 2100 Bioanalyzer (Agilent Technologies, Santa Clara, CA, USA), respectively. The RNA integrity number (RIN) of all samples was greater than 8 and a small RNA peak area was visible on the electropherogram [25].

2.3. miRNA Library Preparation and Sequencing

Libraries (*n* = 16) were prepared and barcoded for sequencing according to Do et al. [10]. Briefly, polyacrylamide gel electrophoresis was used to size separate miRNA libraries from other RNA species. An elution buffer (10 mM Tris-HCl pH 7.5; 50 mM NaCl, 1 mM EDTA) was used to elute the libraries from the gel. Eluted library was concentrated using DNA clean and concentrator-5 (Zymo Research, Irvine, CA, USA). The concentration of purified libraries was evaluated using Picogreen assay (Life Technologies, Waltham, MA, USA) and a Nanodrop 3300 fluorescent spectrophotometer (NanoDrop Technologies), and further confirmed by qPCR using the Kapa Library Quantification Kit for Illumina platforms (KAPA Biosystems, Wilmington, MA, USA). Libraries were multiplexed in equimolar concentrations and sequenced in one lane on an Illumina HiSeq 2500 platform following Illumina's recommended protocol to generate single end data of 50-bases by The Centre for Applied Genomics, The Hospital for Sick Children, Toronto, Canada (http://www.tcag.ca/).

2.4. Small RNA Sequence Data Analysis

Bioinformatics processing of generated small RNA sequences was done as previously described [10,24]. Briefly, the raw sequence data (16 fastq files) was checked for sequencing quality with FastQC program v0.11.3 (http://www.bioinformatics.babraham.ac.uk/projects/fastqc/). Trimming of 3′ and 5′ adaptor sequences, contaminants and repeats was accomplished with Cutadapt v1.2.2 (https://cutadapt.readthedocs.org/). Then, FASTQ Quality Filter tool of FASTX-toolkit (http://hannonlab.cshl.edu/fastx_toolkit/) was used to remove reads having a Phred score <20 for at least 50% of the bases and reads shorter than 18 nucleotides or longer than 30 nucleotides. Clean reads that passed all filtering criteria from the 16 files were parsed into one file and mapped to the bovine genome (UMD3.1) using bowtie 1.0.0 (http://bowtie-bio.sourceforge.net/index.shtml) [26]. Reads that mapped to other RNA species (rRNA, tRNA, snRNA and snoRNA) in the Rfam RNA family database (http://rfam.xfam.org/) or to more than five positions of the genome were removed.

2.5. Identification of Known miRNA and Novel miRNA Discovery

The identification of known miRNAs was performed with miRBase v21 http://www.mirbase.org/) (Kozomara and Griffiths-Jones, 2014), while novel miRNA discovery was achieved with miRDeep2 v2.0.0.8 (https://github.com/rajewsky-lab/mirdeep2) [27]. MiRDeep2 was designed to detect miRNAs from deep sequence reads using a probabilistic algorithm based on the miRNA biogenesis model. The core and quantifier modules of miRDeep2 were applied to discover novel miRNAs in the pooled dataset of all the libraries while the quantifier module was used to profile the detected miRNAs in each library. MiRDeep2 score higher than five was used as cuff point for the identification of novel miRNAs. Subsequently, a threshold of 10 counts per million and present in ≥2 libraries was applied to remove lowly expressed miRNAs. MiRNAs meeting these criteria were further used in downstream analyses including differential expression (DE) analysis.

2.6. Differential miRNA Expression

DeSeq2 (v1.14.1) (https://bioconductor.org/packages/release/bioc/html/DESeq2.html) [28], which implements a negative binomial model, was used to perform differential miRNA expression analysis. Following normalization, normalized counts of miRNAs at D96 were compared with corresponding values on D33. Significant differential miRNA expression between D33 (pre-weaning) and D96 (post-weaning) was defined as having a Benjamini and Hochberg [29] false discovery rate (FDR) or corrected *p*-value < 0.05.

2.7. Predicted Target Genes of miRNAs

In order to investigate the functions of the most highly expressed miRNAs and differently expressed (DE) miRNAs, we firstly predicted their target mRNAs. Perl scripts (targetscan_60.pl and targetscan_61_context_scores.pl) (http://targetscan.org) were used to predict target mRNAs and to calculate their context scores, respectively. Predicted target mRNAs with context + scores above 95th percentile were further used [9,10,30]. The predicted target mRNAs were then filtered against the mRNA transcriptome obtained from ileum tissues of the same animals. Only predicted target genes that were expressed in the mRNA transcriptome of the ileum tissues of the animals [23] were retained for further analysis.

2.8. miRNA–mRNA Co-Expression Analysis and Target Gene Enrichment

For miRNA–mRNA co-expression, the Pearson correlation coefficient between target mRNAs (retained above) and DE miRNAs were calculated. A miRNA-mRNA pair was considered co-expressed if it had a negative and significant correlation value at FDR < 0.05. The mRNAs significantly correlated with miRNAs were then used for downstream target gene ontology and KEGG pathways enrichment using ClueGO (http://apps.cytoscape.org/apps/cluego) [31]. For ClueGO analysis, a hypergeometric test was used for enrichment analyses and Benjamini–Hochberg [29] correction was used for multiple testing correction (FDR < 0.05). Since KEGG pathways enrichment relied on the human database (due to lack of information in bovine), we used a less stringent threshold (uncorrected *p*-value < 0.05) to declare if a pathway was significantly enriched. Interactions between miRNAs and mRNAs were visualized with Cytoscape (http://www.cytoscape.org/) [32].

2.9. Real-Time Quantitative PCR

The method of real-time quantitative PCR was used to validate the expression of four DE (bta-miR-142-5p, miR-146a, miR-24-3p and miR-374b) and two non-DE (bta-miR-486-5p and miR-193b) miRNAs. The same total RNA used in miRNA-sequencing was used. Total RNA was reverse transcribed with Universal cDNA Synthesis Kit II from Exiqon (Exiqon Inc., Woburn, MA, USA), following the manufacturer's instructions. ExiLENT SYBR® Green Master Mix Kit (Exiqon, Woburn, MA, USA) and the miRCURY LNA™ Assay (Exiqon, Woburn, MA, USA) specific for each miRNA listed above were used to perform Quantitative qPCR on a StepOne Plus System (Applied Biosystems, Foster City, CA, USA) according to the manufacturer's instructions. Bta-miR-126-3p was used as endogenous control to assess the expression level of miRNAs using the comparative Ct (ΔΔCt) method. Bta-miR-126-3p was selected as an endogenous control based on its consistent expression throughout all the analyzed samples on D33 and D96.

3. Results

3.1. Identification and Characterization of Ileum miRNAs

MiRNA sequencing of 16 libraries generated a total of 185,458,022 reads. After adaptor trimming, size selection and quality filtering, 150,999,506 (81.4%) reads with length ranging from 18 to 30 nucleotides and having a phred score >20 were retained for analysis (Table S1). Out of this number, 133,698,161 reads (88.5%) mapped to unique positions on the bovine genome (University of Maryland assembly of *B. taurus*, release 3.1; UMD.3.1), 10,661,520 (7.1%) were unmapped, while 1,150,263 (0.8%) mapped to more than five positions and were discarded (Table S1). Mapped reads belonging to other RNA species, tRNA (3,153,316 (2.1%)), rRNA (480,099 (0.3%)), snRNA (236,118 (0.2%)) and snoRNA (1,620,029 (1.1%)) were discarded. The majority of miRNAs retained for further analyses were 22 nucleotides long (Table S2).

Novel miRNAs were considered to have a minimum MiRDeep2 score of five, as shown in Table S3. After removing lowly expressed reads, a total of 307 known and 81 novel miRNAs satisfying the conditions of having at least 10 read counts per million and present in a minimum of two libraries were used for DE analysis (Table S4a,b).

Abundantly expressed miRNAs having >3% of the total read counts on D33 and D96 were bta-miR-143, bta-miR-192, bta-miR-26a and bta-miR-21-5p, while bta-miR-191, bta-miR-10b, bta-miR-148a and bta-miR-10a were highly expressed with >3% of total read counts on D96 (post-weaning) only (Table 1). The 20 commonly highly expressed miRNAs (>1% of total read counts) targeted 2609 unique genes (Tables 1 and S5a). The target genes were significantly enriched in 459 biological processes (BP), 53 cellular components (CC) and 43 molecular function (MF) gene ontology (GO) terms (Table S5b), as well as in 14 KEGG pathways (Table S5c). Single-organism developmental process (FDR = 1.13×10^{-10}), intracellular (FDR = 4.63×10^{-17}), and protein binding (FDR = 1.10×10^{-5}) were the most significantly enriched BP, CC and MF GO terms, respectively (Table 2), while MAPK signaling pathway was the most significantly enriched KEGG pathway (Table 3). Moreover, a novel miRNA, bta-miR-22-24033, was the most highly expressed among novel miRNAs (accounted for 0.3% of total read counts) (Table S4b).

Table 1. The 20 most abundantly expressed miRNAs in ileum tissue of calves.

miRNA	Pre-Weaning (D33)		Post-Weaning (D96)		Both Periods	
	Read Counts	% of Total	Read Counts	% of Total	Read Counts	% of Total
bta-miR-143	16,742,092	24.51	7,468,034	13.65	24,210,126	19.68
bta-miR-192	4,435,941	6.50	5,383,934	9.84	9,819,875	7.98
bta-miR-26a	2,861,953	4.19	4,861,644	8.89	7,723,597	6.28
bta-miR-191	1,691,133	2.48	4,106,915	7.51	5,798,048	4.71
bta-miR-10b	1,693,467	2.48	2,031,077	3.71	3,724,544	3.03
bta-miR-148a	1,702,572	2.49	1,754,282	3.21	3,456,854	2.81
bta-miR-10a	1,413,716	2.07	1,734,580	3.17	3,148,296	2.56
bta-miR-21-5p	4,233,604	6.20	1,673,008	3.06	5,906,612	4.80
bta-miR-99a-5p	1,282,425	1.88	1,452,116	2.65	2,734,541	2.22
bta-miR-215	1,320,560	1.93	1,373,575	2.51	2,694,135	2.19
bta-miR-27b	1,557,472	2.28	1,310,625	2.40	2,868,097	2.33
bta-let-7a-5p	1,729,397	2.53	1,231,252	2.25	2,960,649	2.41
bta-let-7f	1,668,877	2.44	1,226,489	2.24	2,895,366	2.35
bta-miR-125b	923,543	1.35	987,775	1.81	1,911,318	1.55
bta-miR-145	1,152,495	1.69	798,715	1.46	1,951,210	1.59
bta-miR-30e-5p	699,838	1.02	735,220	1.34	1,435,058	1.17
bta-let-7g	1,083,909	1.59	685,389	1.25	1,769,298	1.44
bta-miR-194	1,498,650	2.19	571,343	1.04	2,069,993	1.68
bta-miR-30d	747,419	1.09	560,435	1.02	1,307,854	1.06

Table 2. Enriched gene ontology (GO) terms for target genes of 20 most abundantly expressed miRNAs.

GO Class	GOID	GO Term	*p*-Value	FDR
Biological process	GO:0044767	Single-organism developmental process	4.40×10^{-13}	1.13×10^{-10}
	GO:0044260	Cellular macromolecule metabolic process	7.96×10^{-13}	1.79×10^{-10}
	GO:0044237	Cellular metabolic process	3.29×10^{-12}	6.57×10^{-10}
	GO:0007275	Multicellular organismal development	2.62×10^{-11}	4.71×10^{-9}
	GO:0048731	System development	7.45×10^{-11}	1.22×10^{-8}
	GO:0009888	Tissue development	9.21×10^{-11}	1.38×10^{-8}
	GO:0048856	Anatomical structure development	1.24×10^{-10}	1.72×10^{-8}
	GO:0036211	Protein modification process	8.55×10^{-10}	9.61×10^{-8}
	GO:0030154	Cell differentiation	1.74×10^{-9}	1.74×10^{-7}
Cellular component	GO:0043412	Macromolecule modification	3.53×10^{-9}	3.34×10^{-7}
	GO:0005622	Intracellular	2.57×10^{-20}	4.63×10^{-17}
	GO:0044424	Intracellular part	1.35×10^{-19}	1.22×10^{-16}
	GO:0043227	Membrane-bounded organelle	3.43×10^{-17}	2.06×10^{-14}
	GO:0043231	Intracellular membrane-bounded organelle	1.95×10^{-15}	8.79×10^{-13}
	GO:0043229	Intracellular organelle	3.08×10^{-15}	1.11×10^{-12}
	GO:0005737	Cytoplasm	4.55×10^{-15}	1.36×10^{-12}
	GO:0044422	Organelle part	2.39×10^{-10}	3.07×10^{-8}
	GO:0044446	Intracellular organelle part	3.80×10^{-10}	4.56×10^{-8}
	GO:0044444	Cytoplasmic part	9.28×10^{-10}	9.83×10^{-8}
	GO:0005634	Nucleus	7.71×10^{-8}	4.34×10^{-6}
Molecular function	GO:0005515	Protein binding	3.18×10^{-7}	1.10×10^{-5}
	GO:0019207	Kinase regulator activity	4.91×10^{-6}	1.08×10^{-4}
	GO:0019887	Protein kinase regulator activity	3.01×10^{-5}	4.79×10^{-4}
	GO:0003723	RNA binding	3.24×10^{-5}	5.06×10^{-4}
	GO:0019210	Kinase inhibitor activity	3.78×10^{-5}	5.76×10^{-4}
	GO:0004702	Receptor signaling protein serine/threonine kinase activity	8.23×10^{-5}	1.06×10^{-3}
	GO:0005057	Receptor signaling protein activity	1.94×10^{-4}	2.02×10^{-3}
	GO:0061650	Ubiquitin-like protein conjugating enzyme activity	3.20×10^{-4}	2.96×10^{-3}
	GO:0003700	Transcription factor activity, sequence-specific DNA binding	4.99×10^{-4}	4.17×10^{-3}

Table 3. Enriched KEGG pathways for target genes of 20 most abundantly expressed miRNAs.

KEGG Pathway	*p*-Value	FDR
Cysteine and methionine metabolism	2.56×10^{-3}	4.39×10^{-2}
Amino sugar and nucleotide sugar metabolism	1.73×10^{-3}	3.47×10^{-2}
TGF-beta signaling pathway	9.58×10^{-4}	2.30×10^{-2}
Signaling pathways regulating pluripotency of stem cells	6.67×10^{-4}	2.00×10^{-2}
Pathways in cancer	3.81×10^{-5}	4.57×10^{-3}
Transcriptional misregulation in cancer	1.76×10^{-3}	3.26×10^{-2}
Proteoglycans in cancer	1.39×10^{-4}	8.37×10^{-3}
MAPK signaling pathway	1.22×10^{-5}	2.94×10^{-3}
Cell cycle	3.48×10^{-4}	1.67×10^{-2}
p53 signaling pathway	9.32×10^{-4}	2.49×10^{-2}
Protein processing in endoplasmic reticulum	1.29×10^{-3}	2.81×10^{-2}
ErbB signaling pathway	5.55×10^{-4}	1.90×10^{-2}
FoxO signaling pathway	1.04×10^{-4}	8.35×10^{-3}
Chronic myeloid leukemia	5.21×10^{-4}	2.08×10^{-2}

3.2. Differentially Expressed miRNAs and Downstream Target Gene Enrichment Analyses

A total of 220 miRNAs (104 up-regulated and 116 down-regulated) were significantly DE between D33 (pre-weaning) and D96 (post-weaning) (Figure 1, Table S6a). Bta-miR-374a (FDR = 5.00×10^{29}), bta-miR-15b (FDR = 7.96×10^{24}) and bta-miR-26a (FDR = 1.30×10^{20}) were the most significantly down-regulated miRNAs, while bta-miR-455-5p (FDR = 1.01×10^{23}), bta-miR-210 (FDR = 4.23×10^{20}) and bta-miR-497 (FDR = 9.95×10^{20}) were the most significantly up-regulated miRNAs (Table 4).

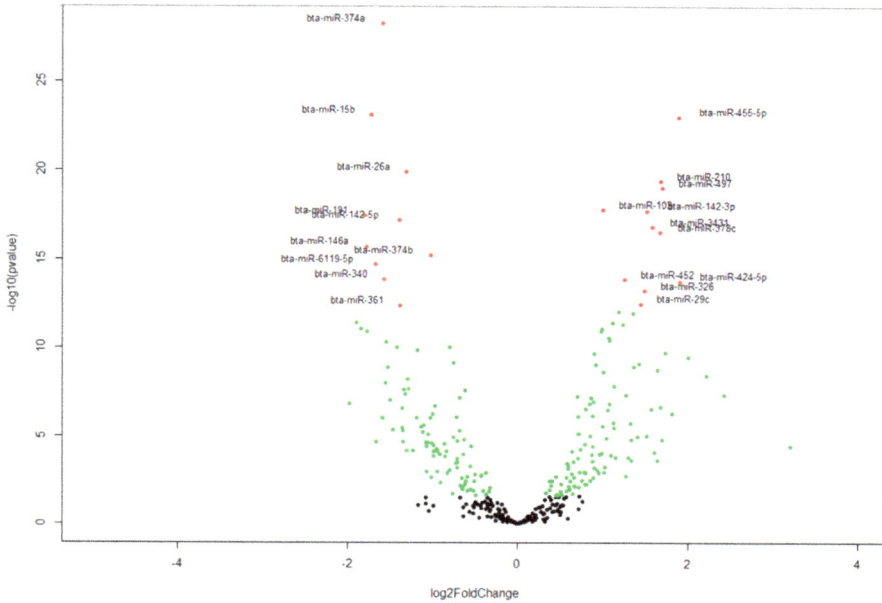

Figure 1. Volcano plot depicting miRNA differential expression results. Each dot represents a miRNA. Green and red dots represent miRNAs significantly differentially expressed at FDR < 0.05 and FDR < 1×10^{11}, respectively. Black dots represent miRNAs that were not differentially expressed. Differential expression analysis was accomplished with DeSeq2.

Table 4. The ten most up- and down-regulated miRNAs between D33 (pre-weaning) and D96 (post-weaning).

miRNA	Base Mean	Log2fold Change	Fold Change	*p*-Value	FDR
bta-miR-374a	8851.83	−1.59	−3.01	5.00×10^{-29}	1.94×10^{-26}
bta-miR-15b	17,755.36	−1.73	−3.32	7.96×10^{-24}	1.31×10^{-21}
bta-miR-26a	494,446.71	−1.32	−2.50	1.30×10^{-20}	1.26×10^{-18}
bta-miR-191	367,869.82	−1.81	−3.51	3.59×10^{-18}	1.55×10^{-16}
bta-miR-142-5p	92,116.08	−1.40	−2.64	6.64×10^{-18}	2.58×10^{-16}
bta-miR-146a	73,690.12	−1.79	−3.45	2.04×10^{-16}	6.08×10^{-15}
bta-miR-374b	4629.32	−1.03	−2.04	6.14×10^{-16}	1.70×10^{-14}
bta-miR-6119-5p	1796.83	−1.68	−3.19	2.08×10^{-15}	5.38×10^{-14}
bta-miR-340	360.85	−1.58	−2.99	1.40×10^{-14}	3.31×10^{-13}
bta-miR-361	9880.75	−1.39	−2.62	4.33×10^{-13}	8.00×10^{-12}
bta-miR-455-5p	282.27	1.89	3.70	1.01×10^{-23}	1.31×10^{-21}
bta-miR-210	486.48	1.67	3.19	4.23×10^{-20}	3.28×10^{-18}
bta-miR-497	692.69	1.70	3.24	9.95×10^{-20}	6.43×10^{-18}
bta-miR-103	39,682.96	1.00	2.00	1.69×10^{-18}	9.35×10^{-17}
bta-miR-142-3p	1742.08	1.52	2.87	1.96×10^{-18}	9.52×10^{-17}
bta-miR-3431	614.77	1.58	2.98	1.58×10^{-17}	5.57×10^{-16}
bta-miR-378c	1268.42	1.67	3.18	3.21×10^{-17}	1.04×10^{-15}
bta-miR-452	393.30	1.26	2.39	1.45×10^{-14}	3.31×10^{-13}
bta-miR-424-5p	301.96	1.90	3.73	1.84×10^{-14}	3.97×10^{-13}
bta-miR-326	378.11	1.49	2.82	6.07×10^{-14}	1.24×10^{-12}

The DE miRNAs (220 miRNAs) were predicted to target 11,691 mRNAs (Table S6b). Using mRNA transcriptome data of the same samples, 1560 mRNAs out of the predicted 11,691 mRNAs, were significantly and negatively correlated with their targeting miRNAs (Table S6c). Bta-miR-2285f had the highest number of target genes (172), while *AGO2* gene was the most popular target for DE miRNAs (targeted by 25 DE miRNAs) (Table S6c). Other common target genes for DE miRNAs were *SLC25A46*, *KCTD13* and *PAXIP1*, each targeted by 9 DE miRNAs (Table S6c). The GO enrichment analyses of the 1560 target genes (significantly and negatively correlated with miRNA) indicated that 158, 26 and 28 of them were significantly enriched in BP-, CC- and MF-GO terms, respectively (Table S7a–c). The most enriched BP-, CC- and MF-GO terms were cellular macromolecule metabolic process (FDR = 9.38×10^{10}), intracellular (FDR = 3.37×10^{19}) and organic cyclic compound binding (FDR = 1.19×10^{4}), respectively (Table 5, Figures 2–4). Moreover, 16 KEGG pathways were significantly enriched for the target genes (1560) of 220 DE miRNAs, and peroxisome (*p* = 0.004) and Hedgehog signaling pathways (*p* = 0.006) were the most significantly enriched (Table S7d, Figure 5). Moreover, among the 1560 target genes negatively correlated with miRNAs, 278 were also significantly DE between D33 and D96 in our previous study (Table S8) [23]. The 278 genes were the targets for 64 DE miRNAs. SOX4 was the most common target, since it was targeted by 6 different miRNAs (bta-miR-191, bta-miR-30e-5p, bta-miR-15-11508, bta-miR-2285f, bta-miR-92b and bta-miR-2285q). Meanwhile, bta-miR-2285f and bta-miR-874 had the highest number of target genes (37 and 28, respectively) (Figure 6).

Table 5. Most significantly enriched gene ontology (GO) terms for target genes of differentially expressed miRNAs.

GO Class	GO ID	GO Term	*p*-Value	FDR
Biological process	GO:0044260	Cellular macromolecule metabolic process	9.38×10^{-10}	5.11×10^{-7}
	GO:0043170	Macromolecule metabolic process	6.49×10^{-10}	7.07×10^{-7}
	GO:0019222	Regulation of metabolic process	9.86×10^{-9}	3.58×10^{-6}
	GO:0048518	Positive regulation of biological process	1.67×10^{-8}	4.55×10^{-6}
	GO:0071704	Organic substance metabolic process	4.25×10^{-8}	7.72×10^{-6}
	GO:0090304	Nucleic acid metabolic process	3.66×10^{-8}	7.97×10^{-6}
	GO:c060255	Regulation of macromolecule metabolic process	9.84×10^{-8}	1.34×10^{-5}
	GO:0044237	Cellular metabolic process	8.67×10^{-8}	1.35×10^{-5}
	GO:0009059	Macromolecule biosynthetic process	1.76×10^{-7}	1.60×10^{-5}
	GO:0048522	Positive regulation of cellular process	1.92×10^{-7}	1.61×10^{-5}
Cellular component	GO:0005622	Intracellular	2.76×10^{-21}	3.37×10^{-19}
	GO:0043229	Intracellular organelle	2.39×10^{-21}	5.85×10^{-19}
	GO:0044424	Intracellular part	1.21×10^{-20}	9.85×10^{-19}
	GO:0043231	Intracellular membrane-bounded organelle	5.54×10^{-18}	3.39×10^{-16}
	GO:0043227	Membrane-bounded organelle	1.80×10^{-17}	8.80×10^{-16}
	GO:0005634	Nucleus	1.41×10^{-12}	5.76×10^{-11}
	GO:0005737	Cytoplasm	1.69×10^{-10}	5.90×10^{-9}
	GO:0005654	Nucleoplasm	1.80×10^{-9}	5.52×10^{-8}
	GO:0044428	Nuclear part	4.96×10^{-9}	1.35×10^{-7}
Molecular function	GO:0044446	Intracellular organelle part	1.06×10^{-8}	2.59×10^{-7}
	GO:0097159	Organic cyclic compound binding	1.19×10^{-6}	1.19×10^{-4}
	GO:1901363	Heterocyclic compound binding	6.87×10^{-7}	1.37×10^{-4}
	GO:0043167	Ion binding	2.32×10^{-5}	1.16×10^{-3}
	GO:0003676	Nucleic acid binding	2.03×10^{-5}	1.36×10^{-3}
	GO:0005515	Protein binding	3.90×10^{-5}	1.56×10^{-3}
	GO:0019207	Kinase regulator activity	5.59×10^{-5}	1.60×10^{-3}
	GO:0019887	Protein kinase regulator activity	5.31×10^{-5}	1.77×10^{-3}
	GO:0043169	Cation binding	1.39×10^{-4}	3.47×10^{-3}
	GO:0003700	Transcription factor activity, sequence-specific DNA binding	1.63×10^{-4}	3.63×10^{-3}
	GO:0019899	Enzyme binding	5.60×10^{-4}	8.61×10^{-3}

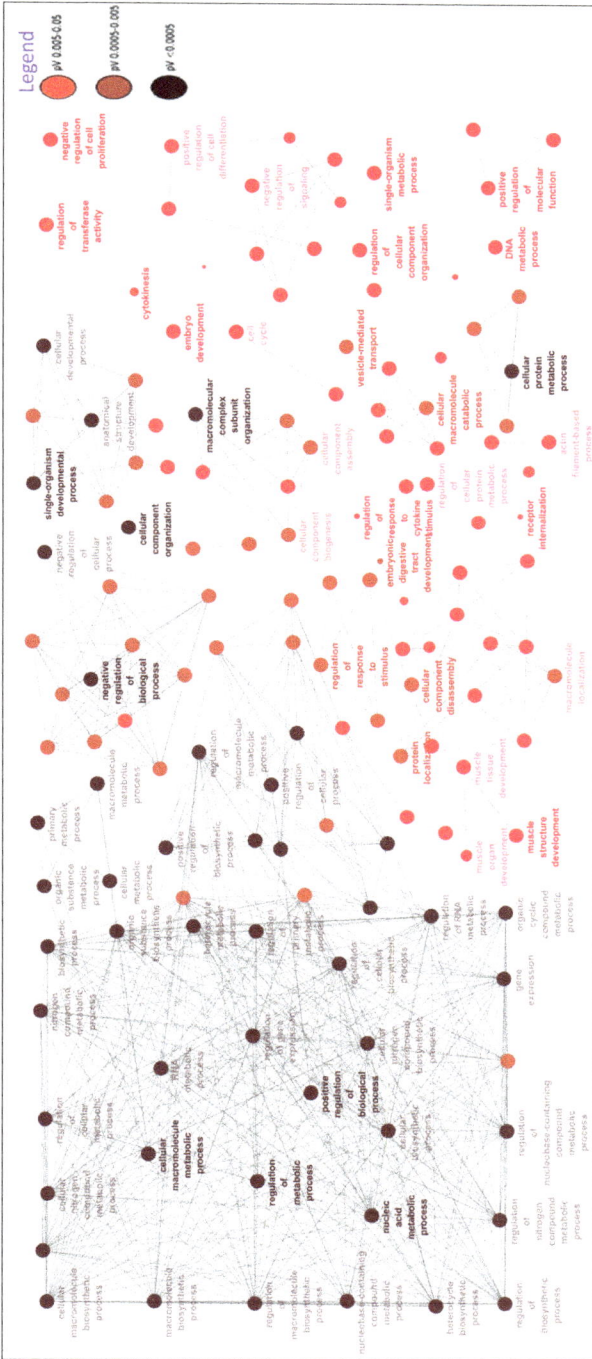

Figure 2. The ClueGO results for biological processes gene ontology terms enrichment for target genes (mRNAs) of differentially expressed miRNAs and relationships between them. The nodes (round shape) represent gene ontology terms, node color represents the level of significance as indicated in the legend, while node size reflects the number of genes enriched in each gene ontology term.

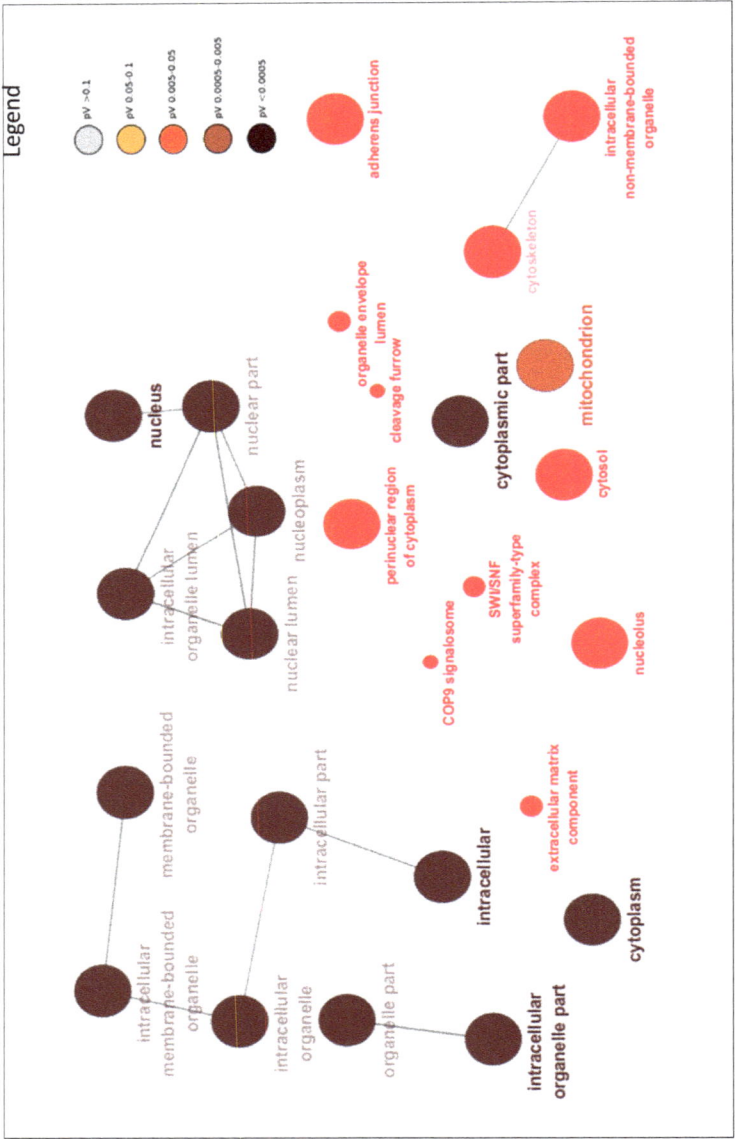

Figure 3. The ClueGO results for cellular processes gene ontology terms enrichment for target genes of differentially expressed miRNAs and relationships between them. The nodes (round shape) represent gene ontology terms, node color represents the level of significance as indicated in the legend, while node size reflects the number of genes enriched in each gene ontology term.

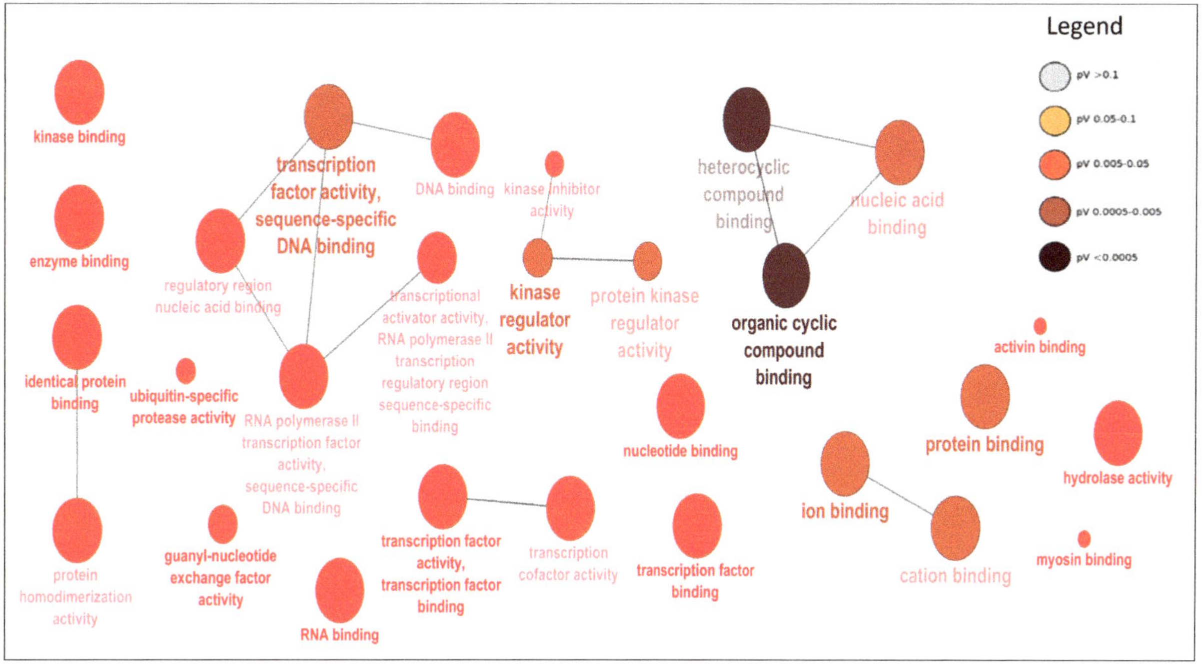

Figure 4. The ClueGO results for molecular functions gene ontology terms enrichment for target genes of differentially expressed miRNAs and relationships between them. The nodes (round shape) represent gene ontology terms, node color represents the level of significance as indicated in the legend, while node size reflects the number of genes enriched in each gene ontology term.

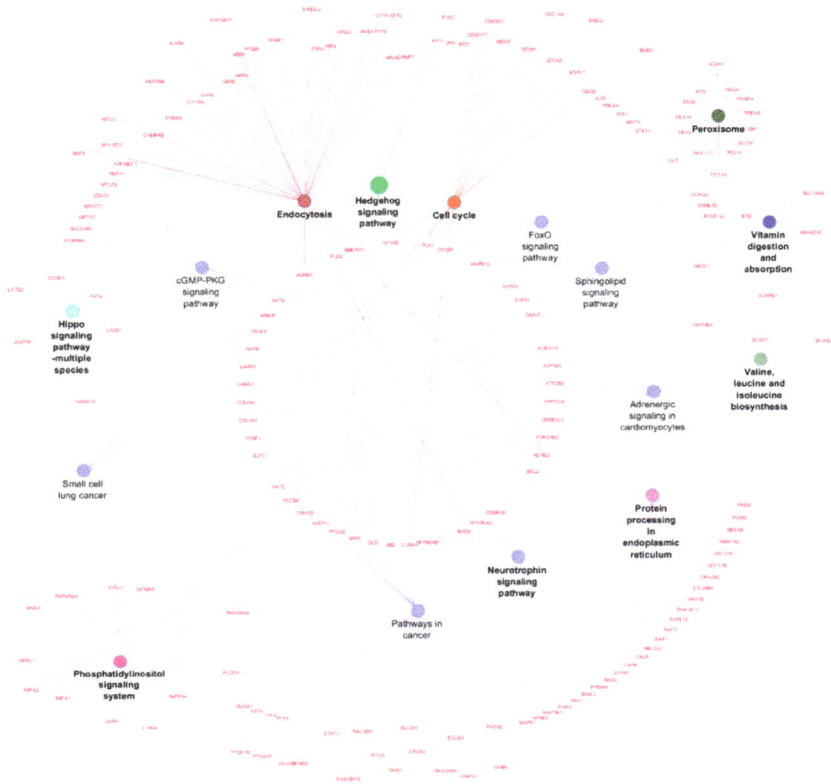

Figure 5. The ClueGO results for KEGG pathways enrichment for target genes of differentially expressed miRNAs and relationships between them. The nodes (round shapes) represent KEGG pathways or genes enriched in the pathways.

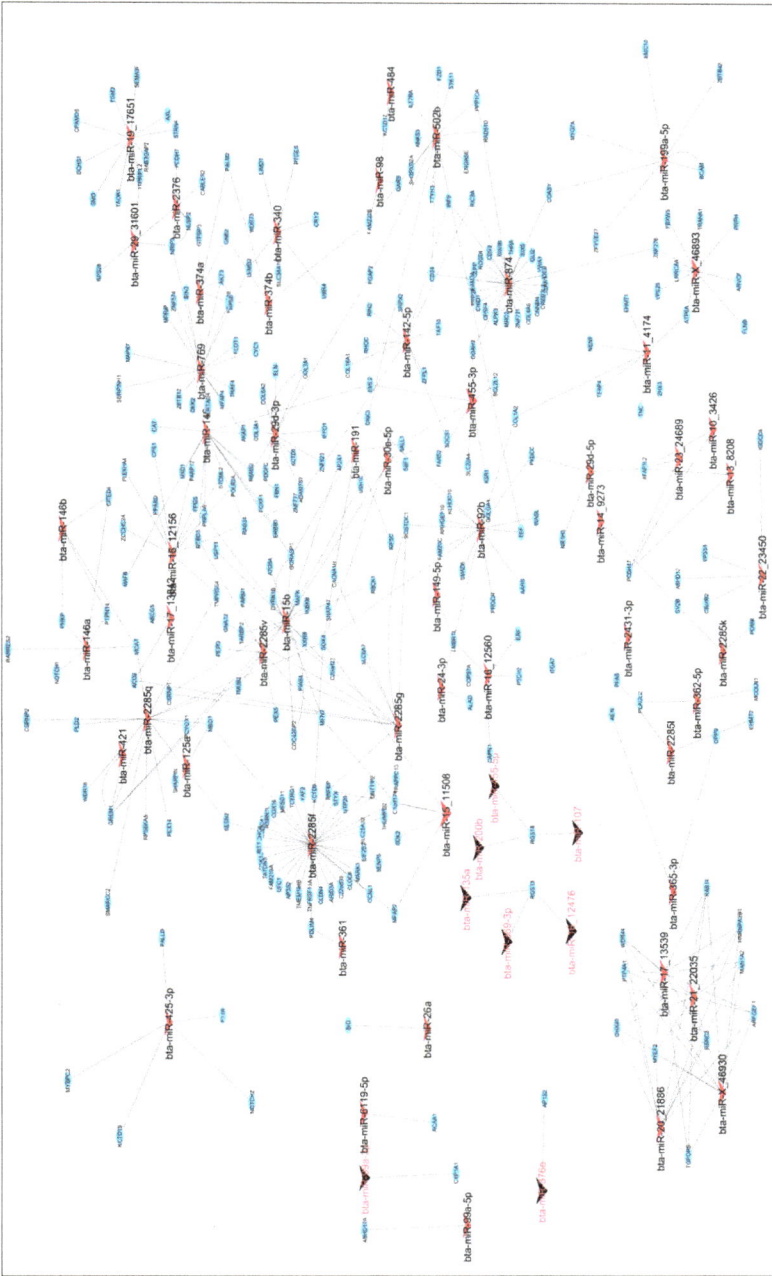

Figure 6. The Cytoscape visualization of the relationships between differentially expressed miRNAs and their target mRNAs. The nodes present either genes (round shape) or miRNAs (V shape). The up- and down-regulated miRNAs are colored red and black, respectively.

3.3. Real Time Quantitative PCR Validation

The RNA-Seq expression of 6 miRNAs was validated by qPCR. Two of them (bta-miR-486-5p and bta-miR-193b) were non DE, while four of them (bta-miR-142-5p, miR-146a, miR-24-3p and miR-374b) were DE between D33 and D96 by RNA-Seq. Observed fold changes for DE miRNAs between both methods were similar, except for the non-DE miRNAs, where an opposite trend was observed (Figure 7).

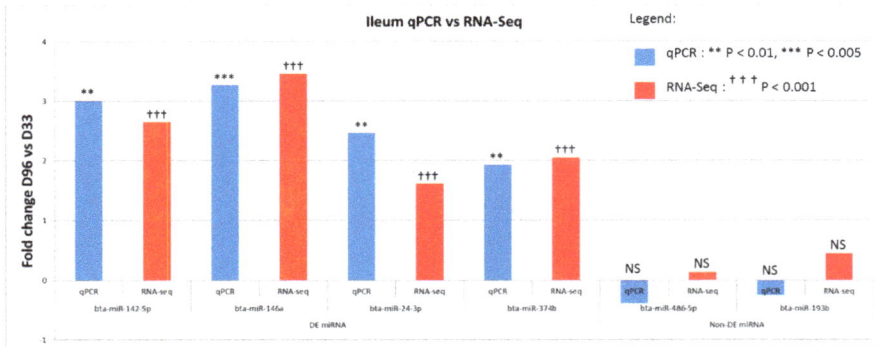

Figure 7. Result of qPCR validation of the expression of miRNAs between day 33 (pre-weaning) and day 96 (post-weaning), and compared with results obtained by miRNA sequencing.

4. Discussion

Physiologically, major metabolic changes that take place in calf gastrointestinal tract following the transition from liquid to solid food are accompanied by rapid changes in gene expression controlled by the signal-mediated coordination of transcriptional and post-transcriptional mechanisms [33]. Previously, we reported that ~20% of bovine genes were significantly DE between the pre-weaning (day 33) and the post-weaning (day 96) period, and enrichment analysis revealed the importance of DE genes in biological processes necessary for the switch in nutrition and developmental stage from the pre-weaning to the post-weaning period [23]. While it is well known that miRNAs are important for the regulation of these processes, little is known about how they participate in the regulation of ileum functions from the pre-weaning to the post-weaning period.

Highly expressed and DE miRNAs identified in this study suggest potential roles in ileum developmental processes during the transition from the pre-weaning to the post-weaning period. However, some highly expressed miRNAs such as bta-miR-21-5p, bta-miR-26a, bta-miR-148a and bta-let-7a-5p (Table 1) were also highly expressed in other tissues such as milk fat [10], milk whey/somatic cells [25], and mammary gland epithelial cells [13]. Interestingly, bta-miR-143, the most highly expressed miRNA, was also among the most abundant miRNAs in bovine testis [34] and reported as the most highly expressed miRNA in ileum tissue of calves at the pre- and post-weaning periods [35]. It was suggested that bta-miR-143 might regulate key genes involved in differentiation of connective tissue cells, the major components of the gut; hence, its high abundance might be important for the regulation of rapid development and growth of the gastrointestinal tract during early life. Indeed, enrichment analysis of target genes of the top 20 abundant miRNAs indicated enrichment in many biological processes and molecular function GO terms (Tables 2 and S5) involved in developmental processes, therefore supporting important roles for highly abundant miRNAs, including bta-miR-143, in these processes. Further supporting evidence was derived from enriched KEGG pathways crucial for cellular development processes such as FoxO signaling pathway, cell cycle, MAPK signaling pathway, and p53 signaling pathway (Table 3). Moreover, we also detected 81 novel miRNAs in this study that were lowly expressed. The most abundant novel miRNA, bta-miR-22_24033, accounted for only 0.3% of the total read counts. Forty-four novel miRNAs were significantly DE between D33 and D96, therefore suggesting roles in the regulation of ileum gene expression during the early period of growth. Nevertheless, novel miRNAs identified in this study will enrich the bovine miRNome as well as enhance knowledge of the potential roles of miRNAs in calf GIT. However, further functional validations to clarify the roles of identified miRNAs in the development of calf gut during the early period of growth are needed.

Bta-miR-374a was the most significantly DE miRNA in this study (Table 4). Bta-miR-374a was found to be differentially expressed between lactating and non-lactating cows [36]. Bta-miR-374a potentially targeted 36 different genes (Table S6b) and some of them might be important for ileum functions such as *EIF2AK4*, *FBXO18*, *GTPBP3* and *GNB2*, etc. *EIF2AK4* is an important transcription factor in host response to infection with pathogenic bacteria associated with Crohn's disease [37]. Moreover, *FBXO18*, *GTPBP3* and *GNB2* have been reported to be significantly DE in calf gastrointestinal tract between the pre- and post-weaning periods [23]. *FBXO18* encodes a member of the F-box protein family with function in phosphorylation-dependent ubiquitination [38]. *FBXO18* is implicated in the regulation of stress-induced apoptosis processes and homologous recombination in familial and sporadic breast cancer [39]. Cells deficient in *FBXO18* were unable to activate the cytotoxic-stress-induced cascade, resulting in increased cell survival [40]. *GTPBP3* is an important gene for mitochondrial functions and a mutation in this gene resulted in defective mitochondrial energy production through oxidative phosphorylation [41]. *GNB2* is important for neuronal apoptosis and was induced by lidocaine in the rat [42]. Nevertheless, the functions of these genes (*FBXO18*, *GTPBP3* and *GNB2*) in the ileum are unknown. Bta-miR-15b, the second most significantly up-regulated miRNA, belongs to the miR-15b family cluster. This miRNA cluster can target cell cycle proteins and the anti-apoptotic *Bcl-2* gene to control cell proliferation and apoptosis [43]. Bta-miR-15b might also play a role in mastitis disease development in cows [44]. Moreover, Liang et al. [35] also reported that bta-miR-15b was significantly DE between 0-day-old and 7-day-old calves and its expression correlated with bacterial population, thus suggesting roles in the regulation of gut development, immune, and digestive functions. Furthermore, bta-miR-15b potentially targeted *IKBKB* (Figure 6), a gene known as an essential molecule for NF-κB signaling pathway [45] with important roles in both innate and acquired immunity [46].

Interestingly, bta-miR-26a, one of the most significantly DE miRNA (Table 4), was also one of the most highly expressed miRNA (Table 1). The human homologue of this miRNA plays important roles in Crohn's disease [15]. In cattle, this miRNA regulates the expression of *PCK1* gene, which is important for semen quality and longevity of Holstein bulls [47]. Bta-miR-26a potentially targeted *BID* gene and their expressions were significantly correlated in this study (Figure 6). *BID* is a pro-apoptotic member of the Bcl-2 protein family with roles in the regulation of apoptosis processes [46]. Therefore, bta-miR-26a might have roles in ileum development via targeting the *BID* gene. Bta-miR-455-5p was the most down-regulated miRNA between D33 and D96. Bta-miR-455-5p was reported to be important for the function of granulosa cells of subordinate and dominant follicles during the early luteal phase of the bovine estrous cycle [48]. In humans, this miRNA homologue down regulated *RAB18* gene in gastric cancer [49]. In fact, bta-miR-455-5p also potentially targeted *RAB18* gene in this study (Figure 6).

Among the top 20 DE miRNAs in this study, bta-miR-142-3p, bta-miR-142-5p, bta-miR-191, bta-miR-146a, bta-miR-210 and bta-miR-424-5p were also found to be DE in calf ileum at the pre-weaning and weaning periods [35]. Some of these miRNAs have been reported to play important roles in immune functions. For example, bta-miR-146a inhibited the mRNA and protein expression levels of *TRAF6* gene and acted as a negative feedback regulator of bovine inflammation and innate immunity through down regulation of the TLR4/TRAF6/NF-κB pathway in bovine mammary epithelial cells [50]. Furthermore, bta-miR-142-5p was important for bovine alveolar macrophage response to *Mycobacterium bovis* infection [51].

As expected, the target genes of DE miRNAs were enriched in important biological process GO-terms related to metabolic processes (such as cellular macromolecule metabolic process, macromolecule metabolic process and regulation of metabolic process) (Table 5 and Figure 2) and molecular function GO-terms related to metabolism of the macromolecule compound (such as organic cyclic compound binding, heterocyclic compound binding, nucleic acid binding and protein binding) (Table 5 and Figure 4), thus suggesting roles in the regulation of these processes. Interestingly, the target genes of DE miRNAs were also enriched in GO-terms like transcription factor activity and sequence-specific DNA binding (FDR = 3.63×10^{-3}) (Table 5), thus suggesting their importance in transcription factor activity. The interaction between miRNAs and transcription factors to regulate gene expression in biological processes is well documented [52,53]. MiRNAs might inhibit transcription factor activities by either directly inhibiting the expression of their encoding genes or by inhibiting other gene(s) that have impact on their activities [53,54]. Previously, we observed that some transcription factors might play important roles in mediating miRNA regulatory functions in cow milk yield and milk component traits [9]. In humans, several miRNAs have been reported to participate in the regulation of intestinal transcription factors; miR-196b inhibited the *GATA6* intestinal transcription factor to control intestinal cell homeostasis and tumorigenesis in colon cancer patients [55], while miR-30 family controlled intestinal epithelial cell proliferation and differentiation by targeting SOX9 (transcription factor) and other genes in ubiquitin ligase pathway [56]. In fact, as mentioned above, the most DE miRNA in this study (bta-miR-374) also potentially targeted a transcription factor (*EIF2AK4*) known to be important for human Crohn's disease [37]. The most important pathway enriched for DE miRNAs target genes was Hedgehog signaling pathway (p = 0.003, Table S7, Figure 5). Hedgehog signaling pathway is important for cell growth, survival and fate, as well as for normal embryonic development [57,58]. This pathway also has multiple patterning functions during mammalian gut development [59]; therefore, it may be important for ileum functions during the early part of life. Another important pathway enriched for target genes of DE miRNAs was peroxisomes pathway (p = 0.006, Table S7, Figure 5). The peroxisomes pathway is crucial for metabolic processes such as fatty acid oxidation, biosynthesis of ether lipids, and free radical detoxification [60]. Since one of the main functions of the ileum is to absorb bile salts, one of the products of fatty acid oxidation, enrichment of the peroxisome pathway supports its role in normal ileum function. Another important role of the ileum is vitamin absorption and the enriched pathway, vitamin digestion and absorption (Table S7 and Figure 5), might reflect the changes in gene expression for different vitamin requirements between the pre- and post-weaning periods. Other enriched pathways also reflect the importance of miRNAs in the regulation of genes involved in lipid metabolism (phosphatidylinositol signaling system and sphingolipid signaling pathway), protein metabolism (valine, leucine and isoleucine biosynthesis and protein processing in endoplasmic reticulum) and intracellular signaling (cGMP-PKG signaling, FoxO signaling and neurotrophin signaling pathway) in the development of the ileum during the early part of life.

5. Conclusions

This is the first integrated miRNA-mRNA analysis characterizing the function of miRNAs in calf ileum during early life. Eighty-one novel miRNAs were identified that will enrich the bovine miRNome repertoire and contribute to the understanding of regulatory processes in calf ileum. This study highlighted potential roles of bta-miR-143, bta-miR-192, bta-miR-26a and bta-miR-21-5p in growth and developmental processes during the transition from the pre-weaning to the post-weaning period. This study also suggested roles for DE miRNAs in metabolic processes, metabolism of the macromolecule compound, transcription factor activities, as well as involvement in pathways related to metabolism (peroxisomes), vitamin digestion and absorption, lipid and protein metabolism, and intracellular signaling (Hedgehog signaling, GMP-PKG signaling, FoxO signaling, neurotrophin signaling pathway). Moreover, several DE miRNAs—DE mRNAs pairs such as bta-miR-374a—*FBXO18*, bta-miR-374a—*GTPBP3* and bta-miR-374a—*GNB2* with potential roles in tissue development, and bta-miR-15b—*IKBKB* with vital roles in immune functions were revealed. This study, therefore, provided insights on miRNA expression and their potential functions in calf ileum development during early life, which might facilitate identification of miRNA biomarkers for growth, nutritional and disease challenges during the pre- and post-weaning periods.

Supplementary Materials: The following are available online at http://www.mdpi.com/2073-4409/7/9/134/s1. Table S1: Mapping statistics of miRNA sequencing reads; Table S2: Length distribution (nt) of miRNA reads; Table S3: Novel miRNAs identified by miRDeep2; Table S4: Novel and known miRNAs and their read count summary; Table S5; The 20 most highly expressed miRNAs and their predicted target genes (a); enriched gene ontology terms (b) and enriched KEGG pathways (c); Table S6; Differentially expressed miRNAs between day 33 (pre-weaning) and day 96 (post-weaning) (a); predicted target genes of DE miRNAs (b) and differentially expressed genes which are both negatively correlated and predicted targets of miRNAs (c); Table S7; Gene ontology and KEGG pathways enriched for target genes of differential expressed miRNAs; Table S8; Differentially expressed genes that were targets of DE miRNAs and also negatively correlated.

Author Contributions: Conceived and designed the experiments, E.M.I.-A.; Performed the experiments, B.E.F. and P.-L.D.; Data analysis/curation, P.-L.D. and D.N.D.; Writing-Original Draft Preparation, D.N.D.; Writing-Review & Editing, E.M.I.-A.; Visualization, D.N.D.; Supervision and Project Administration, E.M.I.-A.; Funding Acquisition, E.M.I.-A.; all authors revised and approved the final draft.

Funding: This work was supported by funding from Agriculture and Agri-Food Canada to Eveline M. Ibeagha-Awemu, grant number J-000218.

Acknowledgments: Authors thank farm staff of Agriculture and Agri-Food Canada's Sherbrooke Research and Development Center for assistance in animal management.

Conflicts of Interest: The authors declare no conflict of interest.

References

1. Ambros, V. The functions of animal micrornas. *Nature* **2004**, *431*, 350–355. [CrossRef] [PubMed]
2. Carthew, R.W.; Sontheimer, E.J. Origins and mechanisms of miRNAs and siRNAs. *Cell* **2009**, *136*, 642–655. [CrossRef] [PubMed]
3. Coutinho, L.L.; Matukumalli, L.K.; Sonstegard, T.S.; Van Tassell, C.P.; Gasbarre, L.C.; Capuco, A.V.; Smith, T.P. Discovery and profiling of bovine micrornas from immune-related and embryonic tissues. *Physiol. Genet.* **2007**, *29*, 35–43. [CrossRef] [PubMed]
4. Ponsuksili, S.; Tesfaye, D.; Schellander, K.; Hoelker, M.; Hadlich, F.; Schwerin, M.; Wimmers, K. Differential expression of mirnas and their target mRNAs in endometria prior to maternal recognition of pregnancy associates with endometrial receptivity for in vivo-and in vitro-produced bovine embryos. *Biol. Reprod.* **2014**, *91*, 1–12. [CrossRef] [PubMed]
5. Hossain, M.; Salilew-Wondim, D.; Schellander, K.; Tesfaye, D. The role of micrornas in mammalian oocytes and embryos. *Anim. Reprod. Sci.* **2012**, *134*, 36–44. [CrossRef] [PubMed]
6. Do, D.N.; Ibeagha-Awemu, E.M. Non-coding RNA roles in ruminant mammary gland development and lactation. In *Current Topics in Lactation*; InTech: Rijeka, Croatia, 2017.
7. Gu, Z.; Eleswarapu, S.; Jiang, H. Identification and characterization of micrornas from the bovine adipose tissue and mammary gland. *FEBS Lett.* **2007**, *581*, 981–988. [CrossRef] [PubMed]
8. Jin, W.; Dodson, M.V.; Moore, S.S.; Basarab, J.A. Characterization of microrna expression in bovine adipose tissues: A potential regulatory mechanism of subcutaneous adipose tissue development. *BMC Mol. Biol.* **2010**, *11*, 29. [CrossRef] [PubMed]
9. Do, D.N.; Dudemaine, P.-L.; Li, R.; Ibeagha-Awemu, E.M. Co-expression network and pathway analyses reveal important modules of mirnas regulating milk yield and component traits. *Int. J. Mol. Sci.* **2017**, *18*, 1560. [CrossRef] [PubMed]
10. Do, D.N.; Li, R.; Dudemaine, P.-L.; Ibeagha-Awemu, E.M. Microrna roles in signalling during lactation: An insight from differential expression, time course and pathway analyses of deep sequence data. *Sci. Rep.* **2017**, *7*. [CrossRef] [PubMed]
11. Jabed, A.; Wagner, S.; McCracken, J.; Wells, D.N.; Laible, G. Targeted microrna expression in dairy cattle directs production of β-lactoglobulin-free, high-casein milk. *Proc. Natl. Acad. Sci. USA* **2012**, *109*, 16811–16816. [CrossRef] [PubMed]

12. Naeem, A.; Zhong, K.; Moisá, S.; Drackley, J.; Moyes, K.; Loor, J. Bioinformatics analysis of microRNA and putative target genes in bovine mammary tissue infected with streptococcus uberis. *J. Dairy Sci.* **2012**, *95*, 6397–6408. [CrossRef] [PubMed]

13. Jin, W.; Ibeagha-Awemu, E.M.; Liang, G.; Beaudoin, F.; Zhao, X.; Guan, L.L. Transcriptome microrna profiling of bovine mammary epithelial cells challenged with *Escherichia coli* or staphylococcus aureusbacteria reveals pathogen directed microRNA expression profiles. *BMC Genet.* **2014**, *15*, 181.

14. Farrell, D.; Shaughnessy, R.G.; Britton, L.; MacHugh, D.E.; Markey, B.; Gordon, S.V. The identification of circulating mirna in bovine serum and their potential as novel biomarkers of early mycobacterium avium subsp paratuberculosis infection. *PLoS ONE* **2015**, *10*, e0134310. [CrossRef] [PubMed]

15. Wu, F.; Zhang, S.; Dassopoulos, T.; Harris, M.L.; Bayless, T.M.; Meltzer, S.J.; Brant, S.R.; Kwon, J.H. Identification of microRNAs associated with ileal and colonic crohn's disease. *Inflamm. Bowel Dis.* **2010**, *16*, 1729–1738. [CrossRef] [PubMed]

16. Paraskevi, A.; Theodoropoulos, G.; Papaconstantinou, I.; Mantzaris, G.; Nikiteas, N.; Gazouli, M. Circulating microRNA in inflammatory bowel disease. *J. Crohn's Colitis* **2012**, *6*, 900–904. [CrossRef] [PubMed]

17. De Souza, H.S.; Fiocchi, C. Immunopathogenesis of ibd: Current state of the art. *Nat. Rev. Gastroenterol. Hepatol.* **2016**, *13*, 13–27. [CrossRef] [PubMed]

18. Dalal, S.R.; Kwon, J.H. The role of microrna in inflammatory bowel disease. *Gastroenterol. Hepatol.* **2010**, *6*, 714.

19. Chapman, C.G.; Pekow, J. The emerging role of miRNAs in inflammatory bowel disease: A review. *Ther. Adv. Gastroenterol.* **2015**, *8*, 4–22. [CrossRef] [PubMed]

20. Liang, G.; Malmuthuge, N.; Griebel, P. Model systems to analyze the role of mirnas and commensal microflora in bovine mucosal immune system development. *Mol. Immunol.* **2015**, *66*, 57–67. [CrossRef] [PubMed]

21. Liang, G.; Malmuthuge, N.; Bao, H.; Stothard, P.; Griebel, P.J. Transcriptome analysis reveals regional and temporal differences in mucosal immune system development in the small intestine of neonatal calves. *BMC Genet.* **2016**, *17*, 602. [CrossRef] [PubMed]

22. Liang, G.; Malmuthuge, N.; Guan, Y.; Ren, Y.; Griebel, P.J.; Guan, L.L. Altered microrna expression and pre-mRNA splicing events reveal new mechanisms associated with early stage mycobacterium avium subspecies paratuberculosis infection. *Sci. Rep.* **2016**, *6*. [CrossRef] [PubMed]

23. Ibeagha-Awemu, E.M.; Do, D.N.; Dudemaine, P.-L.; Fomenky, B.E.; Bissonnette, N. Integration of lncRNA and mrna transcriptome analyses reveals genes and pathways potentially involved in calves' intestinal growth and development during the early weeks of life. *Genes* **2017**. submitted.

24. Do, D.N.; Dudemaine, P.-L.; Fomenky, B.E.; Ibeagha-Awemu, E.M. Integration of miRNA weighted gene co-expression network and miRNA-mRNA co-expression analyses reveals potential regulatory functions of mirnas in calf rumen development. *Genomics* **2018**. [CrossRef] [PubMed]

25. Li, R.; Dudemaine, P.-L.; Zhao, X.; Lei, C.; Ibeagha-Awemu, E.M. Comparative analysis of the mirnome of bovine milk fat, whey and cells. *PLoS ONE* **2016**, *11*, e0154129. [CrossRef] [PubMed]

26. Langmead, B.; Trapnell, C.; Pop, M.; Salzberg, S.L. Ultrafast and memory-efficient alignment of short DNA sequences to the human genome. *Genet. Biol.* **2009**, *10*, R25. [CrossRef] [PubMed]

27. Friedlander, M.R.; Chen, W.; Adamidi, C.; Maaskola, J.; Einspanier, R.; Knespel, S.; Rajewsky, N. Discovering micrornas from deep sequencing data using mirdeep. *Nat. Biotechnol.* **2008**, *26*, 407–415. [CrossRef] [PubMed]

28. Love, M.I.; Huber, W.; Anders, S. Moderated estimation of fold change and dispersion for RNA-seq data with DESeq2. *Genet. Biol.* **2014**, *15*, 550. [CrossRef] [PubMed]

29. Benjamini, Y.; Hochberg, Y. Controlling the false discovery rate: A practical and powerful approach to multiple testing. *J. R. Stat. Soc. Ser. B (Methodol.)* **1995**, *57*, 289–300.

30. Li, R.; Beaudoin, F.; Ammah, A.A.; Bissonnette, N.; Benchaar, C.; Zhao, X.; Lei, C.; Ibeagha-Awemu, E.M. Deep sequencing shows microrna involvement in bovine mammary gland adaptation to diets supplemented with linseed oil or safflower oil. *BMC Genet.* **2015**, *16*, 884. [CrossRef] [PubMed]

31. Bindea, G.; Mlecnik, B.; Hackl, H.; Charoentong, P.; Tosolini, M.; Kirilovsky, A.; Fridman, W.-H.; Pagès, F.; Trajanoski, Z.; Galon, J. Cluego: A cytoscape plug-in to decipher functionally grouped gene ontology and pathway annotation networks. *Bioinformatics* **2009**, *25*, 1091–1093. [CrossRef] [PubMed]

32. Shannon, P.; Markiel, A.; Ozier, O.; Baliga, N.S.; Wang, J.T.; Ramage, D.; Amin, N.; Schwikowski, B.; Ideker, T. Cytoscape: A software environment for integrated models of biomolecular interaction networks. *Genet. Res.* **2003**, *13*, 2498–2504. [CrossRef] [PubMed]

33. Turner, M.; Galloway, A.; Vigorito, E. Noncoding RNA and its associated proteins as regulatory elements of the immune system. *Nat. Immunol.* **2014**, *15*, 484–491. [CrossRef] [PubMed]

34. Huang, J.; Ju, Z.; Li, Q.; Hou, Q.; Wang, C.; Li, J.; Li, R.; Wang, L.; Sun, T.; Hang, S.; et al. Solexa sequencing of novel and differentially expressed micrornas in testicular and ovarian tissues in holstein cattle. *Int. J. Biol. Sci.* **2011**, *7*, 1016–1026. [CrossRef] [PubMed]

35. Liang, G.; Malmuthuge, N.; McFadden, T.B.; Bao, H.; Griebel, P.J.; Stothard, P. Potential regulatory role of micrornas in the development of bovine gastrointestinal tract during early life. *PLoS ONE* **2014**, *9*, e92592. [CrossRef] [PubMed]

36. Li, Z.; Liu, H.; Jin, X.; Lo, L.; Liu, J. Expression profiles of micrornas from lactating and non-lactating bovine mammary glands and identification of mirna related to lactation. *BMC Genet.* **2012**, *13*, 731. [CrossRef] [PubMed]

37. Bretin, A.; Carriere, J.; Dalmasso, G.; Bergougnoux, A.; B'Chir, W.; Maurin, A.C.; Muller, S.; Seibold, F.; Barnich, N.; Bruhat, A.; et al. Activation of the EIF2AK4-EIF2A/eIF2α-ATF4 pathway triggers autophagy response to crohn disease-associated adherent-invasive *Escherichia coli* infection. *Autophagy* **2016**, *12*, 770–783. [CrossRef] [PubMed]

38. Kim, J.; Kim, J.-H.; Lee, S.-H.; Kim, D.-H.; Kang, H.-Y.; Bae, S.-H.; Pan, Z.-Q.; Seo, Y.-S. The novel human DNA helicase hFBH1 is an F-box protein. *J..Biol. Chem.* **2002**, *277*, 24530–24537. [CrossRef] [PubMed]

39. Heyn, H.; Sayols, S.; Moutinho, C.; Vidal, E.; Sanchez-Mut, J.V.; Stefansson Olafur, A.; Nadal, E.; Moran, S.; Eyfjord Jorunn, E.; Gonzalez-Suarez, E.; et al. Linkage of DNA methylation quantitative trait loci to human cancer risk. *Cell Rep.* **2014**, *7*, 331–338. [CrossRef] [PubMed]

40. Laulier, C.; Cheng, A.; Huang, N.; Stark, J.M. Mammalian fbh1 is important to restore normal mitotic progression following decatenation stress. *DNA Repair* **2010**, *9*, 708–717. [CrossRef] [PubMed]

41. Kopajtich, R.; Nicholls Thomas, J.; Rorbach, J.; Metodiev Metodi, D.; Freisinger, P.; Mandel, H.; Vanlander, A.; Ghezzi, D.; Carrozzo, R.; Taylor Robert, W.; et al. Mutations in gtpbp3 cause a mitochondrial translation defect associated with hypertrophic cardiomyopathy, lactic acidosis, and encephalopathy. *Am. J. Hum. Genet.* **2014**, *95*, 708–720. [CrossRef] [PubMed]

42. Tan, Y.; Wang, Q.; Zhao, B.; She, Y.; Bi, X. GNB2 is a mediator of lidocaine-induced apoptosis in rat pheochromocytoma PC12 cells. *NeuroToxicol.* **2016**, *54*, 53–64. [CrossRef] [PubMed]

43. Yue, J.; Tigyi, G. Conservation of miR-15a/16-1 and miR-15b/16-2 clusters. *Mamm. Genome* **2010**, *21*, 88–94. [CrossRef] [PubMed]

44. Chen, L.; Liu, X.; Li, Z.; Wang, H.; Liu, Y.; He, H.; Yang, J.; Niu, F.; Wang, L.; Guo, J. Expression differences of mirnas and genes on NF-κB pathway between the healthy and the mastitis chinese holstein cows. *Gene* **2014**, *545*, 117–125. [CrossRef] [PubMed]

45. Schmid, J.A.; Birbach, A. Iκb kinase β (ikkβ/ikk2/ikbkb)—A key molecule in signaling to the transcription factor NF-κB. *Cytokine Growth Factor Rev.* **2008**, *19*, 157–165. [CrossRef] [PubMed]

46. Pannicke, U.; Baumann, B.; Fuchs, S.; Henneke, P.; Rensing-Ehl, A.; Rizzi, M.; Janda, A.; Hese, K.; Schlesier, M.; Holzmann, K.; et al. Deficiency of innate and acquired immunity caused by an IKBKB mutation. *New Eng. J. Med.* **2013**, *369*, 2504–2514. [CrossRef] [PubMed]

47. Huang, J.; Guo, F.; Zhang, Z.; Zhang, Y.; Wang, X.; Ju, Z.; Yang, C.; Wang, C.; Hou, M.; Zhong, J. PCK1 is negatively regulated by bta-miR-26a, and a single-nucleotide polymorphism in the 3′ untranslated region is involved in semen quality and longevity of holstein bulls. *Mol. Reprod. Dev.* **2016**, *83*, 217–225. [CrossRef] [PubMed]

48. Salilew-Wondim, D.; Ahmad, I.; Gebremedhn, S.; Sahadevan, S.; Hossain, M.M.; Rings, F.; Hoelker, M.; Tholen, E.; Neuhoff, C.; Looft, C. The expression pattern of micrornas in granulosa cells of subordinate and dominant follicles during the early luteal phase of the bovine estrous cycle. *PLoS ONE* **2014**, *9*, e106795. [CrossRef] [PubMed]

49. Liu, J.; Zhang, J.; Li, Y.; Wang, L.; Sui, B.; Dai, D. MiR-455-5p acts as a novel tumor suppressor in gastric cancer by down-regulating rab18. *Gene* **2016**, *592*, 308–315. [CrossRef] [PubMed]

50. Wang, X.P.; Luoreng, Z.M.; Zan, L.S.; Li, F.; Li, N. Bovine miR-146a regulates inflammatory cytokines of bovine mammary epithelial cells via targeting the TRAF6 gene. *J. Dairy Sci.* **2017**, *100*, 7648–7658. [CrossRef] [PubMed]

51. Vegh, P.; Magee, D.A.; Nalpas, N.C.; Bryan, K.; McCabe, M.S.; Browne, J.A.; Conlon, K.M.; Gordon, S.V.; Bradley, D.G.; MacHugh, D.E. Microrna profiling of the bovine alveolar macrophage response to

mycobacterium bovis infection suggests pathogen survival is enhanced by microrna regulation of endocytosis and lysosome trafficking. *Tuberculosis* **2015**, *95*, 60–67. [CrossRef] [PubMed]

52. Martinez, N.J.; Walhout, A.J. The interplay between transcription factors and micrornas in genome-scale regulatory networks. *Bioessays* **2009**, *31*, 435–445. [CrossRef] [PubMed]

53. Hobert, O. Gene regulation by transcription factors and micrornas. *Science* **2008**, *319*, 1785–1786. [CrossRef] [PubMed]

54. Shalgi, R.; Lieber, D.; Oren, M.; Pilpel, Y. Global and local architecture of the mammalian microrna–transcription factor regulatory network. *PLoS Comput. Biol.* **2007**, *3*, e131. [CrossRef] [PubMed]

55. Fantini, S.; Salsi, V.; Reggiani, L.; Maiorana, A.; Zappavigna, V. The miR-196b mirna inhibits the gata6 intestinal transcription factor and is upregulated in colon cancer patients. *Oncotarget* **2017**, *8*, 4747. [CrossRef] [PubMed]

56. Peck, B.C.E.; Sincavage, J.; Feinstein, S.; Mah, A.T.; Simmons, J.G.; Lund, P.K.; Sethupathy, P. MiR-30 family controls proliferation and differentiation of intestinal epithelial cell models by directing a broad gene expression program that includes sox9 and the ubiquitin ligase pathway. *J. Biol. Chem.* **2016**, *291*, 15975–15984. [CrossRef] [PubMed]

57. Mazumdar, T.; DeVecchio, J.; Shi, T.; Jones, J.; Agyeman, A.; Houghton, J.A. Hedgehog signaling drives cellular survival in human colon carcinoma cells. *Cancer Res.* **2011**, *71*, 1092–1102. [CrossRef] [PubMed]

58. Varjosalo, M.; Taipale, J. Hedgehog: Functions and mechanisms. *Genes Dev.* **2008**, *22*, 2454–2472. [CrossRef] [PubMed]

59. Zacharias, W.J.; Madison, B.B.; Kretovich, K.E.; Walton, K.D.; Richards, N.; Udager, A.M.; Li, X.; Gumucio, D.L. Hedgehog signaling controls homeostasis of adult intestinal smooth muscle. *Dev. Biol.* **2011**, *355*, 152–162. [CrossRef] [PubMed]

60. Kim, P.; Hettema, E. Multiple pathways for protein transport to peroxisomes. *J. Mol. Biol.* **2015**, *427*, 1176–1190. [CrossRef] [PubMed]

Review

MicroRNAs as Diagnostic and Prognostic Biomarkers in Ischemic Stroke—A Comprehensive Review and Bioinformatic Analysis

Ceren Eyileten [1,†], Zofia Wicik [2,†], Salvatore De Rosa [3], Dagmara Mirowska-Guzel [1], Aleksandra Soplinska [1], Ciro Indolfi [3,4], Iwona Jastrzebska-Kurkowska [5], Anna Czlonkowska [5] and Marek Postula [1,*]

1   Department of Experimental and Clinical Pharmacology, Medical University of Warsaw, Center for Preclinical Research and Technology CEPT, 02-097 Warsaw, Poland; ceren.eyileten-postula@wum.edu.pl (C.E.); dmirowska@wum.edu.pl (D.M.-G.); ola@soplinska.pl (A.S.)
2   Rheumatology Division, Hospital das Clinicas HCFMUSP, Universidade de Sao Paulo, Sao Paulo, SP 01246-903, Brazil; zofiawicik@gmail.com
3   Division of Cardiology, Department of Medical and Surgical Sciences, "Magna Graecia" University, 88100 Catanzaro, Italy; saderosa@unicz.it (S.D.R.); indolfi@unicz.it (C.I.)
4   URT-CNR, Department of Medicine, Consiglio Nazionale delle Ricerche of IFC, Viale Europa S/N, 88100 Catanzaro, Italy
5   2nd Department of Neurology, Institute of Psychiatry and Neurology, 02-957 Warsaw, Poland; ikurkowska@ipin.edu.pl (I.J.-K.); czlonkow@ipin.edu.pl (A.C.)
*   Correspondence: mpostula@wum.edu.pl; Tel.: +48-22-1166-160; Fax: +48-22-1166-202
†   These authors contributed equally to this work.

Received: 1 November 2018; Accepted: 2 December 2018; Published: 6 December 2018

Abstract: Stroke is the second-most common cause of death worldwide. The pathophysiology of ischemic stroke (IS) is related to inflammation, atherosclerosis, blood coagulation, and platelet activation. MicroRNAs (miRNAs) play important roles in physiological and pathological processes of neurodegenerative diseases and progression of certain neurological diseases, such as IS. Several different miRNAs, and their target genes, are recognized to be involved in the pathophysiology of IS. The capacity of miRNAs to simultaneously regulate several target genes underlies their unique value as diagnostic and prognostic markers in IS. In this review, we focus on the role of miRNAs as diagnostic and prognostic biomarkers in IS. We discuss the most common and reliable detection methods available and promising tests currently under development. We also present original results from bioinformatic analyses of published results, identifying the ten most significant genes (HMGB1, YWHAZ, PIK3R1, STAT3, MAPK1, CBX5, CAPZB, THBS1, TNFRSF10B, RCOR1) associated with inflammation, blood coagulation, and platelet activation and targeted by miRNAs in IS. Additionally, we created miRNA-gene target interaction networks based on Gene Ontology (GO) information derived from publicly available databases. Among our most interesting findings, miR-19a-3p is the most widely modulated miRNA across all selected ontologies and might be proposed as novel biomarker in IS to be tested in future studies.

Keywords: miRNA; bioinformatic analysis; ischemic stroke; miRNA-gene target interaction; network; biomarker; diagnosis; prognosis

---

1. Introduction

Stroke is the second-most common cause of death worldwide, accounting for almost 6.5 million stroke deaths each year. Approximately 795,000 strokes occur in the United States each year. On average, every 40 s, someone in the United States has a stroke, and on average, every 4 min, someone dies of a stroke [1,2]. In Europe, strokes account for 405,000 deaths (9%) in men and 583,000 (13%) deaths in women each year [3]. While hemorrhagic strokes accounts for 15% and a further 5% are due to unknown etiology, approximately 80% of all acute strokes are ischemic, mainly from large vessel occlusion due to either artery-to-artery embolism or cardiac embolism. [4,5]. Ischemic stroke (IS) is characterized by the sudden loss of blood circulation to an area of the brain, resulting in irreversible brain injury and subsequent neurologic deficits occurring already few minutes after the onset of ischemia [6].

MicroRNAs (miRNA) are small, endogenous, single-stranded, noncoding RNA molecules ranging in length from 18–25 nt that are found in eukaryotic cells. They regulate approximately 60% of the mammalian protein coding genes, primarily through the interaction with mRNAs. This effect is exerted by binding to complementary regions of messenger transcripts to repress their translation or less frequently inducing their degradation. Up to now, more than 5000 human miRNAs have been identified, and about 1000 are estimated to exist [7]. Several studies showed that miRNAs can be useful for the treatment of particular diseases. One advantage of miRNA-based therapies is that synthetic mimics and inhibitors can be used to alter endogenous miRNA levels in clinic. This approach has been utilized for cancer and hepatitis C virus therapy [8,9]. Apart from their treatment value, it was shown that expression of miRNAs plays important roles in many physiological and pathological processes, including epigenetics and neurodegenerative diseases and progression of specific neurological diseases [10,11]. It has been shown that some pathological processes of the central nervous system, and that Alzheimer's disease, multiple sclerosis, and stroke are associated to specific alteration of circulating miRNAs [12]. This suggests that circulating miRNAs could be used as clinical biomarkers, to provide a sort of "liquid biopsy" taken from peripheral blood but providing information on pathophysiological processes going on in the brain and underlying stroke [13–15].

Potential advantages of this approach include the early diagnosis of acute cerebrovascular events, as already proven before with coronary artery disease, peripheral arterial disease and ischemia, acute myocardial infarction, myocarditis, heart failure, takotsubo cardiomyopathy in-stent restenosis, diabetes, or platelet dysfunction [7,16–29].

Several conventional methods were used by researchers in order to detect miRNAs such as, northern blotting, quantitative polymerase chain reaction (qPCR), and microarrays. However, each of these methods has its own individual limitations. Northern analysis was a widely used method for miRNA analyses, as it is easy to perform, specificity is relatively high, sequences with even partial homology can be used as hybridization probes, and the cost of running many gels is low once the equipment is set up [30]. However, northern blot technology also has disadvantages. For example, contamination due to radiolabeling, low detection efficiency, and poor sensitivity for oligonucleotide probes may complicate the measurements [31]. RT-PCR and microarrays allow reliable and effective detection for miRNAs. Techniques like northern blot and RT-PCR allow testing for only a few miRNAs per experiment. On the contrary, using miRNA microarray, it is possible to identify more than several hundred differentially regulated miRNAs in a particular disease or condition in one single run [7]. Several new technical developments have been brought about in the field of RT-PCR in order to further improve their efficiency. Among the others, the coupling with microfluidic cards should be mentioned, together with the digital droplet PCR, which allows very high analytical standards, minimizing reaction volumes [32]. Additionally, next-generation-sequencing technology has quickly emerged as the preferred platform for studying and discovering novel circulating miRNAs. It is possible to sequence multiple samples in one time by pooling with next-generation-sequencing and also possible to construct comprehensive expression profiles for every assessed sample, on the other hand potentially large investment in bioinformatics analysis, time and personnel are required [33,34]. Beside these techniques, there are new promising methodologies have been developed to detect miRNAs, for example, sensitive miRNA biosensors such as silicon nanowires, gold nanoparticles, silver nanoclusters, or conducting polymer/carbon nanotube hybrids [35–38].

Since circulating miRNAs can be detected by several techniques in plasma, serum, or whole blood, they have become the focus of interest as potential new diagnostic and prognostic biomarkers and tools for understanding their role in IS. Therefore, we systematically reviewed all available studies assessing the clinical usefulness of circulating miRNAs as diagnostic or prognostic markers in IS. Thus, in this article, we present an overview of the current knowledge on diagnostic and prognostic value of miRNAs related to IS based on human studies and report the results of a quantitative bioinformatic analysis highlighting the most promising miRNAs for clinical application in IS.

2. Circulating miRNAs and Stroke

Acute myocardial ischemia and IS share several common pathophysiological aspects. Previous studies implied that the circulating miRNAs could be exploited as diagnostic and/or prognostic indicators in several cardiovascular diseases (CVD) [27]. However, limited number of researches on role of miRNAs in IS and its relation to clinical outcomes were published.

2.1. PCR-Based Analysis for miRNA Expression

Quantitative reverse transcription PCR (RT-qPCR) technology is the gold standard for gene expression measurement. Several companies have developed qPCR-based assays for the detection of miRNA expression [39]. RT-qPCR technique allowed us to achieve high sensitivity, high specificity, and quantitative data of miRNA expression analysis. However, RT-qPCR is low-throughput profiling for a genome-wide miRNA-profiling assay [40].

In one of the first study, alterations in plasma miRNA levels in patients with IS were compared to healthy controls. The circulating levels of miR-30a, miR-126, and let-7b were analyzed in plasma samples of patients with IS obtained at different time-points both in acute phase (24 h from the onset of symptoms) and recovery phase (at 1 week, 4 weeks, 24 weeks, and 48 weeks after the episode). Circulating miR-30a and miR-126 expression were down regulated in IS patients regardless of etiology in all time points, except the last measurement at 48 weeks when it increase to baseline value. Interestingly, the expression pattern of circulating let-7b in IS patients with large-vessel atherosclerosis (LA) was increased in contrast to what observed in patients with other stroke subtypes (i.e., small-vessel disease (SV), cardioembolic stroke (CE) and undetermined cause (UDN) groups). Follow-up evaluations revealed that all circulating miRNAs levels returned to the baseline 48 weeks after the episode. Moreover, based on the level of three miRNAs authors defined a specific clinical score system that was further correlated with functional status as evaluated with the modified Rankin Scale (mRS), suggesting that miR-30a, miR-126, and let-7b could be used as a reliable marker for the diagnosis of IS (See Tables 1 and 2) [41]. It is particularly interesting that a modulation of miR-30 has also been reported in patients with acute myocardial infarction [42]. If on one hand this strengthens the hypothesis that miRNAs are "smart" biomarkers capable of reflecting the pathophysiological processes underlying CVS, on the other hand it undermines its usefulness as a disease-specific biomarker.

In another study, two miRNAs selected based on literature search, namely let-7e and miR-338, were analyzed in serum from IS patients in three phases of the ischemic event (i.e., acute- 1–7 days, subacute- 8–14 days, recovered- over 15 days) and miRNA levels were compared with 51 healthy volunteers. In this study only let-7e expression levels were significantly higher at all time-points with significantly higher levels in IS patients at the acute stage. Moreover, let-7e level was correlated with serum high sensitivity C-reactive protein (hs-CRP) level (r = 0.67, *p* = 0.033). A prognostic role of both miRNAs failed to be proven as there was no correlation between miRNAs expression level and National Institutes of Health Stroke Scale (NIHSS) scores, which is a widely used tool that determine the severity of a stroke. However, serum let-7e showed a specificity up to 73.4% and a sensitivity of 82.8% in IS patients at the acute stage, whereas serum miR-338 in IS patients showed a specificity up to 53.2% and a sensitivity of 71.9% in the acute stage. Thus, authors suggested that let-7e expression in serum may serve as a useful noninvasive circulating biomarker for the acute stage of IS [43].

Huang et al. sought to evaluate another let-7 family member, let-7e-5p, in two independent case-control IS populations. The results showed that the expression level of let-7e-5p was significantly higher in IS patients than in control subjects. Logistic regression analysis revealed that let-7e-5p expression was associated with an increased risk of IS (adjusted OR, 1.89; 95% CI, 1.61~2.21; $p < 0.001$). Also, diagnostic accuracy of acute phase specific miRNAs were tested through calculating the area under curve (AUC) of receiver operating characteristic (ROC) curves. The addition let-7e-5p to the traditional risk factor model improved the diagnostic potential to an AUC of 0.82 (95% CI, 0.78~0.85). In order to find targets of let-7e both bioinformatics and target gene expression analysis were performed. It showed that let-7e-5p expression was negatively correlated with several genes (ATF2, CASP3, FGFR2, NLK, PTPN7, RASGRP1, and TGFBR1). Therefore, the study showed that let-7e-5p expression is significantly higher in IS patients and is associated with the occurrence of IS. Moreover, authors suggested that let-7e-5p may be involved in the pathogenesis of IS by regulating CASP3 and NLK expression, as two genes enriched in the MAPK signaling pathway [44].

Gong et al. performed study in order to determine prognostic value of several let-7 family members (let-7a, let-7b, let-7c, let-7d, let-7e, let-7f, let-7g, miR-98) and relation to massive cerebral infarction (MCI) within first 48 h from the onset of symptoms. In their report, the expression of let-7f was down regulated in IS with MCI in comparison to healthy controls and IS without MCI, and up regulated in group without MCI at baseline, i.e., 48 h. When comparing relative expression of let-7f between groups with and without hemorrhagic transformation (HT), authors found up-regulation of let-7f in the MCI without HT after two weeks from the baseline. However, when compared MCI with and without HT significant up-regulation of let-7f was found in the first group after two weeks. It is worth to mention that the level of hs-CRP was negatively correlated with the relative expression of let-7f in the MCI group. Another important finding of the study showed that the expression level of let-7f in the MCI without HT is positively correlated with patients status based on Glasgow Coma Scale score and negatively with hs-CRP concentration ($r = -0.88$, $p < 0.0001$). Also in this study target gene expression analysis was performed and the relative expression of let-7f was negatively correlated with interleukin-6 (IL-6) expression in the MCI without HT (48 h and 2 weeks) ($r = -0.40$, $p < 0.001$), but not in the MCI with HT group, which may suggest that the downregulation of let-7f expression in patients with MCI without HT may induce inflammation [45]. The large involvement of let-7 with cerebrovascular disease, as proven by multiple evidences from independent studies is particular interesting. In fact, let-7 expression and availability is influenced by a circular noncoding RNA named CircPVT1, involved in the modulation of cell senescence [46].

Leung et al. investigated and compared plasma concentrations of miR-124-3p and miR-16 as diagnostic markers in acute stroke. Ninety-three patients with IS, 19 patients with hemorrhagic stroke, and 23 healthy controls enrolled in the study. Plasma concentrations of miRNAs were determined by RT- PCR. Median plasma 124-3p concentrations taken within 24 h of symptom onset were higher in hemorrhagic stroke patients than that in IS patients, while median miR-16 concentration in IS patients were higher than that in hemorrhagic stroke patients. Authors concluded that both miR-124-3p and miR-16 are diagnostic markers to discriminate hemorrhagic stroke and IS [47]. Another study aimed to investigate circulating miRNAs namely miR-15a, miR-16, and miR-17-5p [48]. The selection of specific miRNA was based on assumption that miR-15a and miR-16 are increased in the serum of patients with ischemic events and miR-17-5p may be a critical factor in post-stroke adult neurogenesis [49–52]. In this study, Wu et al. evaluated the utility of these three miRNAs in peripheral blood of 106 IS patients and 120 healthy controls as IS serum biomarkers for diagnostic value. Serum levels of miR-15a, miR-16, and miR-17-5p were significantly higher in IS patients compared to control subjects. Serum miR-15a levels showed a significant positive correlation with age. Besides, there was a strong negative correlation between serum miR-16 levels and high-density lipoprotein (HDL) and apolipoprotein A1 (ApoA1). Moreover, in multivariate logistic regression model, it was found that serum miR-17-5p level was a significant and independent predictor for IS (OR 3.968; CI 95% 1.001–14.29, $p = 0.035$). Also, ROC analysis showed that selected miRNAs were useful IS biomarkers (See Table 2) [48]. In line with these findings, miR-16 was recently associated to peripheral ischemia and was suggested as a potential mediator of remote vascular remodeling [19]. Hence, it could reflect systemic vascular dysfunction, rather than specific and acute localization of ischemic disease.

Jin and Xing evaluated 28 miRNAs selected based on previous publications that were described to possess pro-angiogenic or anti-angiogenic properties. Selected miRNAs were evaluated in plasma samples of IS patients and controls taken within 24 h after ischemic events. In the exploring stage performed in 10 patients and 10 controls, 11 differentially expressed miRNAs (DEM) were identified and included into the validating stage. In the second stage, in order to validate the significantly expressed miRNAs, 106 IS patients and 110 controls were analyzed. The expression of miR-126, miR-130a, and miR-378 in plasma decreased in IS patients; however, miR-222, miR-218, and miR-185 plasma levels were increased. At logistic regression analysis they found that miR-126 (OR 0.840; CI 95% 0.766–0.9220), miR-130a (OR 0.885; CI 95% 0.827–0.948), miR-222 (OR 1.064; CI 95% 1.004–1.126), miR-218 (OR 1.138; CI 95% 1.036–1.250), and miR-185 (OR 1.099; CI 95% 1.003–1.205) were independent predictor factors for IS. Moreover, the combined analysis of these five DEMs demonstrated a good diagnostic performance for IS with sensitivity of 87.7%, and specificity of 54.5%. Additionally, miR-126, miR-378, and miR-101 were negatively associated, and miR-222, while miR-218, miR-206 positively associated with NIHSS score what could be used for disease severity management of IS [53].

So far, several studies showed the importance of miR-145 in stroke. In one of the first study, circulating miR-145 was evaluated in 32 IS patients and compared to 14 control participants. The results showed significant up-regulation of circulating miR-145 expression and non-significant downregulation after one month in IS patients ($N = 11$). Authors suggested that miR-145 might be a desirable biomarker in IS [54].

In the next study nine previously reported stroke associated miRNAs (miR-21, miR-23a, miR-29b, miR-124, miR-145, miR-210, miR-221, miR-223 and miR-483-5p) were screened both in 146 IS group and 96 control subjects. In this validation process only miR-145 was significantly up-regulated, but miR-23a and miR-221 were significantly downregulated. Serum miR-23a and miR-221 were moderately negatively correlated with plasma hs-CRP. Moreover, a strong positive correlation existed between serum miR-145 and hs-CRP (r = 0.6713), and a moderate correlation existed between serum miR-145 and IL-6 (r = 0.5896). Interestingly, miR-145 level was positively correlated with infarct volume and NIHSS scores (r = 0.6249 and r = 0.6288, respectively). Finally, the prediction value using both hs-CRP and miR-145 was significantly higher than for hs-CRP alone (See Table 2). In 49 IS patients long term follow-up lasting 2 years was performed and showed that the expression of miR-145 was highest within the acute phase (1 to 7 days) of stroke, but decreased within the recovery phase (1 month, 6 months, and 2 years) in the serum of patients [55]. The abovementioned role of miR-145 was confirmed through bioinformatics analyses (Gene Ontology and Kyoto Encyclopedia of Genes and Genomes enrichment) that included two mRNA and 1 miRNA microarray expression profile data from the Gene Expression Omnibus database. Based on this analysis two miRNAs, namely miR-145 and miR-122, may represent potential biomarkers in IS. Also, three novel miRNAs (miR-99b, miR-542-3p, and miR-455-5p) were deregulated what may suggest their roles in the pathological processes of IS [56].

Contrary to recent studies, Tsai et al. did not find significant modulation of miR-145 in IS. However, they found that miR-21 level was significantly higher, but miR-221 was significantly lower compared to healthy controls. MiR-145 was excluded from further analysis, as there were no significant differences between groups. At multivariable logistic regression analysis both increased miR-21 level (OR 6.16; 95% CI 2.82, 14.64) and decreased miR-221 level (OR 10.38; 95%CI 4.52, 26.45) were associated with increased risk of stroke when added to classical risk factors like age, sex, diabetes, hypertension, smoking, and hyperlipidemia. Also ROC analysis showed that adding both miR-21 and miR-221 into the model substantially improved AUC value for IS prediction. Hence, the authors suggested that miR-21 and miR-221 can be novel biomarkers for stroke, but not miR-145 [57].

Also, Zhou and Zhang identified decreased plasma miR-21 and miR-24 levels in 68 IS patients compared with 21 healthy controls. Moreover, miR-21 expressions correlated with miR-24 level, and both miRNAs negatively correlated with early outcome of IS based on NIHHS score (r = −0.703, *p* < 0.05 for miR-21; r = −0.694, *p* < 0.05 for miR-24) [58].

Previous study analyzed the changes of miR-223 levels in 75 healthy control samples and compared to 79 IS patients within 72 h from the onset of symptoms. It was found that miR-223 expression correlates with stroke subtype (i.e., LA and SA subtype) and negatively correlates with NIHSS scores, but the correlations with infarct volume and insulin-like growth factor 1 (IGF-1) levels were low [59]. Beside miRNA studies in circulating blood miR-223 was detect also in exosomes. Exosomes are 30–100 nm vesicles that cells secrete into extracellular space when multivesicular bodies fuse with cell membrane. MiR-223 was shown to be one of the most highly expressed miRNAs in plasma exosomes of healthy human and results about its function in ischemia injury are inconsistent [60]. Chen et al. found that the level of miR-223 in circulating exosomes was elevated after onset of IS, and exosomal miR-223 expression was positively correlated to NIHSS score. Moreover, stroke patients with poor outcomes based on mRS inclined to have a greater exosomal miR-223 expression. Therefore, increased exosomal miRNA-223 possess both diagnostic and prognostic value for IS [61].

In another study serum exosomal miRNAs, namely miR-9 and miR-124, were evaluated in 65 IS patients and compared with 66 non-stroke volunteers in order to explore their diagnostic and prognostic value. The team found that both miR-9 and miR-124 levels were increased in IS patients. Moreover, both miRNAs expression were positively correlated with NIHSS score (r = 0.7126 and 0.6825 respectively, *p* < 0.01), infarct volume evaluated by magnetic resonance imaging (MRI) (r = 0.6768 and 0.6312, respectively, *p* < 0.01) and serum IL-6 concentration (r = 0.6980 and 0.6550, respectively, *p* < 0.01). Finally, diagnostic value for IS of exosomal miR-9 and miR-124 was confirmed in ROC analysis [62]. Interestingly, another small study done by Liu et al. showed that serum expression of miR-124 was downregulated within the first 24 h after the ischemic event, but there were no differences in miR-9 and miR-219 levels between 31 IS and 11 healthy controls. Also, only miR-124 and miR-9 were negatively correlated with infarct lesion volume (r = −0.423, *p* = 0.022; and r = −0.608, *p* < 0.001), but not with NIHSS score. Besides, a significant negative correlation existed between plasma hs-CRP levels and serum miR-124 and miR-9 levels (r = −0.421, *p* = 0.023; and r = −0.511, *p* = 0.004) [63].

In the latest study, 17 previously reported stroke-associated miRNAs were measured in serum using RT-qPCR. Researchers first evaluated miRNA levels in randomly selected 30 IS patients compared with 30 control participants. MiR-21, miR-145, miR-29b, and miR-146b were significantly increased but miR-23a and miR-221 levels were decreased within 24 h after stroke onset compared with the control group. Further verification was done for miR-21, miR-23a, miR-29b, miR-145, miR-146b, and miR-221 with a larger number of patients. MiR-146b was significantly upregulated in the serum of 128 IS patients compared with 102 control participants. In addition to this, positive correlation was found between upregulated serum miR-146b level and plasma hs-CRP, infarct volume and NIHSS score, and serum IL-6 of patients. Importantly, the combination of plasma hs-CRP and serum miR-146b gained a better sensitivity/specificity for prediction of IS (AUC from 0.782 to 0.863). Besides, it was found that decreased miR-221 level negatively correlated with plasma hs-CRP level but not serum IL-6 level in IS patients. Authors suggested that upregulated serum miR-146b might be a potential diagnostic biomarker for IS evaluation [64].

2.2. MiRNA Profiling and RNA Sequencing Strategy

MiRNA microarrays are commonly used for miRNA expression profiling analysis. They have advantages and disadvantages comparison to RT-qPCR technologies. First of all, they allow us a powerful high-throughput profiling for a genome-wide miRNA profiling assay and the protocol can be easily standardized. Despite, there are some major disadvantages associated with miRNA microarrays. For example, generally a large amount of high-quality RNA samples are needed for the microarray experiments, which is often a major challenge for miRNA expression analysis of clinical samples. Moreover, lower miRNA detection sensitivity and specificity is the main limitation of microarray analysis compared to RT-qPCR analysis [65]. Beside miRNA microarrays, the next-generation-sequencing technology has quickly emerged as the preferred platform for studying circulating miRNA profiling. Comparison to microarray technology RNA-sequencing technology has higher sensitivity and specificity, however high cost is a practical consideration for researchers [66].

In one of the first study using microarray strategy Jickling et al. identified candidate miRNAs that could serve as a diagnostic tool in 24 IS patients. In a group with IS six miRNAs, expression was increased (i.e., miR-122, miR-148a, let-7i, miR-19a, miR-320d, miR-4429), and two miRNAs, expression was decreased (miR-363 and miR-487b) in comparison to healthy controls. By using gene target database, they also evaluated potential gene targets for selected miRNAs and they found two main pathways which are regulated by these miRNAs, i.e., the nuclear factor-κB (NF-κB) and Toll-like receptor signaling pathways, the pathways, which are involved in immune activation, leukocyte extravasation and thrombosis [67].

Using screening technique Li et al. discovered 115 miRNAs that were differentially expressed in IS patients compared to controls. For the further analysis they selected based on well-defined criteria and literature search 13 miRNAs. In order to validate the initial results they performed RT-qPCR of selected miRNAs and found that expression of 8 miRNAs (miR-32-3p, miR-106b-5p, miR-423-5p, miR-451a, miR-1246, miR-1299, miR-3149, and miR-4739) significantly increased, and the expression of 5 miRNAs (miR-224-3p, miR-377-5p, miR-518b, miR-532-5p, and miR-1913) significantly decreased in the serum of IS patients. According to functional assays only upregulated miR-32-3p, miR-106b-5p, and miR-1246, and downregulated miR-532-5p in IS serum might play a vital role in the pathogenesis of IS. Moreover, through bioinformatic analysis, they found that stroke-related genes vascular endothelial growth factor-A (VEGFA), myeloid cell leukemia-1 (Mcl-1), and superoxide dismutase 2 (SOD2) might be the targets of miR-106b, therefore authors suggested that miR-106b may affect multiple pathways such as apoptosis, oxidation, angiogenesis, and neurogenesis in IS [68–71].

MiRNA profiling strategy in the whole blood samples was proposed to analyze the utility of miRNA for the disease progression and stroke subtype evaluation in young patients with IS. This strategy showed that 157 miRNAs were differentially regulated across stroke samples. In total, 138 miRNAs were upregulated, and 19 miRNAs were downregulated. Interestingly, different miRNAs expression was observed between stroke subtypes groups, i.e., 8 miRNAs were downregulated (hsa-let-7f, miR-126, -1259, -142-3p, -15b, -186, -519e, and -768-5p) and 17 miRNAs were upregulated (hsa-let-7e, miR-1184, -1246, -1261, -1275, -1285, -1290, -181a, -25*, -513a-5p, -550, -602, -665, -891a, -933, -939, and -923) across subtypes of stroke. Further analysis showed also the utility of miRNA profiling in prognosis [72].

Profile of circulating miRNA was also evaluated by Sepramaniam et al. In three different cohorts that contained 169 patients with IS. In total 314 miRNAs were detected upon profiling. In the initial step of the analysis they found 105 different miRNAs that were deregulated in stroke cases. Among the 105 miRNAs, 58 were downregulated while 47 were upregulated. Moreover, further analysis help to significantly distinguish the stroke etiology i.e., LA, CE and SV based on 32 miRNAs (let-7a, let-7d*, let-7g, let-7i, miR-126, -130a, -187*, -18a*, -20a, -22*, -26b, -30b, -30c, -30e*, -320b, -320d, -324-5p, -331-3p, -340, -342-3p, -361-5p, -363, -422a, -423-3p, -501-5p, -502-3p, -505*, -574-3p, -675,-886-5p, -92a, and -93*), however detailed analysis was not provided in the study. Finally, the patients were segregated into acute and recovery phase and 26 miRNAs (let-7d*, miR-125b-2*, -1261, -1299, -130a, -1321, -208a, -22*, -23a, -27a*, -320b, -320d, -30c, -340, -422a, -423-3p, -488, -502-5p, -549a, -574-3p, -574-5p, -617, -627, -886-5p, -92a, and -93*) were unique for acute stroke and 16 miRNAs (let-7a, let-7g, miR-129-5p, -192-5p, -196a*, -26b, -30b, -30e*, -370, -381, -493*, -525-5p, -652, -920, -933, and -96) were unique for stroke patients at 7 days after the event ("recovered patients"). Diagnostic accuracy for acute phase of IS tested through calculating the AUC of ROC curves showed that 5 different miRNAs could serve as a potential biomarkers i.e., miR-125b-2*, -27a*, -422a, -488 and -627 [73] (See Figure 1).

Table 1. Human studies evaluating miRNAs as diagnostic/prognostic biomarkers in stroke.

Ref	miRNAs	Sampling/Sampling Time Point	Number of Stroke pts/Controls	Inclusion Criteria	Exclusion Criteria	Stroke Subtype	Prognostic or Diagnostic Value	Regulation of miRNAs	Correlation
Long et al. [41]	miR-30a, miR-126, Let-7b,	Plasma/24 h, 1 w, 4 w, 24 w, 48 w	197/50	First-ever stroke patients with cerebral infarction	Exclusion criteria included TIA, subarachnoid hemorrhage, embolic brain infarction, brain tumors, and cerebrovascular malformation, pulmonary fibrosis, endocrine and metabolic diseases (except type 2 diabetes), inflammatory and autoimmune diseases, and serious chronic diseases, for example, hepatic cirrhosis and renal failure. Cardioembolic stroke and documented atrial fibrillation were also excluded from the study.	LA 51, SA 48, CE 50, UDN 48	Diagnostic value	MiR-30a and miR-126 were downregulated in 24 h, 1 w, 4 w, and 24 w. After 48 w miRNA levels increased to the baseline	No correlation was found between HDL, LDL, triglyceride, systolic and diastolic blood pressures, diabetes and smoking status and miRNAs.
Peng et al. [43]	miR-338, Let-7e	Serum and CSF/1–7 days (acute phase), 8–14 days (subacute phase), over 15 days (recovery)	72/51	Diagnosis of an initial episode of cerebral infarction based on clinical history and MRI results, ages ranging from 55 to 75 years, patient arrival at the hospital after 4.5 h but within 24 h after the event, NIHSS score of 4 to 15, and without hemorrhagic transformation.	Exclusion criteria for all enrolled patients included recurrent stroke, tumors, abnormal renal or liver function, infectious diseases, immune diseases, blood disorders, and psychiatric illness including depression and schizophrenia.	NA	Let-7e may be a biomarker in acute phase/had no prognostic value	Let-7e significantly higher at all time-points in serum. MiR-338 and Let-7e in CSF was upregulated in 8–14 days (subacute).	No correlation was found between NIHSS and Let-7e, correlation was found between CRP levels and Let-7e.
Huang et al. [44]	Let-7e-5p	Whole blood/24 h	Two groups: 44/44, 302/02	First-ever IS patients	Patients with a history of stroke, peripheral arterial occlusive disease or cancer were excluded from this study.	NA	Diagnostic value for Let-7e-5p	Let-7e-5p significantly higher with IS patients than controls	Negatively correlated with ATF2, CASP3, FGFR2, NLK, PTPN7, RASGRP1 and TGFBRI genes.

Table 1. Cont.

Ref	miRNAs	Sampling/Sampling Time Point	Number of Stroke pts/Controls	Inclusion Criteria	Exclusion Criteria	Stroke Subtype	Prognostic or Diagnostic Value	Regulation of miRNAs	Correlation
Gong et al. [45]	Let-7f	Serum/after 48 h and 2 w	88/130	Selection criteria included age >18 years, within 48 h after stroke attack, based on CT or MRI, the patient had an infarct of at least 67% of the middle cerebral artery territory, with or without the additional infarction of the anterior or posterior cerebral artery on the same side.	Exclusion criteria included unconsciousness due to metabolic disturbances or medication, any sedation or surgery, a pre-stroke score on the mRS of more than 2; and the presence of a concurrent serious illness which may affect the patient's outcome, such as severe cardiopulmonary complications.	NA	Prognostic value	Let-7f was downregulated in IS with MCI and upregulated in IS without MCI at 48 h.	Let-7f was negatively correlated with hs-CRP in IS MCI patients, also negatively correlated with IL-6 in MCI without HT
Leung et al. [47]	miR-124-3p and miR-16	Plasma/(≤6 h), (6–24 h)	93 IS + 19 HS/23	Patients aged 18 years old and above were included to the study, HS or IS confirmed by CT scan and/or MRI, who presented within 24 h of symptom onset.	NA	NA	Diagnostic value	MiR-124-3p levels were markedly higher in patients with HS patients compared to IS patients only in cases presenting early (≤6 h), increased miR-16 were found in patients with IS compared to those with HS in patients presenting late after symptom onset (6–24 h)	Plasma concentrations of miR-124-3p, but not miR-16, positively correlated with lesion volume on CT in HS patients; however, both plasma miR-124-3p and miR-16 did not correlate with lesion volume on MRI in IS patients.
Wu et al. [48]	miR-15a, miR-16, miR-17-5p	Serum/before treatment	106/120	The cohort included 55 men and 51 women with a mean age of 64.8 years (range, 39–88 years).	Symptoms indicative of subarachnoid hemorrhage, even if no imaging findings of hemorrhage were found on CT or MRI, intracranial hemorrhage, acute myocardial infarction, critical limb ischemia.	NA	Diagnostic value	MiR-15a, miR-16, and miR-17-5p were significantly higher in IS patients compared to control subjects	MiR-15a was significantly correlated with age, strong negative correlation between miR-16 levels and HDL and ApoA1 was found.

Table 1. *Cont.*

Ref	miRNAs	Sampling/Sampling Time Point	Number of Stroke pts/Controls	Inclusion Criteria	Exclusion Criteria	Stroke Subtype	Prognostic or Diagnostic Value	Regulation of miRNAs	Correlation
Jin F and Xing, J. [53]	miR-126, miR-17-5p, miR-17-3p, miR-18a, miR-19a, miR-20a, miR-19b-1, miR-92a, Let-7b, Let-7i, miR-130a, miR-210, miR-378, miR-296, miR-101, miR-221, miR-222, miR-328, miR-15b, miR-16, miR-26b, miR-27b, miR-218, miR-206, miR-338-3p, miR-497, miR-195a-3p, miR-185	Plasma/24h	106/110	Within 24 h post the onset of symptom, diagnosed with IS according to patient history, laboratory and neurological examination, CT scan, MRI, and/or MRA.	Patients with infection, renal or hepatic failure, hematological malignancies, solid tumors, immunosuppressive therapy, or treatment with thrombolytic therapy were excluded from the study.	NA	Diagnostic value, disease severity management	MiR-126 and miR-130a decreased in the IS patients while miR-222, miR-218, and miR-185 increased in the IS patients.	MiR-126, miR-378, miR-101 negatively, miR-222, miR-218, miR-206 were positively correlated with NIHSS
Gan et al. [54]	miR-145	Whole blood/NA/after one month second sampling (N = 11)	32/14	IS patients between the ages of 18 and 49 years were included. IS was confirmed by either MRI or CT imaging of the brain.	Excluded from the study were subjects with hemorrhage stroke.	NA	No diagnostic and prognostic value was found	Upregulation of miR-145	No correlation was found
Jia et al. [55]	miR-21, miR-23a, miR-29b, miR-124, miR-145, miR-210, miR-221, miR-223, miR-483-5p	Serum/24 h	146/96	Patients with IS within 24 h after symptom onset were included.	Exclusion criteria included under 18 years old, being on thrombolytic or anticoagulant therapies, intracerebral hemorrhage or hemorrhagic transformation, other complicating neurological or neuropsychological diseases, cancer, comorbidity with proinflammatory conditions and clinical signs of infection at any time during the study.	NA	Diagnostic value	MiR-145 was upregulated, miR-23a and miR-221 were significantly downregulated	Positive correlation between miR-145 and hs-CRP, IL-6, infarct volume and NIHSS scores was found, serum miR-23a and miR-221 were moderate negatively correlated with plasma hs-CRP

Table 1. *Cont.*

Ref	miRNAs	Sampling/Sampling Time Point	Number of Stroke pts/Controls	Inclusion Criteria	Exclusion Criteria	Stroke Subtype	Prognostic or Diagnostic Value	Regulation of miRNAs	Correlation
Tsai et al. [57]	miR-145, miR-21, miR-221	Serum/7 days	167/157	Patients with IS based on the World Health Organization criteria. The blood samples from the patients were taken within 7 days of the onset of stroke. Demographic data and histories of hypertension, diabetes, hypercholesterolemia and cigarette smoking were obtained from each study subject.	NA	NA	Diagnostic value	MiR-21 was downregulated, miR-221 was upregulated, no significance for miR-145 was found	MiR-21 expressions was correlated with miR-221 levels.
Zhou et al. [58]	miR-21, miR-24	Plasma/24 h	68/21	ACI participated included in the study. The diagnosis of ACI was conducted based on patient history, lab examination, neurological deficit, MRI and MRA results.	Patients with a history of tumor, immune disease, blood disease, acute infectious disease, renal or liver failure were excluded.	NA	Diagnostic value	MiR-21 and miR-24 were downregulated	MiR-21 expressions were positively correlated with miR-24 level, and negatively correlated with NIHSS score
Chen et al. [61]	miR-223	Exosomes/plasma/less than 72 h	50/33	Stroke patients included in the study. Demography feature, related previous history including hypertension, diabetes mellitus, hyperlipidemia, cardiopathy, associated laboratory test, and imaging information including blood glucose, blood lipid, electrocardiogram, cardiac ultrasonography, carotid artery ultrasonography, MRI, and MRA were also collected for analysis.	Exclusive criteria included recurrent stroke or stroke onset longer than 72 h, renal or liver failure, acute infectious disease, tumor, hematologic disease, and patients who are unable to cooperate with physical examination.	LA 25, SA 17, CE 8	Diagnostic and prognostic value	MiR-223 was upregulated	MiR-223 was positively correlated with NIHSS score

Table 1. *Cont.*

Ref	miRNAs	Sampling/Sampling Time Point	Number of Stroke pts/Controls	Inclusion Criteria	Exclusion Criteria	Stroke Subtype	Prognostic or Diagnostic Value	Regulation of miRNAs	Correlation
Wang et al. [59]	miR-223	Whole blood/less than 72 h	79/75	IS patients included in the study.	The exclusion criteria included recurrent stroke, intracranial tumor, multiple trauma, hematological system diseases, renal or liver failure, acute infectious diseases and other diseases affecting the hemogram. If the time from the onset of stroke symptoms to blood sample collection was longer than 72 h, the patient was excluded.	LA 37, SA 9, CE 5, UN 28	Diagnostic value	MiR-223 in IS patients were greatly increased compared to the control	MiRNA-223 was negatively correlated with NIHSS scores, plasma level of IGF-1 was positively correlated with that of miRNA-223
Ji et al. [62]	miR-9, miR-124	Exosomes/plasma/the mean time of enrollment blood draw was 16.5 h.	65/66	IS patients were recruited after either MRI or CT imaging of the brain.	Patients with intracerebral hemorrhage or unknown etiology were excluded.	NA	Diagnostic and prognostic value	MiR-9 and miR-124 in IS patients were increased compared to the control	The levels of both miR-9 and miR-124 were positively correlated with NIHSS scores, infarct volumes and serum concentrations of IL-6.
Liu et al. [63]	miR-9, miR-124, miR-219	Serum/24 h	31/11	Patients with IS 24 h after symptom onset were enrolled to the study.	Exclusion criteria were being under 18 years old, being on thrombolytic or anticoagulant therapies, intracerebral hemorrhage or hemorrhagic transformation, other complicating neurological or neuropsychological diseases, cancer, comorbidity with proinflammatory conditions, and clinical signs of infection at any time during the study.	NA	Inflammatory value	MiR-124 was downregulated	Both serum miR-124 and miR-9 levels within 24 h were negatively correlated with infarct volume and plasma hs-CRP levels. All three miRNAs were negatively correlated with MMP-9 levels.
Jickling et al. [67]	miR-122, miR-148a, Let-7i, miR-19a, miR-320d, miR-4429, miR-363, miR-487b	Whole blood/NA	24/24	Patients with IS were enrolled in the study. Stroke diagnosis restricted diffusion on brain MRI (positive DWI-MRI).	Patients with infection (current or within 2 weeks of stroke), immunosuppressive therapy, lymphoma, leukemia, or treatment with thrombolytic therapy were excluded from study.	LA 8, SA 8, CE 8	Diagnostic value	In patients with IS, miR-122, miR-148a, let-7i, miR-19a, miR-320d, miR-4429 were decreased and miR-363, miR-487b were increased compared to vascular risk factor controls.	MiRNAs may regulates NF-κB and toll-like receptor signaling pathways, which are involved in immune activation, leukocyte extravasation and thrombosis.

Table 1. Cont.

Ref	miRNAs	Sampling/Sampling Time Point	Number of Stroke pts/Controls	Inclusion Criteria	Exclusion Criteria	Stroke Subtype	Prognostic or Diagnostic Value	Regulation of miRNAs	Correlation
Li et al. [68]	In total 115 miRNAs were screened miR-32-3p, miR-106b-5p, miR-423-5p, miR-451a, miR-1246, miR-1299, miR-3149, miR-4739, miR-224-3p, miR-377-5p, miR-518b, miR-532-5p, miR-1913	Serum/24 h	117/82	IS patients (aged >45) within 24 h after stroke onset were enrolled to the study.	Exclusion criteria included other types of stroke (TIA, subarachnoid hemorrhage, brain tumors, and cerebrovascular malformation); severe systemic diseases, i.e., pulmonary fibrosis, endocrine, and metabolic diseases (except type 2 diabetes); inflammatory and autoimmune diseases; and serious chronic diseases, for example, hepatic cirrhosis and renal failure.	NA	Diagnostic value of upregulated miR-32-3p, miR-106b-5p, miR-1246, and downregulated miR-532-5p	MiR-32-3p, miR-106b-5p, miR-423-5p, miR-451a, miR-1246, miR-1299, miR-3149, miR-4739 were upregulated	MiR-106b may affect multiple pathways such as apoptosis, oxidation, angiogenesis, and neurogenesis in IS.
Tan et al. [72]	in total 157 miRNAs were screened hsa-let-7f, miR-126, -1259, -142-3p, -15b-186, -519e, -768-5p hsa-let-7e, miR-1184, -1246, -1261, -1275, -1285, -1290, -181a, -25*, -513a-5p, -550, -602, -665, -891a, -933, -939, -923	Whole blood / within 6–18 months	19/5	Asian stroke patients between the ages of 18 to 49 were enrolled to the study. Blood samples collected from stroke patients within 6–18 months in time scale from the index stroke. IS was confirmed either with CT or MRI of the brain.	NA	LA 8, SA 3, CE 5, UN 3	Diagnostic value	In total 138 miRNAs were upregulated and in total 19 miRNAs were downregulated. hsa-let-7f, miR-126, -15b, -186, -519e, -768-5p were downregulated, hsa-let-7e, miR-1184, -1246, -1261, -1275, -1285, -1290, -181a, -25*, -513a-5p, -550, -602, -665, -891a, -933, -939, -923 were upregulated.	Among the upregulated miRNAs, the expression of miR-101, -106b, -130a, -144, -18a, -18b, -19a, -19b, -194, -22, -22, -29b, -29c and -363 were the highest for LA (mRS = 3) stroke sample and positively correlated to the profile observed for LA mRS > 2

Table 1. *Cont.*

Ref	miRNAs	Sampling/Sampling Time Point	Number of Stroke pts/Controls	Inclusion Criteria	Exclusion Criteria	Stroke Subtype	Prognostic or Diagnostic Value	Regulation of miRNAs	Correlation
Sepramaniam et al. [73]	In total 314 miRNAs were screened. hsa-let-7e, hsa-miR-125b-2*, hsa-miR-1261, hsa-miR-129-5p, hsa-miR-1321, hsa-miR-135b, hsa-miR-145, hsa-miR-184, hsa-miR-187*, hsa-miR-196a*, hsa-miR-198, hsa-miR-200b*, hsa-miR-210, hsa-miR-214, hsa-miR-220c, hsa-miR-25*, hsa-miR-602, hsa-miR-611, hsa-miR-617, hsa-miR-623, hsa-miR-627, hsa-miR-637, hsa-miR-638, hsa-miR-659, hsa-miR-668, hsa-miR-671-5p, hsa-miR-675, hsa-miR-920, hsa-miR-933, hsa-miR-943, hsa-miR-99a, hsa-let-7a, hsa-let-7b*, hsa-let-7c, hsa-let-7d*, hsa-let-7l, hsa-let-7g, hsa-let-7i,	Whole blood/24 h, 48 h, 7 days, from 2 months to 2 years from stroke onset	169/118	Patients with IS were enrolled to the study. IS was confirmed through either MRI or CT imaging of the brain.	NA	NA	Diagnostic value, potential diagnostic biomarkers; miR-125b-2*, -27a*, -422a, -488 and -627	Among the significant 105 miRNAs, 58 were downregulated while 47 were upregulated. Upregulated miRNAs: hsa-let-7e, hsa-miR-125b-2*, hsa-miR-1261, hsa-miR-129-5p, hsa-miR-1321, hsa-miR-135b, hsa-miR-145, hsa-miR-184, hsa-miR-187*, hsa-miR-196a*, hsa-miR-198, hsa-miR-200b*, hsa-miR-210, hsa-miR-214, hsa-miR-220c, hsa-miR-25*, hsa-miR-602, hsa-miR-611, hsa-miR-617, hsa-miR-623, hsa-miR-627, hsa-miR-637, hsa-miR-638, hsa-miR-659, hsa-miR-668, hsa-miR-671-5p, hsa-miR-675, hsa-miR-920, and downregulated miRNAs: hsa-let-7a, hsa-let-7b*, hsa-let-7c, hsa-let-7d*,	

Table 1. Cont.

Ref	miRNAs	Sampling/Sampling Time Point	Number of Stroke pts/Controls	Inclusion Criteria	Exclusion Criteria	Stroke Subtype	Prognostic or Diagnostic Value	Regulation of miRNAs	Correlation
	hsa-miR-106b*, hsa-miR-126, hsa-miR-1299, hsa-miR-130a, hsa-miR-151-5p, hsa-miR-18a*, hsa-miR-182, hsa-miR-183, hsa-miR-186, hsa-miR-192, hsa-miR-20a, hsa-miR-208a, hsa-miR-22*, hsa-miR-500, hsa-miR-500*, hsa-miR-501-5p, hsa-miR-502-5p, hsa-miR-502-3p, hsa-miR-505*, hsa-miR-532-5p, hsa-miR-574-5p, hsa-miR-574-3p, hsa-miR-576-5p, hsa-miR-625, hsa-miR-629, hsa-miR-652, hsa-miR-7, hsa-miR-886-5p, hsa-miR-92a, hsa-miR-93*, hsa-miR-96							hsa-let-7i, hsa-let-7g, hsa-let-7i, hsa-miR-106b*, hsa-miR-126, hsa-miR-1299, hsa-miR-130a, hsa-miR-151-5p, hsa-miR-18a*, hsa-miR-182, hsa-miR-183, hsa-miR-186, hsa-miR-192, hsa-miR-20a, hsa-miR-208a, hsa-miR-22*, hsa-miR-500, hsa-miR-500*, hsa-miR-501-5p, hsa-miR-502-5p, hsa-miR-502-3p, hsa-miR-505*, hsa-miR-532-5p, hsa-miR-574-5p, hsa-miR-574-3p, hsa-miR-576-5p, hsa-miR-625, hsa-miR-629, hsa-miR-652, hsa-miR-7, hsa-miR-886-5p, hsa-miR-92a, hsa-miR-93*, hsa-miR-96	

Table 1. *Cont.*

Ref	miRNAs	Sampling/Sampling Time Point	Number of Stroke pts/Controls	Inclusion Criteria	Exclusion Criteria	Stroke Subtype	Prognostic or Diagnostic Value	Regulation of miRNAs	Correlation
Wang et al. [74]	Microarray revealed 17 upregulated miRNAs and 103 downregulated miRNAs in MRI(−) acute stroke patients compared with MRI(+) acute stroke patients, 33 upregulated miRNAs and 36 downregulated miRNAs in MRI(+) acute stroke hsa-miR-106b-5P, hsa-miR-4306, hsa-miR-320e hsa-miR-320d	Plasma/0–3 h 23 patients, 3–6 h 37 patients, 6–12 h 31 patients, 12–24 h 45 patients.	136/116	The inclusion criteria consisted of having IS or TIA and having no history of coronary artery disease.	Patients were excluded if they had received intravenous thrombolytic or anticoagulant therapy before the initial blood samples were collected.	LA 60, SA 51, CE 23, UDN 2	Diagnostic value of hsa-miR-106b-5P, hsa-miR-4306, hsa-miR-320e, and hsa-miR-320d	hsa-miR-106b-5p hsa-miR-4306 increased, hsa-miR-320d decreased which are associated with diagnostic value	NA
Tian et al. [75]	hsa-mir-4454, hsa-mir-140-3p, hsa-mir-106b-5p, hsa-mir-25-3p, hsa-mir-16-5p, hsa-mir-223-3p, hsa-mir-484, hsa-mir-130b-3p, hsa-mir-151a-3p, hsa-mir-130a-3p, hsa-mir-93-5p, hsa-mir-107	Plasma/6 h	40/30	Inclusion criteria: time duration from stroke onset to admission was less than 6 h.	Patients with immune disease, trauma, coronary heart disease, organ failure, tumor, and infection were excluded from the study.	LA 10, SA 8, CE 14, UN 7, ODE 1	Diagnostic and prognostic value of miR-16	MiR-140, miR-106b, miR-130a, miR-16, miR-223, miR-93, miR-484, miR-25, miR-130b, miR-107, and miR-151 were upregulated and miR-4454 was downregulated	Plasma concentrations of miR-16 were related to TOAST criteria, OCSP criteria, and the prognosis of HACI patients

Table 1. *Cont.*

Ref	miRNAs	Sampling/Sampling Time Point	Number of Stroke pts/Controls	Inclusion Criteria	Exclusion Criteria	Stroke Subtype	Prognostic or Diagnostic Value	Regulation of miRNAs	Correlation
Tiedt et al. [76]	hsa-let-7b-3p, hsa-let-7d-3p, hsa-let-7f-5p, hsa-let-7i-5p, hsa-miR-1, hsa-miR-16-2-3p, hsa-miR-17-3p, hsa-miR-17-5p, hsa-miR-18a-3p, hsa-miR-18a-5p, hsa-miR-20a-5p, hsa-miR-26b-5p, hsa-miR-92a-3p, hsa-miR-99b-5p, hsa-miR-101-3p, hsa-miR-125a-5p, hsa-miR-125b-5p, hsa-miR-126-5p, hsa-miR-130a-3p, hsa-miR-140-3p, hsa-miR-143-3p, hsa-miR-181a-5p, hsa-miR-193a-5p, hsa-miR-378a-3p, hsa-miR-423-3p, hsa-miR-532-5p, hsa-miR-660-5p, hsa-miR-3158-3p, hsa-miR-3158-5p, hsa-miR-3184-5p, hsa-miR-3688-3p, hsa-miR-3688-5p	Plasma/24 h, 48 h, 72 h, 90th day.	260/160	IS and TIA patients were enrolled to the study.	Patients with active malignant disease, inflammatory or infectious diseases, surgery within the last three months and prior medication with low-molecular or unfractionated heparin within the last month were excluded. Further, for the discovery sample patients with prior use of antiplatelet medication within the last month, a history of myocardial infarction, stroke, or TIA, or signs for silent CNS infarction on neuroimaging were also excluded. For the replication sample, patients with prior medication with low-molecular or unfractionated heparin within the last month were excluded.	LA 61, SA 18, CE 79, UN 96	Diagnostic value of miR-125a-5p, miR-125b-5p, miR-143-3p	MiR-125a-5p, miR-125b-5p, and miR-143-3p were upregulated.	The transformed infarct volumes of IS patients (N = 188) were correlated with expression levels of miR-125a-5p, miR-125b-5p, and miR-143-3p.

Table 1. *Cont.*

Ref	miRNAs	Sampling/Sampling Time Point	Number of Stroke pts/Controls	Inclusion Criteria	Exclusion Criteria	Stroke Subtype	Prognostic or Diagnostic Value	Regulation of miRNAs	Correlation
Mick et al. [77]	ex-RNAs by RNASeq (331 miRNAs, 97 piRNAs, and 43 snoRNAs) for RT-qPCR analysis: miR-877-5p, miR-124-3p, miR-320d, snoRNA SNO1402, hsa-miR-656-3p, hsa-miR-941	Plasma/NA	2763 participants included from (Framingham Heart Study; Offspring Cohort Exam 8), unbiased next-generation sequencing conducted using plasma from 40 participants from the cohort.	Subjects were diagnosed with stroke based on review of medical records, including relevant hospitalizations, and clinic reported events by at least 2 neurologists agreeing on one of the following manifestations: definite cerebrovascular accident, atherothrombotic infarction of the brain, cerebral embolism, intracerebral hemorrhage, or subarachnoid hemorrhage.	NA	NA	NA	Observational study	miR-877-5p, miR-124-3p, and miR-320d) and one snoRNA (SNO1402) were independently associated with prevalent stroke, hsa-miR-656-3p and hsa-miR-941 were significantly associated with incident stroke
Chen et al. [84]	Let-7b, miR-23a, miR-126, miR-15a, miR-16, miR-17-5p, miR-19b, miR-29b, miR-339-5p, miR-21, miR-221, miR-32-3p, miR-106-5p, miR-532-5p, miR-145, miR-146b, miR-210	Serum/24 h	128/102	Patients with IS within 24 h after symptoms were enrolled to the study.	Exclusion criteria were being under 18 years old, being on thrombolytic or anticoagulant therapies, intracerebral hemorrhage or hemorrhagic transformation, other complicating neurological or neuropsychological diseases, cancer, comorbidity with proinflammatory conditions, and clinical signs of infection at any time during the study.	NA	Diagnostic value for miR-146b. There is no significance of the expression of serum miRNAs among 5 IS groups in this study. It was suggested that miR-146b may be a biomarker for IS but not for separating subtypes of IS.	MiR-21, miR-145, miR-29b and miR-146b were significantly upregulated and miR-23a and miR-221 levels were significantly downregulated	Positive significant correlation was found between serum miR-146b level and plasma hs-CRP, infarct volume and NIHSS score and serum IL-6 of patients.

Table 1. *Cont.*

Ref	miRNAs	Sampling/Sampling Time Point	Number of Stroke pts/Controls	Inclusion Criteria	Exclusion Criteria	Stroke Subtype	Prognostic or Diagnostic Value	Regulation of miRNAs	Correlation
Wu et al. [78]	754 miRNAs were screened, 71 miRNAs were upregulated and 49 miRNAs were downregulated in IS patients	Serum/NA	50/50 then it was confirmed with a larger cohort 177 IS, 81 TIA patients and 42 controls	IS and TIA patients were enrolled to the study. The diagnosis of IS was based on the acute occurrence of focal neurological deficit lasting for more than 24 h and was confirmed by the positive findings of brain CT and MRI.	Patients with a history of hemorrhagic infarction, peripheral arterial occlusive diseases, chronic liver/kidney diseases, primary/metastatic neoplasms or other malignant diseases were excluded.	NA	Prognostic value for miR-23b-3p and miR-29b-3p	MiR-23b-3p, miR-29b-3p, miR-181a-5p and miR-21-5p were markedly increased in IS patients. MiR-23b-3p, miR-29b-3p and miR-181a-5p were also significantly elevated in TIA patients	MiR-23b-3p levels in IS patients were positively related with discharge mRS scores, while miR-23b-3p and miR-29b-3p levels in IS patients were negatively related with discharge BI scores.

Abbreviations: CE, cardioembolic stroke; LA, large artery stroke; SA, small artery stroke; ST, stroke types; TP, time point of blood sampling; UDN, stroke due to undetermined cause, ODE, other determined etiologies; miR, microRNA; hs-CRP, high sensitivity C-reactive protein; HDL, high density lipoprotein; LDL, low density lipoprotein; NIHSS, National Institutes of Health Stroke Scale; IL-6, interleukin 6; TOAST, Trial of Org 10,172 in Acute Stroke Treatment; OCSP, Oxfordshire Community Stroke Project; HACI, hyperacute cerebral infarction; mRS, Modified Rankin Scale; BI, Barthel index; CSF, cerebrospinal fluid; CNS, central nervous system; NA, no data; IS, ischemic stroke; ACI, acute cerebral infraction; MCI, massive cerebral infarction; HS, hemorrhagic stroke; MRI, Magnetic resonance imaging; MRA, magnetic resonance angiography; CT, computed tomography; ApoA1, apolipoprotein A1; MMP-9, matrix metallopeptidase 9; NF-κB, Nuclear factor-κB; TIA, transient ischemic attack; IGF-1, Insulin-like growth factor 1; h, hour; w, week.

Table 2. The results of area under the receiver operating characteristic curves from human studies.

Ref	MicroRNA	AUC		Specificity	Sensitivity
Long et al. [41]	miR-30a	24 h	0.91	80%	94%
		1 week	0.91	84%	93%
		4 weeks	0.92	84%	90%
		24 weeks	0.93	84%	92%
	miR-126	24 h	0.92	84%	92%
		1 week	0.94	86%	90%
		4 weeks	0.93	84%	92%
		24 weeks	0.92	82%	92%
	Let-7b	24 h	0.93	84%	92%
		1 week	0.92	84%	90%
		4 weeks	0.92	86%	92%
		24 weeks	0.91	80%	89%
Wu et al. [48]	miR-15a,	0.698		NA	NA
	miR-16,	0.820			
	miR-17-5p	0.784			
	miR-15a + miR-16 + miR-17-5p	0.845			
Peng et al. [43]	Let-7e	0.86		73.4%	82.8%
	miR-338	0.63		53.2%	71.9%
Huang et al. [44]	Let-7e-5p	0.82		NA	NA
Jin F. and Xing J. [53]	miR-126	0.654		NA	NA
	miR-130a	0.642			
	miR-222	0.584			
	miR-218	0.624			
	miR-185	0.601			
	miR-126 + miR-130a + miR-222 + miR-218 + miR-185	0.767			
Chen et al. [61]	miR-223	0.859		78.8%	84%
Sepramaniam et al. [73]		Cohort 1	Cohort 2		
	miR-125b-2*	0.95	0.85	NA	NA
	miR-27a*	0.89	0.86		
	miR-422a	0.92	0.86		
	miR-488	0.87	0.86		
	miR-627	0.84	0.76		
Tian et al. [75]	miR-16 (overall patients)	0.775		87%	69.7%
	miR-16 (LAA patients)	0.952		91.3%	100%
Wang et al. [74]		Total	MRI(+)		
	hsa-miR-106b-5p	0.999	0.962	NA	NA
	hsa-miR-4306	0.877	0.952		
	hsa-miR-320e	0.953	0.981		
	hsa-miR-320d	0.977	0.987		

Table 2. *Cont.*

Ref	MicroRNA	AUC	Specificity	Sensitivity
Jia et al. [55]	CRP	0.794	NA	NA
	CRP+ miR-145	0.896		
	CRP + miR-23a	0.816		
	CRP + miR-221	0.819		
Tsai et al. [57]	miR-21/miR-221 (traditional risk factors)	0.93	NA	NA
Ji et al. [62]	miR-9	0.8026	NA	NA
	miR-124	0.6976		
Tiedt et al. [76]	miR-125a-5p + miR-125b-5p + miR-143-3p	0.663 (IS vs. TIA)		
	miR-125a-5p + miR-125b-5p + miR-143-3p + IL6 + NSE	0.661 (IS vs. TIA)		
Chen et al. [64]	miR146b	0.776	NA	NA
	CRP	0.782		
	IL-6	0.684		
	CRP+miR146b	0.863		
	IL-6+miR146b	0.819		
	CRP+IL-6+miR146b	0.866		

Abbreviations; MiRNA, microRNA; AUC, area under curve; NA, no data; CRP, C-reactive protein; LAA, large artery atherosclerosis; IL-6, interleukin 6; NSE, neuron specific enolase; MRI, magnetic resonance imaging; IS, ischemic stroke; TIA, transient ischemic attack.

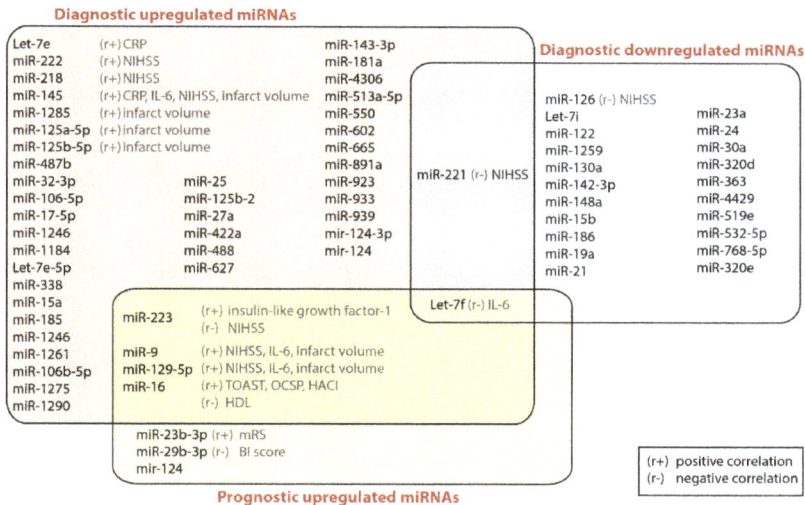

Figure 1. Regulation of diagnostic and prognostic miRNAs serving as biomarkers in ischemic stroke, based on human studies. Abbreviation: miRNA, microRNA; CRP, C-reactive protein; HDL, high density lipoprotein; NIHSS, National Institutes of Health Stroke Scale; IL-6, interleukin 6; TOAST, Trial of Org 10,172 in Acute Stroke Treatment; OCSP, Oxfordshire Community Stroke Project; HACI, hyperacute cerebral infarction; mRS, Modified Rankin Scale; BI, Barthel index. Refs: [41,43–45,47,48,53, 55,57,58,61,62,67,68,72–76,78].

Profiling of miRNAs expression in plasma samples taken from IS patients was performed in 136 individuals with MRI evaluation on admission. In this population 76 patients had already ischemic changes detected with MRI (MRI+). Initial analysis revealed 120 miRNAs in MRI (−) group differentially expressed than in MRI (+) group and 69 miRNAs in MRI (+) in comparison to control group. Seventeen miRNAs were significantly changed (i.e., 10-fold) in plasma samples from both MRI (−) and (+) IS patients compared with control subjects and were further evaluated. However, only two miRNAs (hsa-miR-106b-5p and hsa-miR-4306) showed a gradient of increase, and only two miRNAs (hsa-miR-320e and hsa-miR-320d) showed a gradient of decrease from control patients to MRI(−) and MRI(+) IS patients. Also, ROC analysis performed on data from all IS patients using 4 miRNAs profile allowed to discriminate patients with IS from healthy controls. Thus, the authors showed that hsa-miR-106b-5p, hsa-miR-4306, hsa-miR-320e, and hsa-miR-320d might facilitate the diagnosis and clinical management of IS [74].

Another study aimed to identify specific circulating miRNAs that would facilitate the diagnosis of hyperacute cerebral infarction less than 6 h from the acute event, and validate their usefulness in follow-up up to 3 months after the onset of cerebral infarction. Forty patients with hyperacute cerebral infarction and 30 age-matched healthy volunteers were recruited in this study. Seven hyperacute cerebral infarction and four age-matched healthy volunteers were selected randomly for microarray analysis. Thirty-three patients and 23 controls were selected for RT-qPCR validation. In discovery phase microarray analysis showed 11 miRNAs that were upregulated (miR-140, miR-106b, miR-130a, miR-16, miR-223, miR-93, miR-484, miR-25, miR-130b, miR-107, and miR-151) and 1 miRNA that was downregulated (miR-4454), however in validation phase, only miR-16 was significantly different between patients and the control group. In multivariate logistic regression analysis only miR-16 (OR 1.669, 95% CI 1.071 ± 2.602, *p* = 0.024) was predictive for hyperacute cerebral infarction stroke. Moreover, miR-16 expression was significantly higher in the poor prognosis group than in the good prognosis group during follow-up and the diagnostic accuracy of miR-16 as a biomarker for was confirmed in ROC analysis [75].

In previously published research that consisted from three stages: discovery stage (20 IS patients and 20 healthy control subjects), validation stage (40 IS patients and 40 controls), and replication stage (200 IS patients and 100 controls). RNA sequencing was performed in order to study expression changes of circulating miRNAs. Out of 32 miRNAs, three miRNAs were upregulated in IS patients compared to both controls (i.e., miR-125a-5p, miR-125b-5p, and miR-143-3p) and transient ischemic attacks (TIA) patients. In long-term follow-up lasting up to 90 days both miR-125b-5p and miR-143-3p levels decreased, starting at day two, with no significant difference compared to controls. Expression levels of miR-125a-5p subsequently increased and stayed constantly elevated in comparison to control group. It is worth mentioning that in the studied population, different stroke subtypes were included, and miR-125a-5p, miR-125b-5p, and miR-143-3p were similar across etiological subgroups, i.e., patients with LA, CE, and stroke of undetermined etiology. Moreover, miR-125a-5p, miR-125b-5p, and miR-143-3p differentiated between controls and IS patients with an area AUC of 0.90 (sensitivity: 85.6%; specificity: 76.3%). Importantly, they found the expression levels of miR-125a-5p, miR-125b-5p, and miR-143-3p to be independent of infarct volume and stroke etiology. This finding emphasizes potential utility of these miRNAs as a broadly applicable diagnostic marker for IS. Based on another step of the experiment it was shown that the most important source of these miRNAs are platelets. Moreover, significantly higher concentration of miR-143-3p were found in extracellular microvesicles isolated from IS patients in comparison to healthy controls, but not for miR-125a-5p or miR-125b-5p [76].

One of the largest studies was done by Mick et al. They identify and validate ex-RNAs (extracellular noncoding RNAs) from plasma in participants of the Framingham Heart Study (FHS). The results demonstrate that when studied in a large observational cohort, miRNAs are significantly associated with stroke. In comparison to previously published studies there are some differences making this study highly valuable. First of all it was the largest, unbiased, community-based report of association between plasma circulating ex-RNAs (miRNAs, piRNAs, and snoRNAs) and stroke.

Secondly as a method for ex-RNA discovery RNA sequencing was used that allowed to search not only for miRNAs, but also some other type of non-coding RNA like piRNAs, and snoRNAs. They selected the most abundantly expressed ex-RNAs by RNASeq (331 miRNAs, 97 piRNAs, and 43 snoRNAs) for RT-qPCR analysis in the entire FHS Offspring Cohort. Based on proposed strategy 3 miRNAs (miR-877-5p, miR-124-3p, and miR-320d) and one snoRNA (SNO1402) were independently associated with prevalent stroke. Moreover, two other miRNAs i.e., miR-656-3p and miR-941 were significantly associated with incident stroke risk adjusting for each other and potentially confounding clinical variables. However, such a strategy did not allow analysis of the relation between IS subtypes and miRNAs [77].

Wu et al. evaluated the expression patterns of specific miRNAs in TIA patients. In their study, 754 miRNAs were initially screened by the TaqMan Low Density Array (TLDA) in two pooled serum samples from 50 IS patients and 50 controls. After miRNA profiling, significantly changed miRNAs validated by RT-qPCR in the same cohort and further confirmed in another larger cohort including 177 IS, 81 TIA patients, and 42 controls. Consequently, TLDA screening showed that 71 miRNAs were upregulated, and 49 miRNAs were downregulated in IS patients. RT-qPCR validation confirmed that serum levels of miR-23b-3p, miR-29b-3p, miR-181a-5p, and miR-21–5p were markedly increased in IS patients. Strikingly, miR-23b-3p, miR-29b-3p, and miR-181a-5p were also significantly elevated in TIA patients. Logistic regression and ROC curve analyses showed that these changed miRNAs may function as predictive and discriminative biomarkers for IS and TIA. Thus, authors suggested that, their distinctive expression signatures may contribute to assessing neurological deficit severity of IS and subsequent stroke risk after TIA [78].

Summing up the above reported results, accumulating evidence indicates that circulating miRNAs might be used as an innovative diagnostic and prognostic biomarker and potential novel therapeutic target through its potential roles in inflammatory processes. However, to better evaluate the diagnostic and prognostic role of miRNAs, more studies are required to investigate the intricate interactions between the miRNAs and their target genes in IS. Results of the studies presented in this review should be interpreted with consciousness. As we discussed above studies found controversy findings. The discrepancy of the results might be the reason of demographic differences between populations, heterogeneity of populations, various cohort sizes and study designs. It is necessary to conduct further studies to validate the current hypotheses and closely determine the association between various miRNAs and their contribution to cardiovascular diseases development.

3. Future Perspectives for Using miRNAs in Diagnosis and Prognosis in Ischemic Stroke

The use of circulating miRNAs as disease biomarkers in stroke is a very attractive and promising concept. In fact, they might offer several advantages over traditional disease biomarkers: (i) they can reflect specific cellular pathophysiological alterations; (ii) experimental evidence suggests that they could potentially indicate the specific etiology of stroke; (iii) since specific miRNAs showed significant modulation well before the development of the acute and irreversible cerebral damage, they could potentially allow early stroke diagnosis and/or the identification of subjects at risk before they develop an acute stroke.

Nevertheless, a number of challenges still need to be addressed before circulating miRNAs could enter the clinical arena: (a) only few studies have tested their clinical usefulness, most of which are quite small and present methodological flaws; (b) the large variation observed for miRNAs within the same patients group could make the identification of diagnostic cutoffs particularly challenging; (c) most of the available studies that evaluated the clinical usefulness of circulating miRNAs in stroke were performed in Asian populations and their results could therefore not completely apply to other ethnicities; (d) most studies did not assess the impact of other clinical variables on the prognostic potential, which could be additive in some cases and thus potentially helpful if used to correct/adjust biomarker values; (e) finally, the current standard method for the reliable measurement of circulating miRNAs is RT-PCR, which presents some limitations for the use as clinical biomarker in the setting of an acute disease. Although several alternative methodological approaches are being tested, a reliable, fast and cheaper analytical methodology has not been established, yet.

4. Bioinformatics Analysis

We performed bioinformatics analysis in order to identify the most commonly regulated circulating miRNA among the selected studies that are involved in inflammation, blood coagulation, and platelet activation. The miRNAs list was narrowed down by searching human studies literature. R programming was used to build a miRNA-Gene-Network based on the interactions of the miRNAs and target genes and their functions. TargetScan and MirTarBase databases were used to predict the target genes of the miRNAs [79,80]. To identify the genes associated with inflammation process; platelet activation; blood coagulation we performed a screening of the Gene Ontology (GO) terms for the presence of the key words using the biomaRt package in R [81]. Key words used for screening the GO terms are given in the Supplementary Materials (See Table S1). Electronic database Pubmed and Scopus was searched between December 2017 and March 2018, and original studies were reviewed to evaluate the potential diagnostic or/and prognostic role of circulating miRNAs associated with IS. Review articles and meta-analyses were also investigated, and their secondary references were examined for possible inclusion. Our search was limited to human studies only and did not exclude studies on the basis of ethnicity of study participants. Search terms comprised of the following search syntax: "Search ("micrornas" [MeSH Terms] OR "mir" [MeSH Terms] OR "mirna" [MeSH Terms] OR "circulating miRNA" [MeSH Terms] OR "circulating microRNA" [MeSH Terms]) AND ("ischemic stroke" [MeSH Terms] OR "stroke" [All Fields]) AND ("diagnostic" [MeSH Terms] OR "diagnosis" [All Fields]) OR ("prognostic" [MeSH Terms] OR "prognosis" [All Fields])" Filters: Humans. A total of 52 records were identified after duplicates removal. Titles and abstracts were screened by two independent operators, with exclusion of 27 records for any of the following reasons: (a) they were not related to the specific research question ($n = 11$); (b) they did not present original data ($n = 14$); or they were not human studies ($n = 2$). Finally, 25 articles were selected to be used in this review.

MiRNA-target interaction network was constructed using R programming and visualized using Cytoscape software v 3.6.1 [82]. miRNA-target interactions used for constructing the network were obtained from TargetScan and MirTarBase databases. In the next step we selected miRNAs targeting at least 5 genes associated with specific GO process (inflammation process; platelet activation; blood coagulation), and target genes regulated by at least 5 miRNAs (See Figure S1b). In the third step we subsetted miRNAs regulating at least 5 target genes from the network created in the first step. Final networks were created by using circular layout and direct force layout (See Figure S1c). Circular networks were sorted by the degree of connections between miRNAs and their targeted genes. It enabled to retrieve top miRNAs and top shared targets involved in analyzed GO process (See Figures 2a, 3a, and 4a). To retrieve gene–gene interactions between our top targets we used interactome datasets from String app v 1.2.2 for Cytoscape [83].

The pathophysiology of IS is related with inflammation, blood coagulation process and platelet activation. In course of thromboembolic stroke, activated platelets orchestrate a thrombo-inflammatory cascade by promoting thrombus formation and growth, activating leukocytes, and potentiating cerebral endothelium injury. In line with this aim, we focused on miRNAs involved in inflammation, blood coagulation, and platelet activation in our bioinformatics analysis in order to identify to account for all potential pathophysiological mechanisms underlying the development of IS. In fact, the study that were selected included different subtypes of IS, such as thromboembolic, LA, SA, and UDN. We adopted such broad criteria is the intention to identify a pattern of possibly few miRNAs that could help to recognize the different underlying etiologies and/or main pathophysiological mechanisms. Using a similar approach, we were recently able to identify as few as two circulating miRNAs to be used for the differential diagnosis of heart failure of ischemic origin from other forms of non-ischemic origin [22].

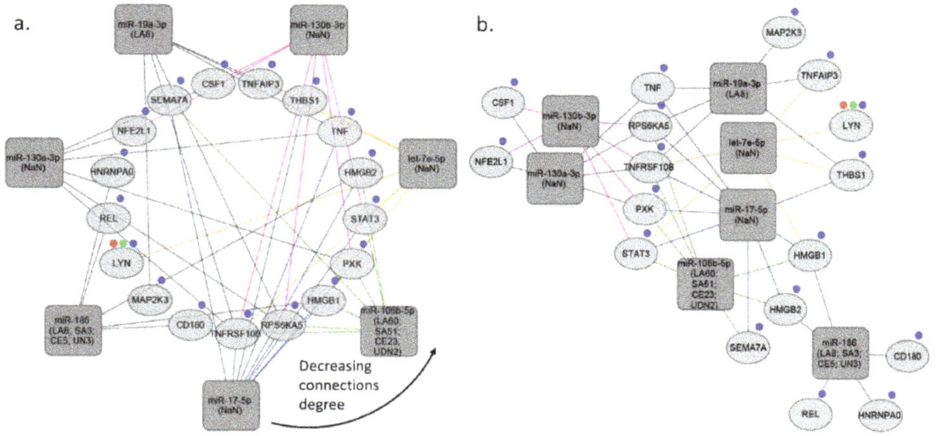

Figure 2. miRNA-target gene networks based on inflammatory response. (**a**) Inflammatory response-network sorted by the degree of connections, (**b**) Inflammatory response-interaction network. The rectangles indicate the stroke type miRNAs, the ellipses indicate target genes. Red, green and blue marks represent specific GO process-blood coagulation, platelet activation, and inflammation process, respectively. LA, large artery stroke; SA, small artery stroke; UDN, stroke due to undetermined cause; CE, cardioembolic stroke; NaN, no data.

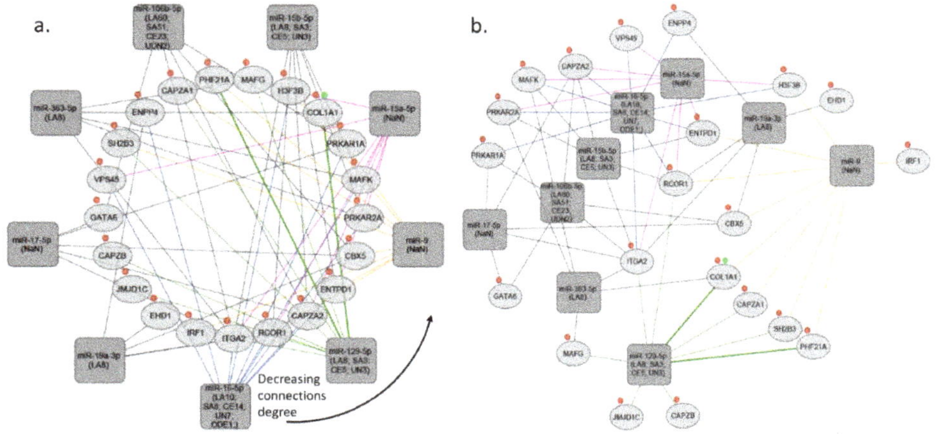

Figure 3. miRNA-target gene networks based on blood coagulation. (**a**) Blood coagulation- network sorted by the degree of connections, (**b**) Blood coagulation-interaction network. The rectangles indicate the stroke type miRNAs, the ellipses indicate target genes. Red and green marks represent specific GO process-blood coagulation and platelet activation, respectively. LA, large artery stroke; SA, small artery stroke; UDN, stroke due to undetermined cause; CE, cardioembolic stroke; NaN, no data; ODE, other determined etiology; UN, determined cause.

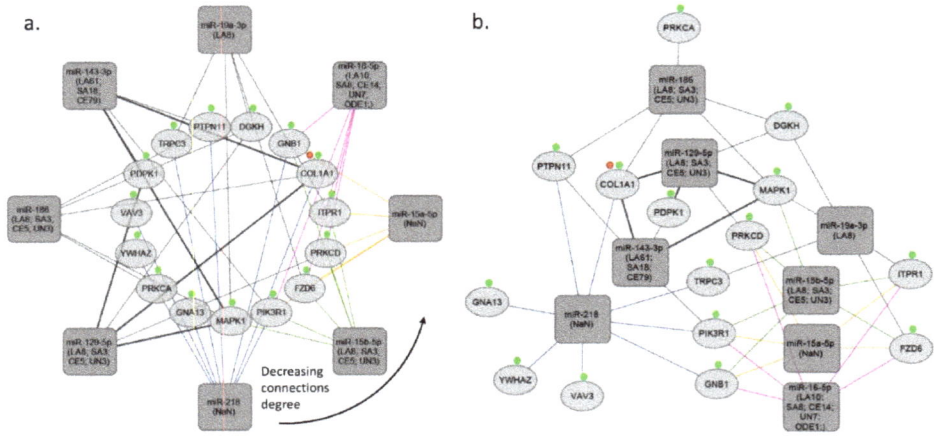

Figure 4. miRNA-target gene networks based on platelet activation. (**a**) Platelet activation-network sorted by the degree of connections, (**b**) Platelet activation-interaction network. The rectangles indicate the stroke type miRNAs, the ellipses indicate target genes. Red and green marks represent specific GO process-blood coagulation and platelet activation, respectively. LA, large artery stroke; SA, small artery stroke; UDN, stroke due to undetermined cause; CE, cardioembolic stroke; NaN, no data; ODE, other determined etiology; UN, determined cause.

Our analysis showed that 2 common miRNAs namely, miR-17-5p and miR-106b-5p may regulate both inflammatory response and blood coagulation. 1 common miRNA, which is miR-186 may regulate both inflammatory response and platelet activation. Addition to this, we found that, 4 common miRNAs namely, miR-15b-5p, miR-15a-5p, miR-16-5p, and miR-129-5p may regulate both blood coagulation and platelet activation. Interestingly, 1 common miRNA, miR-19a-3p was found, that is involved in inflammatory response, blood coagulation, and platelet activation (See Figure 5b). Thus, we found that miR-19a-3p can be the crucial miRNA which regulates all of these three processes. MiR-19a is a member of the miR-17-92 cluster, and its importance was shown in breast cancer cells [84]. Besides, it was found that members of miR-17-92 cluster are involved in coronary artery disease and vascular functions, including ischemia responses and angiogenesis [85–87]. Liu et al. investigated the function of the miR17-92 cluster in adult neural progenitor cells after experimental stroke. They found that stroke substantially up-regulated miR17-92 cluster expression in neural progenitor cells of the adult mouse and inhibition of individual members of the miR17-92 cluster, such as miR-19a, suppressed cell proliferation and increased apoptosis [52]. As we discussed, Jinkling et al. demonstrate that, low circulating miR-19a levels in the blood were found in patients with IS compared to controls [67]. This can be the reason of potential cell protective effect of miR-19a in ischemia. It is worth underlining that only two studies out of 25 analyzed miR-19a-3p, and the only study which used miRNA profiling was found the significant relation with IS. There are several reasons that could help to understand such a discrepancy; (i) different inclusion and exclusion criteria implemented in each study, (ii) differences in population origin, (ii) lack of homogeneity of IS population due to inclusion of different IS subtypes. These weak points should be considered, and better-designed studies should confirm the actual biological role and clinical usefulness of miR-19a-3p in IS. Additionally, in our analysis we demonstrated ten the most significant genes targeted by miRNAs associated with blood coagulation, platelet activation, and inflammation process in IS, namely HMGB1, YWHAZ, PIK3R1, STAT3, MAPK1, CBX5, CAPZB, THBS1, TNFRSF10B, RCOR1 (High-mobility group box 1- HMGB1; tyrosine 3 monooxygenase/tryptophan 5 monooxygenase activating protein, zeta polypeptide-YWHAZ; phosphoinositide-3-kinase regulatory subunit 1- PIK3R1; signal transducers and activators of transcription 3- STAT3; mitogen-activated protein kinase 1- MAPK1; chromobox homolog 5- CBX5; capping actin protein of muscle Z-line subunit beta- CAPZB; thrombospondin-1- THBS1; TNF receptor superfamily member 10b- TNFRSF10B; REST corepresor 1- RCOR1) (See Figure 5a).

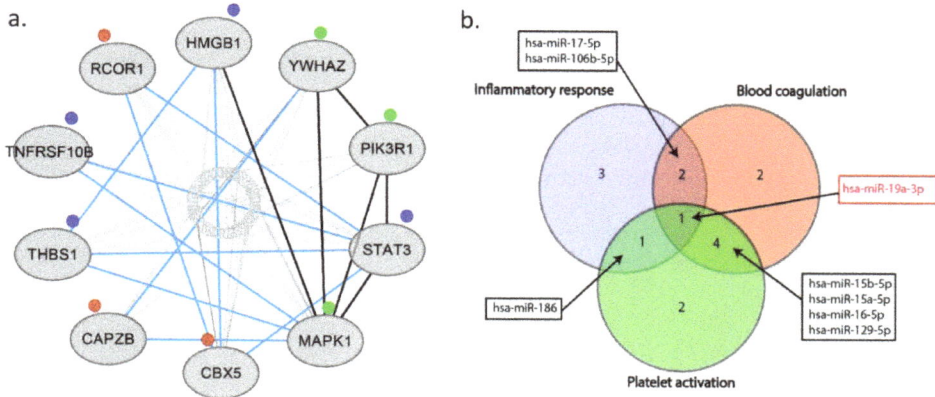

Figure 5. Most important genes targeted by miRNAs and overlapped miRNAs in GO process in IS. (**a**) Top ten genes targeted by miRNAs associated with IS. The circle in the middle indicates the stroke type miRNAs, the ellipses indicate target genes. Red, green and blue marks represent specific GO process-blood coagulation, platelet activation, and inflammation process, respectively. Black edges represent the high confidence connections between genes, blue edges represent low confidence connections. (**b**) Overlapped miRNAs in inflammatory response, blood coagulation, and platelet activation. 2 common miRNAs observed in inflammatory response and blood coagulation namely, miR-17-5p miR-106b-5p. 1 common miRNA observed in inflammatory response and platelet activation namely, miR-186. 4 common miRNAs observed in blood coagulation and platelet activation namely, miR-15b-5p, miR-15a-5p, miR-16-5p, miR-129-5p. 1 common miRNA observed in inflammatory response, blood coagulation and platelet activation namely, miR-19a-3p. miR, microRNA.

HMGB1 is released from necrotic brain tissue and its differential redox forms attract and activate immune cells after ischemic brain injury. Its concentrations correlate with disease severity and outcome after brain injury [88]. Our results support the importance of HMGB in IS. The role of YWHAZ gene was shown in several neurodegenerative diseases such as, Huntington's disease, Alzheimer's disease, and amyotrophic lateral sclerosis [89–91]. As we found in our bioinformatic analysis a similar study showed the importance of YWHAZ gene in CE [92]. Besides, the role of PIK3R1 gene in stroke is yet to be published. On the other hand, one study showed that PIK3R1 is the target of miR-221 in endothelial progenitor cells (EPCs). EPCs assist angiogenesis and have been linked to ischemia-related disorders, including coronary artery disease [93]. Abovementioned, human studies showed that, miR-221 is a potential diagnostic biomarker in IS, thus the role of PIK3R1 and relation with miR-221 in IS should be confirmed by future studies [55,57,64]. As we showed the relation of STAT3 with IS in our analysis, lately Liang et al. reviewed several significant cerebral ischemic and HS treatments that target the STAT3 signaling pathway, including pharmacological and physical therapies [94]. Additionally, Bam et al. investigated miRNA profile in the peripheral blood mononuclear cells of IS patients. Further, they performed bioinformatic analysis and they showed that pro-inflammatory genes like STAT3, IL12A, and IL12B are some of the highly predicted targets for the dysregulated miRNAs [95]. Moreover, studies showed that MAPK can be an important regulator in ischemic and hemorrhagic cerebral vascular disease and raising the possibility that it might be a drug discovery target for stroke. In our bioinformatics analysis we confirmed previous results that showed the importance of MAPK1 gene and its signaling pathway in IS [44,68,96]. CBX5 is a member of the heterochromatin protein 1 (HP1) family, HPs are closely related to the development of major vascular injuries, such as atherosclerosis and hypertension. To the best of our knowledge, only one study showed that association of serum levels of antibodies against CBX5 could potentially represent useful tools for the diagnosis of TIA and predicting the onset of acute-phase of cerebral infarction [97]. In our analysis we showed that CBX5 can be related with IS targeted by miRNAs. THBS1 is a potent regulator of angiogenesis and previous studies showed the importance of THBS1 in stroke [98,99]. Gao et al. showed that plasma THBS1 protein concentrations are elevated and are highly associated with long-term outcome of IS [98]. Based on our bioinformatic analysis we also demonstrated the relation between THBS1 gene and IS targeted by miRNAs. It was shown that decreased cardiac CAPZ protein protects hearts against acute ischemia-reperfusion injury [100]. However, there is no published study describing the role of CAPZB in IS. Similarly, the role of RCOR1 and TNFRSF10B genes in stroke is yet to be published. In our analysis for the first time we found that CAPZB, RCOR1, and TNFRSF10B may have association with IS targeted by miRNAs.

5. Conclusions

The present article provides an up-to-date and comprehensive overview of all circulating miRNAs studied so far as diagnostic or prognostic markers in IS. As we deeply reviewed above, studies showed the impact of more than 1000 different miRNAs on IS, and their association with biochemical and hematological parameters in stroke patients. These studies clarify the value of miRNAs in IS the ischemic pathophysiological process and help us to better understand processes involved in IS pathophysiology. Through a bioinformatic analysis, we associated key biological function and signaling pathways related to the miRNAs with most prominent differential expression pattern in IS compared to controls. Using this approach, we found miR-19a-3p as the single miRNA involved in all three main biological processes selected for the bioinformatic analysis: inflammation, blood coagulation and platelet activation. Furthermore, results of bioinformatics analysis shows the following most significant genes targeted by miRNAs in IS: HMGB1, YWHAZ, PIK3R1, STAT3, MAPK1, CBX5, CAPZB, THBS1, TNFRSF10B, and RCOR1. Moreover, it is worth mentioning that CRP and IL-6 are the most common targets within the inflammatory ontology. This latter finding is particularly interesting, as both CRP and IL-6 are well-known prognostic markers in atherosclerotic vascular disease [101]. In addition, CRP might represent a key link between vascular inflammation and arterial thrombosis or platelet activation [102–104]. Finally, the finding that multiple miRNAs are correlated to inflammation supports the use of a panel of multiple miRNAs.

In conclusion, since many known biomarkers related with IS are not disease specific and have been also associated with other types of brain injury, novel and specific diagnostic/prognostic biomarkers in IS are needed. In this context, circulating miRNAs are very promising biomarkers in stroke. Since specific miRNAs showed significant modulation well before the development of the acute and irreversible cerebral damage, they may potentially allow early stroke diagnosis and/or the identification of subjects at risk before they develop an acute stroke, however future studies are needed.

Supplementary Materials: The following are available online at http://www.mdpi.com/2073-4409/7/12/249/ s1.

Author Contributions: C.E.: Analysis of literature search, collection, analysis, and interpretation of data, preparing manuscript, graphical preparation, acceptance of the final version of the manuscript, preparation of the revision. Z.W.: Bioinformatic analysis, graphical preparation, acceptance of the final version of the manuscript, preparation of the revision. S.D.R.: Preparing manuscript, acceptance of the final version of the manuscript, preparation of the revision. D.M.-G.: Analysis of literature search, acceptance of the final version of the manuscript. A.S.: Analysis of literature search, preparation of the revision. C.I.: Analysis of literature search, acceptance of the final version of the manuscript. I.J.-K.: Analysis of literature search, acceptance of the final version of the manuscript. A.C.: Acceptance of the final version of the manuscript, critical revision of the article. M.P.: Analysis of literature search, Collection and interpretation of data, preparing manuscript, acceptance of the final version of the manuscript, preparation of the revision. All authors agreed to be accountable for all aspects of the work in ensuring that questions related to the accuracy or integrity of any part of the work are appropriately investigated and resolved.

Funding: The work was supported financially as part of the research grant 'Preludium' from the National Science Center, Poland (grant number 2017/25/N/NZ5/00545) and internal funding of the Department of Experimental and Clinical Pharmacology, Medical University of Warsaw, Centre for Preclinical Research and Technology CEPT, Warsaw, Poland.

Conflicts of Interest: The authors declare that there are no conflicts of interest.

References

1. Benjamin, E.J.; Blaha, M.J.; Chiuve, S.E.; Cushman, M.; Das, S.R.; Deo, R.; de Ferranti, S.D.; Floyd, J.; Fornage, M.; Gillespie, C.; et al. American Heart Association Statistics Committee and Stroke Statistics Subcommittee. Heart Disease and Stroke Statistics-2017 Update: A Report from the American Heart Association. *Circulation* **2017**, *135*, 146–603. [CrossRef] [PubMed]
2. De Rosa, S.; Sievert, H.; Sabatino, J.; Polimeni, A.; Sorrentino, S.; Indolfi, C. Percutaneous Closure Versus Medical Treatment in Stroke Patients with Patent Foramen Ovale: A Systematic Review and Meta-analysis. *Ann. Intern. Med.* **2018**, *168*, 343–350. [CrossRef] [PubMed]
3. *European Cardiovascular Disease Statistics*, 2017th ed.; European Heart Network: Brussels, Belgium, 2017.
4. Luengo-Fernandez, R.; Gray, A.M.; Bull, L.; Welch, S.; Cuthbertson, F.; Rothwell, P.M. Quality of life after TIA and stroke: Ten-year results of the Oxford Vascular Study. *Neurology* **2013**, *81*, 1588–1595. [CrossRef] [PubMed]
5. Beal, C.C. Gender and stroke symptoms: A review of the current literature. *J. Neurosci. Nurs.* **2010**, *42*, 80–87. [CrossRef] [PubMed]
6. Chen, Y.; Xiao, Y.; Lin, Z.; Xiao, X.; He, C.; Bihl, J.C.; Zhao, B.; Ma, X.; Chen, Y. The Role of Circulating Platelets Microparticles and Platelet Parameters in Acute Ischemic Stroke Patients. *J. Stroke Cerebrovasc. Dis.* **2015**, *24*, 2313–32320. [CrossRef] [PubMed]
7. Pordzik, J.; Pisarz, K.; De Rosa, S.; Jones, A.D.; Eyileten, C.; Indolfi, C.; Malek, L.; Postula, M. The Potential Role of Platelet-Related microRNAs in the Development of Cardiovascular Events in High-Risk Populations, Including Diabetic Patients: A Review. *Front. Endocrinol.* **2018**, *9*, 74. [CrossRef] [PubMed]
8. Janssen, H.L.; Reesink, H.W.; Lawitz, E.J.; Zeuzem, S.; Rodriguez-Torres, M.; Patel, K.; van der Meer, A.J.; Patick, A.K.; Chen, A.; Zhou, Y.; et al. Treatment of HCV infection by targeting microRNA. *N. Engl. J. Med.* **2013**, *368*, 1685–1694. [CrossRef] [PubMed]
9. Bader, A.G. miR-34—A microRNA replacement therapy is headed to the clinic. *Front. Genet.* **2012**, *3*, 120. [CrossRef] [PubMed]
10. Bushati, N.; Cohen, S.M. MicroRNAs in neurodegeneration. *Curr. Opin. Neurobiol.* **2008**, *18*, 292–296. [CrossRef] [PubMed]
11. Nelson, P.T.; Wang, W.X.; Rajeev, B.W. MicroRNAs (miRNAs) in neurodegenerative diseases. *Brain Pathol.* **2008**, *18*, 130–138. [CrossRef] [PubMed]
12. Lee, H.-J. Exceptional stories of microRNAs. *Exp. Biol. Med.* **2013**, *238*, 339–343. [CrossRef] [PubMed]
13. De Rosa, S.; Indolfi, C. Circulating microRNAs as Biomarkers in Cardiovascular Diseases. In *Experientia Supplementum*; Springer: Basel, Switzerland, 2015; Volume 106, pp. 139–149.
14. Di Ieva, A.; Butz, H.; Niamah, M.; Rotondo, F.; De Rosa, S.; Sav, A.; Yousef, G.M.; Kovacs, K.; Cusimano, M.D. MicroRNAs as biomarkers in pituitary tumors. *Neurosurgery* **2014**, *75*, 181–189, discussion 188–189. [CrossRef] [PubMed]
15. De Rosa, S.; Curcio, A.; Indolfi, C. Emerging role of microRNAs in cardiovascular diseases. *Circ. J.* **2014**, *78*, 567–575. [CrossRef] [PubMed]
16. Fichtlscherer, S.; De Rosa, S.; Fox, H.; Schwietz, T.; Fischer, A.; Liebetrau, C.; Weber, M.; Hamm, C.W.; Röxe, T.; Müller-Ardogan, M.; et al. Circulating microRNAs in patients with coronary artery disease. *Circ. Res.* **2010**, *107*, 677–684. [CrossRef] [PubMed]
17. De Rosa, R.; De Rosa, S.; Leistner, D.; Boeckel, J.N.; Keller, T.; Fichtlscherer, S.; Dimmeler, S.; Zeiher, A.M. Transcoronary Concentration Gradient of microRNA-133a and Outcome in Patients with Coronary Artery Disease. *Am. J. Cardiol.* **2017**, *120*, 15–24. [CrossRef] [PubMed]
18. Sluijter, J.P.; Doevendans, P.A. Circulating microRNA profiles for detection of peripheral arterial disease: Small new biomarkers for cardiovascular disease. *Circ. Cardiovasc. Genet.* **2013**, *6*, 441–443. [CrossRef]

19. Sorrentino, S.; Iaconetti, C.; De Rosa, S.; Polimeni, A.; Sabatino, J.; Gareri, C.; Passafaro, F.; Mancuso, T.; Tammè, L.; Mignogna, C.; et al. Hindlimb Ischemia Impairs Endothelial Recovery and Increases Neointimal Proliferation in the Carotid Artery. *Sci. Rep.* **2018**, *8*, 761. [CrossRef]

20. De Rosa, S.; Fichtlscherer, S.; Lehmann, R.; Assmus, B.; Dimmeler, S.; Zeiher, A.M. Transcoronary concentration gradients of circulating microRNAs. *Circulation* **2011**, *124*, 1936–1944. [CrossRef]

21. Corsten, M.F.; Dennert, R.; Jochems, S.; Kuznetsova, T.; Devaux, Y.; Hofstra, L.; Wagner, D.R.; Staessen, J.A.; Heymans, S.; Schroen, B. Circulating MicroRNA-208b and MicroRNA-499 reflect myocardial damage in cardiovascular disease. *Circ. Cardiovasc. Genet.* **2010**, *3*, 499–506. [CrossRef]

22. De Rosa, S.; Eposito, F.; Carella, C.; Strangio, A.; Ammirati, G.; Sabatino, J.; Abbate, F.G.; Iaconetti, C.; Liguori, V.; Pergola, V.; et al. Transcoronary concentration gradients of circulating microRNAs in heart failure. *Eur. J. Heart Fail.* **2018**, *20*, 1000–1010. [CrossRef]

23. Tijsen, A.J.; Creemers, E.E.; Moerland, P.D.; de Windt, L.J.; van der Wal, A.C.; Kok, W.E.; Pinto, Y.M. MiR423-5p as a circulating biomarker for heart failure. *Circ. Res.* **2010**, *106*, 1035–1039. [CrossRef] [PubMed]

24. Jaguszewski, M.; Osipova, J.; Ghadri, J.R.; Napp, L.C.; Widera, C.; Franke, J.; Fijalkowski, M.; Nowak, R.; Fijalkowska, M.; Volkmann, I.; et al. A signature of circulating microRNAs differentiates takotsubo cardiomyopathy from acute myocardial infarction. *Eur. Heart J.* **2014**, *35*, 999–1006. [CrossRef] [PubMed]

25. Gareri, C.; De Rosa, S.; Indolfi, C. MicroRNAs for Restenosis and Thrombosis After Vascular Injury. *Circ. Res.* **2016**, *118*, 1170–1184. [CrossRef] [PubMed]

26. Polimeni, A.; De Rosa, S.; Indolfi, C. Vascular miRNAs after balloon angioplasty. *Trends Cardiovasc. Med.* **2013**, *23*, 9–14. [CrossRef] [PubMed]

27. De Rosa, S.; Arcidiacono, B.; Chiefari, E.; Brunetti, A.; Indolfi, C.; Foti, D.P. Type 2 Diabetes Mellitus and Cardiovascular Disease: Genetic and Epigenetic Links. *Front. Endocrinol.* **2018**, *9*, 2. [CrossRef] [PubMed]

28. Zampetaki, A.; Kiechl, S.; Drozdov, I.; Willeit, P.; Mayr, U.; Prokopi, M.; Mayr, A.; Weger, S.; Oberhollenzer, F.; Bonora, E.; et al. Plasma microRNA profiling reveals loss of endothelial miR-126 and other microRNAs in type 2 diabetes. *Circ. Res.* **2010**, *107*, 810–817. [CrossRef] [PubMed]

29. Carino, A.; De Rosa, S.; Sorrentino, S.; Polimeni, A.; Sabatino, J.; Caiazzo, G.; Torella, D.; Spaccarotella, C.; Mongiardo, A.; Strangio, A.; et al. Modulation of Circulating MicroRNAs Levels during the Switch from Clopidogrel to Ticagrelor. *Biomed. Res. Int.* **2016**, *2016*, 3968206. [CrossRef] [PubMed]

30. Streit, S.; Michalski, C.W.; Erkan, M.; Kleeff, J.; Friess, H. Northern blot analysis for detection and quantification of RNA in pancreatic cancer cells and tissues. *Nat. Protoc.* **2009**, *4*, 37–43. [CrossRef] [PubMed]

31. Várallyay, E.; Burgyán, J.; Havelda, Z. MicroRNA detection by northern blotting using locked nucleic acid probes. *Nat. Protoc.* **2008**, *3*, 190–196. [CrossRef] [PubMed]

32. Ferraro, D.; Champ, J.; Teste, B.; Serra, M.; Malaquin, L.; Viovy, J.L.; de Cremoux, P.; Descroix, S. Microfluidic platform combining droplets and magnetic tweezers: Application to HER2 expression in cancer diagnosis. *Sci. Rep.* **2016**, *6*, 25540. [CrossRef] [PubMed]

33. Eminaga, S.; Christodoulou, D.C.; Vigneault, F.; Church, G.M.; Seidman, J.G. Quantification of microRNA expression with next-generation sequencing. *Curr. Protoc. Mol. Biol.* **2013**, *103*, 4–17. [CrossRef] [PubMed]

34. Chugh, P.; Dittmer, D.P. Potential Pitfalls in microRNA Profiling. *Wiley Interdiscip. Rev. RNA* **2012**, *3*, 601–616. [CrossRef] [PubMed]

35. Yang, H.; Hui, A.; Pampalakis, G.; Soleymani, L.; Liu, F.F.; Sargent, E.H.; Kelley, S.O. Direct, electronic microRNA detection for the rapid determination of differential expression profiles. *Angew. Chem. Int. Ed. Engl.* **2009**, *48*, 8461–8464. [CrossRef] [PubMed]

36. Roy, S.; Soh, J.H.; Gao, Z. A microfluidic-assisted microarray for ultrasensitive detection of miRNA under an optical microscope. *Lab Chip* **2011**, *11*, 1886–1894. [CrossRef] [PubMed]

37. Dong, H.; Jin, S.; Ju, H.; Hao, K.; Xu, L.P.; Lu, H.; Zhang, X. Trace and label-free microRNA detection using oligonucleotide encapsulated silver nanoclusters as probes. *Anal. Chem.* **2012**, *84*, 8670–8674. [CrossRef] [PubMed]

38. Tran, H.V.; Piro, B.; Reisberg, S.; Tran, L.D.; Duc, H.T.; Pham, M.C. Label-free and reagentless electrochemical detection of microRNAs using a conducting polymer nanostructured by carbon nanotubes: Application to prostate cancer biomarker miR-141. *Biosens. Bioelectron.* **2013**, *49*, 164–169. [CrossRef] [PubMed]

39. Chen, C.; Ridzon, D.A.; Broomer, A.J.; Zhou, Z.; Lee, D.H.; Nguyen, J.T.; Barbisin, M.; Xu, N.L.; Mahuvakar, V.R.; Andersen, M.R.; et al. Real-time quantification of microRNAs by stem-loop RT-PCR. *Nucleic Acids Res.* **2005**, *33*, e179. [CrossRef] [PubMed]

40. Kang, K.; Peng, X.; Luo, J.; Gou, D. Identification of circulating miRNA biomarkers based on global quantitative real-time PCR profiling. *J. Anim. Sci. Biotechnol.* **2012**, *3*, 4. [CrossRef]

41. Long, G.; Wang, F.; Li, H.; Yin, Z.; Sandip, C.; Lou, Y.; Wang, Y.; Chen, C.; Wang, D.W. Circulating miR-30a, miR-126 and let-7b as biomarker for ischemic stroke in humans. *BMC Neurol.* **2013**, *13*, 178. [CrossRef]

42. Li, Y.; Maegdefessel, L. My heart will go on-beneficial effects of anti-miR-30 after myocardial infarction. *Ann. Transl. Med.* **2016**, *4*, 144. [CrossRef]

43. Peng, G.; Yuan, Y.; Wu, S.; He, F.; Hu, Y.; Luo, B. MicroRNA let-7e is a potential circulating biomarker of acute stage ischemic stroke. *Transl. Stroke Res.* **2015**, *6*, 437–445. [CrossRef] [PubMed]

44. Huang, S.; Lv, Z.; Guo, Y.; Li, L.; Zhang, Y.; Zhou, L.; Yang, B.; Wu, S.; Zhang, Y.; Xie, C.; et al. Identification of blood Let-7e-5p as a biomarker for ischemic stroke. *PLoS ONE* **2016**, *11*, e0163951. [CrossRef] [PubMed]

45. Gong, Z.; Zhao, S.; Zhang, J.; Xu, X.; Guan, W.; Jing, L.; Liu, P.; Lu, J.; Teng, J.; Peng, T.; et al. Initial research on the relationship between let-7 family members in the serum and massive cerebral infarction. *J. Neurol. Sci.* **2016**, *361*, 150–157. [CrossRef] [PubMed]

46. Panda, A.C.; Grammatikakis, I.; Kim, K.M.; De, S.; Martindale, J.L.; Munk, R.; Yang, X.; Abdelmohsen, K.; Gorospe, M. Identification of senescence-associated circular RNAs (SAC-RNAs) reveals senescence suppressor CircPVT1. *Nucleic Acids Res.* **2017**, *45*, 4021–4035. [CrossRef] [PubMed]

47. Leung, L.Y.; Chan, C.P.; Leung, Y.K.; Jiang, H.L.; Abrigo, J.M.; Wang, D.F.; Chung, J.S.; Rainer, T.H.; Graham, C.A. Comparison of miR-124-3p and miR-16 for early diagnosis of hemorrhagic and ischemic stroke. *Clin. Chim. Acta* **2014**, *433*, 139–144. [CrossRef] [PubMed]

48. Wu, J.; Du, K.; Lu, X. Elevated expressions of serum miR-15a, miR-16, and miR-17-5p are associated with acute ischemic stroke. *Int. J. Clin. Exp. Med.* **2015**, *8*, 21071–21079.

49. Spinetti, G.; Fortunato, O.; Caporali, A.; Shantikumar, S.; Marchetti, M.; Meloni, M.; Descamps, B.; Floris, I.; Sangalli, E.; Vono, R.; et al. MicroRNA-15a and microRNA-16 impair human circulating proangiogenic cell functions and are increased in the proangiogenic cells and serum of patients with critical limb ischemia. *Circ. Res.* **2013**, *112*, 335–346. [CrossRef]

50. Liu, L.F.; Liang, Z.; Lv, Z.R.; Liu, X.H.; Bai, J.; Chen, J.; Chen, C.; Wang, Y. MicroRNA-15a/b are upregulated in response to myocardial ischemia/reperfusion injury. *J. Geriatr. Cardiol.* **2012**, *9*, 28–32.

51. Yin, K.J.; Fan, Y.; Hamblin, M.; Zhang, J.; Zhu, T.; Li, S.; Hawse, J.R.; Subramaniam, M.; Song, C.Z.; Urrutia, R.; et al. KLF11 mediates PPARgamma cerebrovascular protection in ischaemic stroke. *Brain* **2013**, *136*, 1274–1287. [CrossRef]

52. Liu, X.S.; Chopp, M.; Wang, X.L.; Zhang, L.; Hozeska-Solgot, A.; Tang, T.; Kassis, H.; Zhang, R.L.; Chen, C.; Xu, J.; et al. MicroRNA-17-92 cluster mediates the proliferation and survival of neural progenitor cells after stroke. *J. Biol. Chem.* **2013**, *288*, 12478–12488. [CrossRef]

53. Jin, F.; Xing, J. Circulating pro-angiogenic and anti-angiogenic microRNA expressions in patients with acute ischemic stroke and their association with disease severity. *Neurol. Sci.* **2017**, *38*, 2015–2023. [CrossRef] [PubMed]

54. Gan, C.S.; Wang, C.W.; Tan, K.S. Circulatory microRNA-145 expression is increased in cerebral ischemia. *Genet. Mol. Res.* **2012**, *11*, 147–152. [CrossRef] [PubMed]

55. Jia, L.; Hao, F.; Wang, W.; Qu, Y. Circulating miR-145 is associated with plasma high-sensitivity C-reactive protein in acute ischemic stroke patients. *Cell. Biochem. Funct.* **2015**, *33*, 314–319. [CrossRef] [PubMed]

56. He, W.; Chen, S.; Chen, X.; Li, S.; Chen, W. Bioinformatic analysis of potential microRNAs in ischemic stroke. *J. Stroke Cerebrovasc. Dis.* **2016**, *25*, 1753–1759. [CrossRef] [PubMed]

57. Tsai, P.C.; Liao, Y.C.; Wang, Y.S.; Lin, H.F.; Lin, R.T.; Juo, S.H. Serum microRNA-21 and microRNA-221 as potential biomarkers for cerebrovascular disease. *J. Vasc. Res.* **2013**, *50*, 346–354. [CrossRef] [PubMed]

58. Zhou, J.; Zhang, J. Identification of miRNA-21 and miRNA-24 in plasma as potential early stage markers of acute cerebral infarction. *Mol. Med. Rep.* **2014**, *10*, 971–976. [CrossRef] [PubMed]

59. Wang, Y.; Zhang, Y.; Huang, J.; Chen, X.; Gu, X.; Wang, Y.; Zeng, L.; Yang, G.Y. Increase of circulating miR-223 and insulin-like growth factor-1 is associated with the pathogenesis of acute ischemic stroke in patients. *BMC Neurol.* **2014**, *14*, 77. [CrossRef] [PubMed]

60. Chevillet, J.R.; Kang, Q.; Ruf, I.K.; Briggs, H.A.; Vojtech, L.N.; Hughes, S.M.; Cheng, H.H.; Arroyo, J.D.; Meredith, E.K.; Gallichotte, E.N.; et al. Quantitative and stoichiometric analysis of the microRNA content of exosomes. *Proc. Natl. Acad. Sci. USA* **2014**, *111*, 14888–14893. [CrossRef] [PubMed]

61. Chen, Y.; Song, Y.; Huang, J.; Qu, M.; Zhang, Y.; Geng, J.; Zhang, Z.; Liu, J.; Yang, G.Y. Increased circulating exosomal miRNA-223 is associated with acute ischemic stroke. *Front. Neurol.* **2017**, *8*, 57. [CrossRef]

62. Ji, Q.; Ji, Y.; Peng, J.; Zhou, X.; Chen, X.; Zhao, H.; Xu, T.; Chen, L.; Xu, Y. Increased brain-specific miR-9 and miR-124 in the serum exosomes of acute ischemic stroke patients. *PLoS ONE* **2016**, *11*, e0163645. [CrossRef]

63. Liu, Y.; Zhang, J.; Han, R.; Liu, H.; Sun, D.; Liu, X. Downregulation of serum brain specific microRNA is associated with inflammation and infarct volume in acute ischemic stroke. *J. Clin. Neurosci.* **2015**, *22*, 291–295. [CrossRef] [PubMed]

64. Chen, Z.; Wang, K.; Huang, J.; Zheng, G.; Lv, Y.; Luo, N.; Liang, M.; Huang, L. Upregulated serum miR-146b serves as a biomarker for acute ischemic stroke. *Cell. Physiol. Biochem.* **2018**, *45*, 397–405. [CrossRef] [PubMed]

65. Wang, Y.; Barbacioru, C.; Hyland, F.; Xiao, W.; Hunkapiller, K.L.; Blakem, J.; Chan, F.; Gonzalez, C.; Zhang, L.; Samaha, R.R. Large scale real-time PCR validation on gene expression measurements from two commercial long-oligonucleotide microarrays. *BMC Genom.* **2006**, *7*, 59. [CrossRef]

66. Moldovan, L.; Batte, K.E.; Trgovcich, J.; Wisler, J.; Marsh, C.B.; Piper, M. Methodological challenges in utilizing miRNAs as circulating biomarkers. *J. Cell. Mol. Med.* **2014**, *18*, 371–390. [CrossRef]

67. Jickling, G.C.; Ander, B.P.; Zhan, X.; Noblett, D.; Stamova, B.; Liu, D. MicroRNA expression in peripheral blood cells following acute ischemic stroke and their predicted gene targets. *PLoS ONE* **2014**, *9*, e99283. [CrossRef]

68. Li, P.; Teng, F.; Gao, F.; Zhang, M.; Wu, J.; Zhang, C. Identification of circulating microRNAs as potential biomarkers for detecting acute ischemic stroke. *Cell. Mol. Neurobiol.* **2015**, *35*, 433–447. [CrossRef]

69. Zeng, L.; He, X.; Wang, Y.; Tang, Y.; Zheng, C.; Cai, H.; Liu, J.; Wang, Y.; Fu, Y.; Yang, G.Y. MicroRNA-210 overexpression induces angiogenesis and neurogenesis in the normal adult mouse brain. *Gene Ther.* **2014**, *21*, 37–43. [CrossRef]

70. Ouyang, Y.B.; Xu, L.; Yue, S.; Liu, S.; Giffard, R.G. Neuroprotection by astrocytes in brain ischemia: Importance of microRNAs. *Neurosci. Lett.* **2014**, *565*, 53–58. [CrossRef]

71. Ouyang, Y.B.; Giffard, R.G. MicroRNAs affect BCL-2 family proteins in the setting of cerebral ischemia. *Neurochem. Int.* **2014**, *77C*, 2–8. [CrossRef]

72. Tan, K.S.; Armugam, A.; Sepramaniam, S.; Lim, K.Y.; Setyowati, K.D.; Wang, C.W.; Jeyaseelan, K. Expression profile of MicroRNAs in young stroke patients. *PLoS ONE* **2009**, *4*, e7689. [CrossRef]

73. Sepramaniam, S.; Tan, J.R.; Tan, K.S.; DeSilva, D.A.; Tavintharan, S.; Woon, F.P.; Wang, C.W.; Yong, F.L.; Karolina, D.S.; Kaur, P.; et al. Circulating microRNAs as biomarkers of acute stroke. *Int. J. Mol. Sci.* **2014**, *15*, 1418–1432. [CrossRef] [PubMed]

74. Wang, W.H.; Guan, S.; Zhang, L.Y.; Lei, S.; Zeng, Y.J. Circulating microRNAs as novel potential biomarkers for early diagnosis of acute stroke in humans. *J. Stroke Cerebrovasc. Dis.* **2014**, *23*, 2607–2613. [CrossRef]

75. Tian, C.; Li, Z.; Yang, Z.; Huang, Q.; Liu, J.; Hong, B. Plasma microRNA-16 is a biomarker for diagnosis, stratification, and prognosis of hyperacute cerebral infarction. *PLoS ONE* **2016**, *11*, e0166688. [CrossRef]

76. Tiedt, S.; Prestel, M.; Malik, R.; Schieferdecker, N.; Duering, M.; Kautzky, V.; Stoycheva, I.; Böck, J.; Northoff, B.H.; Klein, M.; et al. RNA-Seq identifies circulating miR-125a-5p, miR-125b-5p, and miR-143-3p as potential biomarkers for acute ischemic stroke. *Circ. Res.* **2017**, *121*, 970–980. [CrossRef]

77. Mick, E.; Shah, R.; Tanriverdi, K.; Murthy, V.; Gerstein, M.; Rozowsky, J.; Kitchen, R.; Larson, M.G.; Levy, D.; Freedman, J.E. Stroke and circulating extracellular RNAs. *Stroke* **2017**, *48*, 828–834. [CrossRef] [PubMed]

78. Wu, J.; Fan, C.L.; Ma, L.J.; Liu, T.; Wang, C.; Song, J.X.; Lv, Q.S.; Pan, H.; Zhang, C.N.; Wang, J.J. Distinctive expression signatures of serum microRNAs in ischaemic stroke and transient ischaemic attack patients. *Thromb. Haemost.* **2017**, *117*, 992–1001. [CrossRef]

79. Agarwal, V.; Bell, G.W.; Nam, J.; Bartel, D.P. Predicting effective microRNA target sites in mammalian mRNAs. *eLife* **2015**, *4*, e05005. [CrossRef] [PubMed]

80. Chou, C.H.; Shrestha, S.; Yang, C.D.; Chang, N.W.; Lin, Y.L.; Liao, K.W.; Huang, W.C.; Sun, T.H.; Tu, S.J.; Lee, W.H.; et al. miRTarBase update 2018: A resource for experimentally validated microRNA-target interactions. *Nucleic Acids Res.* **2018**, *46*, D296–D302. [CrossRef]

81. Durinck, S.; Spellman, P.; Birney, E.; Huber, W. Mapping identifiers for the integration of genomic datasets with the R/Bioconductor package biomaRt. *Nat. Protoc.* **2009**, *4*, 1184–1191. [CrossRef]

82. Shannon, P.; Markiel, A.; Ozier, O.; Baliga, N.S.; Wang, J.T.; Ramage, D.; Amin, N.; Schwikowski, B.; Ideker, T. Cytoscape: A software environment for integrated models of biomolecular interaction networks. *Genome Res.* **2003**, *13*, 2498–2504. [CrossRef]

83. Szklarczyk, D.; Morris, J.H.; Cook, H.; Kuhn, M.; Wyder, S.; Simonovic, M.; Santos, A.; Doncheva, N.T.; Roth, A.; Bork, P.; et al. The STRING database in 2017: Quality-controlled protein-protein association networks, made broadly accessible. *Nucleic Acids Res.* **2017**, *45*, D362–D368. [CrossRef]

84. Zhang, X.; Yu, H.; Lou, J.R.; Zheng, J.; Zhu, H.; Popescu, N.I.; Lupu, F.; Lind, S.E.; Ding, W.Q. MicroRNA-19 (miR-19) regulates tissue factor expression in breast cancer cells. *J. Biol. Chem.* **2011**, *286*, 1429–1435. [CrossRef]

85. Weber, M.; Baker, M.B.; Patel, R.S.; Quyyumi, A.A.; Bao, G.; Searles, C.D. MicroRNA expression profile in CAD patients and the impact of ACEI/ARB. *Cardiol. Res. Pract.* **2011**, *2011*, 532915. [CrossRef]

86. Doebele, C.; Bonauer, A.; Fischer, A.; Scholz, A.; Reiss, Y.; Urbich, C.; Hofmann, W.K.; Zeiher, A.M.; Dimmeler, S. Members of the microRNA-17-92 cluster exhibit a cell-intrinsic antiangiogenic function in endothelial cells. *Blood* **2010**, *115*, 4944–4950. [CrossRef]

87. Bonauer, A.; Carmona, G.; Iwasaki, M.; Mione, M.; Koyanagi, M.; Fischer, A.; Burchfield, J.; Fox, H.; Doebele, C.; Ohtani, K.; et al. MicroRNA-92a controls angiogenesis and functional recovery of ischemic tissues in mice. *Science* **2009**, *324*, 1710–1713. [CrossRef]

88. Singh, V.; Roth, S.; Veltkamp, R.; Liesz, A. HMGB1 as a key mediator of immune mechanisms in ischemic stroke. *Antioxid. Redox Signal.* **2016**, *24*, 635–651. [CrossRef]

89. Diamanti, D.; Lahiri, N.; Tarditi, A.; Magnoni, L.; Fondelli, C.; Morena, E.; Malusa, F.; Pollio, G.; Diodato, E.; Tripepi, G.; et al. Reference genes selection for transcriptional profiling in blood of HD patients and R6/2 mice. *J. Huntingtons Dis.* **2013**, *2*, 185–200. [CrossRef]

90. Dervishi, I.; Gozutok, O.; Murnan, K.; Gautam, M.; Heller, D.; Bigio, E.; Ozdinler, P.H. Protein-protein interactions reveal key canonical pathways, upstream regulators, interactome domains, and novel targets in ALS. *Sci. Rep.* **2018**, *8*, 14732. [CrossRef]

91. Miller, J.A.; Oldham, M.C.; Geschwind, D.H. A systems level analysis of transcriptional changes in Alzheimer's disease and normal aging. *J. Neurosci.* **2008**, *28*, 1410–1420. [CrossRef]

92. Wu, C.C.; Chen, B.S. Key immune events of the pathomechanisms of early cardioembolic stroke: Multi-database mining and systems biology approach. *Int. J. Mol. Sci.* **2016**, *17*, 305. [CrossRef]

93. Chang, T.Y.; Huang, T.S.; Wang, H.W.; Chang, S.J.; Lo, H.H.; Chiu, Y.L.; Wang, Y.L.; Hsiao, C.D.; Tsai, C.H.; Chan, C.H.; et al. miRNome traits analysis on endothelial lineage cells discloses biomarker potential circulating microRNAs which affect progenitor activities. *BMC Genom.* **2014**, *15*, 802. [CrossRef]

94. Liang, Z.; Wu, G.; Fan, C.; Xu, J.; Jiang, S.; Yan, X.; Di, S.; Ma, Z.; Hu, W.; Yang, Y. The emerging role of signal transducer and activator of transcription 3 in cerebral ischemic and hemorrhagic stroke. *Prog. Neurobiol.* **2016**, *137*, 1–16. [CrossRef]

95. Bam, M.; Yang, X.; Sen, S.; Zumbrun, E.E.; Dennis, L.; Zhang, J.; Nagarkatti, P.S.; Nagarkatti, M. Characterization of dysregulated miRNA in peripheral blood mononuclear cells from ischemic stroke patients. *Mol. Neurobiol.* **2018**, *55*, 1419–1429. [CrossRef]

96. Sun, J.; Nan, G. The mitogen-activated protein kinase (MAPK) signaling pathway as a discovery target in stroke. *J. Mol. Neurosci.* **2016**, *59*, 90–98; Erratum in: *J. Mol. Neurosci.* **2016**, *59*, 430. [CrossRef]

97. Wang, H.; Zhang, X.M.; Tomiyoshi, G.; Nakamura, R.; Shinmen, N.; Kuroda, H.; Kimura, R.; Mine, S.; Kamitsukasa, I.; Wada, T.; et al. Association of serum levels of antibodies against MMP1, CBX1, and CBX5 with transient ischemic attack and cerebral infarction. *Oncotarget* **2017**, *9*, 5600–5613. [CrossRef]

98. Gao, J.B.; Tang, W.D.; Wang, H.X.; Xu, Y. Predictive value of thrombospondin-1 for outcomes in patients with acute ischemic stroke. *Clin. Chim. Acta* **2015**, *450*, 176–180. [CrossRef]

99. Cevik, O.; Baykal, A.T.; Sener, A. Platelets proteomic profiles of acute ischemic stroke patients. *PLoS ONE* **2016**, *11*, e0158287. [CrossRef]

100. Yang, F.H.; Pyle, W.G. Reduced cardiac CapZ protein protects hearts against acute ischemia-reperfusion injury and enhances preconditioning. *J. Mol. Cell. Cardiol.* **2012**, *52*, 761–772. [CrossRef]

101. Ridker, P.M.; Everett, B.M.; Thuren, T.; MacFadyen, J.G.; Chang, W.H.; Ballantyne, C.; Fonseca, F.; Nicolau, J.; Koenig, W.; Anker, S.D.; et al. CANTOS Trial Group. Antiinflammatory Therapy with Canakinumab for Atherosclerotic Disease. *N. Engl. J. Med.* **2017**, *377*, 1119–1131. [CrossRef]
102. Cirillo, P.; Golino, P.; Calabrò, P.; Calì, G.; Ragni, M.; De Rosa, S.; Cimmino, G.; Pacileo, M.; De Palma, R.; Forte, L.; et al. C-reactive protein induces tissue factor expression and promotes smooth muscle and endothelial cell proliferation. *Cardiovasc. Res.* **2005**, *68*, 47–55. [CrossRef]
103. De Rosa, S.; Cirillo, P.; Pacileo, M.; Di Palma, V.; Paglia, A.; Chiariello, M. Leptin stimulated C-reactive protein production by human coronary artery endothelial cells. *J. Vasc. Res.* **2009**, *46*, 609–617. [CrossRef] [PubMed]
104. Cirillo, P.; Golino, P.; Calabrò, P.; Ragni, M.; Forte, L.; Piro, O.; De Rosa, S.; Pacileo, M.; Chiariello, M. Activated platelets stimulate tissue factor expression in smooth muscle cells. *Thromb. Res.* **2003**, *112*, 51–57. [CrossRef] [PubMed]

Review

MicroRNAs as Biomarkers in Amyotrophic Lateral Sclerosis

Claudia Ricci *, Carlotta Marzocchi and Stefania Battistini

Department of Medical, Surgical and Neurological Sciences, University of Siena, 53100 Siena, Italy;
carlottamarzocchi@libero.it (C.M.), stefania.battistini@unisi.it (S.B.)
* Correspondence: claudia.ricci@unisi.it; Tel.: +39-0577-233142

Received: 25 October 2018; Accepted: 17 November 2018; Published: 20 November 2018

Abstract: Amyotrophic lateral sclerosis (ALS) is an incurable and fatal disorder characterized by the progressive loss of motor neurons in the cerebral cortex, brain stem, and spinal cord. Sporadic ALS form accounts for the majority of patients, but in 1–13.5% of cases the disease is inherited. The diagnosis of ALS is mainly based on clinical assessment and electrophysiological examinations with a history of symptom progression and is then made with a significant delay from symptom onset. Thus, the identification of biomarkers specific for ALS could be of a fundamental importance in the clinical practice. An ideal biomarker should display high specificity and sensitivity for discriminating ALS from control subjects and from ALS-mimics and other neurological diseases, and should then monitor disease progression within individual patients. microRNAs (miRNAs) are considered promising biomarkers for neurodegenerative diseases, since they are remarkably stable in human body fluids and can reflect physiological and pathological processes relevant for ALS. Here, we review the state of the art of miRNA biomarker identification for ALS in cerebrospinal fluid (CSF), blood and muscle tissue; we discuss advantages and disadvantages of different approaches, and underline the limits but also the great potential of this research for future practical applications.

Keywords: amyotrophic lateral sclerosis (ALS); biomarker; microRNA; cerebrospinal fluid (CSF); muscle biopsy; circulating miRNAs

---

1. Introduction

Amyotrophic lateral sclerosis (ALS), the most common adult-onset neurodegenerative disorder, is an incurable and invariably fatal condition characterized by the progressive loss of motor neurons in the motor cortex, brain stem, and spinal cord [1]. Motor neurons are selectively affected by degeneration and death, however the collective evidence is that ALS is non-cell autonomous, but rather pathogenesis and disease progression depend on the active participation of non-neuronal neighboring cells such as microglia, astrocytes, muscle and T cells [2,3]. Motor neuron degeneration causes progressive weakness of limb, thoracic, abdominal, and bulbar muscles.

During the early stages of the disease symptoms may vary depending on dysfunction of upper motor neurons (UMN) in the motor cortex (resulting in hyperreflexia, extensor plantar response, and increased muscle tone), or lower motor neuron (LMN) in the brainstem and spinal cord (leading to generalized weakness, muscle atrophy, hyporeflexia, fasciculations, and muscle cramps) [1]. Patients with bulbar onset ALS usually develop slurred and nasal speech and difficulty chewing or swallowing. Bulbar onset occurs less frequently than limb involvement, and accounts for about 25% of ALS cases. During the disease course, most cases show the presence of both LMN and UMN signs affecting spinal and brainstem regions [4]. Death, mainly due to bulbar dysfunction and respiratory insufficiency, occurs within 2–4 years of first symptoms; however, a small group of patients with ALS may survive for 10 or more years [5].

1.1. Epidemiology and Genetic Factors

The incidence of ALS is 2.1 per 100,000 persons per year, with an estimated prevalence of 5.4 cases per 100,000 population [6]. Based on data collected by population-based registers, the incidence of ALS increases after the age of 40, shows a peak in the late 60s or early 70s, and then displays a fast decline [7]. The reported male to female ratio varies widely with the age: a sex ratio of 2 or higher is observed for younger patients, while it appears to decrease towards 1 when the proportion of older patients increases [8]. Over the years, several environmental and lifestyle risk factors have been suggested as potential contributors to the cause of ALS. Nevertheless, no conclusive data are yet available, and further studies are required to identify exogenous risk factors of ALS [7,9].

Most cases (around 90%) are classified as sporadic ALS (SALS), since they are not associated with a documented family history. In 1–13% of patients the disease is inherited and defined as familial ALS (FALS), most frequently with a Mendelian dominant inheritance and high penetrance, even though pedigrees with recessive inheritance or incomplete penetrance have been described [10]. The mean age of onset for FALS is 46 years and for SALS is 56 years. In familial ALS, age of onset displays a Gaussian distribution, whereas an age-dependent incidence characterizes sporadic ALS [4]. Disease with an onset prior to 25 years of age is defined as "juvenile ALS" [11]. Apart from the mean age of onset, sporadic and familial forms are clinically indistinguishable suggesting a common pathogenesis.

Several genes have been associated with pathogenesis of ALS. The most common ALS causative genes include chromosome 9 open reading frame 72 (*C9orf72*), Cu2+/Zn2+ superoxide dismutase (*SOD1*), TAR DNA-binding protein 43 (*TARDBP*), and RNA binding protein FUS (*FUS*) [12–14], but a lot of other genes have been associated with the disease [15]. Notably, the mutated genes in ALS encode for proteins with very distinct functions in the cell. However, interestingly many ALS-linked genes, particularly *TARDBP* and *FUS*, are involved in RNA metabolism, including microRNA (miRNA) processing [16,17].

1.2. Diagnosis and Treatment

There is no objective laboratory test able to provide the diagnosis of ALS, which remains mainly based on clinical assessment, electrophysiological examinations, and exclusion of conditions that can mimic ALS. The certainty level of the diagnosis of ALS may be classified into different categories by clinical and laboratory assessments based on El Escorial criteria [18].

Currently, riluzole and edaravone represent the only drugs approved by the FDA for ALS, providing however a limited improvement in survival [5]. The most significant benefit of riluzole is observed after intervention in the early stages of the disease [19]. Thus, an early diagnosis of ALS could provide the most effective results. Since diagnosis of ALS relies on clinical symptoms, and the time from the first symptoms to diagnosis is about 12 months, there is a delay hindering a successful therapy [5]. This phenomenon underlies the importance of the development of screening tests able to detect the disease in early stages.

2. Role of Biomarkers in ALS

In the last years, research has been focused on the identification of potential biological markers to use in diagnostic procedure and clinical practice.

According to the National Institutes of Health Biomarkers Definitions Working Group, a biomarker is defined as "a characteristic that is objectively measured and evaluated as an indicator of normal biological processes, pathogenic processes, or pharmacologic responses to a therapeutic intervention" [20]. Biomarkers can be classified into three general categories: (1) diagnostic biomarkers, which are used for differential diagnosis; (2) prognostic biomarkers, which can differentiate a good or a bad outcome of the disease; and (3) predictive biomarkers, which are utilized for assessing whether a treatment may be effective for a specific patient or not.

In the case of ALS, biomarkers would allow an earlier and more accurate diagnosis, with the opportunity to start an earlier treatment able to modify the disease course. They could help the classification/stratification of ALS patients, monitor the disease progression and identify patients who will respond better to a particular drug. Biomarkers can also provide a valuable tool for the identification of new therapeutic approaches and drive patients' enrollment in clinical trials. Furthermore, they may represent a link between the results obtained in animal models and the human patients, providing insight on potential therapeutic targets.

Over the last two decades, intensive work has been carried out to find consistent biomarkers for ALS. Several candidates involved in excitotoxicity, oxidative stress, neuroinflammation, metabolic dysfunction, and neurodegeneration processes have been explored [21], but, unfortunately, none of these biomarkers has been currently translated into a practical diagnostic tool.

3. miRNAs as Biomarkers

Recently, among the different categories of potential biomarkers, miRNAs have aroused great interest in several fields of research. miRNAs are short (about 22 nucleotides in length) non-coding RNA molecules that play an important role as endogenous regulators of gene expression acting at the post-transcriptional level. miRNAs are synthesized from primary miRNAs, which are transcribed in the nucleus. Primary miRNAs are processed into pre-miRNAs by Drosha and then exported to the cytoplasm. Pre-miRNAs are eventually processed by the Dicer complex, resulting in mature miRNAs, which form RNA-induced silencing complexes [22]. miRNAs have a tissue-specific expression and this knowledge can help to better understand a normal and a disease development of the respective tissue [23]. miRNAs are known to play important roles in many physiological and pathological processes, including tumorigenesis [24], metabolism [25], immune function [26], and several neurodegenerative disorders [27], such as Parkinson's disease, Alzheimer's disease, Huntington's disease [28] and also ALS [29].

miRNAs have several intrinsic characteristics that make them promising as biomarkers. An ideal biomarker should display high sensitivity, specificity, and predictive power. miRNAs have been shown to have high specificity, and, in particular in cancer research, where a plethora of publications has been generated, it has been demonstrated that miRNA expression profiles differ among cancer types according to diagnosis and developmental stage of the tumor, with a better resolution than traditional gene expression analysis [30]. Moreover, unlike other RNA classes, miRNAs are remarkably stable and therefore can be robustly measured in many biological body fluids including plasma, tears, saliva and cerebrospinal fluid [31]. Indeed, miRNAs appear resistant to boiling, repeated freeze-thawing cycles, pH changes, and fragmentation by chemical or enzymes [32–34]. Furthermore, recent evidence indicates that miRNAs can be detected in biological fluids and can be used to "capture" changes in the cells of origin, including neurons [35].

In addition to these general considerations, several findings suggest a specific involvement of miRNAs in ALS. For example, the loss of Dicer is sufficient to cause progressive degeneration of spinal motor neurons [36]; in addition, a global down-regulation of miRNAs is a frequent molecular denominator for multiple forms of human ALS [37]. Moreover, a common theme for several ALS-related genes is a role in RNA processing pathways [38]. FUS facilitates co-transcriptional Drosha recruitment to specific miRNA loci [39] and TARDBP participate to miRNA biogenesis as a component of both Drosha and Dicer complexes [16].

miRNA Detection

During the last decade, the development of methods for detecting miRNAs has risen to become a very attractive area of research. Although miRNAs have characteristics that made them suitable biomarkers, the detection of these molecules is challenging due to their intrinsic characteristics including small size, sequence similarity among various members, low level and tissue-specific or developmental stage-specific expression. Two approaches commonly used in the research of miRNAs as biomarkers, including studies in the area of neurodegenerative diseases and in particular in ALS, are reported below.

(1) Measurement of hundreds of miRNAs in specimens from patients with a pathology of interest and from control subjects using profiling methods, such as microarray, quantitative Real-Time Polymerase Chain Reaction (qRT-PCR)-based array, quantitative nCounter or Next Generation Sequencing (NGS), with subsequent validation of identified miRNAs by qRT-PCR;

(2) Analysis of selected miRNA(s) already known as related to specific tissues, cell types, or gene expression pathways. In this case, the number of miRNA(s) to be tested is limited, which makes the use of individual qRT-PCR appropriate, increasing sensitivity and reproducibility of the analysis.

Among the profiling methods, microarray is a powerful high-throughput widely used tool that screens large numbers of miRNAs analyzing simultaneously several samples processed in parallel in a single experiment [40]. An alternative method is deep-sequencing, which relays on NGS machines that can process millions of sequence reads in parallel in just a few days [41,42]. Sequence reads are processed by bioinformatics analysis, which identifies both known and novel miRNAs in the data sets, and perform a relative quantification using a digital approach [43]. Finally, qRT-PCR arrays can also be used to detect multiple miRNAs at the same time [44]. This approach is able to detect miRNAs in very low copy number [45]. This is an important aspect, since large amounts of RNA from clinical samples can be difficult to obtain. Other advantages of qRT-PCR-based techniques used in routine diagnostic are sensitivity, specificity, speed and simplicity [46]. Of note, potential biomarkers selected by array-based analysis need to be confirmed by qRT-PCR, due to high variability and low reproducibility of results obtained from these techniques [47].

A critical issue in qRT-PCR analysis is the data normalization approach. Normalization refers to adjusting for variations in data that are due to known factors (usually technical factors) and not related to the biological differences that are being investigated, and that could otherwise lead to inaccurate quantification. For this reason, stable normalizers are needed, but identifying such molecules is challenging, and it is often necessary to select them on a case-by-case basis [48]. Normalization to reference invariant miRNAs [49] is effective in many cases, but this approach requires that the reference miRNA is not influenced by the condition being studied. Exogenous spike-in controls added to samples during the miRNAs extraction may be used to compensate the variability caused by extraction efficiency and possible presence of inhibitors [50]. The combined use of two or more normalizers usually allows reducing experimental variability and improving reliability of the analysis.

4. miRNAs as Biomarkers for ALS

The first paper about miRNAs as biomarkers for ALS in human samples was published by De Felice and colleagues in 2012 [51]. Since then, a large number of studies have been performed on cerebrospinal fluid (CSF), blood and muscle biopsies from ALS patients. See Figure 1 for a schematic workflow of identification of miRNA-based biomarkers.

	qRT-PCR	Microarray	NGS	nCounter
Sensitivity	+++	+/++	++	++
Specificity	+++	+/++	+++	++
Throughput	+/++	+++	+++	++
Identification of novel miRNAs	no	no	yes	no
Validation data	no	yes	yes	yes
Normalization	yes	no	no	yes

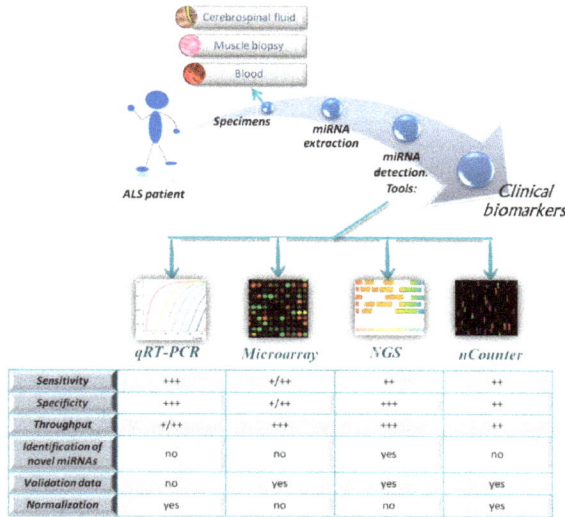

Figure 1. MicroRNA (miRNA)-based biomarkers in amyotrophic lateral sclerosis (ALS) patients. Schematic workflow to identify possible miRNAs as biomarkers starting from ALS patients' sample using different quantitative approaches. The comparison among the common characteristics of miRNA detection platforms is summarized in the figure. Sensibility, specificity and throughput are classified as follows: +++ (very high), ++ (moderate), +/++ (moderate to low) and + (low). Abbreviations: qRT-PCR, quantitative Real-Time Polymerase Chain Reaction; NGS, Next Generation Sequencing.

4.1. miRNAs in Cerebrospinal Fluid

Cerebrospinal fluid (CSF) is the fluid that bathes the central nervous system (CNS) and, due to this direct interaction, represents a potentially ideal source for identifying biomarkers for ALS. miRNAs present in CSF can mirror CNS physiological and pathological conditions representing more sensitive biomarkers of brain changes than those present in other biofluids [35]. The presence of miRNAs in CSF was first demonstrated by Cogswell and colleagues [52]. The authors reported that the amount of miRNAs secreted or excreted from other organs to CSF is very limited and that the major source of miRNAs detected in CSF are immune cells present in this biofluid. In addition, other studies showed that the miRNAs present in CSF derived also from neurons [53].

CSF samples are obtained by lumbar puncture, a procedure used for diagnostic purposes to confirm ALS diagnosis and exclude other pathologies, as inflammatory nerve conditions. Lumbar puncture, however, represents an invasive procedure, that cannot be repeated during the disease course for ethical implications. Thus, analysis of miRNAs in CSF is not suitable to identify biomarkers to follow disease progression.

Up to date, five studies have been published about the identification of miRNAs as biomarkers in CSF from ALS patients. Results are shown in Table 1.

The first three studies in Table 1 selected a limited set of miRNAs to analyze: 43 miRNAs found up-regulated in SOD1 spinal cord CD39+ microglia and splenic Ly6Chi monocytes [54], a group of *TARDBP* binding miRNAs [55], or one selected miRNA, over-expressed in ALS blood leucocytes [56], respectively. The other two studies performed a miRNA expression profiling, using qRT-PCR [57] or small RNA sequencing (NGS) [58]. In both profiling studies, results were validated by qRT-PCR for each single miRNA. While Benigni and colleagues found eight out of fourteen miRNAs as significantly deregulated, Waller and coworkers failed to confirm statistically significant differences in miRNA expression [57,58].

Table 1. Deregulated microRNAs (miRNAs) in cerebrospinal fluid (CSF) of amyotrophic lateral sclerosis (ALS) patients compared to healthy controls.

miRNA (Hsa-miR)	miRNA Expression Change	No. of Specimens	miRNA Detection Approach	Ref.
150, 99b, 146a	↑ in SALS	SALS: 10 FALS: 5 HCs: 10	qRT-PCR	[54]
27b, 328, 532-3p	↑ in SALS and FALS			
132-5p, 132-3p, 143-3p	↓	SALS: 22 HCs: 24	qRT-PCR	[55]
143-5p, 574-5p	↑			
338-3p	↑	SALS: 10 HCs: 10	qRT-PCR	[56]
181a-5p	↑	SALS: 24 HCs: 24	qRT-PCR	[57]
21-5p, 195-5p, 148-3p, 15b-5p, let7a-5p, let7b-5p, let7f-5p	↓			
124-3p, 127-3p, 143-3p, 125b-2-3p, 9-5p, 27b-3p	↑	SALS: 32 HCs: 10 NCs: 6	NGS	[58]
486-5p, let7f-5p, 16-5p, 28-3p, 146a-3p, 150-5p, 378a-3p, 142-5p, 92a-5p	↓			

Abbreviations: Ref., Reference; ↑/↓, up-regulated/down-regulated; SALS, sporadic amyotrophic lateral sclerosis patients; FALS, familial amyotrophic lateral sclerosis patients; HCs, healthy controls; NCs, neurological disease control subjects (multiple sclerosis); qRT-PCR, quantitative Real-Time Polymerase Chain Reaction; NGS, Next Generation Sequencing.

A common feature observed by the authors is an overall down-regulation of miRNAs in CSF samples from ALS patients [55,57,58], in agreement with other studies showing that the majority of deregulated miRNAs in tissues from ALS models and ALS patients are down-regulated [59]. This could suggest a general default in RNA metabolism in ALS [38].

In general, however, these studies highlight a wide heterogeneity among miRNAs significantly deregulated. A possible explanation could be the variability in terms of experimental approach and technical procedures and the reduced number of CSF samples analyzed in each study.

Some authors evaluated the correlation between CSF and serum miRNA expression levels. A significant positive correlation between expression levels in CSF and serum from ALS patients was found for miR-338-3p [56] and miR-143-3p [55]. However, the amount of most miRNAs was independently regulated between the two biofluids at individual level. This suggests that CSF miRNAs do not simply reflect the usually more abundant serum miRNAs, and changes in the serum do not necessarily reproduce alterations of CSF levels [55].

It should be noted that, among the different body fluids, the lowest abundance of miRNAs appears in CSF. Thus, it is possible that some potentially promising and informative miRNAs, identified both in vivo and in vitro ALS models, are below the limit of detection of the available methods of analysis. For example, the miR-218, a motor neurons-enriched miRNA, has been found increased in CSF of ALS rodent models: its expression correlates with the number of remaining spinal motor neurons and is responsive to motor neuron sparing therapy [60]. miR-218 could thus represent a potential biomarker to assess drug effects on motor neurons during clinical trials in ALS patients. However, at the present time, this miRNA is detectable only in some CSF samples, and thus a comparison between ALS patients and controls is not possible.

An approach to overcome the technical limits due to low abundance of several miRNAs in CSF could be to focus on those miRNAs found up-regulated in ALS patients. Among these, miR-338-3p seems to be very promising, since it has been reported as consistently upregulated in CSF, serum and leukocytes from ALS patients [56]. miR-338-3p is involved in several molecular pathways and could contribute to ALS pathogenesis through different modalities, such as neurodegeneration and apoptosis. Recent evidence suggests that miR-338 participates in the control of neuroblast apoptosis and in neuroblastoma pathogenesis [61] and it is able to suppress neuroblastoma proliferation, invasion and migration [62]. Interestingly, also another miRNA found up-regulated in CSF from ALS patients, miR181a-5p, has been proposed as an anti-oncomir, which acts as a tumor suppressor in normal tissues, promoting growth inhibition and apoptosis [63]. These findings suggest that these up-regulated miRNAs are involved in ALS pathogenetic process through apoptotic mechanisms responsible for cell death.

In order to increase diagnostic accuracy, up-regulated miRNAs can be used in combination with other miRNAs, identified as down-regulated. Benigni and colleagues reported that the ratios of miR-181a-5p/miR-15b-5p and miR-181a-5p/miR-21-5p considerably increased the specificity with a slight decrease in sensitivity compared with each individual miRNA [57]. A wider use of this strategy could allow improvements in the performance of identified biomarkers and should be taken into account for future studies, as further discussed in the following paragraphs.

4.2. Circulating miRNAs

The use of blood samples in the diagnostic routine presents several advantages. Blood specimens are easy to obtain, process and store, and the samples required for the analysis can be collected without using invasive procedures for the patients. The lack of ethical implications as compared with CSF and muscle biopsy makes it possible to repeat the blood draw during the disease progression. Since miRNAs circulate in the blood in a highly stable form, this may facilitate the procedure of storage and conservation and increase the flexibility of the analysis.

Blood-based biomarkers may originate from the CNS through a transfer between the blood and CSF at the blood–CSF barrier [64,65], suggesting that the same biomarkers could be present in both biofluids. They may be generated also by other organs and tissues affected during ALS, such as degenerating muscles or peripheral blood cells. Therefore, blood can represent an excellent biofluid for discovery and validation of biomarkers for ALS [66]. On the other hand, miRNAs present in blood can reflect other pathophysiological conditions concurrent but not directly related to ALS disease (e.g. inflammatory status, response to pharmacological treatments, etc.), which may represent confounding factors.

Several studies on circulating miRNAs as potential biomarkers for ALS have been published. The findings from such studies are summarized in Table 2.

Table 2. Deregulated circulating microRNAs (miRNAs) in amyotrophic lateral sclerosis (ALS) patients compared to healthy controls.

miRNA (Has-miR) Expression Change	Source	miRNA Detection Approach	No. of Specimens for miRNAs Validation	Ref.
↑: 338-3p	Leukocytes	Microarray→ miRNAs validation with qRT-PCR	SALS: 14 HCs: 14	[51]
↑: 27a, 155, 142-5p, 223, 30b, 532-3p	Monocytes (CD14+ CD16-)	Nanostring nCounter [1] → miRNAs validation with qRT-PCR	SALS: 22 FALS: 4 HCs: 24	[54]
↓: 132-3p, 132-5p, 143-3p, 143-5p, let-7b	Serum	Nine *TARDBP* binding miRNAs and miR-9-5p → qRT-PCR	SALS: 22 HCs: 24	[55]

Table 2. *Cont.*

miRNA (Has-miR) Expression Change	Source	miRNA Detection Approach	No. of Specimens for miRNAs Validation	Ref.
↑: 206, 106b	Serum	Microarray [1] → miRNAs validation with qRT-PCR	SALS: 12 HCs: 12	[67]
↑: 338-3p	Leukocytes and serum	miR-338-3p → qRT-PCR	SALS: 10 HCs: 10	[56]
↓ in FALS/SALS: 4745-5p, 3665, 4530 ↓ in FALS: 1915-3p	Serum	Microarray → miRNAs validation with qRT-PCR	FALS: 23 HCs: 24 SALS: 14 HCs: 14	[68]
↓ in FALS/SALS: 1825 ↓ in SALS: 1234-3p	Serum	Microarray→ miRNAs validation with qRT-PCR	SALS: 20 HCs: 20 FALS: 13 HCs: 13	[69]
↑: 4649-5p ↓: 4299	Plasma	Microarray → miRNAs validation with qRT-PCR	SALS: 48 HCs: 47	[70]
↓: 183, 193b, 451, 3935	Leukocytes	Microarray → miRNAs validation with qRT-PCR	SALS: 83 HCs: 61	[71]
↑: 424, 206	Plasma	Microarray [2] → miRNAs validation with qRT-PCR	SALS: 39 HCs: 39	[72]
↑: 206, 133a,133b ↓: 146a, 149*, 27a	Serum	Preselected myo-miRNAs, inflammatory and angiogenic miRNA → qRT-PCR	SALS: 14 HCs: 8	[73]
↑: 206 Deregulated MicroRN pairs: 206/338-3p 9*/129-3p 335-5p/338-3p	Plasma	Thirty seven brain-enriched and inflammation-associated microRNAs → qRT-PCR	ALS: 50 HCs: 50	[74]
↑ [†]: 1, 133a-3p, 133b, 144-5p, 192-3p, 195-5p, 19a-3p ↓ [†]: let-7d-3p, 320a, 320b, 320c, 425-5p, 139-5p	Serum	qRT-PCR array	SALS: 20 FALS: 3 HCs: 30 NCs: 103	[75]
↑: 206, 143-3p ↓: 374b-5p	Serum	qRT-PCR array → miRNAs validation with qRT-PCR	SALS: 23 CRL: 22	[76]
↑: 9, 338, 638, 663a, 124a, 451a, 132, 206, let-7b	Leukocytes	Preselected 10 miRNAs → miRNAs validation with qRT-PCR	SALS: 84 HCs: 27	[77]
↑: 142-3p ↓: 1249-3p	Serum	Microarray [1] → miRNAs validation with qRT-PCR	SALS: 20 HCs: 20	[78]
↓: 27a-3p	Serum exosomes	miR-27a-3p → qRT-PCR	ALS: 10 HCs: 20	[79]
↓: let-7a-5p, let-7d-5p, let-7f-5p, let-7g-5p, let-7i-5p, 103a-3p, 106b-3p, 128-3p, 130a-3p, 130b-3p, 144-5p, 148a-3p, 148b-3p, 15a-5p, 15b-5p, 151a-5p, 151b, 16-5p, 182-5p, 183-5p, 186-5p, 22-3p, 221-3p, 223-3p, 23a-3p, 26a-5p, 26b-5p, 27b-3p, 28-3p, 30b-5p, 30c-5p, 342-3p, 425-5p, 451a, 532-5p, 550a-3p, 584-5p, 93-5p	Whole blood	NGS → qRT-PCR	SALS: 50 HCs: 15	[80]

[1] analysis carried out on samples from transgenic mice; [2] analysis carried out on samples of ALS patients' skeletal muscle biopsies; [†], miRNAs deregulated in ALS patients compared to healthy controls and neurological controls (including multiple sclerosis and Alzheimer's disease patients). Abbreviations: Ref., Reference; ↑/↓, up-regulated/down-regulated; SALS, sporadic amyotrophic lateral sclerosis patients; FALS, familial amyotrophic lateral sclerosis patients; HCs, healthy controls; NCs, neurological controls; qRT-PCR, quantitative Real-Time Polymerase Chain Reaction; NGS, Next Generation Sequencing.

As reported in Table 2, several studies have identified numerous potential miRNA biomarkers in peripheral blood from ALS patients, however their results rarely overlap with each other. This high discrepancy in the identified miRNAs is probably associated with the variability of quantification methods, miRNA normalizers used, number of samples included, clinical features of patients, and also with the differences in selected source of miRNAs (serum, plasma, leukocytes, and whole blood). Another possible reason for the poor reproducibility of results may be the high level of heterogeneity in miRNA profiles of SALS patients in comparison to FALS patients. Freischmidt and colleagues initially reported a signature of 22 miRNAs significantly down-regulated in FALS and presymptomatic mutation carriers [68]. Subsequently, the same authors replicated the analysis of these miRNAs in a larger SALS sample group using identical technical procedures, and found only 2 miRNAs significantly down-regulated in all SALS patients. A more accurate analysis of results revealed that around 60% of SALS patients shared a serum miRNA fingerprint with genetic cases, while the remaining around 40% of patients were evenly distributed among control samples. The absence of FALS-like miRNA patterns in these patients may mirror a higher impact of exogenous factors and possibly a lower and/or different genetic influence in a subgroup of SALS patients [69].

Interestingly, the miRNA expression profiles derived from the study performed by Freischmidt and colleagues [68] were re-elaborated applying principal component analysis (PCA)-based unsupervised feature extraction (FE), another analysis approach [81]. The authors identified a total of 51 deregulated miRNAs, 27 down-regulated and 24 up-regulated in ALS patients in comparison with healthy controls. Applying the linear discriminant analysis (LDA) to these selected miRNAs, overall accuracy was 0.66 including healthy controls, ALS mutation carriers, FALS and SALS patients. Of note, excluding SALS patients, LDA was able to successfully discriminate healthy controls, ALS mutation carriers and FALS patients, with an accuracy rising up to 0.84, confirming as the heterogeneity of SALS group can introduce a wider variability in circulating miRNA profiles.

Among the studies published until now, a largely used approach is miRNA profiling on blood samples from ALS patients and controls, carried out by microarray [51,68–71], PCR-array [75,76] and NGS [80]. Other studies performed analysis on specific miRNAs, selected from data previously reported in the literature [55,56,73,74,77]. In other cases, the first step of the research was a microarray analysis on samples from transgenic mice [54,67,78] or skeletal muscle biopsies from ALS patients [68], followed by validation of miRNAs found deregulated in the first step of analysis.

Only one study analyzed miRNA expression specifically in serum exosomes [79]. Exosomes are double lipid vesicles secreted by a variety of cells and widespread in the peripheral body fluid. They can reflect physiological and pathological changes of the cells of origin, representing potential new biomarkers for disease diagnosis [82]. miRNAs are enriched in exosomes, and the exosome membrane structure can protect them from degradation by RNA enzymes. The authors investigated the expression of only miR-27a-3p, previously reported as present in myoblast-derived exosomes [83], and found a down-regulation of this miRNA in ALS patients, suggesting that miRNA exosome analysis could represent a future perspective for ALS biomarker identification.

Despite a poor overlapping among the miRNAs identified as deregulated in ALS, some circulating miRNAs seem to be particularly promising as potential biomarkers in ALS patients. Table 3 summarizes these miRNAs, reported as de-regulated in two or more papers.

As shown in the Table 3, some common pathways emerge: some miRNAs are involved in neurodegeneration and apoptosis (miR-338, miR-142, miR-183 and let-7d), other miRNAs act at muscle level (miR-206, miR-133a, miR-133b and miR-27a). In particular, miR-206, miR-133a and miR-133b are myo-miRNAs, molecules specifically expressed in striated muscle and involved in muscle proliferation, repair and regeneration. Their expression levels change during the process of myogenesis, development, atrophy, degeneration, and myopathies [84]. The more recurrent result is an up-regulation of circulating miR-206 in ALS patients. miR-206 is a human skeletal muscle-specific miRNA that promotes the formation of new neuromuscular junctions following nerve injury, and therefore plays a crucial role in the reinnervation process [85]. In miR-206 knock-out mice, delayed and incomplete muscular reinnervation was observed in comparison to those animals that expressed miR-206. In addition, high expression levels of miR-206 were found in a mouse model of ALS, and its under-expression was associated with a faster progression of the disease [86]. A consensus for higher expression levels of this miRNA in ALS patients compared to controls was reported by several authors [67,72–74,76,77]. Although miR-206 seems to represent a valid circulating biomarker for ALS, it is still to define whether the elevated expression of this miRNA is the result of the disease or its cause.

Table 3. The most promising circulating microRNAs (miRNAs) detected as potential biomarkers in amyotrophic lateral sclerosis (ALS) patients.

miRNAs (Has-miR)	miRNA Change	Role in ALS	Ref.
206	↑	Myo-miRNA: muscle proliferation, repair and regeneration. It promotes neuromuscular connectivity and enhances reinnervation	[67,72–74,76,77]
338	↑	Involvement in different pathways such as apoptosis, neurodegeneration, and/or glutamate clearance	[51,67,74,77]
133a	↑	Myo-miRNA: muscle proliferation, repair and regeneration	[73,75]
133b	↑	Myo-miRNA: muscle proliferation, repair and regeneration	[73,75]
142	↑	miRNA predicted to target a specific set of genes associated to the pathophysiology of ALS, including *TARDBP* and *C9orf72*.	[54,78]
183	↓	miRNA involved in neurodegenerative signaling pathway, including PI3K-Akt and MAPK pathway. miR-183/mTOR pathway contributes to spinal muscular atrophy pathology	[71,80]
27a	↓	miRNA involved in muscle growth, myoblast proliferation acting on myostatin. It is present in myoblast-derived exosomes	[73,79]
let-7d	↓	Involvement in apoptosis by the Hippo signaling pathway	[75,80]

Abbreviations: Ref., Reference; ↑/↓, up-regulated/down-regulated.

While all the works performed a comparison between samples from ALS patients and healthy controls, only a subset of them included also samples from patients affected by other neurological disorders. Neurological controls comprised Parkinson's disease [56,70,71,74], Alzeihmer's Disease [56,69,74,75], Huntington's disease [56,69,71], Multiple Sclerosis [54,75] and ALS-mimic conditions [76]. The use of neurological controls can help to discriminate whether identified miRNAs are really specific for ALS or are common features linked to neurodegenerative processes. For example, the comparison of miRNA expression between ALS and Parkinson's disease patients suggested that miR-183 might be specific for SALS, whereas miR-451 and miR-3935 might be more general biomarkers linked to neurodegenerative disorders [71]. In addition, the inclusion of an ALS-mimic patients' group may contribute to identify miRNA biomarkers to use in the differential diagnosis in the early stages of the disease.

Only a part of the studies performed until now investigated the potential correlations among miRNA expression levels and ALS clinical features, sometimes in longitudinal studies, measuring miRNA levels in the same ALS patient over time [70,72,75,76,80]. In some case this analysis failed to find any association [67,69,77], in other cases specific correlations were reported. Some authors described associations of miRNA expression levels with ALS site of onset [70,74,76,80], ALS Functional Rating Scale-revised (ALSFRS-R) and/or vital capacity (VC) [70,75,78,80], Medical Research Council (MRC) sumscore [72] and with the disease progression rate [72,80]. Only two studies investigated the possible associations of specific serum miRNAs with riluzole treatment, failing to identify any correlation [75,76]. Such results must be anyway considered with caution, since the number of subjects included in every group is limited. They need to be confirmed in larger cohorts of ALS patients, to really define the role of miRNA expression in ALS clinical presentation and progression. From this perspective, it would be very important that, after the identification of potential miRNA biomarkers, more longitudinal studies were performed, to evaluate if these miRNAs could be used as prognostic indicators.

As already mentioned for CSF studies, also in serum the analysis of combinations of several miRNAs has shown a higher accuracy than single miRNAs in discriminating ALS from healthy controls or other neurological disorders [71,74,75]. A very interesting approach is reported by Sheinerman and colleagues, who developed a strategy based on miRNA pairs, consisting of one miRNA enriched in synapses of a brain region affected by the disease and another miRNA enriched in a different brain region or cell type. The use of the pair of miRNA derived from the same organ allowed decreasing potential overlap with pathologies of other organs and reducing also inter-individual variability. The authors demonstrated that, combining two or three effective miRNA pairs into a single miRNA classifier, they could achieve a greater accuracy in discriminating ALS both from healthy controls and patients affected by other neurological disorders [74]. Thus, in the future studies it should be considered that, while the deregulation of a single miRNA can be a feature common to several neurological diseases, panels of deregulated miRNAs, or combinations of them, may result highly specific for ALS and represent a signature for this disease.

Finally, a relevant aspect of the use of miRNAs as ALS biomarkers is their potential of identifying the disease in very early stages, also before any clinical manifestation. In their work, Freischmidt and colleagues showed that a specific subset of miRNAs, reduced in the serum of patients with familial and sporadic ALS, was reduced also in presymptomatic carriers of pathogenic ALS mutations. Moreover, the down-regulation was largely independent of the underlying disease gene and was stronger in patients with familial ALS than in pre-manifest mutation carriers, suggesting that alterations of miRNA profiles could be progressive when comparing the pre-manifest and manifest phase of the disease [68]. If confirmed, these findings may be of fundamental importance for the development of screening tests able to detect ALS in early asymptomatic stages and for future preventive therapeutic strategies before the occurrence of clinically evaluable symptoms.

4.3. miRNAs in Muscle Biopsies

Skeletal muscle is another potential source for the identification of candidate miRNA biomarkers. In the last years, it has become evident that ALS does not affect only motor neurons but also other cell types, including striated muscle, which play an active role in the disease pathogenesis. Before the clinical onset and during the disease progression, the affected skeletal muscle of ALS patients attempts to restore function by futile cycles of reinnervation and denervation [87]. Eventually, persistent muscle wasting exceeds the ability to repair and consequently the atrophy process starts. Due to the crucial role of the skeletal muscle in ALS pathogenesis, recent studies have focused their research on the identification of specific muscle miRNAs in ALS tissues, which could potentially be use as prognostic biomarkers of disease. Moreover, miRNAs identified in skeletal muscle of ALS patients could be used as biomarkers also in plasma or serum, where they can be released by the affected tissues. This strategy seems to be particularly interesting, since muscle biopsy is unfortunately an invasive practice and cannot be proposed for longitudinal studies to follow disease progression.

Several studies focused on analysis of myo-miRNAs, including miR-1, miR-133a, miR-133b, miR-206, miR-208a, miR-208b, miR-499, and miR-486 [88]. Most of them explored the role of these miRNAs in mouse models (for a review see [89]), but only few studies investigated the role of these molecules as possible markers in muscle biopsies of patients with ALS, due to the rarity and difficulty to obtain this kind of samples. miRNAs found deregulated in muscle biopsies from ALS patients compared to healthy control subjects are shown in Table 4.

Most studies focused on the expression of myo-miRNAs [90–93]; only in some cases also other miRNAs were included, for example miRNAs related to inflammation/angiogenesis [93] or selected by microarray [72] or NGS approaches [94,95]. Overall, results are sometimes contrasting and poorly reproducible. These non-concordant finding could be attributed to different types of muscle used for biopsy, discordance among the samples in terms of inclusion criteria of patients (age, gender, evolution of disease, onset) and different techniques and internal control molecules used to assess miRNA expression levels.

Table 4. Deregulated miRNAs in skeletal muscle biopsies of amyotrophic lateral sclerosis (ALS) patients compared to healthy controls.

miRNA (Hsa-miR)	miRNA Expression Change	Type of Muscle	No. of Muscle Biopsies	miRNA Detection Approach	Ref.
206	↑	Deltoid, anconeus	FALS: 1 SALS: 10 HCs: 6	mir-206 → qRT-PCR	[90]
23a, 29b, 206, 455, 31	↑	Vastus lateralis	ALS: 14 HCs: 10	Myo-miRNAs and miRNAs dysregulated in human muscle disease → qRT-PCR	[91]
1, 26a, 133a, 455	↓	Vastus lateralis	ALS: 5 HCs: 7	Myo-miRNAs → qRT-PCR	[92]
424, 214, 206	↑	Biceps brachii	ALS: 5 HCs: 5	Microarray → miRNAs validation with qRT-PCR	[72]
1, 206, 133a, 133b, 27a, 155, 146a, 221	↑	Quadriceps femoris	SALS: 13 HCs: 5	Inflammatory/angiogenic miRNAs and myo-miRNAs → qRT-PCR	[93]
1, 10b-5p, 100-5p, 133a-3p, 133b-3p	↓	Biceps, deltoid, tibialis anterior, vastus lateralis	ALS: 19 HCs: 9	NGS [1] and qRT-PCR [1] → qRT-PCR	[94]
100-5p, 10a, 125a-5p, 133a-1/-2-3p, 362, 500a-3p, 542-5p, 99a-5p	↑	Vastus lateralis	FALS: 2 SALS: 9 HCs: 11	NGS	[95]
1303-3p, 150-5p, 26a-1/-2-5p, 486-1/-2-5p,	↓				

[1] analysis carried out on samples from transgenic mice. Abbreviations: Ref., Reference; ↑/↓, up-regulated/down-regulated; SALS, sporadic amyotrophic lateral sclerosis patients; FALS, familial amyotrophic lateral sclerosis patients; HCs, healthy controls; qRT-PCR, quantitative Real-Time Polymerase Chain Reaction; NGS, Next Generation Sequencing.

Some authors performed also a correlation analysis among miRNA expression and ALS clinical features. Table 5 reports miRNAs altered in tissue of specific stratified ALS patients' groups analyzed in comparison to control subjects.

In addition, in other papers the associations with clinical variables were analyzed comparing groups of patients to each other. Stratifying ALS patients, an up-regulation of myo-miRNAs (miR -206, miR-133a, miR-133b and miR-27a) and of inflammatory miRNAs (miR-155, miR-146a and miR -221) was discovered in ALS patients with earlier age at onset (<55 years) and longer disease duration [93]. Moreover, significantly higher expression levels of the same myo-miRNAs and inflammatory miRNAs were detected in male than in female. This gender difference has been hypothesized to be related to a difference in hormonal regulation, implying a slower disease progression in women [93]. In another paper, miR-29c, miR-208b and miR-499 were reported as increased in patients with slow disease course [96]. Expression data were analyzed in patients categorized into "early" and "late" based on disease duration at the moment of biopsy (more or less one year). miR-9 and miR-206 significantly increased in the early patients' group and, of note, miR-206 inversely correlated with the time from symptoms onset to muscle biopsy, indicating an early response to denervation in skeletal muscle [96].

Table 5. Deregulated miRNAs in skeletal muscle biopsies of specific amyotrophic lateral sclerosis (ALS) patients' groups analyzed in comparison to healthy controls.

miRNA (Hsa-miR)	miRNA Expression Change in Specific ALS Patients' Group	Type of Muscle	No. of Muscle Biopsies	miRNA Detection Approach	Ref.
133a, 29c, 9, 208b	↑in ALS slow group [1]		FALS: 3 SALS: 11 HCs: 24	Eleven skeletal muscle related miRNAs → qRT-PCR	[96]
1, 208b	↓ in ALS rapid group [2]	Deltoid and quadriceps	Slow group [1]: 6 Rapid group [2]: 5		
133a, 133b, 206, 29c, 9, 155, 23a	↑ in early stage group [3]		Early group [3]: 4 Late group [4]:9		
100-5p, 199a-1/-2, 199b-3p, 27a-5p, 3607-3p, 424-5p, 450a-1/-2-5p, 450b-5p, 501-3p, 502-3p, 542-5p, 660-5p	↑ in higher disease severity [5]	Vastus lateralis	Higher disease group [5]: 7 HCs: 11	NGS	[95]
1303-3p, 133a-1/-2-3p, 150-5p, 378, 486-1/-2-5p, 502-3p, 855-3p	↓ in higher disease severity [5]				

[1], ALS slow group (≥4 years of disease progression without requiring respiratory supports); [2], ALS rapid group (<4 years of disease progression without respiratory supports or death occurring <4 years from symptoms onset); [3], early stage group (less than one year from symptom onset to muscle biopsy); [4], late stage group (more than one year from symptom onset to muscle biopsy); [5], group of patients with higher disease severity. Abbreviations: Ref., Reference; ↑/↓, up-regulated/down-regulated; SALS, sporadic amyotrophic lateral sclerosis patients; FALS, familial amyotrophic lateral sclerosis patients; HCs, healthy controls; qRT-PCR, quantitative Real-Time Polymerase Chain Reaction; NGS, Next Generation Sequencing.

Although the results are often inconsistent among different studies, some trends in miRNAs deregulation seem to emerge. One of the most interesting miRNA is miR-133a, which was found to be up-regulated in human ALS tissues [93,95,96], particularly in patients with slow disease progression and in biopsies obtained before one year from the symptom onset [96]. At the same time, a significant reduction of this miRNA was present in a specific ALS patients' group with higher disease severity [95], suggesting changes in its expression during the disease progression. In contrast, however, other studies detected a down-regulation of miR-133a in human biopsies, as reported also in mice [92,94]. At the moment, the strongest data are those concerning miR-206. Indeed, the mechanisms responsible for the increase of this miRNA seem to be conserved in the skeletal muscle of mouse models and in that from ALS patients, and the up-regulation described in both cases is an ALS-specific response to the denervation. miR-206 was found significantly up-regulated in muscle samples from ALS patients compared to control subjects [72,90,91,93], similarly to what observed in blood samples, strengthening the role of this miRNA as potential biomarker for ALS. Of note, miR-206 showed an increased trend in muscle biopsies from long-term survivor patients, even though below the statistically significance [90]. De Andrade and colleagues [72] reported that this miRNA was over-expressed both in plasma and skeletal muscle of patients with ALS, but the over-expression was not progressive during the follow-up. They supposed that miR-206 expression increased early in the disease course, reaches a plateau and then begins to fall. In agreement with this hypothesis, an up-regulation of miR-206 was described in muscle biopsies from ALS patients within one year from the clinical onset, becoming less evident as the disease progresses to a later stage [96]. Finally, Si and collaborators reported a non-significant upward trend in miR-206 in muscle samples from ALS patients compared to controls. This result, however, was correlated to a high standard error for this miRNA due to the variability among samples. Moreover, the authors reported a significant inverse correlation between this miRNA and the muscle power of the biopsied muscle, hypothesizing that it could be a marker of disease activity. This finding highlights the importance to associate miR-206 levels with muscle-specific clinical assessment rather than overall clinical status [94].

Although a concordant miRNA signature have not been identified yet in ALS patient muscle biopsies, these findings show that miRNAs could be useful prognostic markers to better understand the course of disease. In particular, the identification of a specific muscular miRNA profile through multicenter studies, able to increase the statistical power of the analysis, could lead to a stratification of the patients in order to identify prognostic biomarkers to use as indicator of disease progression, facilitating the clinical management of patients.

5. Conclusions and Future Perspectives

Despite the intense research activity of the last years, the use of miRNAs as biomarkers for diagnosis of ALS and clinical management of patients is still in an early stage of development. Several interesting data have been obtained so far, with important insights into the disease processes. However, results achieved in different studies are most of the time conflicting and poorly reproducible, making it difficult to unequivocally identify which miRNA(s) may be selected as biomarker in clinical practice. In order to overcome these limits, some improvements in the research approach should be taken into account.

First of all, one factor strongly complicating the comparison among data reported by different research groups is the wide range of methods used for the identification of potential miRNA biomarkers and the different techniques for miRNA measurement and data normalization. A common acceptance of certain guidelines, standard research protocols, and strong methods of statistical analysis of miRNAs will be important in the future to achieve reliable biomarkers.

Another critical issue is the relatively small number of patients included in the studies performed until now. Results are often interesting, but they need to be verified in larger cohorts of ALS patients. It would be really important that those miRNAs, which have shown initial promise, were validated in independent laboratories and/or in multicenter collaborations. In addition, since ALS is a highly heterogeneous disease, replication studies should increase the number of patients stratifying them based on clinical and genetic features, in order to obtain a better assessment of the potential associations among miRNAs and these variables.

Further, in several studies miRNA levels of ALS patients have been compared only to those of control subjects not affected by neurological disorders. This approach may bring to the identification of miRNAs able to successfully differentiate patients from healthy control subjects, but these miRNAs are often associated with common pathologic processes of neurodegeneration and are not specific for ALS. It will be of fundamental importance to extend the comparison to patients affected by other neurodegenerative diseases, in particular ALS-mimic disorders, to evaluate the specificity of deregulated miRNAs for ALS.

One of the more interesting approaches to miRNA biomarker identification is the use of a complex set of biomarkers, or combinations or ratios of biomarkers from different pathogenic pathways, rather than the employ of a single marker. This strategy has been shown to increase the sensitivity and/or specificity of potential ALS biomarkers and to contain more exhaustive diagnostic information, and should be more widely used in future researches.

At the same time, when possible, future studies should try to combine data obtained from multiple source of sample (blood, CSF, muscle) of the same patient. Up to date, only few studies have performed this kind of analysis, and their results are quite conflicting. However, an extensive analysis of correlations among different samples could be helpful to obtain more informative data and improve patients' stratification. In addition, for circulating miRNAs, it would be important to perform longitudinal studies on a large number of patients, in order to identify potential biomarkers of disease progression, and evaluate their role as prognostic indicators.

In conclusion, miRNAs constitute very promising biomarkers for ALS, but there is still much work to be done to validate and use them in clinical routine. The ultimate objective is to include these biomarkers in all phases of ALS management, from the diagnosis to the clinical trials, and, in perspective, to the identification of future therapeutic approaches.

Author Contributions: Writing—review and editing, C.R. and C.M.; supervision, S.B.

Funding: This work was funded by the Italian Ministry for Instruction, University and Research (FFARB 2017), fund-holder S.B.

Conflicts of Interest: The authors declare no conflict of interest.

References

1. Rowland, L.P.; Shneider, N.A. Amyotrophic lateral sclerosis. *N. Engl. J. Med.* **2001**, *344*, 1688–1700. [CrossRef] [PubMed]
2. Rothstein, J.D. Current hypotheses for the underlying biology of amyotrophic lateral sclerosis. *Ann. Neurol.* **2009**, *65*, S3–S9. [CrossRef] [PubMed]
3. Ilieva, H.; Polymenidou, M.; Cleveland, D.W. Non-cell autonomous toxicity in neurodegenerative disorders: ALS and beyond. *J. Cell Biol.* **2009**, *187*, 761–772. [CrossRef] [PubMed]
4. Wijesekera, L.C.; Leigh, P.N. Amyotrophic lateral sclerosis. *Orphanet J. Rare Dis.* **2009**, *4*, 3. [CrossRef] [PubMed]
5. Brown, R.H., Jr.; Al-Chalabi, A. Amyotrophic lateral sclerosis. *N. Engl. J. Med.* **2017**, *377*, 162–172. [CrossRef] [PubMed]
6. Chiò, A.; Logroscino, G.; Traynor, B.J.; Collins, J.; Simeone, J.C.; Goldstein, L.A.; White, L.A. Global epidemiology of amyotrophic lateral sclerosis: A systematic review of the published literature. *Neuroepidemiology* **2013**, *41*, 118–130. [CrossRef] [PubMed]
7. Logroscino, G.; Traynor, B.J.; Hardiman, O.; Chio', A.; Couratier, P.; Mitchell, J.D.; Swingler, R.J.; Beghi, E. Descriptive epidemiology of amyotrophic lateral sclerosis: New evidence and unsolved issues. *J. Neurol. Neurosurg. Psychiatry* **2008**, *79*, 6–11. [CrossRef] [PubMed]
8. Manjaly, Z.R.; Scott, K.M.; Abhinav, K.; Wijesekera, L.; Ganesalingam, J.; Goldstein, L.H.; Janssen, A.; Dougherty, A.; Willey, E.; Stanton, B.R.; et al. The sex ratio in amyotrophic lateral sclerosis: A population based study. *Amyotroph. Lateral Scler.* **2010**, *11*, 439–442. [CrossRef] [PubMed]
9. Sutedja, N.A.; Fischer, K.; Veldink, J.H.; van der Heijden, G.J.; Kromhout, H.; Heederik, D.; Huisman, M.H.; Wokke, J.J.; van den Berg, L.H. What we truly know about occupation as a risk factor for ALS: A critical and systematic review. *Amyotroph. Lateral Scler.* **2009**, *10*, 295–301. [CrossRef] [PubMed]
10. Andersen, P.M. Amyotrophic lateral sclerosis associated with mutations in the CuZn superoxide dismutase gene. *Curr. Neurol. Neurosci. Rep.* **2006**, *6*, 37–46. [CrossRef] [PubMed]
11. Ben Hamida, M.; Hentati, F.; Ben Hamida, C. Hereditary motor system diseases (chronic juvenile amyotrophic lateral sclerosis). Conditions combining a bilateral pyramidal syndrome with limb and bulbar amyotrophy. *Brain* **1990**, *113*, 347–363. [CrossRef] [PubMed]
12. Andersen, P.M.; Al-Chalabi, A. Clinical genetics of amyotrophic lateral sclerosis: What do we really know? *Nat. Rev. Neurol.* **2011**, *7*, 603–615. [CrossRef] [PubMed]
13. Bigio, E.H.; Weintraub, S.; Rademakers, R.; Baker, M.; Ahmadian, S.S.; Rademaker, A.; Weitner, B.B.; Mao, Q.; Lee, K.H.; Mishra, M.; Ganti, R.A.; Mesulam, M.M. Frontotemporal lobar degeneration with TDP-43 proteinopathy and chromosome 9p repeat expansion in C9ORF72: Clinicopathologic correlation. *Neuropathology* **2013**, *33*, 122–133. [CrossRef] [PubMed]
14. Rizzo, F.; Riboldi, G.; Salani, S.; Nizzardo, M.; Simone, C.; Corti, S.; Hedlund, E. Cellular therapy to target neuroinflammation in amyotrophic lateral sclerosis. *Cell. Mol. Life Sci.* **2014**, *71*, 999–1015. [CrossRef] [PubMed]
15. Volk, A.E.; Weishaupt, J.H.; Andersen, P.M.; Ludolph, A.C.; Kubisch, C. Current knowledge and recent insights into the genetic basis of amyotrophic lateral sclerosis. *Med. Genet.* **2018**, *30*, 252–258. [CrossRef] [PubMed]

16. Kawahara, Y.; Mieda-Sato, A. TDP-43 promotes microRNA biogenesis as a component of the drosha and dicer complexes. *Proc. Natl. Acad. Sci. USA* **2012**, *109*, 3347–3352. [CrossRef] [PubMed]

17. Lagier-Tourenne, C.; Polymenidou, M.; Cleveland, D.W. TDP-43 and FUS/TLS: Emerging roles in RNA processing and neurodegeneration. *Hum. Mol. Genet.* **2010**, *19*, R46–R64. [CrossRef] [PubMed]

18. Brooks, B.R.; Miller, R.G.; Swash, M.; Munsat, T.L.; World Federation of Neurology Research Group on Motor Neuron Diseases. El escorial revisited: Revised criteria for the diagnosis of amyotrophic lateral sclerosis. *Amyotroph. Lateral Scler. Other Motor. Neuron Disord.* **2000**, *1*, 293–299. [CrossRef] [PubMed]

19. Zoing, M.C.; Burke, D.; Pamphlett, R.; Kiernan, M.C. Riluzole therapy for motor neurone disease: An early Australian experience (1996–2002). *J. Clin. Neurosci.* **2006**, *13*, 78–83. [CrossRef] [PubMed]

20. Biomarkers Definitions Working Group. Biomarkers and surrogate endpoints: Preferred definitions and conceptual framework. *Clin. Pharmacol. Ther.* **2001**, *69*, 89–95. [CrossRef] [PubMed]

21. Robelin, L.; Gonzalez De Aguilar, J.L. Blood biomarkers for amyotrophic lateral sclerosis: Myth or reality? *Biomed. Res. Int.* **2014**, *2014*, 10. [CrossRef] [PubMed]

22. O'Brien, J.; Hayder, H.; Zayed, Y.; Peng, C. Overview of microrna biogenesis, mechanisms of actions, and circulation. *Front. Endocrinol. Lausanne* **2018**, *9*, 402. [CrossRef] [PubMed]

23. Ludwig, N.; Leidinger, P.; Becker, K.; Backes, C.; Fehlmann, T.; Pallasch, C.; Rheinheimer, S.; Meder, B.; Stähler, C.; Meese, A.; et al. Distribution of miRNA expression across human tissues. *Nucleic Acids Res.* **2016**, *44*, 3865–3877. [CrossRef] [PubMed]

24. Mocellin, S.; Pasquali, S.; Pilati, P. Oncomirs: From tumor biology to molecularly targeted anticancer strategies. *Mini Rev. Med. Chem.* **2009**, *9*, 70–80. [CrossRef] [PubMed]

25. Aumiller, V.; Förstemann, K. Roles of microRNAs beyond development—metabolism and neural plasticity. *Biochim. Biophys. Acta.* **2008**, *1779*, 692–696. [CrossRef] [PubMed]

26. Carissimi, C.; Fulci, V.; Macino, G. MicroRNAs: Novel regulators of immunity. *Autoimmun. Rev.* **2009**, *8*, 520–524. [CrossRef] [PubMed]

27. Bushati, N.; Cohen, S.M. MicroRNAs in neurodegeneration. *Curr. Opin. Neurobiol.* **2008**, *18*, 292–296. [CrossRef] [PubMed]

28. Rajgor, D. Macro roles for microRNAs in neurodegenerative diseases. *Noncoding RNA Res.* **2018**, *3*, 154–159. [CrossRef] [PubMed]

29. Rinchetti, P.; Rizzuti, M.; Faravelli, I.; Corti, S. MicroRNA Metabolism and dysregulation in amyotrophic lateral sclerosis. *Mol. Neurobiol.* **2018**, *55*, 2617–2630. [CrossRef] [PubMed]

30. Lu, J.; Getz, G.; Miska, E.A.; Alvarez-Saavedra, E.; Lamb, J.; Peck, D.; Sweet-Cordero, A.; Ebert, B.L.; Mak, R.H.; Ferrando, A.A.; et al. MicroRNA expression profiles classify human cancers. *Nature.* **2005**, *435*, 834–838. [CrossRef] [PubMed]

31. Weber, J.A.; Baxter, D.H.; Zhang, S.; Huang, D.Y.; Huang, K.H.; Lee, M.J.; Galas, D.J.; Wang, K. The microRNA spectrum in 12 body fluids. *Clin. Chem.* **2010**, *56*, 1733–1741. [CrossRef] [PubMed]

32. Mitchell, P.S.; Parkin, R.K.; Kroh, E.M.; Fritz, B.R.; Wyman, S.K.; Pogosova-Agadjanyan, E.L.; Peterson, A.; Noteboom, J.; O'Briant, K.C.; Allen, A.; et al. Circulating microRNAs as stable blood-based markers for cancer detection. *Proc. Natl. Acad. Sci. USA* **2008**, *105*, 10513–10518. [CrossRef] [PubMed]

33. Cortez, M.A.; Bueso-Ramos, C.; Ferdin, J.; Lopez-Berestein, G.; Sood, A.K.; Calin, G.A. MicroRNAs in body fluids-the mix of hormones and biomarkers. *Nat. Rev. Clin. Oncol.* **2011**, *8*, 467–477. [CrossRef] [PubMed]

34. Mo, M.H.; Chen, L.; Fu, Y.; Wang, W.; Fu, S.W. Cell-free circulating miRNA biomarkers in cancer. *J. Cancer* **2012**, *3*, 432–448. [CrossRef] [PubMed]

35. Rao, P.; Benito, E.; Fischer, A. MicroRNAs as biomarkers for CNS disease. *Front. Mol. Neurosci.* **2013**, *6*, 39. [CrossRef] [PubMed]

36. Haramati, S.; Chapnik, E.; Sztainberg, Y.; Eilam, R.; Zwang, R.; Gershoni, N.; McGlinn, E.; Heiser, P.W.; Wills, A.M.; Wirguin, I.; et al. MiRNA malfunction causes spinal motor neuron disease. *Proc. Natl. Acad. Sci. USA* **2010**, *107*, 13111–13116. [CrossRef] [PubMed]

37. Emde, A.; Eitan, C.; Liou, L.L.; Libby, R.T.; Rivkin, N.; Magen, I.; Reichenstein, I.; Oppenheim, H.; Eilam, R.; Silvestroni, A.; et al. Dysregulated miRNA biogenesis downstream of cellular stress and als-causing mutations: A new mechanism for ALS. *EMBO J.* **2015**, *34*, 2633–2651. [CrossRef] [PubMed]

38. Strong, M.J. The evidence for altered RNA metabolism in amyotrophic lateral sclerosis (ALS). *J. Neurol. Sci.* **2010**, *288*, 1–12. [CrossRef] [PubMed]

39. Morlando, M.; Dini Modigliani, S.; Torrelli, G.; Rosa, A.; Di Carlo, V.; Caffarelli, E.; Bozzoni, I. FUS stimulates microRNA biogenesis by facilitating co-transcriptional drosha recruitment. *EMBO J.* **2012**, *31*, 4502–4510. [CrossRef] [PubMed]

40. Li, W.; Ruan, K. MicroRNA detection by microarray. *Anal. Bioanal. Chem.* **2009**, *394*, 1117–1124. [CrossRef] [PubMed]

41. Motameny, S.; Wolters, S.; Nürnberg, P.; Schumacher, B. Next generation sequencing of miRNAs–strategies, resources and methods. *Genes* **2010**, *1*, 70–84. [CrossRef] [PubMed]

42. Friedländer, M.R.; Chen, W.; Adamidi, C.; Maaskola, J.; Einspanier, R.; Knespel, S.; Rajewsky, N. Discovering MicroRNAs from deep sequencing data using miRDeep. *Nat. Biotechnol.* **2008**, *26*, 407–415. [CrossRef] [PubMed]

43. Creighton, C.J.; Reid, J.G.; Gunaratne, P.H. Expression profiling of microRNAs by deep sequencing. *Brief. Bioinform.* **2009**, *10*, 490–497. [CrossRef] [PubMed]

44. Mestdagh, P.; Feys, T.; Bernard, N.; Guenther, S.; Chen, C.; Speleman, F.; Vandesompele, J. High-throughput stem-loop RT-qPCR miRNA expression profiling using minute amounts of input RNA. *Nucleic Acids Res.* **2008**, *36*, e143. [CrossRef] [PubMed]

45. Schmittgen, T.D.; Jiang, J.; Liu, Q.; Yang, L. A high-throughput method to monitor the expression of microRNA precursors. *Nucleic Acids Res.* **2004**, *32*, e43. [CrossRef] [PubMed]

46. Murphy, J.; Bustin, S.A. Reliability of real-time reverse-transcription PCR in clinical diagnostics: Gold standard or substandard? *Expert. Rev. Mol. Diagn.* **2009**, *9*, 187–197. [CrossRef] [PubMed]

47. De Planell-Saguer, M.; Rodicio, M.C. Detection methods for microRNAs in clinic practice. *Clin. Biochem.* **2013**, *46*, 869–878. [CrossRef] [PubMed]

48. Schwarzenbach, H.; da Silva, A.M.; Calin, G.; Pantel, K. Data normalization strategies for microRNA quantification. *Clin. Chem.* **2015**, *61*, 1333–1342. [CrossRef] [PubMed]

49. Peltier, H.J.; Latham, G.J. Normalization of microRNA expression levels in quantitative RT-PCR assays: Identification of suitable reference RNA targets in normal and cancerous human solid tissues. *RNA* **2008**, *14*, 844–852. [CrossRef] [PubMed]

50. Kroh, E.M.; Parkin, R.K.; Mitchell, P.S.; Tewari, M. Analysis of circulating microRNA biomarkers in plasma and serum using quantitative reverse transcription-PCR (qRT-PCR). *Methods* **2010**, *50*, 298–301. [CrossRef] [PubMed]

51. De Felice, B.; Guida, M.; Guida, M.; Coppola, C.; De Mieri, G.; Cotrufo, R. A miRNA signature in leukocytes from sporadic amyotrophic lateral sclerosis. *Gene* **2012**, *508*, 35–40. [CrossRef] [PubMed]

52. Cogswell, J.P.; Ward, J.; Taylor, I.A.; Waters, M.; Shi, Y.; Cannon, B.; Kelnar, K.; Kemppainen, J.; Brown, D.; Chen, C.; et al. Identification of miRNA changes in Alzheimer's disease brain and CSF yields putative biomarkers and insights into disease pathways. *J. Alzheimers Dis.* **2008**, *14*, 27–41. [CrossRef] [PubMed]

53. Alexandrov, P.N.; Dua, P.; Hill, J.M.; Bhattacharjee, S.; Zhao, Y.; Lukiw, W.J. microRNA (miRNA) speciation in Alzheimer's disease (AD) cerebrospinal fluid (CSF) and extracellular fluid (ECF). *Int. J. Biochem. Mol. Biol.* **2012**, *3*, 365–373. [PubMed]

54. Butovsky, O.; Siddiqui, S.; Gabriely, G.; Lanser, A.J.; Dake, B.; Murugaiyan, G.; Doykan, C.E.; Wu, P.M.; Gali, R.R.; Iyer, L.K.; et al. Modulating inflammatory monocytes with a unique microRNA gene signature ameliorates murine ALS. *J. Clin. Invest.* **2012**, *122*, 3063–3087. [CrossRef] [PubMed]

55. Freischmidt, A.; Müller, K.; Ludolph, A.C.; Weishaupt, J.H. Systemic dysregulation of TDP-43 binding microRNAs in amyotrophic lateral sclerosis. *Acta Neuropathol. Commun.* **2013**, *1*, 42. [CrossRef] [PubMed]

56. De Felice, B.; Annunziata, A.; Fiorentino, G.; Borra, M.; Biffali, E.; Coppola, C.; Cotrufo, R.; Brettschneider, J.; Giordana, M.L.; Dalmay, T.; et al. MiR-338-3p is over-expressed in blood, CFS, serum and spinal cord from sporadic amyotrophic lateral sclerosis patients. *Neurogenetics* **2014**, *15*, 243–253. [CrossRef] [PubMed]

57. Benigni, M.; Ricci, C.; Jones, A.R.; Giannini, F.; Al-Chalabi, A.; Battistini, S. Identification of miRNAs as potential biomarkers in cerebrospinal fluid from amyotrophic lateral sclerosis patients. *Neuromolecular Med.* **2016**, *18*, 551–560. [CrossRef] [PubMed]

58. Waller, R.; Wyles, M.; Heath, P.R.; Kazoka, M.; Wollff, H.; Shaw, P.J.; Kirby, J. Small RNA sequencing of sporadic amyotrophic lateral sclerosis cerebrospinal fluid reveals differentially expressed miRNAs related to neural and glial activity. *Front. Neurosci.* **2018**, *11*, 731. [CrossRef] [PubMed]

59. Paez-Colasante, X.; Figueroa-Romero, C.; Sakowski, S.A.; Goutman, S.A.; Feldman, E.L. Amyotrophic lateral sclerosis: mechanisms and therapeutics in the epigenomic era. *Nat. Rev. Neurol.* **2015**, *11*, 266–279. [CrossRef] [PubMed]

60. Hoye, M.L.; Koval, E.D.; Wegener, A.J.; Hyman, T.S.; Yang, C.; O'Brien, D.R.; Miller, R.L.; Cole, T.; Schoch, K.M.; Shen, T.; et al. MicroRNA profiling reveals marker of motor neuron disease in ALS models. *J. Neurosci.* **2017**, *37*, 5574–5586. [CrossRef] [PubMed]

61. Kos, A.; Olde Loohuis, N.F.; Wieczorek, M.L.; Glennon, J.C.; Martens, G.J.; Kolk, S.M.; Aschrafi, A. A potential regulatory role for intronic microRNA-338-3p for its host gene encoding apoptosis-associated tyrosine kinase. *PLoS ONE* **2012**, *7*, e31022. [CrossRef] [PubMed]

62. Chen, X.; Pan, M.; Han, L.; Lu, H.; Hao, X.; Dong, Q. miR-338-3p suppresses neuroblastoma proliferation, invasion and migration through targeting PREX2a. *FEBS Lett.* **2013**, *587*, 3729–3737. [CrossRef] [PubMed]

63. Conti, A.; Aguennouz, M.; La Torre, D.; Tomasello, C.; Cardali, S.; Angileri, F.F.; Maio, F.; Cama, A.; Germanò, A.; Vita, G.; et al. miR-21 and 221 upregulation and miR-181b downregulation in human grade II-IV astrocytic tumors. *J. Neurooncol.* **2009**, *93*, 325–332. [CrossRef] [PubMed]

64. Johanson, C.E.; Stopa, E.G.; McMillan, P.N. The blood-cerebrospinal fluid barrier: Structure and functional significance. *Methods Mol. Biol.* **2011**, *686*, 101–131. [CrossRef] [PubMed]

65. Spector, R.; Robert Snodgrass, S.; Johanson, C.E. A balanced view of the cerebrospinal fluid composition and functions: Focus on adult humans. *Exp. Neurol.* **2015**, *273*, 57–68. [CrossRef] [PubMed]

66. Vu, L.T.; Bowser, R. Fluid-based biomarkers for amyotrophic lateral sclerosis. *Neurotherapeutics* **2017**, *14*, 119–134. [CrossRef] [PubMed]

67. Toivonen, J.M.; Manzano, R.; Oliván, S.; Zaragoza, P.; García-Redondo, A.; Osta, R. MicroRNA-206: A potential circulating biomarker candidate for amyotrophic lateral sclerosis. *PLoS ONE* **2014**, *9*, e89065. [CrossRef] [PubMed]

68. Freischmidt, A.; Müller, K.; Zondler, L.; Weydt, P.; Volk, A.E.; Božič, A.L.; Walter, M.; Bonin, M.; Mayer, B.; von Arnim, C.A.; et al. Serum microRNAs in patients with genetic amyotrophic lateral sclerosis and pre-manifest mutation carriers. *Brain* **2014**, *137*, 2938–2950. [CrossRef] [PubMed]

69. Freischmidt, A.; Müller, K.; Zondler, L.; Weydt, P.; Mayer, B.; von Arnim, C.A.; Hübers, A.; Dorst, J.; Otto, M.; Holzmann, K.; et al. Serum microRNAs in sporadic amyotrophic lateral sclerosis. *Neurobiol. Aging* **2015**, *36*, 2660.e15–2660.e20. [CrossRef] [PubMed]

70. Takahashi, I.; Hama, Y.; Matsushima, M.; Hirotani, M.; Kano, T.; Hohzen, H.; Yabe, I.; Utsumi, J.; Sasaki, H. Identification of plasma microRNAs as a biomarker of sporadic amyotrophic lateral Sclerosis. *Mol. Brain* **2015**, *8*, 67. [CrossRef] [PubMed]

71. Chen, Y.; Wei, Q.; Chen, X.; Li, C.; Cao, B.; Ou, R.; Hadano, S.; Shang, H.F. Aberration of miRNAs expression in leukocytes from sporadic amyotrophic lateral sclerosis. *Front. Mol. Neurosci.* **2016**, *9*, 69. [CrossRef] [PubMed]

72. De Andrade, H.M.; de Albuquerque, M.; Avansini, S.H.; de S Rocha, C.; Dogini, D.B.; Nucci, A.; Carvalho, B.; Lopes-Cendes, I.; França, M.C., Jr. MicroRNAs-424 and 206 are potential prognostic markers in spinal onset amyotrophic lateral sclerosis. *J. Neurol. Sci.* **2016**, *368*, 19–24. [CrossRef] [PubMed]

73. Tasca, E.; Pegoraro, V.; Merico, A.; Angelini, C. circulating microRNAs as biomarkers of muscle differentiation and atrophy in ALS. *Clin. Neuropathol.* **2016**, *35*, 22–30. [CrossRef] [PubMed]

74. Sheinerman, K.S.; Toledo, J.B.; Tsivinsky, V.G.; Irwin, D.; Grossman, M.; Weintraub, D.; Hurtig, H.I.; Chen-Plotkin, A.; Wolk, D.A.; McCluskey, L.F.; et al. Circulating brain-enriched microRNAs as novel biomarkers for detection and differentiation of neurodegenerative diseases. *Alzheimers Res. Ther.* **2017**, *9*, 89. [CrossRef] [PubMed]

75. Raheja, R.; Regev, K.; Healy, B.C.; Mazzola, M.A.; Beynon, V.; Von Glehn, F.; Paul, A.; Diaz-Cruz, C.; Gholipour, T.; Glanz, B.I.; et al. Correlating serum micrornas and clinical parameters in amyotrophic lateral sclerosis. *Muscle Nerve* **2018**, *58*, 261–269. [CrossRef] [PubMed]

76. Waller, R.; Goodall, E.F.; Milo, M.; Cooper-Knock, J.; Da Costa, M.; Hobson, E.; Kazoka, M.; Wollff, H.; Heath, P.R.; Shaw, P.J. Serum miRNAs miR-206, 143–3p and 374b-5p as potential biomarkers for amyotrophic lateral sclerosis (ALS). *Neurobiol. Aging* **2017**, *55*, 123–131. [CrossRef] [PubMed]

77. Vrabec, K.; Boštjančič, E.; Koritnik, B.; Leonardis, L.; Dolenc Grošelj, L.; Zidar, J.; Rogelj, B.; Glavač, D.; Ravnik-Glavač, M. Differential expression of several miRNAs and the host genes AATK and DNM2 in leukocytes of sporadic ALS patients. *Front. Mol. Neurosci.* **2018**, *11*, 106. [CrossRef] [PubMed]

78. Matamala, J.M.; Arias-Carrasco, R.; Sanchez, C.; Uhrig, M.; Bargsted, L.; Matus, S.; Maracaja-Coutinho, V.; Abarzua, S.; van Zundert, B.; Verdugo, R. Genome-wide circulating microRNA expression profiling reveals potential biomarkers for amyotrophic lateral sclerosis. *Neurobiol. Aging* **2018**, *64*, 123–138. [CrossRef] [PubMed]

79. Xu, Q.; Zhao, Y.; Zhou, X.; Luan, J.; Cui, Y.; Han, J. Comparison of the extraction and determination of serum exosome and miRNA in serum and the detection of miR-27a-3p in serum exosome of ALS patients. *Intractable Rare Dis. Res.* **2018**, *7*, 13–18. [CrossRef] [PubMed]

80. Liguori, M.; Nuzziello, N.; Introna, A.; Consiglio, A.; Licciulli, F.; D'Errico, E.; Scarafino, A.; Distaso, E.; Simone, I.L. Dysregulation of MicroRNAs and target genes networks in peripheral blood of patients with sporadic amyotrophic lateral sclerosis. *Front. Mol. Neurosci.* **2018**, *11*, 288. [CrossRef] [PubMed]

81. Taguchi, Y.H.; Wang, H. Exploring microRNA biomarker for amyotrophic lateral sclerosis. *Int. J. Mol. Sci.* **2018**, *19*, 1318. [CrossRef] [PubMed]

82. Bang, C.; Thum, T. Exosomes: New players in cell-cell communication. *Int. J. Biochem. Cell. Biol.* **2012**, *44*, 2060–2064. [CrossRef] [PubMed]

83. Cui, Y.; Luan, J.; Li, H.; Zhou, X.; Han, J. Exosomes derived from mineralizing osteoblasts promote ST2 cell osteogenic differentiation by alteration of microRNA expression. *FEBS Lett.* **2016**, *590*, 185–192. [CrossRef] [PubMed]

84. Sharma, M.; Juvvuna, P.K.; Kukreti, H.; McFarlane, C. Mega roles of microRNAs in regulation of skeletal muscle health and disease. *Front. Physiol.* **2014**, *5*, 239. [CrossRef] [PubMed]

85. Ma, G.; Wang, Y.; Li, Y.; Cui, L.; Zhao, Y.; Zhao, B.; Li, K. MiR-206, a key modulator of skeletal muscle development and disease. *Int. J. Biol. Sci.* **2015**, *11*, 345–352. [CrossRef] [PubMed]

86. Williams, A.H.; Valdez, G.; Moresi, V.; Qi, X.; McAnally, J.; Elliott, J.L.; Bassel-Duby, R.; Sanes, J.R.; Olson, E.N. MicroRNA-206 delays ALS progression and promotes regeneration of neuromuscular synapses in mice. *Science* **2009**, *326*, 1549–1554. [CrossRef] [PubMed]

87. Loeffler, J.P.; Picchiarelli, G.; Dupuis, L.; Gonzalez De Aguilar, J.L. The role of skeletal muscle in amyotrophic lateral sclerosis. *Brain Pathol.* **2016**, *26*, 227–236. [CrossRef] [PubMed]

88. Horak, M.; Novak, J.; Bienertova-Vasku, J. Muscle-specific microRNAs in skeletal muscle development. *Dev. Biol.* **2016**, *410*, 1–13. [CrossRef] [PubMed]

89. Di Pietro, L.; Lattanzi, W.; Bernardini, C. Skeletal muscle MicroRNAs as key players in the pathogenesis of amyotrophic lateral sclerosis. *Int. J. Mol. Sci.* **2018**, *19*, 1534. [CrossRef] [PubMed]

90. Bruneteau, G.; Simonet, T.; Bauché, S.; Mandjee, N.; Malfatti, E.; Girard, E.; Tanguy, M.L.; Behin, A.; Khiami, F.; Sariali, E.; et al. Muscle histone deacetylase 4 upregulation in amyotrophic lateral sclerosis: potential role in reinnervation ability and disease progression. *Brain* **2013**, *136*, 2359–2368. [CrossRef] [PubMed]

91. Russell, A.P.; Wada, S.; Vergani, L.; Hock, M.B.; Lamon, S.; Léger, B.; Ushida, T.; Cartoni, R.; Wadley, G.D.; Hespel, P.; et al. Disruption of skeletal muscle mitochondrial network genes and miRNAs in amyotrophic lateral sclerosis. *Neurobiol. Dis.* **2013**, *49*, 107–117. [CrossRef] [PubMed]

92. Jensen, L.; Jørgensen, L.H.; Bech, R.D.; Frandsen, U.; Schrøder, H.D. Skeletal muscle remodelling as a function of disease progression in amyotrophic lateral sclerosis. *Biomed. Res. Int.* **2016**, *2016*, 5930621. [CrossRef] [PubMed]

93. Pegoraro, V.; Merico, A.; Angelini, C. Micro-RNAs in ALS muscle: Differences in gender, age at onset and disease duration. *J. Neurol. Sci.* **2017**, *380*, 58–63. [CrossRef] [PubMed]

94. Si, Y.; Cui, X.; Crossman, D.K.; Hao, J.; Kazamel, M.; Kwon, Y.; King, P.H. Muscle microRNA signatures as biomarkers of disease progression in amyotrophic lateral sclerosis. *Neurobiol. Dis.* **2018**, *114*, 85–94. [CrossRef] [PubMed]

95. Kovanda, A.; Leonardis, L.; Zidar, J.; Koritnik, B.; Dolenc-Groselj, L.; Ristic Kovacic, S.; Curk, T.; Rogelj, B. Differential expression of microRNAs and other small RNAs in muscle tissue of patients with ALS and healthy age-matched controls. *Sci. Rep.* **2018**, *8*, 5609. [CrossRef] [PubMed]

96. Di Pietro, L.; Baranzini, M.; Berardinelli, M.G.; Lattanzi, W.; Monforte, M.; Tasca, G.; Conte, A.; Logroscino, G.; Michetti, F.; Ricci, E.; et al. Potential therapeutic targets for ALS: MIR206, MIR208b and MIR499 are modulated during disease progression in the skeletal muscle of patients. *Sci. Rep.* **2017**, *7*, 9538. [CrossRef] [PubMed]

cells

MDPI

Article

Exploring MicroRNA Biomarkers for Parkinson's Disease from mRNA Expression Profiles

Y-h. Taguchi [1] and Hsiuying Wang [2],*

[1] Department of Physics, Chuo University, 1-13-27 Kasuga, Bunky-ku, Tokyo 112-8551, Japan;
     tag@granular.com
[2] Institute of Statistics, National Chiao Tung University, Hsinchu 30010, Taiwan
*   Correspondence: wang@stat.nctu.edu.tw; Tel.: +886-3-5712121(ext. 56813)

Received: 30 October 2018; Accepted: 4 December 2018; Published: 5 December 2018

Abstract: Parkinson's disease (PD) is a chronic, progressive neurodegenerative disease characterized by both motor and nonmotor features. The diagnose of PD is based on a review of patients' signs and symptoms, and neurological and physical examinations. So far, no tests have been devised that can conclusively diagnose PD. In this study, we explore both microRNA and gene biomarkers for PD. Microarray gene expression profiles for PD patients and healthy control are analyzed using a principal component analysis (PCA)-based unsupervised feature extraction (FE). 244 genes are selected to be potential gene biomarkers for PD. In addition, we implement these genes into Kyoto Encyclopedia of Genes and Genomes (KEGG) pathways, and find that the 15 microRNAs (miRNAs), hsa-miR-92a-3p, 16-5p, 615-3p, 877-3p, 100-5p, 320a, 877-5p, 23a-3p, 484, 23b-3p, 15a-5p, 324-3p, 19b-3p, 7b-5p and 505-3p, significantly target these 244 genes. These miRNAs are shown to be significantly related to PD. This reveals that both selected genes and miRNAs are potential biomarkers for PD.

Keywords: biomarker; gene; microRNA; parkinson's disease

1. Introduction

Parkinson's disease (PD), first described by Dr. James Parkinson in 1817, is a chronic, progressive neurodegenerative disease characterized by both motor and nonmotor features [1]. PD motor symptoms such as shaking, rigidity, and slowness of movement are caused by the loss of striatal dopaminergic neurons [2]. The nonmotor symptoms of PD include sleep disorders, depression, and cognitive changes [3–5].

The incidence and prevalence of PD increase with age. So far, it is problematic to conclusively diagnose PD because of the lack of a reference standard test [6]. The diagnose of PD is based on a review of patients' signs and symptoms, and neurological and physical examinations. Resting tremor, cogwheel rigidity, and bradykinesia are three "cardinal signs" of PD, and postural instability, a late finding in PD, is the fourth cardinal sign of PD [6].

Owing to the lack of a standard test for diagnosing PD, genetic testing of mutations in disease-causing genes may be one of a helpful way to diagnose familial Parkinson's disease (fPD) and sporadic Parkinson's disease (sPD). A number of PD disease-causing genes have been discovered and debated for both physicians and patients regarding diagnostic and presymptomatic genetic testing of PD in the clinic [7].

A common form of monogenic PD with dominant inheritance is caused by mutations in the gene for leucine-rich repeat kinase 2 (LRRK2) [8,9]; H-Synuclein (SNCA), which is a presynaptic neuronal protein, is linked genetically and neuropathologically to PD [10]; mutations in Parkin are the second most common known cause of PD [11]; mutations in DJ-1 and PTEN Kinase 1 (PINK1) can cause PD [11]; there is a significant inverse association of the ubiquitin carboxy-terminal hydrolase L1

(UCHL1) S18Y variant with PD [12]; and variants in glucocerebrosidase (GBA) and SNCA influence PD risk [13].

In addition to neurological and physical examinations and genetic testing, the gene expression differences between PD and healthy controls can be used as a potential prognosis of PD. Several studies have explored gene biomarkers for PD [14,15]. In this study, we explore gene and microRNA (miRNA) biomarkers for PD.

2. Materials and Methods

2.1. Gene Expression

We downloaded three mRNA expression profiles of PD and healthy control from Gene Expression Omnibus (GEO) under the GEO ID, GSE20295, GSE20163, and GSE20164. For all three profiles, raw data files (CEL files) (Supplementary file S1) were read with the function ReadAffy of the package Affy (Affymetrix; Thermo Fisher Scientific, Inc., Waltham, MA, USA) in R. Loaded CEL files were treated by mas5 function in the affy package. Then, write.exprs function was used to output normalized mRNA expression profiles. The tissue of the data was substantia nigra (see Supplementary file S1). Integrating the three datasets, we had 35 normal control and 25 PD patients' substantia nigra mRNA expression profiles.

2.2. Principal Component Analysis Based Unsupervised Feature Extraction

We applied principal component analysis (PCA)-based unsupervised feature extraction (FE) to mRNA expression profiles in order to select mRNAs that were expressed distinctly between controls and PD patients (see Supplementary file S2 for more details). This method has been successful in identifying potential biomarkers for other neurological disorders [16,17]. PCA was applied to only mRNAs identified by PCA-based unsupervised feature extraction (FE). PC loadings and PC scores were attributed to samples and genes, respectively. Furthermore, PC loadings are associated with the distinction between controls and PD patients; PD patients and controls were differentiated using linear discriminant analysis (LDA) with these selected PC loadings (see Supplementary file S2 for more details).

2.3. Validation of Obtained mRNAs

In order to validate the obtained mRNAs, we computed the area under the curve (AUC) of the receiver operating characteristic curve (ROC). Gene symbols associated with mRNAs selected by PCA-based unsupervised FE were uploaded to Enrichr [18,19], and various enriched biological terms were identified (see Supplementary file S4 for more details).

3. Results

255 probes (Supplementary file S3) were identified using PCA-based unsupervised FE. PCA was applied to 255 probes, and PC loadings attributed to samples were computed. The fourth PC loadings turned out to be associated with distinction between PD patients and controls. Table 1 shows the confusion table obtained by the linear discriminant analysis (LDA) using the fourth PC loading. Sensitivity of PD is 0.88, precision of PD is 0.73, F1 score (F-measure) is 0.8, and accuracy is 0.80. AUC is 0.95. Odds ratio is 20.5. p value computed by Fisher's exact test is 2.5×10^{-6}. All of these evaluations suggest that the fourth PC loadings can significantly discriminate controls and PD patients.

Although we found 255 probes that successfully discriminate PD patients from controls, biological evaluation of obtained probes is important. Two hundred and forty four gene symbols corresponding to these 255 probes (See Supplementary Materials) were uploaded to Enrichr for biological evaluation. There turned out to be many biological terms enriched in these gene symbols. Table 2 shows the top-ranked five categories in "Disease Perturbations from GEO down" of Enrichr (full list is available in Supplementary Material S4). The first four among these five categories are the PD. Since the GEO datasets used in Enrichr are not the same datasets used in our study, PCA based unsupervised FE successfully identified genes downregulated in PD patients.

Table 1. Confusion table obtained by linear discriminant analysis (LDA) using the fourth principal component (PC) loading obtained by principal component analysis (PCA) using 255 mRNAs identified by PCA-based unsupervised feature extraction (FE). Rows: true, columns: prediction.

	Control	PD
Control	24	8
PD	3	22

Table 2. Top ranked five categories in "Disease Perturbations from Gene Expression Omnibus (GEO) down" of Enrichr.

Term	Overlap	p-Value	Adjusted p-Value
Parkinson's disease DOID-14330 human GSE19587 sample 740	65/207	5.02×10^{-83}	4.18×10^{-80}
Parkinson's disease DOID-14330 human GSE19587 sample 1080	56/167	5.88×10^{-73}	1.60×10^{-70}
Parkinson's disease DOID-14330 human GSE19587 sample 496	73/361	3.90×10^{-78}	1.59×10^{-75}
Parkinson's disease DOID-14330 human GSE7621 sample 940	67/365	2.96×10^{-68}	6.06×10^{-66}
Dystonia C0393593 human GSE3064 sample 329	62/317	1.06×10^{-64}	1.74×10^{-62}

As for "Disease Perturbations from GEO up" of Enrichr, PD is less enriched, but there are still seven PD experiments with a significant p-value (Table 3, full list in Supplementary Material S4).

Table 3. Seven PD expression profiles in "Disease Perturbations from GEO up" where 244 gene symbols are enriched.

Term	Overlap	p-Value	Adjusted p-value
Parkinson's disease DOID-14330 human GSE19587 sample 741	33/158	5.14×10^{-35}	1.22×10^{-33}
Parkinson's disease DOID-14330 human GSE7621 sample 940	35/235	1.05×10^{-31}	1.74×10^{-30}
Parkinson's disease DOID-14330 human GSE7621 sample 941	38/342	1.55×10^{-29}	2.19×10^{-28}
Parkinson's disease DOID-14330 human GSE19587 sample 1080	37/433	1.17×10^{-24}	8.78×10^{-24}
Parkinson's disease DOID-14330 human GSE6613 sample 788	26/274	1.24×10^{-18}	5.50×10^{-18}
Parkinson's disease DOID-14330 human GSE19587 sample 496	15/239	1.03×10^{-8}	2.15×10^{-8}

The next biological term investigated is KEGG pathway (Table 4). Although PD was not top ranked, it is the eighth most significant enriched KEGG pathway. Tables 1–4 suggest that the identified 244 genes are significantly related to PD.

Table 4. Top ranked 10 KEGG pathways enriched in 244 identified gene symbols.

Term	Overlap	p-Value	Adjusted p-Value
Ribosome_Homo sapiens_hsa03010	28/137	1.68×10^{-29}	2.92×10^{-27}
Phagosome_Homo sapiens_hsa04145	16/154	1.72×10^{-12}	1.49×10^{-10}
Synaptic vesicle cycle_Homo sapiens_hsa04721	10/63	4.11×10^{-10}	2.38×10^{-8}
Pathogenic Escherichia coli infection_Homo sapiens_hsa05130	9/55	2.38×10^{-9}	1.04×10^{-7}
Gap junction_Homo sapiens_hsa04540	10/88	1.18×10^{-8}	4.11×10^{-7}
Mineral absorption_Homo sapiens_hsa04978	8/51	2.68×10^{-8}	7.76×10^{-7}
Oxidative phosphorylation_Homo sapiens_hsa00190	10/133	6.09×10^{-7}	1.51×10^{-5}
Parkinson's disease_Homo sapiens_hsa05012	10/142	1.11×10^{-6}	2.42×10^{-5}
Vibrio cholerae infection_Homo sapiens_hsa05110	6/51	8.84×10^{-6}	1.71×10^{-4}
GABAergic synapse_Homo sapiens_hsa04727	7/88	2.15×10^{-5}	3.73×10^{-4}

In this regard, we found that 15 miRNAs significantly target 244 gene symbols by checking "miRTarBase 2017" in Enrichr (Table 5); 113 gene symbols out of 244 gene symbols were targeted by either of 15 miRNAs. These 15 miRNAs were reported to be related to PD (Table 5). Thus, we should consider these 15 miRNAs as key regulators of PD instead of 244 gene symbols. In reality, miRNAs are generally suggested as key regulator of PD [20].

Table 5. "miRTarBase 2017" in Enrichr when 244 gene symbols were uploaded.

Term	Overlap	p-Value	Adjusted p-Value	Reference
hsa-miR-92a-3p	37/1404	1.41×10^{-8}	2.71×10^{-5}	[21]
hsa-miR-16-5p	37/1555	1.93×10^{-7}	1.85×10^{-4}	[22]
hsa-miR-615-3p	25/891	1.38×10^{-6}	8.85×10^{-4}	[23]
hsa-miR-877-3p	19/606	5.92×10^{-6}	2.28×10^{-3}	[24]
hsa-miR-100-5p	12/250	5.37×10^{-6}	2.28×10^{-3}	[25]
hsa-miR-320a	18/584	1.33×10^{-5}	4.25×10^{-3}	[26]
hsa-miR-877-5p	11/235	1.68×10^{-5}	4.63×10^{-3}	[24]
hsa-miR-23a-3p	11/249	2.88×10^{-5}	6.91×10^{-3}	[25]
hsa-miR-484	22/890	4.37×10^{-5}	9.33×10^{-3}	[25]
hsa-miR-23b-3p	12/322	6.55×10^{-5}	1.26×10^{-2}	[27]
mmu-miR-15a-5p	15/499	9.42×10^{-5}	1.65×10^{-2}	[25]
hsa-miR-324-3p	12/338	1.04×10^{-4}	1.66×10^{-2}	[28]
mmu-miR-19b-3p	11/310	2.03×10^{-4}	3.00×10^{-2}	[20]
mmu-miR-7b-5p	13/438	3.13×10^{-4}	4.02×10^{-2}	[20]
hsa-miR-505-3p	9/222	2.93×10^{-4}	4.02×10^{-2}	[29]

4. Discussion

More Brain Synapse-Related Biological Terms Are Enriched

Although in the previous section, we notice that downregulated genes and upregulated genes in PD GEO datasets and in PD KEGG pathway are enriched in the 244 identified gene symbols, more detailed analyses give us more convincing insight about the relationship between these 244 gene symbols and PD. Other than PD, some KEGG pathways in Table 4 are also related to PD. For example, as for ribosome that is top-ranked in Table 4, ribosomal protein s15 phosphorylation was reported to mediate LRRK2 neurodegeneration in Parkinson's disease [30] As for phagosome that was the second top-ranked, LRRK2 was reported to be a negative regulator of Mycobacterium tuberculosis phagosome maturation in macrophages [31]. As for synaptic vesicle cycle that was the third top-ranked, synaptic vesicle trafficking was reported to be related to Parkinson's disease [32]. These results suggest that 244 gene symbols are expected to be closely related to PD progression mechanisms.

Although LRRK2 itself was not included in 244 gene symbols, significant number of genes obtained from "Single Gene Perturbations from GEO up/down" affected by LRKK2 KO/KI (knocked out/knocked in) were included in 244 gene symbols (Tables 6 and 7). This also supports that 244 gene symbols are related to mechanisms of PD.

Other than PD specificity, it is important to check if 244 genes are specifically upregulated in brain/synapse, since, otherwise, genes identified are not convincing. Tables 8 and 9 show that 244 genes are highly brain tissue-specific, while Table 10 shows that 244 genes are overlapped with genes downregulated in other tissues than brain (full list is in Supporting Materials).

Table 6. Top ranked four LRRK2 KO/KI experiments in "Single Gene Perturbations from GEO up".

Name	Overlap	p-Value	Adjusted p-Value
LRRK2 Gly2019Ser (G2019S) mutation knockin human GSE36321 sample 1688	21/335	1.03×10^{-11}	4.82×10^{-11}
LRRK2 mutant human GSE33298 sample 2039	16/309	4.94×10^{-8}	1.55×10^{-7}
LRRK2 dominant negative mutation-G2019S homozygous human GSE33298 sample 1743	12/280	1.68×10^{-5}	4.14×10^{-5}
LRRK2 dominant negative mutation-G2019S homozygous human GSE33298 sample 1741	12/337	1.01×10^{-4}	2.33×10^{-4}

Table 7. Top ranked two LRRK2 KO/KI experiments in "Single Gene Perturbations from GEO down".

Name	Overlap	p-Value	Adjusted p-Value
LRRK2 Gly2019Ser (G2019S) mutation knockin human GSE36321 sample 1688	24/265	8.38×10^{-17}	9.78×10^{-16}
LRRK2 dominant negative mutation-G2019S heterozygous human GSE33298 sample 1739	9/282	1.61×10^{-3}	3.56×10^{-3}

Table 8. Top ranked five gene expression profiles in drug matrix whose altered genes are associated with 244 gene symbols.

Term	Overlap	p-Value	Adjusted p-Value
Oxcarbazepine-1600-mg/kg-in_CMC-Rat-Brain-3d-dn	31/369	1.66×10^{-20}	1.31×10^{-16}
Carbachol-15-mg/kg_in_Water-Rat-Brain-3d-up	26/318	4.96×10^{-17}	7.82×10^{-14}
Piracetam-2500_mg/kg_in_CMC-Rat-Brain-5d-up	27/325	7.89×10^{-18}	2.56×10^{-14}
Theophylline-225_mg/kg_in_Water-Rat-Brain-3d-dn	25/314	3.87×10^{-16}	5.08×10^{-13}
Tramadol-114_mg/kg_in_Water-Rat-Brain-5d-dn	26/315	3.93×10^{-17}	7.75×10^{-14}

Table 9. Top ranked five gene expression profiles in "Genotype-Tissue Expression (GTEx) Tissue Sample Gene Expression Profiles up" whose altered genes are associated with 244 gene symbols.

Term	Overlap	p-Value	Adjusted p-Value
GTEX-X585-0011-R2B-SM-46MVF_brain_male_50-59_years	81/1895	1.63×10^{-33}	4.72×10^{-30}
GTEX-WHSE-0011-R2A-SM-3P5ZL_brain_male_20-29_years	71/1660	7.67×10^{-29}	1.11×10^{-25}
GTEX-X261-0011-R8A-SM-4E3I5_brain_male_50-59_years	70/1878	8.35×10^{-25}	8.08×10^{-22}
GTEX-N7MT-0011-R10A-SM-2I3E1_brain_female_60-69_years	70/1918	2.88×10^{-24}	2.09×10^{-21}
GTEX-TSE9-0011-R8A-SM-3DB7R_brain_female_60-69_years	62/1548	2.52×10^{-23}	1.47×10^{-20}

Table 10. Top ranked five gene expression profiles in "GTEx Tissue Sample Gene Expression Profiles down" whose altered genes are associated with 244 gene symbols.

Term	Overlap	p-Value	Adjusted p-Value
GTEX-S4Q7-1226-SM-4AD5I_testis_male_20-29_years	22/329	8.84×10^{-13}	2.36×10^{-9}
GTEX-U4B1-1526-SM-4DXSL_testis_male_40-49_years	20/282	3.48×10^{-12}	3.09×10^{-9}
GTEX-UPK5-1426-SM-4JBHH_liver_male_40-49_years	79/3879	2.06×10^{-12}	2.74×10^{-9}
GTEX-OHPM-2126-SM-3LK75_testis_male_50-59_years	26/525	6.48×10^{-12}	4.07×10^{-9}
GTEX-S7PM-0626-SM-4AD4Q_testis_male_60-69_years	34/911	7.63×10^{-12}	4.07×10^{-9}

Thus, we can conclude that 244 gene symbols are not only PD-specific but also brain tissue-specific. This also suggests that we successfully identified gene-related PD mechanisms. In spite of successfully identifications, it is not very easy to investigate as many as 244 gene symbols one by one. It is better to find a more limited number of factors that regulate 244 gene symbols.

For the selected 15 miRNAs in Table 5, we confirmed our results by comparing other studies. hsa-miR-92a appeared as novel hub miR in both regulatory and co-expression network, indicating its strong functional role in PD; GBA deficiency is associated with PD, and miR-16-5p has been shown to correspond to enhanced GBA protein levels [22,33]; and both PD and HD (Huntington's disease) are neurodegenerative and caused by protein inclusions. miR-615-3p was identified as differentially expressed in HD prefrontal cortex compared to non-neurological disease controls, and hsa-miR-615-3p was identified up-regulated in HD [23]; miR-100, miR-23a, and miR484 were identified to be PD-related miRNAs [25]; miR-320a was identified as a PD-related miRNA [26]; tumor necrosis factor-alpha (TNF-α), a pro-inflammatory cytokine, was elevated in blood, CSF, and striatum regions of the brain in PD patients, and hsa-miR-23a, hsa-miR-23b, and hsa-miR-320a significantly decreased in the presence of TNF-α [27]; mmu-miR-15a-5p, mmu-miR-19b-3p, and mmu-miR-7b-5p miR-7 were shown to downregulate the inflammatory response in cellular in vitro and/or in vivo PD neurotoxic models, and miR-19b was shown to be downregulated in the prodromal stage of alpha-synucleinopathies and pinpointed this miRNA as a potential biomarker also for PD and DLB [20]; and miR-505-3p was identified as a PD-predictive miRNA by microarrays [29].

5. Conclusions

PD is a long-term degenerative disorder that usually affects elderly people. Since there are no standard tests that can conclusively diagnose PD, biological biomarkers can help in early diagnosis. In this study, PD-related mRNAs are selected using a PCA-based, unsupervised FE method, and miRNA biomarkers of PD are explored based on these selected mRNAs. Two hundred and forty-four genes and 15 miRNAs are identified to be related to PD. Biological evidence shows the selected genes and miRNAs are potential PD biomarkers. However, since the tissues used in this study are substantia nigra from postmortem brain, it needs to be further verified whether these selected biomarkers can be used as blood/serum PD biomarkers.

Supplementary Materials: The following are available online at http://www.mdpi.com/2073-4409/7/12/245/s1, File S1: samples for PD and control, File S2: Mathematical details of PCA based unsupervised FE, File S3: the selected genes, File S4: "miRTarBase 2017" in Enrichr when 244 gene symbols were uploaded

Author Contributions: Conceptualization, H.W.; Methodology, Y-h.T.; Formal Analysis, Y-h.T.; Writing, Y-h.T. and H.W.

Funding: This research was funded by Japan Society for the Promotion of Science under the grant number KAKENHI 17K00417 and the Ministry of Science and Technology 107-2118-M-009-002-MY2, Taiwan.

Conflicts of Interest: The authors declare no conflict of interest.

References

1. DeMaagd, G.; Philip, A. Parkinson's disease and its management: Part 1: Disease entity, risk factors, pathophysiology, clinical presentation, and diagnosis. *Pharm. Ther.* **2015**, *40*, 504.
2. Drui, G.; Carnicella, S.; Carcenac, C.; Favier, M.; Bertrand, A.; Boulet, S.; Savasta, M. Loss of dopaminergic nigrostriatal neurons accounts for the motivational and affective deficits in Parkinson's disease. *Mol. Psychiatry* **2014**, *19*, 358–367. [CrossRef] [PubMed]
3. Neikrug, A.B.; Maglione, J.E.; Liu, L.; Natarajan, L.; Avanzino, J.A.; Corey-Bloom, J.; Palmer, B.W.; Loredo, J.S.; Ancoli-Israel, S. Effects of sleep disorders on the non-motor symptoms of Parkinson disease. *J. Clin. Sleep Med.* **2013**, *9*, 1119–1129. [CrossRef] [PubMed]
4. Marsh, L. Depression and Parkinson's disease: Current knowledge. *Curr. Neurol. Neurosci. Rep.* **2013**, *13*, 409. [CrossRef] [PubMed]
5. Watson, G.S.; Leverenz, J.B. Profile of cognitive impairment in Parkinson's disease. *Brain Pathol.* **2010**, *20*, 640–645. [CrossRef] [PubMed]
6. Levine, C.B.; Fahrbach, K.R.; Siderowf, A.D.; Estok, R.P.; Ludensky, V.M.; Ross, S.D. Diagnosis and treatment of Parkinson's disease: A systematic review of the literature. *Evid. Rep. Technol. Assess.* **2003**, *57*, 1–4.
7. Tan, E.-K.; Jankovic, J. Genetic testing in Parkinson disease: Promises and pitfalls. *Arch. Neurol.* **2006**, *63*, 1232–1237. [CrossRef]

8. Zimprich, A.; Biskup, S.; Leitner, P.; Lichtner, P.; Farrer, M.; Lincoln, S.; Kachergus, J.; Hulihan, M.; Uitti, R.J.; Calne, D.B.; et al. Mutations in LRRK2 cause autosomal-dominant parkinsonism with pleomorphic pathology. *Neuron* **2004**, *44*, 601–607. [CrossRef]

9. Gasser, T. Usefulness of genetic testing in PD and PD trials: A balanced review. *J. Park. Dis.* **2015**, *5*, 209–215. [CrossRef]

10. Stefanis, L. α-Synuclein in Parkinson's disease. *Cold Spring Harbor Perspect. Med.* **2012**, *2*, a009399. [CrossRef]

11. Dawson, T.M.; Dawson, V.L. The role of parkin in familial and sporadic Parkinson's disease. *Mov. Disord.* **2010**, *25*, S32–S39. [CrossRef] [PubMed]

12. Tan, E.K.; Puong, K.Y.; Fook-Chong, S.; Chua, E.; Shen, H.; Yuen, Y.; Pavanni, R.; Wong, M.-C.; Puvan, K.; Zhao, Y. Case-control study of UCHL1 S18Y variant in Parkinson's disease. *Mov. Disord.* **2010**, *25*, S32–S39. [CrossRef] [PubMed]

13. Davis, A.A.; Andruska, K.M.; Benitez, B.A.; Racette, B.A.; Perlmutter, J.S.; Cruchaga, C. Variants in GBA, SNCA, and MAPT influence Parkinson disease risk, age at onset, and progression. *Neurobiol. Aging* **2016**, *37*, 209. e1–209. e7. [CrossRef] [PubMed]

14. Pinho, R.; Guedes, L.C.; Soreq, L.; Lobo, P.P.; Mestre, T.; Coelho, M.; Rosa, M.M.; Gonçalves, N.; Wales, P.; Mendes, T. Gene expression differences in peripheral blood of Parkinson's disease patients with distinct progression profiles. *PLoS ONE* **2016**, *11*, e0157852. [CrossRef]

15. Aguiar, P.M.C.; Severino, P. Biomarkers in Parkinson Disease: global gene expression analysis in peripheral blood from patients with and without mutations in PARK2 and PARK8. *Einstein* **2010**, *8*, 291–297. [CrossRef] [PubMed]

16. Taguchi, Y.H.; Wang, H. Exploring microRNA Biomarker for Amyotrophic Lateral Sclerosis. *Int. J. Mol. Sci.* **2018**, *19*, 1318. [CrossRef] [PubMed]

17. Taguchi, Y.H.; Wang, H. Genetic association between amyotrophic lateral sclerosis and cancer. *Genes* **2017**, *8*, 243. [CrossRef]

18. Kuleshov, M.V.; Jones, M.R.; Rouillard, A.D.; Fernandez, N.F.; Duan, Q.; Wang, Z.; Koplev, S.; Jenkins, S.L.; Jagodnik, K.M.; Lachmann, A.; et al. Enrichr: A comprehensive gene set enrichment analysis web server 2016 update. *Nucleic Acids Res.* **2016**, *44*, W90–W97. [CrossRef]

19. Chen, E.Y.; Tan, C.M.; Kou, Y.; Duan, Q.; Wang, Z.; Meirelles, G.V.; Clark, N.R.; Ma'ayan, A. Enrichr: Interactive and collaborative HTML5 gene list enrichment analysis tool. *BMC Bioinf.* **2013**, *14*, 128. [CrossRef]

20. Leggio, L.; Vivarelli, S.; L'Episcopo, F.; Tirolo, C.; Caniglia, S.; Testa, N.; Marchetti, B.; Iraci, N. microRNAs in Parkinson's disease: From pathogenesis to novel diagnostic and therapeutic approaches. *Int. J. Mol. Sci.* **2017**, *18*, 2698. [CrossRef]

21. Chatterjee, P.; Bhattacharyya, M.; Bandyopadhyay, S.; Roy, D. Studying the system-level involvement of microRNAs in Parkinson's disease. *PLoS ONE* **2014**, *9*, e93751. [CrossRef] [PubMed]

22. Hoss, A.G.; Labadorf, A.; Beach, T.G.; Latourelle, J.C.; Myers, R.H. microRNA profiles in Parkinson's disease prefrontal cortex. *Front. Aging Neurosci.* **2016**, *8*, 36. [CrossRef] [PubMed]

23. Hoss, A.G.; Kartha, V.K.; Dong, X.; Latourelle, J.C.; Dumitriu, A.; Hadzi, T.C.; Macdonald, M.E.; Gusella, J.F.; Akbarian, S.; Chen, J.F.; et al. MicroRNAs located in the Hox gene clusters are implicated in huntington's disease pathogenesis. *PLoS Genet.* **2014**, *10*, e1004188. [CrossRef] [PubMed]

24. Sibley, C.R.; Seow, Y.; Curtis, H.; Weinberg, M.S.; Wood, M.J. Silencing of Parkinson's disease-associated genes with artificial mirtron mimics of miR-1224. *Nucleic Acids Res.* **2012**, *40*, 9863–9875. [CrossRef] [PubMed]

25. Chen, L.; Yang, J.; Lu, J.; Cao, S.; Zhao, Q.; Yu, Z. Identification of aberrant circulating miRNAs in Parkinson's disease plasma samples. *Brain Behav.* **2018**, *8*, e00941. [CrossRef]

26. Heman-Ackah, S.M.; Hallegger, M.; Rao, M.S.; Wood, M.J. RISC in PD: The impact of microRNAs in Parkinson's disease cellular and molecular pathogenesis. *Front. Mol. Neurosci.* **2013**, *6*, 40. [CrossRef]

27. Prajapati, P.; Sripada, L.; Singh, K.; Bhatelia, K.; Singh, R.; Singh, R. TNF-α regulates miRNA targeting mitochondrial complex-I and induces cell death in dopaminergic cells. *Biochim. Biophys. Acta Mol. Basis Dis.* **2015**, *1852*, 451–461. [CrossRef]

28. Vallelunga, A.; Ragusa, M.; Di Mauro, S.; Iannitti, T.; Pilleri, M.; Biundo, R.; Weis, L.; Di Pietro, C.; De Iuliis, A.; Nicoletti, A. Identification of circulating microRNAs for the differential diagnosis of Parkinson's disease and Multiple System Atrophy. *Front. Cell. Neurosci.* **2014**, *8*, 156. [CrossRef]

29. Khoo, S.K.; Petillo, D.; Kang, U.J.; Resau, J.H.; Berryhill, B.; Linder, J.; Forsgren, L.; Neuman, L.A.; Tan, A.C. Plasma-based circulating MicroRNA biomarkers for Parkinson's disease. *J. Park. Dis.* **2012**, *2*, 321–331.

30. Martin, I.; Kim, J.W.; Lee, B.D.; Kang, H.C.; Xu, J.C.; Jia, H.; Stankowski, J.; Kim, M.S.; Zhong, J.; Kumar, M.; et al. Ribosomal protein s15 phosphorylation mediates LRRK2 neurodegeneration in Parkinson's disease. *Cell* **2014**, *157*, 472–485. [CrossRef]

31. Härtlova, A.; Herbst, S.; Peltier, J.; Rodgers, A.; Bilkei-Gorzo, O.; Fearns, A.; Dill, B.D.; Lee, H.; Flynn, R.; Cowley, S.A. LRRK2 is a negative regulator of Mycobacterium tuberculosis phagosome maturation in macrophages. *EMBO J.* **2018**, *37*, e98694. [CrossRef] [PubMed]

32. Esposito, G.; Clara, F.A.; Verstreken, P. Synaptic vesicle trafficking and Parkinson's disease. *Dev. Neurobiol.* **2012**, *72*, 134–144. [CrossRef] [PubMed]

33. Siebert, M.; Westbroek, W.; Chen, Y.-C.; Moaven, N.; Li, Y.; Velayati, A.; Saraiva-Pereira, M.L.; Martin, S.E.; Sidransky, E. Identification of miRNAs that modulate glucocerebrosidase activity in Gaucher disease cells. *RNA Biol.* **2014**, *11*, 1291–1300. [CrossRef] [PubMed]

Article

Inferring Novel Autophagy Regulators Based on Transcription Factors and Non-Coding RNAs Coordinated Regulatory Network

Shuyuan Wang [1,†], Wencan Wang [1,†], Qianqian Meng [1], Shunheng Zhou [2], Haizhou Liu [2], Xueyan Ma [1], Xu Zhou [1], Hui Liu [1], Xiaowen Chen [1,*] and Wei Jiang [2,*]

[1] College of Bioinformatics Science and Technology, Harbin Medical University, Harbin 150081, China; bioccwsy@163.com (S.W.); wangwencan1314@163.com (W.W.); mqq1992hmu@163.com (Q.M.); 18345550297@163.com (X.M.); biomathzx@163.com (X.Z.); liuhui870320@163.com (H.L.)

[2] College of Automation Engineering, Nanjing University of Aeronautics and Astronautics, Nanjing 211106, China; zhoushunheng@163.com (S.Z.); liuhaizhou2015@126.com (H.L.)

* Correspondence: hrbmucxw@163.com (X.C.); weijiang@nuaa.edu.cn (W.J.)

† These authors contributed equally to this work.

Received: 8 October 2018; Accepted: 30 October 2018; Published: 2 November 2018

Abstract: Autophagy is a complex cellular digestion process involving multiple regulators. Compared to post-translational autophagy regulators, limited information is now available about transcriptional and post-transcriptional regulators such as transcription factors (TFs) and non-coding RNAs (ncRNAs). In this study, we proposed a computational method to infer novel autophagy-associated TFs, micro RNAs (miRNAs) and long non-coding RNAs (lncRNAs) based on TFs and ncRNAs coordinated regulatory (TNCR) network. First, we constructed a comprehensive TNCR network, including 155 TFs, 681 miRNAs and 1332 lncRNAs. Next, we gathered the known autophagy-associated factors, including TFs, miRNAs and lncRNAs, from public data resources. Then, the random walk with restart (RWR) algorithm was conducted on the TNCR network by using the known autophagy-associated factors as seeds and novel autophagy regulators were finally prioritized. Leave-one-out cross-validation (LOOCV) produced an area under the curve (AUC) of 0.889. In addition, functional analysis of the top 100 ranked regulators, including 55 TFs, 26 miRNAs and 19 lncRNAs, demonstrated that these regulators were significantly enriched in cell death related functions and had significant semantic similarity with autophagy-related Gene Ontology (GO) terms. Finally, extensive literature surveys demonstrated the credibility of the predicted autophagy regulators. In total, we presented a computational method to infer credible autophagy regulators of transcriptional factors and non-coding RNAs, which would improve the understanding of processes of autophagy and cell death and provide potential pharmacological targets to autophagy-related diseases.

Keywords: autophagy regulator; transcriptional factor; non-coding RNA; regulatory network; RWR algorithm

1. Introduction

Autophagy is a process of cytoplasmic degradation that is essential in homeostasis and stress-response, as well as in protein degradation and organelles turnover [1]. The regulation of autophagy is critical in human health and disease. Both its insufficient and overdriven activity can disturb the body functions, including causing cancers. For example, autophagy deficiency causes oxidative stress and genome instability which is a known cause of cancer initiation and progression [2] and up-regulation of autophagy in RAS-transformed cancer cells promotes their growth, survival, tumorigenesis invasion, and metastasis [3]. The process of autophagy involves multiple kinds of regulators, including autophagy-related (*ATG*) genes, ATG proteins and non-coding RNAs (ncRNAs). For instance, the autophagy database archived a list of 582 experimentally demonstrated ATG proteins [4] and Wu et al. provided a comprehensive bioinformatics resource to dissect ncRNA-mediated autophagy interactions [5]. In addition, regulation of autophagy by targeting autophagy regulators is a promising strategy for cancer therapy [6]. For example, temsirolimus could significantly prolong progression-free survival of mantle cell lymphoma (MCL) patients by inhibiting the mechanistic target of rapamycin (mTOR) protein, a post-translational autophagy regulator [7].

Currently, post-translational autophagy regulators, such as ATG proteins, are well known while limited information is available about transcriptional and post-transcriptional regulators, such as transcription factors (TFs) and ncRNAs [8]. Inferring that novel transcriptional and post-transcriptional autophagy regulators will help to dissect the autophagy regulation mechanisms and provide possible pharmacological targets to regulate autophagy. The TFs and ncRNAs coordinated regulatory (TNCR) network has demonstrated its power as a tool to study biological issues such as regulatory pathways in human diseases, classifiers for drug resistance and so on [9–11]. For example, Liang et al. performed deconvolution on the transcriptional network and demonstrated that BACH1 was the master regulator of breast cancer bone metastasis [12]. Wang et al. identified disease-related regulatory cascades by dissecting the TF and miRNA regulatory network, which helped understand the pathogenesis [13]. Recently, lncRNAs were found to be targeted by miRNAs and functioned as miRNA sponges to attenuate the inhibition ability of miRNAs to mRNAs. Furthermore, lncRNAs were also shown to play crucial roles in the regulation of gene expression at transcriptional and post-transcriptional levels [14,15]. Thus, lncRNAs introduce an extra layer of complexity to the TNCR network, enhancing the analytical ability of the regulatory network.

In this study, we proposed a computational method to predict novel autophagy-associated TFs, miRNAs and lncRNAs based on the TNCR network. First, experimentally verified transcriptional and post-transcriptional regulatory relationships among TFs, miRNAs and lncRNAs were collected and a comprehensive regulatory network was constructed. Next, the known autophagy-associated TFs, miRNAs and lncRNAs were gathered from public data resources. The random walk with restart (RWR) algorithm was implemented on the regulatory network to prioritize autophagy regulators. Leave-one-out cross-validation (LOOCV) achieved an area under the curve (AUC) of 0.889. Functional enrichment analyses and extensive literature surveys demonstrated the credibility of predicted regulators. Altogether, we presented a computational method of inferring credible autophagy regulators and we believed that this would help improve the understanding of the autophagy regulation mechanisms.

2. Materials and Methods

2.1. Construction of a Comprehensive TNCR Network

We integrated five types of experimentally verified transcriptional and post-transcriptional regulatory relationships among TFs, miRNAs and lncRNAs, including TF-miRNA, TF-lncRNA, miRNA-lncRNA, miRNA-TF, lncRNA-TF. The TFs regulations of miRNAs were downloaded from the database TransmiR, which manually surveyed literature and recorded experimentally supported TF-miRNA regulation [16]. The TFs regulations of lncRNAs were obtained from the database ChIPBase, which decoded the transcriptional regulation of lncRNAs from ChIP-seq data in diverse tissues and cell lines [17]. Here, only TF-lncRNA regulations that were identified in more than 20 datasets were retained. In order to improve the credibility of the regulations, we also used the TRANSFAC Match program to assure transcription factor binding sites (TFBS) in lncRNA sequences [18] using minimum false-positive profiles of vertebrate high quality matrices. The final TF-lncRNA regulations were obtained by intersecting the ChIPBase data source with the TRANSFAC results. The miRNAs regulations of TFs were integrated from two databases, miRecords [19] and miRTarBase [20]. Both of these two databases collected experimentally validated miRNA-target interactions, and we retained the union set of the relationships presented in these two databases. The miRNAs regulations of lncRNAs were derived from LncBase v2 which provided experimentally supported and in silico predicted miRNA recognition elements (MREs) on lncRNAs [21]. We retained the interactions presented in the experimental module and the prediction scores should have been equal to or greater than 0.95. The lncRNAs regulations of TFs were downloaded from LncReg [22] and LncRNA2Target [23]. The database LncReg collected validated lncRNA-associated regulatory entries while LncRNA2Target curated differentially expressed genes after the lncRNA knockdown or overexpression. We kept the union set of the lncRNAs regulations of TFs which were provided by these two databases. Integrating all of the above regulations, we constructed a comprehensive TNCR network.

2.2. Collection of Known Autophagy Regulators

The known autophagy-associated TFs, miRNAs and lncRNAs were collected from public data resources. We first obtained human genes in autophagy related Gene Ontology (GO) terms from the AmiGO-2 database. Next, we downloaded the human autophagy-associated genes from the autophagy database [4], a multifaceted online resource providing information on genes and proteins related to autophagy across several eukaryotic species. The union set of these two gene sets were regarded as known autophagy-associated genes. As for autophagy-associated miRNAs and lncRNAs, we resorted to the database ncRDeathDB, a comprehensive bioinformatics resource archiving ncRNA-associated cell death interactions and picked up the autophagy-associated miRNAs and lncRNAs [5]. All the autophagy-associated genes, miRNAs and lncRNAs we obtained were mapped onto the TNCR network, and the intersections were regarded as seeds for RWR algorithm.

2.3. Prioritization of Novel Autophagy Regulators with the RWR Method

We performed the RWR method on the constructed TNCR network to prioritize novel autophagy regulators. The RWR method simulates a random walker that starts on given seed nodes and transits randomly from the current node to neighboring nodes in the network with the restart probability to teleport to the start nodes. Here, the known autophagy regulators were used as seed nodes. We denoted P_0 as the initial probability vector and P_t as a vector in which the i-th element held the probability of finding the random walker at node i in step t. Let α be the restart probability of the random walk in each step at the source nodes. W denotes the probability transition matrix and is derived from the adjacency matrix of the TNCR network. The formula is defined as:

$$w(i,j) = \begin{cases} A(i,j) / \sum_j A(i,j), & if \ \sum_j A(i,j) \neq 0 \\ 0, & otherwise \end{cases} \tag{1}$$

where $w(i,j)$ represents the element in the probability transition matrix, and $A(i,j)$ represents the element in the adjacency matrix. The probability vector in step $t+1$ can be described as follows:

$$p_{t+1} = (1-\alpha)wp_{t+1} + \alpha p_0 \tag{2}$$

Based upon the previous work, the restart probability (α) was set as 0.5, and the initial probability (P_0) of each seed node was set as $1/n$ (where n is the number of seed autophagy regulators) while the initial probability of all non-seed nodes was set as zero [24,25]. With the iteration steps going on, the probability of the RWR algorithm will become stable. We defined the stable probability as P_∞ when the difference between P_t and P_{t+1} was less than 10^{-10}. The stable probability of P_∞ can be used as a measure of proximity to the seed regulators. If $P_\infty(node_i) > P_\infty(node_j)$, then $node_i$ will be in closer proximity to the seed regulators in the regulatory network than $node_j$. As a result, all candidate nodes in the regulatory network can be ranked according to P_∞ and the top ranked elements can be expected to have a high probability of being associated with autophagy.

2.4. Functional Analysis for Predicted Autophagy Regulators

To demonstrate the credibility of the proposed prediction method, we performed functional analysis for the predicted autophagy regulators. We first retrieved the top 100 ranked regulator candidates (excluding seeds), including TFs, miRNAs and lncRNAs, and performed separately the functional enrichment analyses. For the obtained TFs, we used DAVID to perform GO and Kyoto Encyclopedia of Genes and Genomes (KEGG) pathway enrichment analysis [26]. For the obtained miRNAs, we collected the experimentally verified miRNA targets from the miRecords [19] and miRTarBase [20]; we then used the union set of the miRNA targets to perform GO and KEGG pathway enrichment analysis with DAVID. For the obtained lncRNAs, we utilized the recently developed function annotation tool of non-coding RNA (FARNA), a knowledgebase of inferred functions of human ncRNA transcripts, to implement function annotation analysis. We searched the FARNA database by using each obtained lncRNA, and retrieved promoter-associated transcription factors and transcription co-factors for the lncRNA. Then, all the obtained transcription factors and transcription co-factors were inputted into DAVID to perform GO and KEGG pathway enrichment analysis. In addition, we also performed GO enrichment analysis for the known autophagy-associated TFs, miRNAs and lncRNAs separately, as described above. The union set of the significant GO categories were considered as the autophagy related GO terms. All these DAVID analyses adopted the same criteria that the biological process (BP) category was used for GO analysis, and the significance of enrichment was set at p-value < 0.05. Finally, we calculated the functional similarity scores between the GO terms enriched in the predicted autophagy regulators and the autophagy related GO terms. The computational procedure was implemented using R package GOSemSim [27] and the rcmax method was chosen as a combined method for aggregating multiple GO terms. We also performed 1000 random tests to evaluate the significance of obtained functional similarity scores. In each random test, we randomly chose the same number of GO terms as in the real situation and calculated the functional similarity scores as above. The statistical p-value was calculated as the ratio of random functional similarity scores higher than the real functional similarity score.

3. Results

3.1. Characteristics of the TNCR Network

In this study, we integrated five types of experimentally verified transcriptional and post-transcriptional regulatory relationships from public data resources and constructed a comprehensive TNCR network (see Materials and Methods for details). The TNCR network comprised of 4529 edges, including 155 TFs, 681 miRNAs and 1332 lncRNAs (Figure 1A, Supplementary Table S1). To get an overview of the TNCR network, we examined the degree distribution of the network. As shown in Figure 1B, most nodes (50.4%) had degree one and few nodes had a high degree. In addition, the power-law distribution of the forms $y = 327.4 \times 10^{-1.31}$ $\left(R^2 = 0.823\right)$, $y = 157.4 \times 10^{-1.19}$ $\left(R^2 = 0.773\right)$ and $y = 224.4 \times 10^{-1.36}$ $\left(R^2 = 0.774\right)$ were fitted for degree, out-degree and in-degree respectively. These results indicated that the TNCR network satisfied approximate scale-free topology which is the common feature of most biological networks [28]. Next, we further investigated the in-degree and out-degree distributions for TFs, miRNAs and lncRNAs, respectively (Figure 1C). In general, few nodes had very high degrees and many had low degrees, regardless of TFs, miRNAs or lncRNAs in-degree and out-degree. Furthermore, TFs had a higher median in-degree and out-degree than miRNAs and lncRNAs, which meant that TFs more likely acted as hubs in the TNCR network.

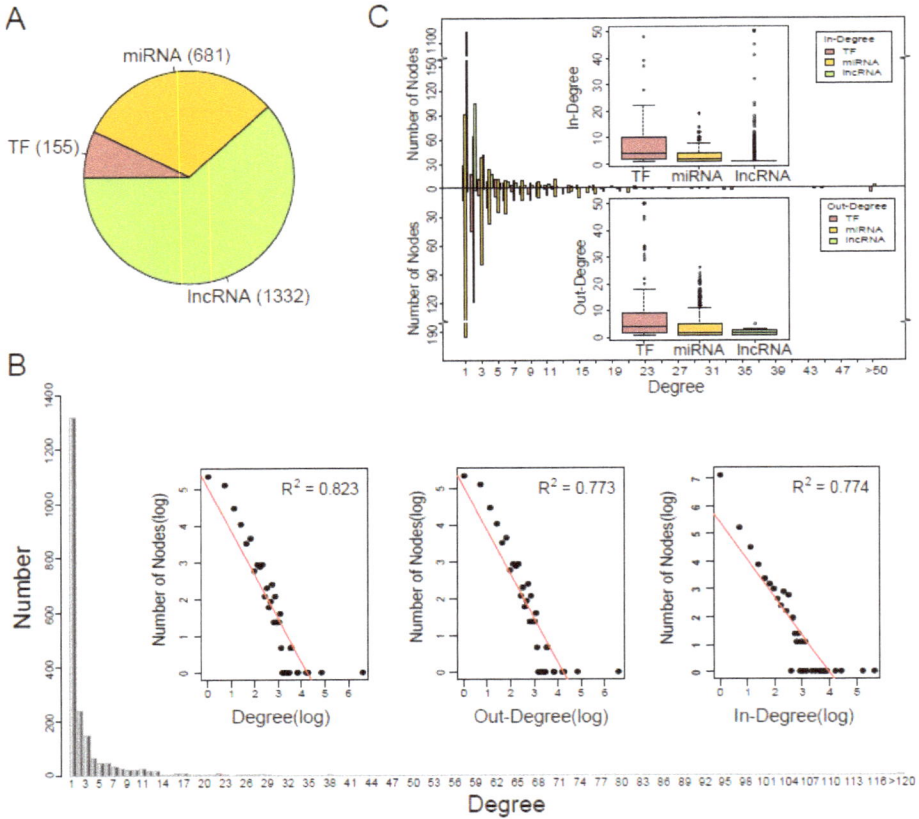

Figure 1. Characteristics of the TFs and ncRNAs coordinated regulatory (TNCR) network. (**A**) Proportion of transcription factor (TF), microRNA (miRNA) and long non-coding RNA (lncRNA) in the TNCR network. (**B**) Degree distribution of all nodes in the TNCR network and the log-log plots for the degree, out-degree and in-degree distributions of all nodes. (**C**) In-degree and out-degree distributions of TFs, miRNAs and lncRNAs in the TNCR network.

3.2. Performance Evaluation of the Proposed Method

By integrating data from AmiGO-2, the autophagy database and the ncRDeathDB, we obtained 1222 known autophagy regulators in total (Supplementary Table S2). After mapping these regulators onto the TNCR network, we finally got 178 autophagy regulators as seeds, including 25 TFs, 152 miRNAs and 1 lncRNAs (Supplementary Table S3). By performing the RWR method on the TNCR network with the seeds, we finally prioritized novel autophagy regulators.

In order to evaluate the performance of our method for inferring autophagy regulators, we performed LOOCV analysis. Each known autophagy regulator was left out in turn as the test case and the other known autophagy regulators were taken as seeds. All the other nodes in the TNCR network were regarded as candidate autophagy regulators. Sensitivity and specificity were calculated for each threshold. Finally, a receiver operating characteristic (ROC) curve was plotted by varying the threshold and then the value of the AUC was calculated. Our method, tested on already known autophagy regulators, achieved an AUC of 0.889 (Figure 2), exhibiting excellent performance. Here, the TNCR network incorporated three kinds of regulators (TFs, miRNAs and lncRNAs) and five kinds of regulations (TF-miRNA, TF-lncRNA, miRNA-lncRNA, miRNA-TF and lncRNA-TF). To demonstrate the effectivity and reliability of the TNCR network, we compared the performance of partial TNCR networks. The AUCs were calculated for a TNCR-ML network (miRNAs and lncRNAs only) and a TNCR-TM network (TFs and miRNAs only) separately by performing LOOCV (the TNCR-TL network (TFs and lncRNAs only) was not analyzed because of missing seed regulators). The AUCs were 0.697 and 0.544 respectively, which were lower than those using the TNCR network (Figure 2). To further determine whether the results of the cross validation might have been generated by chance, we performed randomization tests. The seeds were generated randomly from candidate nodes in all three networks and the AUC values were calculated by performing LOOCV, as above. The AUC values under randomized tests were much lower than those in real situations (0.530, 0.549 and 0.519, respectively, for these three conditions), confirming the valid and reliable performance of autophagy regulator seeds in our method (Figure 2). We also performed RWR on 1000 degree-preserving randomized TNCR networks and the average value of the AUCs was calculated. As shown in Figure 2, the result based on the real TNCR network and the real seed nodes performed best.

The prioritization of all candidate autophagy regulators is provided in Supplementary Table S4. The top 100 ranked candidate regulators, including 55 TFs, 19 miRNAs and 26 lncRNAs, were further validated by literature mining, in which 52 regulators had been verified to be associated with autophagy in published papers (Supplementary Table S5). For example, the fifth ranked regulator MYC was recently proved to mitigate its oncogenic activity by chaperone-mediated autophagy (CMA) regulation [29] and the ninth ranked regulator XIST was determined to increase autophagy activity in non-small-cell lung cancer by regulation of ATG7 [30]. The extensive literature surveys demonstrated the feasibility of our method to predict autophagy regulators.

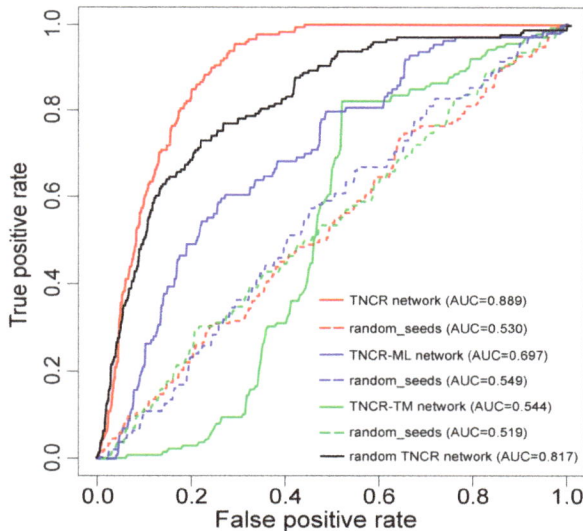

Figure 2. Receiver operating characteristic (ROC) curves and area under the curve (AUC) values for the random walk with restart (RWR) method on the whole, partial and random TNCR networks with real seeds and random seeds. The ROC curves were plotted and AUC values were calculated separately by leave-one-out cross-validation (LOOCV) for the TNCR network, TNCR-ML (miRNAs and lncRNAs only) network, TNCR-TM (TFs and miRNAs only) network and the random TNCR network with real and random seeds.

3.3. Functional Characteristics of Predicted Autophagy Regulators

The top 100 ranked candidate autophagy regulators were retrieved, including 55 TFs, 19 miRNAs and 26 lncRNAs (Supplementary Table S4), then the functional analyses were performed separately for these predicted autophagy regulators (see Materials and Methods for details). The top 20 significantly enriched GO terms and KEGG pathways for TFs are shown in Figure 3. We observed that some cell death related GO terms, such as cell cycle arrest and negative regulation of cell proliferation, were enriched by these top ranked TFs. Several significantly enriched KEGG pathways were also related to cell death, for instance, cell cycle and adherens junction. In addition, some cancer related pathways, such as colorectal cancer, prostate cancer and thyroid cancer, were also enriched, indicating that the autophagy regulators played important roles in cancer. This was consistent with previous studies [11,31,32]. The top 20 significantly enriched GO terms and KEGG pathways by the top ranked miRNAs and lncRNAs are shown in Figure S1 and Figure S2. Similar to the top ranked TFs, the cell death related GO terms and KEGG pathways, such as apoptotic process and cell proliferation, were also enriched by top ranked miRNAs and lncRNAs. Cancer related pathways, such as pancreatic cancer and small cell lung cancer, were enriched by top ranked miRNAs and lncRNAs. We observed that there were obvious overlaps among GO terms and KEGG pathways enriched by top ranked TFs, miRNAs and lncRNAs (Figures 3 and 4A). All of the significantly enriched GO terms and KEGG pathways (*p*-value < 0.05) for top ranked TFs, miRNAs and lncRNAs were shown in Supplemental Table S6.

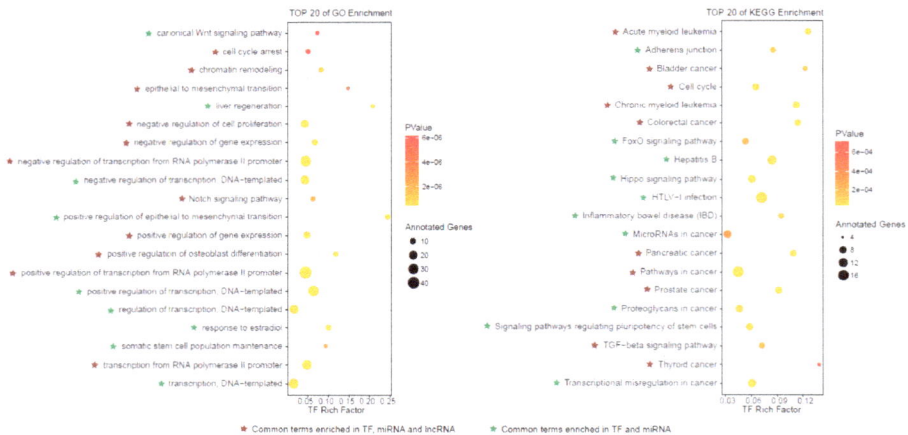

Figure 3. The top 20 Gene Ontology (GO) enrichment and Kyoto Encyclopedia of Genes and Genomes (KEGG) enrichment results for top ranked TFs. The common enriched GO terms and KEGG pathways among top ranked TFs, miRNAs and lncRNAs are marked.

To further evaluate the top ranked regulators associated with autophagy, we compared the GO terms enriched by the top 100 ranked regulators with those enriched by known autophagy-associated factors (including protein-coding genes, miRNAs and lncRNAs). As shown in Figure 4A, the numbers of overlapping enriched GO terms among top-ranked TFs, miRNAs, lncRNAs and known autophagy-associated factors were high (the significantly enriched GO terms for known autophagy-associated factors were shown in Supplemental Table S7). We calculated the functional similarity scores between the GO terms enriched by the top 100 ranked regulators and the autophagy related GO terms. The functional similarity scores between the autophagy related GO terms and those enriched by top ranked TFs, miRNAs, lncRNAs were 0.970, 0.978 and 0.949, respectively. The random functional similarity scores for each kind of regulators, which were calculated by randomly choosing the same number of GO terms as in the real situation, were significantly lower than the real scores (Figure 4B, Figure S3). All these *p*-values were less than 2.2×10^{-16} (see Materials and Methods for details). This meant that the top ranked regulators were significantly associated with autophagy. The functional characteristics of the top ranked regulators indicated that our method was capable of identifying novel autophagy regulators.

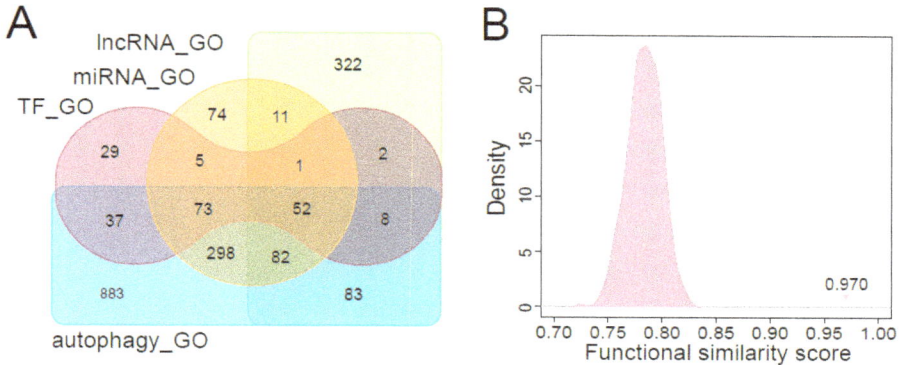

Figure 4. Evaluation of the top ranked regulators associated with autophagy. (**A**) Venn plot for the GO functional annotation comparison among the top ranked TFs, miRNAs, lncRNAs and the known autophagy-associated factors. (**B**) Distribution of random functional similarity scores for the top ranked TFs and the autophagy-associated factors. The triangle indicates the true functional similarity score for top ranked TFs and the known autophagy-associated factors.

4. Discussion

Autophagy is an intracellular catabolic process for maintaining homeostasis and involved systematic regulation at post-translational, transcriptional, and post-transcriptional levels [33]. Both its insufficient and overdriven functions can disturb intracellular homeostasis [34]. Thus, the regulation of autophagy is critical for body cells normal function. Although the knowledge of autophagy regulation is making certain progress, the landscape of autophagy regulators is far from completeness. In addition, autophagy demonstrates a promising therapeutic target in several pathologies [35]. Thus, identification of novel autophagy regulators is beneficial to targeted therapy of complex human diseases. Regulatory networks provide global views of the transmission of genetic information, and are proved to be powerful tools for studying biological issues. In this study, we conducted a computational method to infer novel autophagy regulators based on the regulatory network. We first constructed a comprehensive regulatory TNCR network that incorporated transcriptional and post-transcriptional regulators, including TFs, miRNAs and lncRNAs. Network topological analysis revealed that the degree distribution of the TNCR network approximately followed the power-law distribution. Then, the candidate autophagy regulators were ranked by implementing the RWR method on the TNCR network using the known autophagy regulators as seed nodes. The AUC values determined by LOOCV achieved 0.889, demonstrating the high credibility of our method for recovering known autophagy regulators. Furthermore, functional enrichment analyses revealed that the predicted autophagy regulators were associated with cell death related functional categories such as negative regulation of cell proliferation, cell death and cell cycle arrest. Significantly high functional semantic similarity scores were obtained between the obtained GO terms and the autophagy related GO terms. In addition, extensive literature surveys demonstrated that the top ranked regulators were verified to have associations with autophagy. All these results indicate that our approach is effective in inferring transcriptional and post-transcriptional autophagy regulators and that it would help to improve the understanding of the autophagy regulation mechanisms.

In the past several years, the landscape of TNCR networks has been described elaborately [12,13]. Several experimentally verified transcriptional and post-transcriptional regulatory databases have been developed, such as TransmiR [16], ChIPBase [17], miRTarBase [20] and so on. However, the exhaustive transcriptional and post-transcriptional regulatory relationships still need further elucidation. For example, the characterization of lncRNAs regulation of TFs is still at a primary level [36]. Furthermore, the competing endogenous RNA (ceRNA) relationships involved in TFs, miRNAs and lncRNAs provide further complex regulations among transcriptional and post-transcriptional factors which should be considered in the future analysis of TNCR network [37]. Our approach in this study was based on the general regulatory network TNCR; however, the autophagy plays tissue-specific and double-edged roles in the cellular homeostasis and survival. We believe that the performance of our approach would be improved if we use the data of a specific cancer. In addition, the comprehensiveness of seeds is critical for the performance of the RWR algorithm [38]. Currently, protein-coding regulators of the autophagic machinery are relatively well known, while few studies have been conducted on the non-coding RNA regulators, especially lncRNAs. With the abundance of research of autophagy related regulators, we will obtain comprehensive seed autophagy regulators, and provide more credible, verifiable autophagy regulators.

Supplementary Materials: The following are available online at http://www.mdpi.com/2073-4409/7/11/194/s1.

Author Contributions: Conceptualization, X.C. and W.J.; data curation, Q.M. and X.Z.; formal analysis, S.W. and W.W.; funding acquisition, S.W. and W.J.; investigation, S.W.; methodology, S.Z.; supervision, W.J.; validation, Q.M., H.L. and H.L.; visualization, W.W. and X.M.; writing—original draft, S.W.; writing—review and editing W.J.

Funding: This work was supported by the National Natural Science Foundation of China [61571169]; the Fundamental Research Funds for the Central Universities [NE2018101]; the Harbin medical university scientific research innovation fund [2017JCZX52].

Conflicts of Interest: The authors declare no conflicts of interest.

References

1. Mizushima:, N. Autophagy: Process and function. *Genes Dev.* **2007**, *21*, 2861–2873. [CrossRef] [PubMed]
2. Mathew, R.; Kongara, S.; Beaudoin, B.; Karp, C.M.; Bray, K.; Degenhardt, K.; Chen, G.; Jin, S.; White, E. Autophagy suppresses tumor progression by limiting chromosomal instability. *Genes Dev.* **2007**, *21*, 1367–1381. [CrossRef] [PubMed]
3. Lock, R.; Kenific, C.M.; Leidal, A.M.; Salas, E.; Debnath, J. Autophagy-dependent production of secreted factors facilitates oncogenic ras-driven invasion. *Cancer Discov.* **2014**, *4*, 466–479. [CrossRef] [PubMed]
4. Homma, K.; Suzuki, K.; Sugawara, H. The autophagy database: An all-inclusive information resource on autophagy that provides nourishment for research. *Nucleic Acids Res.* **2011**, *39*, D986–D990. [CrossRef] [PubMed]
5. Wu, D.; Huang, Y.; Kang, J.; Li, K.; Bi, X.; Zhang, T.; Jin, N.; Hu, Y.; Tan, P.; Zhang, L.; et al. Ncrdeathdb: A comprehensive bioinformatics resource for deciphering network organization of the ncrna-mediated cell death system. *Autophagy* **2015**, *11*, 1917–1926. [CrossRef] [PubMed]
6. Janku, F.; McConkey, D.J.; Hong, D.S.; Kurzrock, R. Autophagy as a target for anticancer therapy. *Nat. Rev. Clin. Oncol.* **2011**, *8*, 528–539. [CrossRef] [PubMed]
7. Galimberti, S.; Petrini, M. Temsirolimus in the treatment of relapsed and/or refractory mantle cell lymphoma. *Cancer Manag. Res.* **2010**, *2*, 181–189. [CrossRef] [PubMed]
8. Turei, D.; Foldvari-Nagy, L.; Fazekas, D.; Modos, D.; Kubisch, J.; Kadlecsik, T.; Demeter, A.; Lenti, K.; Csermely, P.; Vellai, T.; et al. Autophagy regulatory network-a systems-level bioinformatics resource for studying the mechanism and regulation of autophagy. *Autophagy* **2015**, *11*, 155–165. [CrossRef] [PubMed]
9. Jiang, W.; Zhang, Y.; Meng, F.; Lian, B.; Chen, X.; Yu, X.; Dai, E.; Wang, S.; Liu, X.; Li, X.; et al. Identification of active transcription factor and mirna regulatory pathways in Alzheimer's disease. *Bioinformatics* **2013**, *29*, 2596–2602. [CrossRef] [PubMed]
10. Jiang, W.; Mitra, R.; Lin, C.C.; Wang, Q.; Cheng, F.; Zhao, Z. Systematic dissection of dysregulated transcription factor-mirna feed-forward loops across tumor types. *Brief. Bioinform.* **2016**, *17*, 996–1008. [CrossRef] [PubMed]

11. Dai, E.; Wang, J.; Yang, F.; Zhou, X.; Song, Q.; Wang, S.; Yu, X.; Liu, D.; Yang, Q.; Dai, H.; et al. Accurate prediction and elucidation of drug resistance based on the robust and reproducible chemoresponse communities. *Int. J. Cancer* **2018**, *142*, 1427–1439. [CrossRef] [PubMed]

12. Liang, Y.; Wu, H.; Lei, R.; Chong, R.A.; Wei, Y.; Lu, X.; Tagkopoulos, I.; Kung, S.Y.; Yang, Q.; Hu, G.; et al. Transcriptional network analysis identifies bach1 as a master regulator of breast cancer bone metastasis. *J. Biol. Chem.* **2012**, *287*, 33533–33544. [CrossRef] [PubMed]

13. Wang, S.; Li, W.; Lian, B.; Liu, X.; Zhang, Y.; Dai, E.; Yu, X.; Meng, F.; Jiang, W.; Li, X. Tmrec: A database of transcription factor and mirna regulatory cascades in human diseases. *PloS. ONE* **2015**, *10*, e0125222. [CrossRef] [PubMed]

14. Martens, J.A.; Laprade, L.; Winston, F. Intergenic transcription is required to repress the saccharomyces cerevisiae *ser3* gene. *Nature* **2004**, *429*, 571–574. [CrossRef] [PubMed]

15. Carrieri, C.; Cimatti, L.; Biagioli, M.; Beugnet, A.; Zucchelli, S.; Fedele, S.; Pesce, E.; Ferrer, I.; Collavin, L.; Santoro, C.; et al. Long non-coding antisense rna controls uchl1 translation through an embedded sineb2 repeat. *Nature* **2012**, *491*, 454–457. [CrossRef] [PubMed]

16. Wang, J.; Lu, M.; Qiu, C.; Cui, Q. Transmir: A transcription factor-microrna regulation database. *Nucleic Acids Res.* **2010**, *38*, D119–122. [CrossRef] [PubMed]

17. Yang, J.H.; Li, J.H.; Jiang, S.; Zhou, H.; Qu, L.H. Chipbase: A database for decoding the transcriptional regulation of long non-coding rna and microrna genes from chip-seq data. *Nucleic Acids Res.* **2013**, *41*, D177–D187. [CrossRef] [PubMed]

18. Matys, V.; Kel-Margoulis, O.V.; Fricke, E.; Liebich, I.; Land, S.; Barre-Dirrie, A.; Reuter, I.; Chekmenev, D.; Krull, M.; Hornischer, K.; et al. Transfac and its module transcompel: Transcriptional gene regulation in eukaryotes. *Nucleic Acids Res.* **2006**, *34*, D108–D110. [CrossRef] [PubMed]

19. Xiao, F.; Zuo, Z.; Cai, G.; Kang, S.; Gao, X.; Li, T. Mirecords: An integrated resource for microrna-target interactions. *Nucleic Acids Res.* **2009**, *37*, D105–D110. [CrossRef] [PubMed]

20. Chou, C.H.; Shrestha, S.; Yang, C.D.; Chang, N.W.; Lin, Y.L.; Liao, K.W.; Huang, W.C.; Sun, T.H.; Tu, S.J.; Lee, W.H.; et al. Mirtarbase update 2018: A resource for experimentally validated microrna-target interactions. *Nucleic Acids Res.* **2018**, *46*, D296–D302. [CrossRef] [PubMed]

21. Paraskevopoulou, M.D.; Vlachos, I.S.; Karagkouni, D.; Georgakilas, G.; Kanellos, I.; Vergoulis, T.; Zagganas, K.; Tsanakas, P.; Floros, E.; Dalamagas, T.; et al. Diana-lncbase v2: Indexing microrna targets on non-coding transcripts. *Nucleic Acids Res.* **2016**, *44*, D231–D238. [CrossRef] [PubMed]

22. Zhou, Z.; Shen, Y.; Khan, M.R.; Li, A. Lncreg: A reference resource for lncrna-associated regulatory networks. *Database (Oxford)* **2015**, *2015*. [CrossRef] [PubMed]

23. Jiang, Q.; Wang, J.; Wu, X.; Ma, R.; Zhang, T.; Jin, S.; Han, Z.; Tan, R.; Peng, J.; Liu, G.; et al. Lncrna2target: A database for differentially expressed genes after lncrna knockdown or overexpression. *Nucleic Acids Res.* **2015**, *43*, D193–D196. [CrossRef] [PubMed]

24. Li, Y.; Patra, J.C. Genome-wide inferring gene-phenotype relationship by walking on the heterogeneous network. *Bioinformatics* **2010**, *26*, 1219–1224. [CrossRef] [PubMed]

25. Chen, X.; Shi, H.; Yang, F.; Yang, L.; Lv, Y.; Wang, S.; Dai, E.; Sun, D.; Jiang, W. Large-scale identification of adverse drug reaction-related proteins through a random walk model. *Sci. Rep.* **2016**, *6*, 36325. [CrossRef] [PubMed]

26. Huang da, W.; Sherman, B.T.; Lempicki, R.A. Systematic and integrative analysis of large gene lists using david bioinformatics resources. *Nat. Protoc.* **2009**, *4*, 44–57. [CrossRef] [PubMed]

27. Yu, G.; Li, F.; Qin, Y.; Bo, X.; Wu, Y.; Wang, S. Gosemsim: An r package for measuring semantic similarity among go terms and gene products. *Bioinformatics* **2010**, *26*, 976–978. [CrossRef] [PubMed]

28. Barabasi, A.L.; Oltvai, Z.N. Network biology: Understanding the cell's functional organization. *Nat. Rev. Genet.* **2004**, *5*, 101–113. [CrossRef] [PubMed]

29. Gomes, L.R.; Menck, C.F.M.; Cuervo, A.M. Chaperone-mediated autophagy prevents cellular transformation by regulating myc proteasomal degradation. *Autophagy* **2017**, *13*, 928–940. [CrossRef] [PubMed]

30. Sun, W.; Zu, Y.; Fu, X.; Deng, Y. Knockdown of lncrna-xist enhances the chemosensitivity of nsclc cells via suppression of autophagy. *Oncol. Rep.* **2017**, *38*, 3347–3354. [CrossRef] [PubMed]

31. Yu, T.; Guo, F.; Yu, Y.; Sun, T.; Ma, D.; Han, J.; Qian, Y.; Kryczek, I.; Sun, D.; Nagarsheth, N.; et al. Fusobacterium nucleatum promotes chemoresistance to colorectal cancer by modulating autophagy. *Cell* **2017**, *170*, 548–563. [CrossRef] [PubMed]

32. Plantinga, T.S.; Tesselaar, M.H.; Morreau, H.; Corssmit, E.P.; Willemsen, B.K.; Kusters, B.; van Engen-van Grunsven, A.C.; Smit, J.W.; Netea-Maier, R.T. Autophagy activity is associated with membranous sodium iodide symporter expression and clinical response to radioiodine therapy in non-medullary thyroid cancer. *Autophagy* **2016**, *12*, 1195–1205. [CrossRef] [PubMed]

33. Das, G.; Shravage, B.V.; Baehrecke, E.H. Regulation and function of autophagy during cell survival and cell death. *Cold Spring Harb. Perspect. Biol.* **2012**, *4*, a008813. [CrossRef] [PubMed]

34. Thorburn, A. Autophagy and its effects: Making sense of double-edged swords. *PLoS biology* **2014**, *12*, e1001967. [CrossRef] [PubMed]

35. Rubinsztein, D.C.; Codogno, P.; Levine, B. Autophagy modulation as a potential therapeutic target for diverse diseases. *Nat. Rev. Drug Discov.* **2012**, *11*, 709–730. [CrossRef] [PubMed]

36. Kopp, F.; Mendell, J.T. Functional classification and experimental dissection of long noncoding rnas. *Cell* **2018**, *172*, 393–407. [CrossRef] [PubMed]

37. Tay, Y.; Rinn, J.; Pandolfi, P.P. The multilayered complexity of cerna crosstalk and competition. *Nature* **2014**, *505*, 344–352. [CrossRef] [PubMed]

38. Zhang, S.W.; Shao, D.D.; Zhang, S.Y.; Wang, Y.B. Prioritization of candidate disease genes by enlarging the seed set and fusing information of the network topology and gene expression. *Mol. Biosyst.* **2014**, *10*, 1400–1408. [CrossRef] [PubMed]

![cells logo] *cells*

Review

MicroRNAs in Cardiac Autophagy: Small Molecules and Big Role

Teng Sun [1], Meng-Yang Li [2], Pei-Feng Li [2] and Ji-Min Cao [1,*]

[1] Key Laboratory of Cellular Physiology, Ministry of Education, Department of Physiology, Shanxi Medical University, Taiyuan 030001, China; tengsun@qdu.edu.cn
[2] Institute for Translational Medicine, Qingdao University, Qingdao 266021, China; limengyang@qdu.edu.cn (M.-Y.L.); peifeng@ioz.ac.cn (P.-F.L.)
* Correspondence: caojimin@126.com; Tel.: +86-351-413-5246

Received: 5 July 2018; Accepted: 9 August 2018; Published: 11 August 2018

Abstract: Autophagy, which is an evolutionarily conserved process according to the lysosomal degradation of cellular components, plays a critical role in maintaining cell homeostasis. Autophagy and mitochondria autophagy (mitophagy) contribute to the preservation of cardiac homeostasis in physiological settings. However, impaired or excessive autophagy is related to a variety of diseases. Recently, a close link between autophagy and cardiac disorders, including myocardial infarction, cardiac hypertrophy, cardiomyopathy, cardiac fibrosis, and heart failure, has been demonstrated. MicroRNAs (miRNAs) are a class of small non-coding RNAs with a length of approximately 21–22 nucleotides (nt), which are distributed widely in viruses, plants, protists, and animals. They function in mediating the post-transcriptional gene silencing. A growing number of studies have demonstrated that miRNAs regulate cardiac autophagy by suppressing the expression of autophagy-related genes in a targeted manner, which are involved in the pathogenesis of heart diseases. This review summarizes the role of microRNAs in cardiac autophagy and related cardiac disorders. Furthermore, we mainly focused on the autophagy regulation pathways, which consisted of miRNAs and their targeted genes.

Keywords: microRNAs; autophagy; mitophagy; cardiac diseases; biomarker

---

1. Overview of Autophagy and MicroRNAs

1.1. Autophagy

Autophagy is an evolutionarily conserved process of the lysosome-dependent degradation of cytoplasm components and damaged organelles, such as endoplasmic reticulum, peroxisomes, and mitochondria [1]. Factors that induce cellular senility and stress, including infections, toxics, hypoxia, and nutrient starvation, can induce autophagy, which plays a role in protecting cells and maintaining homeostasis. When it is stimulated by pathological factors, autophagy is either disrupted or contributes to autophagic cell death. Abnormal autophagy is related to a variety of pathological disorders. Autophagy involves the following steps: autography induction, vesicle nucleation, vesicle elongation and autophagosomes formation, and retrieval and fusion between autophagosomes and lysosomes [2]. Mitochondria produce energy in the form of ATP and play important roles in cellular homeostasis, signaling, apoptosis, autophagy, and metabolism. Damaged and dysfunctional mitochondria are deleterious to cells and can even lead to various types of diseases [3–8]. Therefore, quality control of mitochondria is crucial. Autophagy is the main form of mitochondria degradation, which is a process that is called mitophagy [3,4]. Mitophagy is an autophagic response that specifically targets damaged and hence potentially cytotoxic mitochondria. Putative kinase 1 (PINK1)-Parkin axis is a well-known regulation pathway in mitophagy [9].

Furthermore, Bcl-2 interacting protein 3 (BNIP3) [10], Bcl2-like protein 13 (Bcl2-L-13) [11], and FUN14 domain containing protein1 (FUNDC1) [12] are reported to be involved in mitophagy regulation [3].

Autophagy is widely implicated in cardiac homeostasis in health and diseases, such as myocardial infarction, myocardial hypertrophy, cardiac fibrosis, cardiomyopathy, and heart failure [13–17]. Autophagy positively or negatively regulates myocardial infarction and myocardial hypertrophy. The modulation of autophagy processes affects cardiac function [13,18–20]. Cardiac fibrosis and heart failure are associated with excessive autophagy. The suppression of autophagy could attenuate fibrosis and improve heart function [16,17]. Additionally, autophagy is implicated in cardiomyopathy. It is reported that autophagy is suppressed in high-fat diet-induced obesity cardiomyopathy [15,21], while autophagy is activated in doxorubicin-induced cardiomyopathy and cardiotoxicity [22].

1.2. MicroRNAs

MicroRNAs (miRNAs) are small non-coding RNAs with a length of approximately 21–22 nucleotides (nt), which are distributed widely in viruses, plants, protists, and animals. They function to mediate post-transcriptional gene silencing. MiRNAs usually exhibit a high conservation level and a very low rate of evolution [23,24]. Mature miRNAs are generated via a two-step processing by Drosha and Dicer. The primary transcripts with a length of hundreds of nucleotides, which are named Pri-miRNAs, are initially generated in the nucleus. The Drasha complex cleaves pri-miRNA into pre-miRNA. Pre-miRNA is transported to the cytoplasm by Exportin-5. In the cytoplasm, pre-miRNA is processed by RNase III Dicer. Eventually, one arm is generated as the mature miRNA with a length of about 22 nt, while the other one is degraded [25–28]. The mature miRNAs are assembled into the ribonucleoprotein (RNP) complex, which are called miRNPs or miRISCs. By interacting with the 3′-untranslated region (UTR), 5′-untranslated region (UTR), or coding sequence (CDS) region of the target genes, miRNAs mediate gene silencing at the post-transcriptional or translational levels. Argonaute proteins (AGOs), which are the key components of miRNPs, compete with translation initiation factors (e.g., elf4E) by binding to their m7G cap to inhibit translation initiation. AGOs also recruit the 60S ribosomal subunit binding protein elF6 to prevent the binding of the 60S subunit, which triggers translation initiation. At the post-initiation stage, either miRNAs or AGOs become associated with polysomes to play a supposed role in elongating inhibition or the dropping off of the ribosome. Moreover, the miRNPs could recruit the CCR4-NOT complex to mediate the deadenylation of mRNAs during the translation process [29,30]. Based on their function in mediating gene silencing, miRNAs are deeply linked to numerous biological processes and related diseases [18,31,32].

2. MicroRNAs Regulate the Core Autophagy Signaling Cascades

Accumulated research has indicated that miRNAs play critical roles in the autophagy processes. Autophagy proceeds in several successive stages, including induction, vesicle nucleation, vesicle elongation, and maturation [31,33]. MiRNAs are deeply implicated in the regulation of the main autophagy signaling cascades [31,34], which usually occurs through modulating the expression of autophagy-related genes (Figure 1).

Autophagy induction is mainly regulated by the unc-51-like kinase (ULK) complex, which is composed of ULK1/2, autophagy-related gene 13 (ATG13), ATG101, and focal adhesion kinase family interacting protein with a mass of 200 kDa (FIP200). The mammalian target of rapamycin (mTOR) is one of the most important upstream regulators for the induction of autophagy. Under normal conditions, the mammalian target of rapamycin complex 1 (mTORC1) interacts with and inactivates the ULK1/2. Once exposed to autophagy-induced factors, such as myocardial ischemia, anoxia, or other myocardial damage stimuli, mTORC1 is dissociated from the ULK complex, which dephosphorylates the ULK1/2. The activated ULK1/2 subsequently phosphorylates ATG13 and TIP200, which leads to the initiation of autophagy [33,34]. It was reported that miR-372, miR-17-5p, and miR-106a inhibit autophagy by down-regulating ULK1 [35–37]. MiR-885-3p was found to directly target ULK2 mRNA to control the cell viability and autophagy [38]. ULK2 is suppressed by miR-26b in a targeted manner, which interrupts the autophagy initiation [39,40]. Additionally, miR-20a and miR-20b negatively regulate autophagy via targeting FIP200 [41].

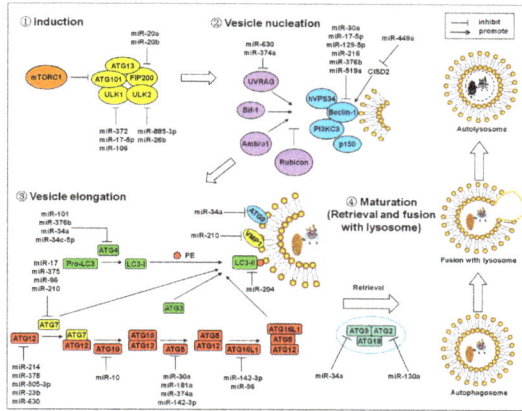

Figure 1. MicroRNAs regulate the core autophagy signaling cascades. Autophagy proceeds in four successive stages including induction, vesicle nucleation, vesicle elongation, and maturation. MicroRNAs are implicated in the processes. See text for detailed explanations. Arrows represent the promotion effect. T bars represent the inhibition effect.

Vesicle nucleation is the step that forms the double membrane-bond vesicles, which are called autophagosomes. The activation of the PI3K complex, which is composed of PI3KC3, hVPS34, Beclin-1, and p150, triggers the vesicle nucleation. In addition, several binding partners including Ambra1, Bif-1, UV irradiation resistance-associated gene (UVRAG), and Rubicon positively or negatively regulate this complex. In response to ample nutrients, the Bcl-2 family anti-apoptotic proteins interact with Beclin-1 and inhibit autophagy [31,42]. It was reported that miR-630 and miR-374a could repress the expression of UVRAG [43]. Regarding the PI3K complex, miR-30a, miR-17-5p, miR-129-5p, miR-216, miR-376b, and miR-519a prevent the autophagy process through directly targeting Beclin-1 [44–48]. MiR-449a indirectly promotes Beclin-1-mediated autophagy via suppressing the CDGSH iron sulfur domain 2 (CISD2) [49]. MiR-449 or miR-146a targeting Bcl-2 is also involved in autophagy regulation [50,51]. ATG12 system and LC3 system, the two ubiquitin-like protein conjugation pathways, are responsible for the vesicle elongation. In the ATG12 pathway, ATG12 interacts with ATG5 by binding ATG7 and ATG10 successively. The ATG12-ATG5 complex finally conjugates with ATG16L1 to form a large multimeric complex. In the LC3 pathway, ATG4 cleaves pro-LC3 to LC3-I. LC3-I is subsequently conjugated to phosphatidylethanolamine (PE) under the catalysis of ATG7 and E2-like ligase ATG3. Mediated by the ATG16L1-ATG5-ATG12 complex, which functions as an E3 ligase, LC3-I is processed to the phagophore membrane-bound LC3-II [31,33]. These two conjugation pathways could be regulated by microRNAs. The ATG12-mediated autophagy is targetedly inhibited by miR-214, miR-378, miR-505-3p, miR-23b, and miR-630 [43,52–55]. MiR-181a/miR-30a/miR-374a targeting ATG5 [43,56,57], miR-142-3p targeting ATG5/ATG16L [58], miR-96 targeting ATG7/ATG16L1 [59], miR-210/miR-17/miR-375 targeting ATG7 [60–62], and miR-20 targeting ATG10 [63] have been demonstrated to regulate autophagy. Additionally, the down-regulation of ATG4B is mediated by miR-34a/34c-5p, which suppresses the rapamycin-induced autophagy [64]. ATG4C and ATG4D are silenced by miR-376b and miR-101, respectively [48,65]. MiR-204 prevents autophagy by targeting LC3-II [66]. VMP1, a critical transmembrane protein for phagophore formation, is a direct target of miR-210 [67].

The maturation of autophagy includes retrieval and fusion with lysosomes. The process is poorly studied in mammals and humans. The retrieval step is mainly mediated by the ATG9-ATG2-AGT18 complex, which recruits lipids and proteins to the growing phagophore. Lastly, autophagosomes fuse to lysosomes, forming autolysosomes where the inner membrane of the former autophagosome and the engulfed cargo are degraded by acid hydrolases [31,68,69]. ATG9A is a direct target of miR-34a in autophagy maturation step [70]. MiR-130a interferes with the ATG9-ATG2-AGT18 complex formation via down-regulating ATG2B [71].

3. The Role of MicroRNAs in Cardiac Autophagy

The importance of autophagy for the preservation of cardiac homeostasis in physiological settings has been demonstrated. When they are subjected to the cardiomyocyte-specific deletion of autophagy-related gene 5 (Atg5), which is required for optimal autophagic responses, mice will develop cardiac hypertrophy, left ventricular dilatation, contractile dysfunction, and premature death. This is accompanied by a disorganized sarcomere structure, mitochondrial misalignment, and aggregation [72–74]. MiR-19a-3p/19b-3p inhibits autophagy in cardiomyocytes by targeting the TGF-β II-Atg 5 pathway [16]. MiR-33 reduces lipid droplet catabolism by negatively regulating Atg5 and miR-214-3p, which reduces the oxidized low-density lipoprotein-initiated autophagy by directly targeting Atg5 mRNA. This ultimately contributes to atherosclerosis [75,76]. In human aortic smooth muscle cells, the forced expression of miR-221/222 silences phosphatase and tensin homolog deleted on chromosome ten (PTEN) and subsequently activates the Akt signaling. This eventually inhibits autophagy by down-regulating the expression of LC3II and ATG5, as well as by elevating the expression of SQSTM1/p62 [77]. MiR-30b disrupts autophagy in vascular smooth muscle cells by decreasing autophagy-related genes, such as Atg5 and LC3II [78]. MiR-212 and miR-132 inhibit autophagy by negatively regulating the pro-autophagic transcription factor forkhead box O3 (FOXO3). The mice engineered to overexpress miR-212 and miR-132 in cardiomyocytes prematurely succumb because of pathological cardiac hypertrophy and heart failure [20,79]. Another critical microRNA that regulates cardiac autophagy is miR-199a. MiR-199a represses autophagy by indirectly activating the mechanistic target of rapamycin complex 1 (mTORC1), which leads to cardiac hypertrophy. In this model, tissue degeneration could be partially reversed by a potent autophagy inducer, rapamycin [14]. In addition, the silencing of miR-143 promotes the autophagy of the c-kit$^+$ cardiac progenitor cells in response to oxidative stress. Autophagy-related gene 7 (Atg7) is identified as the target gene of miR-143 in the autophagy pathway [80].

Mitophagy, which is responsible for the quality control of mitochondria to match the metabolic or developmental demands, is critical for cardiac homeostasis. Mitophagy is usually executed by a series of mitochondria functional molecules, such as Bcl2 interacting protein 3 (BNIP3L, best known as NIX), FUNDC1, Parkin, PTEN-induced putative kinase 1 (PINK 1), and mitofusin 2 (MFN2) [81–84]. MicroRNAs have been found to be involved in the mitophagy flux. MiR-137 markedly inhibits mitophagy without affecting global autophagy in response to hypoxia in neurocytes. A further exploration of the mechanism shows that miR-137 targets two mitophagy receptors, FUNDC1 and NIX, and downregulates their expression. The suppression of mitophagy mediated by miR-137 could be reversed by the forced expression of FUNDC1 and NIX [85]. MiR-27a and miR-27b prevent mitophagic influx by suppressing PINK1 expression at the translational level, which subsequently decreases ubiquitin phosphorylation, Parkin translocation, and LC3 II accumulation in damaged mitochondria. Furthermore, this inhibits the lysosomal degradation of the damaged mitochondria in neurocytes [86]. MiR-181a blocks the colocalization of mitochondria and autophagosomes/lysosomes by targeting Parkin. The overexpression of miR-181a inhibits the mitochondrial uncoupling agents, which induces mitophagy without affecting global autophagy. In contrast, the knockdown of miR-181 accelerates the mitophagy in neuroblastoma cells [87]. The role of microRNAs in cardiac mitophagy has been rarely reported. In the cardiac system, miR-410 inhibits the excessive mitophagy from cardiac ischemia/reperfusion injuries by modulating the heat shock protein B1 activity via a direct interaction with the 3′-untranslated region of high-mobility group box 1 protein [88].

The degradation of damaged cells, organelles, or intracellular components by autophagy has been demonstrated to prevent the aging of cardiomyocytes. It was found that the senescent myocardium is associated with impaired autophagy. The induction of autophagy could reverse age-dependent cardiac hypertrophy and diastolic dysfunction in old mice [89,90]. During the processes, AMP-activated protein kinase (AMPK) strongly promotes an autophagy flux, which consists of unc-51 like autophagy activating kinase (ULK) 1, Beclin 1, and phosphatidylinositol 3-kinase catalytic subunit type 3 (PIK3C3) [72,91–93]. MiR-20a and miR-106b reduce autophagy via the suppression of the ULK1 expression in myoblasts. The cells with the knockdown of endogenous miR-20a and miR-106b exhibit a normal autophagy activity [94]. MiR-19a-3p/19b-3p targeting the Beclin 1 pathway has been demonstrated to inhibit autophagy in cardiomyocytes [16]. MiR-221/222 inhibits autophagy by indirectly regulating Beclin 1 [77]. In vascular smooth muscle cells, Beclin 1 is a target of miR-30b to mediate autophagy [78]. MiR-155 plays a beneficial role in the aging myocardium by negatively regulating the autophagy-inhibitor, mTOR [95].

In summary, autophagy and mitophagy are deeply implicated in the cardiac homeostasis processes, including heart development, compensation, and aging, which could be widely regulated by microRNAs.

4. MicroRNAs in Autophagy-Related Heart Diseases

Autophagy is widely involved in cardiac physiology, and impaired autophagy or abnormal autophagy usually contribute to cardiac pathological disorders [96]. Recent research has suggested a close link between autophagy and heart diseases, including myocardial infarction [18], ischemia-reperfusion injury [66], myocardial hypertrophy [97], cardiac fibrosis [16], cardiomyopathy [98], and heart failure [17]. During the above processes, a variety of microRNAs have been demonstrated to play a critical role (Figure 2). MicroRNAs participate in regulating heart diseases through modulating the cardiac autophagic cell death. Many critical autophagy-related genes, such as ATG7, ATG9, LC3, p62, Beclin-1, AMPK, mTOR, BAG3, TNFα, and PARP-1, have been demonstrated to be direct or indirect targets of microRNAs in the autophagy-signaling pathways, which are involved in the pathogenesis of heart diseases [14,18,19,99–105].

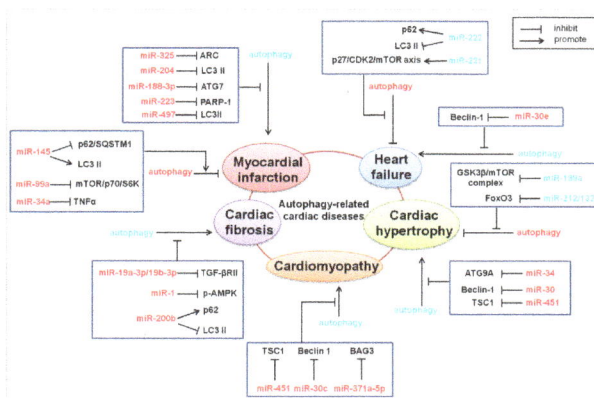

Figure 2. Summary of miRNAs regulating cardiac autophagy and related heart diseases. The molecular mechanisms of the miRNAs regulating the autophagy and autophagy-related heart diseases are shown in the figure. See text for detailed explanations. Arrows represent the promotion effect. T bars represent the inhibition effect. Red represents the negative regulation in diseases. Blue represents the positive regulation in diseases.

4.1. Myocardial Infarction

Myocardial infarction (MI) is pathologically defined as myocardial cell death due to prolonged ischemia or anoxia, which is the most severe manifestation of coronary heart disease. Cardiac cell death plays a decisive role in the pathogenesis of MI, because of the terminal differentiation and loss of regenerative ability of cardiomyocytes [106]. Autophagic cell death has been demonstrated to greatly contribute to the pathogenesis of MI. As a class of important autophagy regulators, microRNAs promote or inhibit MI by mediating the autophagy pathway.

A famous microRNA that is involved in autophagic cell death and MI is miR-325. MiR-325 is upregulated both in the cardiomyocytes treated with anoxia/reoxygenation, and in the hearts subjected to ischemia/reperfusion injury. The cardiomyocyte-specific overexpression of miR-325 potentiates excessive autophagic responses and myocardial infarct sizes, whereas the knockdown of miR-325 inhibited the autophagic cell death and MI. The apoptosis repressor with caspase recruit domain (ARC) is identified as the downstream mediator and the transcription factor, E2F1, as the upstream regulator of miR-325 in the autophagy program. A novel autophagic regulating model composed of E2F1, miR-325, and ARC in MI has been clarified in the literature [107]. The outstanding role of miR-30e in autophagic cell death and related MI has been demonstrated. MiR-30e protects hearts from MI. The cardioprotective mechanism of miR-30e has been explored. The level of miR-30e is dramatically decreased in the animal models of MI. The silencing of miR-30e significantly inhibits cellular apoptosis by modulating the apoptosis-related gene, Bax, and caspase-3, and meanwhile activates the autophagic flux and Notch1/Hes1/Akt signaling pathway. The effect of the knockdown of miR-30e on apoptosis and oxidative stress damage could be reversed significantly by autophagy inhibitor 3-methyladenine [108]. MiR-34a could inhibit autophagic cell death in the hearts that are subjected to ischemia/reperfusion via regulating tumor necrosis factor α (TNFα), thereby reducing myocardial injury [104]. MiR-145 plays a cardioprotective effect in myocardial infarction by promoting cardiac autophagy. Rabbits administered with miR-145 mimics exhibit a significantly smaller infarct size and improved cardiac function than that of the control group upon ischemia/reperfusion injury. Further study shows that miR-145 promotes autophagic cell death in cardiomyocytes by directly targeting fibroblast growth factor receptor substrate 2 (FRS2) mRNA, and it subsequently accelerates the transition of LC3B I to II and down-regulates p62/SQSTM1, which inhibits MI [19]. MiR-223 is up-regulated in rat hearts that have undergone coronary ligation. The overexpression of miR-223 inhibits excessive cardiac autophagy. Further mechanistic study reveals that miR-223 protects cardiomyocytes from excessive autophagy via the Akt/mTOR pathway by targeting poly (ADP-ribose) polymerase 1 (PARP-1) [105]. MiR-188-3p is down-regulated in the myocardial infarction model. The forced expression of miR-188-3p suppresses autophagic cell death by targeting ATG7, which subsequently reduces the infarct sizes and improves cardiac function [18]. MiR-204 could inhibit autophagy by targeting LC3 II upon ischemia-reperfusion injury [66]. MiR-99a plays a cardioprotective role in the post-MI left ventricle remodeling by preventing cell apoptosis and increasing autophagy via an mTOR/p70/S6K pathway, which improves both the cardiac function and survival rate in a murine model of MI [109]. Myocardial miR-497 is dramatically down-regulated in the murine hearts subjected to MI, and in the cardiomyocytes subjected to hypoxia/reoxygenation. The overexpression of miR-497 induces apoptosis by targeting the anti-apoptosis gene, Bcl-2, and inhibits autophagy by targeting the autophagy gene, LC3BII, while the silencing of miR-497 exhibits the opposite effect in response to MI [110].

4.2. Cardiac Hypertrophy

A growing number of studies have revealed a critical pathogenic role of the altered activity of autophagy in cardiac hypertrophy. It was reported that cardiac autophagy was increased in a maladaptive manner in the hearts that were subjected to a pressure overload. However, some other studies have suggested that myocardial autophagy is insufficient, which results from a chronic pressure overload and contributes to maladaptive cardiac remodeling and heart failure [111]. MicroRNAs have been demonstrated to participate in regulating autophagic cell death in cardiac hypertrophy. In angiotensin-induced cardiac hypertrophy, miR-30 and miR-34 play an anti-hypertrophy role by preventing autophagy [97,99]. MiR-451 has been demonstrated to be involved in suppressing the autophagosome formation by targeting tuberous sclerosis complex1 (TSC1), which inhibits abnormal autophagy upon hypertrophic stimuli. By controlling the autophagy process, the ectopic overexpression of miR-451 attenuates cardiac hypertrophy, while the knockdown of miR-451 accelerates hypertrophy [112]. The MiR-212/132 family promotes pathological cardiac hypertrophy and heart failure by directly targeting the anti-hypertrophic and pro-autophagic transcription factor, FoxO3. The MiR-212/132 null mice do not frequently develop hypertrophy and heart failure, whereas the overexpression of miR-212/132 leads to the hyperactivation of pro-hypertrophic calcineurin/NFAT signaling and an impaired autophagic response in response to starvation [20]. Cardiac specific miR-199a transgenic mice suffer cardiac hypertrophy, accompanied with decreased autophagy levels. The enhancement of autophagy by the forced expression of Atg5 attenuates the hypertrophic effects of overexpression of miR-199a on cardiomyocytes. In exploring the molecular mechanism, miR-199a has been demonstrated to target the glycogen synthase kinase 3β (GSK3β)/mTOR complex signaling pathway in modulating autophagy and cardiac hypertrophy [14].

4.3. Cardiac Fibrosis

A link between autophagy and cardiac fibroblasts has been demonstrated recently. However, the exploration on the molecular mechanism, including the role of microRNAs in autophagy-related cardiac fibrosis, is insufficient. It was reported that miR-19a-3p/19b-3p could inhibit epithelial mesenchymal transition, extracellular matrix production, and the invasion of human cardiac fibroblasts by targeting TGF-β RII. Moreover, the enhancement of autophagy rescues the inhibition effect of miR-19a-3p/19b-3p on cardiac fibroblasts. These results suggest that miR-19a-3p/19b-3p exhibits an anti-fibroblast effect through regulating cardiac autophagy [16]. MiR-1 promotes high glucose-induced cardiac fibrosis by down-regulating pro-autophagic p-AMPK [102]. MiR-200b controls cardiac fibroblast autophagy during cardiac fibrosis. The expression level of miR-200b is decreased in the cardiac fibrosis model. The MiR-200b mimic inhibits the LC3BII/I ratio, increases p62, and alleviates cardiac fibroblast autophagy, whereas the knockdown of miR-200b exhibits an opposite effect [113].

4.4. Cardiomyopathy

MicroRNA-mediated autophagy is involved in the pathogenesis of several cardiomyopathy. Until now, miR-30c, miR-371a-5p, and miR-451 have been precisely demonstrated to regulate cardiac autophagy in cardiomyopathy [98,103,112]. The depletion of miR-30c enhances autophagic cell death in diabetic hearts. The overexpression of miR-30c inhibits autophagy in diabetic hearts and subsequently improves cardiac function and structure in the diabetic mice model. In exploring the molecular mechanisms, it was found that miR-30c suppressed autophagy in diabetic cardiomyopathy via targeting BECN1, through direct binding to BECN1 3′ UTR [98]. Bcl-2-associated athanogene 3 (BAG3), an autophagy pathway mediator, is a direct target of miR-371a-5p in healthy donors. Mutations and polymorphisms, including a frequent nucleotide change g2252c in the BAG3 3′-untranslated region (3′-UTR) of Takotsubo patients, leads to the loss of binding to miR-371a-5p, which probably contributes to the Takotsubo cardiomyopathy (TTC) pathogenesis [103]. MiR-451 is down-regulated in the heart tissues from hypertrophic cardiomyopathy (HCM) patients. The ectopic overexpression of miR-451 inhibits the autophagosome formation through targetedly suppressing tuberous sclerosis complex 1 (TSC1), and subsequently decreasing the size of the cardiomyocytes [112].

4.5. Heart Failure

Emerging evidence confirms that restoring autophagy, which improves bulk protein degradation, is proved to be beneficial in heart failure [5]. The cardioprotective effect of autophagy on heart failure could be mediated by microRNAs. MiR-221 has emerged as a representative regulator of autophagy-mediated heart failure. Research suggests that the cardiac-specific overexpression of miR-221 in mice leads to cardiac dysfunction and heart failure, and reduces the autophagic flux by inhibiting the autophagic vesicle formation in the meanwhile. Further study showed that miR-221 inhibits autophagy and promotes heart failure by modulating the p27/CDK2/mTOR axis [17]. MiR-222, which shares the same gene cluster with miR-221, exhibits a potential function in autophagic cell death and heart failure. Transgenic mice with a cardiac-specific expression of miR-222 significantly develop heart failure and are accompanied with autophagy inhibition. MiR-221 downregulates LC3 II, upregulates p62, and activates the mTOR pathway, all of which are critical autophagy regulators [100]. Another microRNA regulating cardiac autophagy in heart failure is miR-30e. MiR-30e inhibits the autophagy of cardiomyocytes by down-regulating the Beclin-1 expression, and subsequently mediates the cardioprotection of the angiotensin-converting enzyme 2 (ACE2) in the rats with Dox-induced heart failure [101].

In summary, microRNAs targeting the autophagy signaling pathways in cardiac disorders have been demonstrated by numerous evidence. The autophagic cell death mediated by microRNAs plays a critical role in the pathogenesis of heart diseases.

5. Conclusions and Perspectives

As outlined in this review, miRNAs function as pro- or anti-regulators in cardiac autophagy, by targeting extensive signaling pathways. This knowledge has improved our understanding of the role and mechanism of autophagy in cardiac diseases. Mitophagy is one critical part of the autophagy processes, which plays a decisive role in cardiac physiology and pathology. However, studies on the role of microRNAs in cardiac mitophagy are greatly insufficient, and thus more explorations need to be carried out in the future. Considering that, usually, one miRNA targets several genes or several miRNAs target one gene in autophagy regulation, the crosslink between the different pathways needs to be further explored. Autophagy is an evolutionarily conserved self-protective process and plays a critical role in maintaining cell homeostasis. However, excessive autophagy usually contributes to the pathogenesis of a variety of diseases, including cardiac disorders. Up until now, it has been a mystery as to how beneficial autophagy turns into a harmful autophagic cell death. With respect to MI as well as cardiac hypertrophy and heart failure, the miRNAs-regulated autophagy probably functions as the promoter or inhibitor in the pathogenesis. For instance, miR-145 plays a cardioprotective effect in MI by promoting cardiac autophagy [19], while miR-188-3p inhibits MI via inactivating the autophagy pathway [18]. MiR-451 functions as a cardiac hypertrophy inhibitor by activating autophagy [112], while the knockdown of miR-199a suppresses cardiac hypertrophy by promoting autophagic cell death [14]. MiR-221 inhibits autophagy and promotes heart failure [100], while miR-30e plays a cardioprotective role in heart failure by modulating autophagy [101]. Whether miRNA-regulated autophagy plays the role of an angel or devil in cardiac pathology, remains controversial. Cardiac cell death is the cytological basis of cardiac disorders. Apart from autophagy, several other types of cell death, such as apoptosis, are involved in the cardiac pathology. In the exploration of miRNA regulating cardiac autophagy, the apoptosis pathway is in crosslink with the autophagy pathway. For example, miR-223, miR-99a, and miR-30e regulate both the cardiac autophagy process and the apoptosis process [101,105,109]. Therefore, research focusing on the relationship of different types of cardiac cell death is necessary in the future. The important role of microRNAs in regulating cardiac autophagy has been demonstrated by accumulated evidence from studies. However, there are still many fuzzy and controversial issues that need to be explored in the future.

Author Contributions: Conceptualization, T.S. and J.-M.C.; Methodology, T.S. and M.-Y.L.; Software, T.S. and M.-Y.L.; Validation, T.S. and J.-M.C.; Formal Analysis, T.S., P.-F.L. and J.-M.C.; Investigation, T.S., M.-Y.L. and P.-F.L.; Resources, T.S.; Data Curation, T.S. and J.-M.C.; Writing-Original Draft Preparation, T.S., M.-Y.L. and J.-M.C.; Writing-Review & Editing, T.S. and J.-M.C.; Visualization, T.S. and J.-M.C.; Supervision, T.S. and J.-M.C.; Project Administration, T.S. and J.-M.C.; Funding Acquisition, T.S. and J.-M.C.

Acknowledgments: This work was supported by the China Postdoctoral Science Foundation (2016M592134), the Qingdao Postdoctoral Application Research Project (2016074), the National Natural Science Foundation of China (81670313), and the fund for Shanxi '1331 Project' Key Subjects Construction (1331KSC).

Conflicts of Interest: The authors declare no conflict of interest in this review.

Abbreviations

ACE2	angiotensin-converting enzyme 2
AGOs	argonaute proteins
AMPK	AMP- activated protein kinase
ARC	apoptosis repressor with caspase recruit domain
ATG	autophagy-related gene
BAG3	Bcl-2-associated athanogene 3
Bcl2-L-13	Bcl2-like protein 13
BNIP3	Bcl-2 interacting protein 3
BNIP3L	Bcl2 interacting protein 3 like
CDS	coding sequence
CISD2	CDGSH iron sulfur domain 2
FIP200	focal adhesion kinase family interacting protein of 200 kDa
FOXO3	forkhead box O3
FUNDC1	FUN14 domain containing protein 13
GSK3β	glycogen synthase kinase 3β
HCM	hypertrophic cardiomyopathy
MFN2	mitofusion2
miRNAs	microRNAs
mTOR	mammalian target of rapamycin
mTORC1	mTOR complex 1
PARP-1	poly (ADP-ribose) polymerase 1
PE	phosphatidylethanolamine
PIK3C3	phosphatidylinositol 3-kinase catalytic subunit type 3
PINK1	putative kinase 1
PTEN	phosphatase and tensin homolog deleted on chromosome ten
TSC1	tuberous sclerosis complex 1
TTC	Takatsubo cardiomyopathy
ULK	Unc-51-like kinase
UTR	untranslated region
UVRAG	UV irradiation resistance-associated gene

References

1. He, C.; Klionsky, D.J. Regulation mechanisms and signaling pathways of autophagy. *Annu. Rev. Genet.* **2009**, *43*, 67–93. [CrossRef] [PubMed]
2. Legakis, J.E.; Yen, W.L.; Klionsky, D.J. A cycling protein complex required for selective autophagy. *Autophagy* **2007**, *3*, 422–432. [CrossRef] [PubMed]
3. Kiriyama, Y.; Nochi, H. Intra- and intercellular quality control mechanisms of mitochondria. *Cells* **2017**, *7*, 1. [CrossRef] [PubMed]
4. Harper, J.W.; Ordureau, A.; Heo, J.M. Building and decoding ubiquitin chains for mitophagy. *Nat. Rev. Mol. Cell Biol.* **2018**, *19*, 93–108. [CrossRef] [PubMed]
5. Ghosh, R.; Pattison, J.S. Macroautophagy and chaperone-mediated autophagy in heart failure: The known and the unknown. *Oxid. Med. Cell. Longev.* **2018**, *2018*. [CrossRef] [PubMed]
6. Liu, H.; Dai, C.; Fan, Y.; Guo, B.; Ren, K.; Sun, T.; Wang, W. From autophagy to mitophagy: The roles of p62 in neurodegenerative diseases. *J. Bioenerg. Biomembr.* **2017**, *49*, 413–422. [CrossRef] [PubMed]
7. Yan, C.; Li, T.S. Dual role of mitophagy in cancer drug resistance. *Anticancer Res.* **2018**, *38*, 617–621. [PubMed]
8. Herst, P.M.; Rowe, M.R.; Carson, G.M.; Berridge, M.V. Functional mitochondria in health and disease. *Front. Endocrinol.* **2017**, *8*. [CrossRef] [PubMed]

9. Cummins, N.; Gotz, J. Shedding light on mitophagy in neurons: What is the evidence for pink1/parkin mitophagy in vivo? *Cell. Mol. Life Sci.* **2018**, *75*, 1151–1162. [CrossRef] [PubMed]

10. Zhang, J.; Liu, L.; Xue, Y.; Ma, Y.; Liu, X.; Li, Z.; Li, Z.; Liu, Y. Endothelial monocyte-activating polypeptide-ii induces bnip3-mediated mitophagy to enhance temozolomide cytotoxicity of glioma stem cells via down-regulating mir-24-3p. *Front. Mol. Neurosci.* **2018**, *11*, 92. [CrossRef] [PubMed]

11. Murakawa, T.; Yamaguchi, O.; Hashimoto, A.; Hikoso, S.; Takeda, T.; Oka, T.; Yasui, H.; Ueda, H.; Akazawa, Y.; Nakayama, H.; et al. Bcl-2-like protein 13 is a mammalian atg32 homologue that mediates mitophagy and mitochondrial fragmentation. *Nat. Commun.* **2015**, *6*, 7527. [CrossRef] [PubMed]

12. Zhou, H.; Zhu, P.; Wang, J.; Zhu, H.; Ren, J.; Chen, Y. Pathogenesis of cardiac ischemia reperfusion injury is associated with ck2alpha-disturbed mitochondrial homeostasis via suppression of fundc1-related mitophagy. *Cell Death Differ.* **2018**, *25*, 1080. [CrossRef] [PubMed]

13. Liu, C.Y.; Zhang, Y.H.; Li, R.B.; Zhou, L.Y.; An, T.; Zhang, R.C.; Zhai, M.; Huang, Y.; Yan, K.W.; Dong, Y.H.; et al. Lncrna caif inhibits autophagy and attenuates myocardial infarction by blocking *p53*-mediated myocardin transcription. *Nat. Commun.* **2018**, *9*, 29. [CrossRef] [PubMed]

14. Li, Z.; Song, Y.; Liu, L.; Hou, N.; An, X.; Zhan, D.; Li, Y. Mir-199a impairs autophagy and induces cardiac hypertrophy through mtor activation. *Cell Death Differ.* **2017**, *24*, 1205–1213. [CrossRef] [PubMed]

15. Wang, S.; Wang, C.; Turdi, S.; Richmond, K.L.; Zhang, Y.; Ren, J. ALDH2 protects against high fat diet-induced obesity cardiomyopathy and defective autophagy: Role of CaM kinase II, histone H3K9 methyltransferase SUV39H, Sirt1, and PGC-1α deacetylation. *Int. J. Obes.* **2018**, *42*, 1073–1087. [CrossRef] [PubMed]

16. Zou, M.; Wang, F.; Gao, R.; Wu, J.; Ou, Y.; Chen, X.; Wang, T.; Zhou, X.; Zhu, W.; Li, P.; et al. Autophagy inhibition of hsa-miR-19a-3p/19b-3p by targeting TGF-β R II during TGF-β1-induced fibrogenesis in human cardiac fibroblasts. *Sci. Rep.* **2016**, *6*, 24747. [CrossRef] [PubMed]

17. Su, M.; Wang, J.; Wang, C.; Wang, X.; Dong, W.; Qiu, W.; Wang, Y.; Zhao, X.; Zou, Y.; Song, L.; et al. MicroRNA-221 inhibits autophagy and promotes heart failure by modulating the p27/CDK2/mTOR axis. *Cell Death Differ.* **2015**, *22*, 986–999. [CrossRef] [PubMed]

18. Wang, K.; Liu, C.Y.; Zhou, L.Y.; Wang, J.X.; Wang, M.; Zhao, B.; Zhao, W.K.; Xu, S.J.; Fan, L.H.; Zhang, X.J.; et al. APF lncRNA regulates autophagy and myocardial infarction by targeting miR-188-3p. *Nat. Commun.* **2015**, *6*, 6779. [CrossRef] [PubMed]

19. Higashi, K.; Yamada, Y.; Minatoguchi, S.; Baba, S.; Iwasa, M.; Kanamori, H.; Kawasaki, M.; Nishigaki, K.; Takemura, G.; Kumazaki, M.; et al. MicroRNA-145 repairs infarcted myocardium by accelerating cardiomyocyte autophagy. *Am. J. Physiology-Heart Circ. Physiol.* **2015**, *309*, H1813–H1826. [CrossRef] [PubMed]

20. Ucar, A.; Gupta, S.K.; Fiedler, J.; Erikci, E.; Kardasinski, M.; Batkai, S.; Dangwal, S.; Kumarswamy, R.; Bang, C.; Holzmann, A.; et al. The miRNA-212/132 family regulates both cardiac hypertrophy and cardiomyocyte autophagy. *Nat. Commun.* **2012**, *3*, 1078. [CrossRef] [PubMed]

21. Chen, Z.; Li, Y.; Wang, Y.; Qian, J.; Ma, H.; Wang, X.; Jiang, G.; Liu, M.; An, Y.; Ma, L.; et al. Cardiomyocyte-restricted low density lipoprotein receptor-related protein 6 (LRP6) deletion leads to lethal dilated cardiomyopathy partly through Drp1 signaling. *Theranostics* **2018**, *8*, 627–643. [CrossRef] [PubMed]

22. Carresi, C.; Musolino, V.; Gliozzi, M.; Maiuolo, J.; Mollace, R.; Nucera, S.; Maretta, A.; Sergi, D.; Muscoli, S.; Gratteri, S.; et al. Anti-oxidant effect of bergamot polyphenolic fraction counteracts doxorubicin-induced cardiomyopathy: Role of autophagy and c-kit[pos]CD45[neg]CD31[neg] cardiac stem cell activation. *J. Mol. Cell. Cardiol.* **2018**, *119*, 10–18. [CrossRef] [PubMed]

23. Ambros, V. The functions of animal microRNAs. *Nature* **2004**, *431*, 350–355. [CrossRef] [PubMed]

24. Bartel, D.P. Micrornas: Genomics, biogenesis, mechanism, and function. *Cell* **2004**, *116*, 281–297. [CrossRef]

25. Lee, Y.; Ahn, C.; Han, J.; Choi, H.; Kim, J.; Yim, J.; Lee, J.; Provost, P.; Radmark, O.; Kim, S.; et al. The nuclear RNase III drosha initiates microRNA processing. *Nature* **2003**, *425*, 415–419. [CrossRef] [PubMed]

26. Han, J.; Lee, Y.; Yeom, K.H.; Nam, J.W.; Heo, I.; Rhee, J.K.; Sohn, S.Y.; Cho, Y.; Zhang, B.T.; Kim, V.N. Molecular basis for the recognition of primary micrornas by the Drosha-DGCR8 complex. *Cell* **2006**, *125*, 887–901. [CrossRef] [PubMed]

27. Lund, E.; Guttinger, S.; Calado, A.; Dahlberg, J.E.; Kutay, U. Nuclear export of microRNA precursors. *Science* **2004**, *303*, 95–98. [CrossRef] [PubMed]

28. Hutvagner, G.; McLachlan, J.; Pasquinelli, A.E.; Balint, E.; Tuschl, T.; Zamore, P.D. A cellular function for the RNA-interference enzyme Dicer in the maturation of the let-7 small temporal RNA. *Science* **2001**, *293*, 834–838. [CrossRef] [PubMed]

29. Filipowicz, W.; Bhattacharyya, S.N.; Sonenberg, N. Mechanisms of post-transcriptional regulation by microRNAs: Are the answers in sight? *Nat. Rev. Genet.* **2008**, *9*, 102–114. [CrossRef] [PubMed]

30. Eulalio, A.; Huntzinger, E.; Izaurralde, E. Getting to the root of miRNA-mediated gene silencing. *Cell* **2008**, *132*, 9–14. [CrossRef] [PubMed]

31. Su, Z.; Yang, Z.; Xu, Y.; Chen, Y.; Yu, Q. Micrornas in apoptosis, autophagy and necroptosis. *Oncotarget* **2015**, *6*, 8474–8490. [CrossRef] [PubMed]

32. Llorens, F.; Thune, K.; Marti, E. Regional and subtype-dependent miRNA signatures in sporadic Creutzfeldt-Jakob disease are accompanied by alterations in miRNA silencing machinery and biogenesis. *PloS Pathog.* **2018**, *14*, e1006802. [CrossRef] [PubMed]

33. Zhou, L.; Ma, B.; Han, X. The role of autophagy in angiotensin II induced pathological cardiac hypertrophy. *J. Mol. Endocrinol.* **2016**, *57*, R143–R152. [CrossRef] [PubMed]

34. Frankel, L.B.; Lund, A.H. Microrna regulation of autophagy. *Carcinogenesis* **2012**, *33*, 2018–2025. [CrossRef] [PubMed]

35. Chen, H.; Zhang, Z.; Lu, Y.; Song, K.; Liu, X.; Xia, F.; Sun, W. Downregulation of UKL1 by microRNA-372 inhibits the survival of human pancreatic adenocarcinoma cells. *Cancer Sci.* **2017**, *108*, 1811–1819. [CrossRef] [PubMed]

36. Duan, X.; Zhang, T.; Ding, S.; Wei, J.; Su, C.; Liu, H.; Xu, G. MicroRNA-17-5p modulates bacille calmette-guerin growth in RAW264.7 cells by targeting ULK1. *PloS ONE* **2015**, *10*, e0138011. [CrossRef] [PubMed]

37. Rothschild, S.I.; Gautschi, O.; Batliner, J.; Gugger, M.; Fey, M.F.; Tschan, M.P. MicroRNA-106a targets autophagy and enhances sensitivity of lung cancer cells to Src inhibitors. *Lung Cancer* **2017**, *107*, 73–83. [CrossRef] [PubMed]

38. Huang, Y.; Chuang, A.Y.; Ratovitski, E.A. Phospho-ΔNp63α/miR-885–3p axis in tumor cell life and cell death upon cisplatin exposure. *Cell Cycle* **2011**, *10*, 3938–3947. [CrossRef] [PubMed]

39. Li, Z.; Li, J.; Tang, N. Long noncoding RNA Malat1 is a potent autophagy inducer protecting brain microvascular endothelial cells against oxygen-glucose deprivation/reoxygenation-induced injury by sponging miR-26b and upregulating ULK2 expression. *Neuroscience* **2017**, *354*, 1–10. [CrossRef] [PubMed]

40. John Clotaire, D.Z.; Zhang, B.; Wei, N.; Gao, R.; Zhao, F.; Wang, Y.; Lei, M.; Huang, W. Mir-26b inhibits autophagy by targeting ULK2 in prostate cancer cells. *Biochem. Biophys. Res. Commun.* **2016**, *472*, 194–200. [CrossRef] [PubMed]

41. Li, S.; Qiang, Q.; Shan, H.; Shi, M.; Gan, G.; Ma, F.; Chen, B. MiR-20a and miR-20b negatively regulate autophagy by targeting RB1CC1/FIP200 in breast cancer cells. *Life Sci.* **2016**, *147*, 143–152. [CrossRef] [PubMed]

42. Cheng, Y.; Ren, X.; Hait, W.N.; Yang, J.M. Therapeutic targeting of autophagy in disease: Biology and pharmacology. *Pharmacol. Rev.* **2013**, *65*, 1162–1197. [CrossRef] [PubMed]

43. Huang, Y.; Guerrero-Preston, R.; Ratovitski, E.A. Phospho-ΔNp63α-dependent regulation of autophagic signaling through transcription and micro-RNA modulation. *Cell Cycle* **2012**, *11*, 1247–1259. [CrossRef] [PubMed]

44. Chen, J.; Yu, Y.; Li, S.; Liu, Y.; Zhou, S.; Cao, S.; Yin, J.; Li, G. Micro RNA-30a ameliorates hepatic fibrosis by inhibiting Beclin1-mediated autophagy. *J. Cell. Mol. Med.* **2017**, *21*, 3679–3692. [CrossRef] [PubMed]

45. Hou, W.; Song, L.; Zhao, Y.; Liu, Q.; Zhang, S. Inhibition of Beclin-1-mediated autophagy by microRNA-17-5p enhanced the radiosensitivity of glioma cells. *Oncol. Res. Featur. Preclin. Clin. Cancer Ther.* **2017**, *25*, 43–53. [CrossRef] [PubMed]

46. Zhao, K.; Zhang, Y.; Kang, L.; Song, Y.; Wang, K.; Li, S.; Wu, X.; Hua, W.; Shao, Z.; Yang, S.; et al. Methylation of microRNA-129-5p modulates nucleus pulposus cell autophagy by targeting Beclin-1 in intervertebral disc degeneration. *Oncotarget* **2017**, *8*, 86264–86276. [CrossRef] [PubMed]

47. Zhang, X.; Shi, H.; Lin, S.; Ba, M.; Cui, S. MicroRNA-216a enhances the radiosensitivity of pancreatic cancer cells by inhibiting beclin-1-mediated autophagy. *Oncol. Rep.* **2015**, *34*, 1557–1564. [CrossRef] [PubMed]

48. Korkmaz, G.; le Sage, C.; Tekirdag, K.A.; Agami, R.; Gozuacik, D. miR-376b controls starvation and mTOR inhibition-related autophagy by targeting ATG4C and BECN1. *Autophagy* **2012**, *8*, 165–176. [CrossRef] [PubMed]

49. Sun, A.G.; Meng, F.G.; Wang, M.G. Cisd2 promotes the proliferation of glioma cells via suppressing beclin1mediated autophagy and is targeted by microRNA449a. *Mol. Med. Rep.* **2017**, *16*, 7939–7948. [CrossRef] [PubMed]

50. Han, R.; Ji, X.; Rong, R.; Li, Y.; Yao, W.; Yuan, J.; Wu, Q.; Yang, J.; Yan, W.; Han, L.; et al. MiR-449a regulates autophagy to inhibit silica-induced pulmonary fibrosis through targeting Bcl2. *J. Mol. Med.* **2016**, *94*, 1267–1279. [CrossRef] [PubMed]

51. Chen, G.; Gao, X.; Wang, J.; Yang, C.; Wang, Y.; Liu, Y.; Zou, W.; Liu, T. Hypoxia-induced microRNA-146a represses Bcl-2 through Traf6/IRAK1 but not Smad4 to promote chondrocyte autophagy. *Biol. Chem.* **2017**, *398*, 499–507. [CrossRef] [PubMed]

52. Hu, J.L.; He, G.Y.; Lan, X.L.; Zeng, Z.C.; Guan, J.; Ding, Y.; Qian, X.L.; Liao, W.T.; Ding, Y.Q.; Liang, L. Inhibition of ATG12-mediated autophagy by miR-214 enhances radiosensitivity in colorectal cancer. *Oncogenesis* **2018**, *7*, 16. [CrossRef] [PubMed]

53. Tan, D.; Zhou, C.; Han, S.; Hou, X.; Kang, S.; Zhang, Y. MicroRNA-378 enhances migration and invasion in cervical cancer by directly targeting autophagy-related protein 12. *Mol. Med. Rep.* **2018**, *17*, 6319–6326. [CrossRef] [PubMed]

54. Yang, K.; Yu, B.; Cheng, C.; Cheng, T.; Yuan, B.; Li, K.; Xiao, J.; Qiu, Z. *Mir505-3p* regulates axonal development via inhibiting the autophagy pathway by targeting *Atg12*. *Autophagy* **2017**, *13*, 1679–1696. [CrossRef] [PubMed]

55. Sun, L.; Liu, A.; Zhang, J.; Ji, W.; Li, Y.; Yang, X.; Wu, Z.; Guo, J. miR-23b improves cognitive impairments in traumatic brain injury by targeting ATG12-mediated neuronal autophagy. *Behav. Brain Res.* **2018**, *340*, 126–136. [CrossRef] [PubMed]

56. Yang, J.; He, Y.; Zhai, N.; Ding, S.; Li, J.; Peng, Z. MicroRNA-181a inhibits autophagy by targeting Atg5 in hepatocellular carcinoma. *Front. Biosci. (Landmark edition)* **2018**, *23*, 388–396.

57. Fu, X.T.; Shi, Y.H.; Zhou, J.; Peng, Y.F.; Liu, W.R.; Shi, G.M.; Gao, Q.; Wang, X.Y.; Song, K.; Fan, J.; et al. MicroRNA-30a suppresses autophagy-mediated anoikis resistance and metastasis in hepatocellular carcinoma. *Cancer Lett.* **2018**, *412*, 108–117. [CrossRef] [PubMed]

58. Zhang, K.; Chen, J.; Zhou, H.; Chen, Y.; Zhi, Y.; Zhang, B.; Chen, L.; Chu, X.; Wang, R.; Zhang, C. Pu.1/microRNA-142-3p targets ATG5/ATG16L1 to inactivate autophagy and sensitize hepatocellular carcinoma cells to sorafenib. *Cell Death Dis.* **2018**, *9*, 312. [CrossRef] [PubMed]

59. Gan, J.; Cai, Q.; Qu, Y.; Zhao, F.; Wan, C.; Luo, R.; Mu, D. MiR-96 attenuates status epilepticus-induced brain injury by directly targeting Atg7 and Atg16l1. *Sci. Rep.* **2017**, *7*, 10270. [CrossRef] [PubMed]

60. Wang, C.; Zhang, Z.Z.; Yang, W.; Ouyang, Z.H.; Xue, J.B.; Li, X.L.; Zhang, J.; Chen, W.K.; Yan, Y.G.; Wang, W.J. MiR-210 facilitates ECM degradation by suppressing autophagy via silencing of ATG7 in human degenerated NP cells. *Biomed. Pharmacother.* **2017**, *93*, 470–479. [CrossRef] [PubMed]

61. Comincini, S.; Allavena, G.; Palumbo, S.; Morini, M.; Durando, F.; Angeletti, F.; Pirtoli, L.; Miracco, C. MicroRNA-17 regulates the expression of ATG7 and modulates the autophagy process, improving the sensitivity to temozolomide and low-dose ionizing radiation treatments in human glioblastoma cells. *Cancer Biol. Ther.* **2013**, *14*, 574–586. [CrossRef] [PubMed]

62. Liu, L.; Shen, W.; Zhu, Z.; Lin, J.; Fang, Q.; Ruan, Y.; Zhao, H. Combined inhibition of EGFR and c-ABL suppresses the growth of fulvestrant-resistant breast cancer cells through miR-375-autophagy axis. *Biochem. Bioph. Res. Commun.* **2018**, *498*, 559–565. [CrossRef] [PubMed]

63. He, W.; Cheng, Y. Inhibition of miR-20 promotes proliferation and autophagy in articular chondrocytes by PI3K/AKT/mTOR signaling pathway. *Biomed. Pharmacother.* **2018**, *97*, 607–615. [CrossRef] [PubMed]

64. Wu, Y.; Dai, X.; Ni, Z.; Yan, X.; He, F.; Lian, J. The downregulation of ATG4B mediated by microRNA-34a/34c-5p suppresses rapamycin-induced autophagy. *Irani. J. Basic Med. Sci.* **2017**, *20*, 1125–1130.

65. Frankel, L.B.; Wen, J.; Lees, M.; Hoyer-Hansen, M.; Farkas, T.; Krogh, A.; Jaattela, M.; Lund, A.H. MicroRNA-101 is a potent inhibitor of autophagy. *EMBO J.* **2011**, *30*, 4628–4641. [CrossRef] [PubMed]

66. Xiao, J.; Zhu, X.; He, B.; Zhang, Y.; Kang, B.; Wang, Z.; Ni, X. MiR-204 regulates cardiomyocyte autophagy induced by ischemia-reperfusion through LC3-II. *J. Biomed. Sci.* **2011**, *18*, 35. [CrossRef] [PubMed]

67. Ying, Q.; Liang, L.; Guo, W.; Zha, R.; Tian, Q.; Huang, S.; Yao, J.; Ding, J.; Bao, M.; Ge, C.; et al. Hypoxia-inducible microRNA-210 augments the metastatic potential of tumor cells by targeting vacuole membrane protein 1 in hepatocellular carcinoma. *Hepatology* **2011**, *54*, 2064–2075. [CrossRef] [PubMed]

68. Rotter, D.; Rothermel, B.A. Targets, trafficking, and timing of cardiac autophagy. *Pharmacol. Res.* **2012**, *66*, 494–504. [CrossRef] [PubMed]

69. Wang, Z.V.; Rothermel, B.A.; Hill, J.A. Autophagy in hypertensive heart disease. *J. Biol. Chem.* **2010**, *285*, 8509–8514. [CrossRef] [PubMed]

70. Yang, J.; Chen, D.; He, Y.; Melendez, A.; Feng, Z.; Hong, Q.; Bai, X.; Li, Q.; Cai, G.; Wang, J.; et al. MiR-34 modulates caenorhabditis elegans lifespan via repressing the autophagy gene atg9. *Age* **2013**, *35*, 11–22. [CrossRef] [PubMed]

71. Kovaleva, V.; Mora, R.; Park, Y.J.; Plass, C.; Chiramel, A.I.; Bartenschlager, R.; Dohner, H.; Stilgenbauer, S.; Pscherer, A.; Lichter, P.; et al. MircoRNA-130a targets ATG2b and DICER1 to inhibit autophagy and trigger killing of chronic lymphocytic leukemia cells. *Cancer Res.* **2012**, *72*, 1763–1772. [CrossRef] [PubMed]

72. Bravo-San Pedro, J.M.; Kroemer, G.; Galluzzi, L. Autophagy and mitophagy in cardiovascular disease. *Circ. Res.* **2017**, *120*, 1812–1824. [CrossRef] [PubMed]

73. Nakai, A.; Yamaguchi, O.; Takeda, T.; Higuchi, Y.; Hikoso, S.; Taniike, M.; Omiya, S.; Mizote, I.; Matsumura, Y.; Asahi, M.; et al. The role of autophagy in cardiomyocytes in the basal state and in response to hemodynamic stress. *Nat. Med.* **2007**, *13*, 619–624. [CrossRef] [PubMed]

74. Taneike, M.; Yamaguchi, O.; Nakai, A.; Hikoso, S.; Takeda, T.; Mizote, I.; Oka, T.; Tamai, T.; Oyabu, J.; Murakawa, T.; et al. Inhibition of autophagy in the heart induces age-related cardiomyopathy. *Autophagy* **2010**, *6*, 600–606. [CrossRef] [PubMed]

75. Ouimet, M.; Ediriweera, H.; Afonso, M.S.; Ramkhelawon, B.; Singaravelu, R.; Liao, X.; Bandler, R.C.; Rahman, K.; Fisher, E.A.; Rayner, K.J.; et al. MicroRNA-33 regulates macrophage autophagy in atherosclerosis. *Arterioscleros. Thromb. Vasc. Biol.* **2017**, *37*, 1058–1067. [CrossRef] [PubMed]

76. Wang, J.; Wang, W.N.; Xu, S.B.; Wu, H.; Dai, B.; Jian, D.D.; Yang, M.; Wu, Y.T.; Feng, Q.; Zhu, J.H.; et al. MicroRNA-214-3p: A link between autophagy and endothelial cell dysfunction in atherosclerosis. *Acta Physiol.* **2018**, *222*, e12973. [CrossRef] [PubMed]

77. Li, L.; Wang, Z.; Hu, X.; Wan, T.; Wu, H.; Jiang, W.; Hu, R. Human aortic smooth muscle cell-derived exosomal miR-221/222 inhibits autophagy via a PTEN/Akt signaling pathway in human umbilical vein endothelial cells. *Biochem. Biophys. Res. Commun.* **2016**, *479*, 343–350. [CrossRef] [PubMed]

78. Wang, J.; Sun, Y.T.; Xu, T.H.; Sun, W.; Tian, B.Y.; Sheng, Z.T.; Sun, L.; Liu, L.L.; Ma, J.F.; Wang, L.N.; et al. MicroRNA-30b regulates high phosphorus level-induced autophagy in vascular smooth muscle cells by targeting BECN1. *Cell. Physiol. Biochem.* **2017**, *42*, 530–536. [CrossRef] [PubMed]

79. Pietrocola, F.; Izzo, V.; Niso-Santano, M.; Vacchelli, E.; Galluzzi, L.; Maiuri, M.C.; Kroemer, G. Regulation of autophagy by stress-responsive transcription factors. *Semin. Cancer Biol.* **2013**, *23*, 310–322. [CrossRef] [PubMed]

80. Ma, W.; Ding, F.; Wang, X.; Huang, Q.; Zhang, L.; Bi, C.; Hua, B.; Yuan, Y.; Han, Z.; Jin, M.; et al. By targeting *Atg7* microRNA-143 mediates oxidative stress-induced autophagy of c-kit⁺ mouse cardiac progenitor cells. *EBioMedicine* **2018**, *32*, 182–191. [CrossRef] [PubMed]

81. Dorn, G.W., 2nd. Mitochondrial pruning by Nix and BNip3: An essential function for cardiac-expressed death factors. *J. Cardiovasc. Transl. Res.* **2010**, *3*, 374–383. [CrossRef] [PubMed]

82. Gong, G.; Song, M.; Csordas, G.; Kelly, D.P.; Matkovich, S.J.; Dorn, G.W., 2nd. Parkin-mediated mitophagy directs perinatal cardiac metabolic maturation in mice. *Science* **2015**, *350*, aad2459. [CrossRef] [PubMed]

83. Andres, A.M.; Hernandez, G.; Lee, P.; Huang, C.; Ratliff, E.P.; Sin, J.; Thornton, C.A.; Damasco, M.V.; Gottlieb, R.A. Mitophagy is required for acute cardioprotection by simvastatin. *Antioxid. Redox Signal.* **2014**, *21*, 1960–1973. [CrossRef] [PubMed]

84. Billia, F.; Hauck, L.; Konecny, F.; Rao, V.; Shen, J.; Mak, T.W. PTEN-inducible kinase 1 (PINK1)/Park6 is indispensable for normal heart function. *Proc. Natl. Acad. Sci. USA* **2011**, *108*, 9572–9577. [CrossRef] [PubMed]

85. Li, W.; Zhang, X.; Zhuang, H.; Chen, H.G.; Chen, Y.; Tian, W.; Wu, W.; Li, Y.; Wang, S.; Zhang, L.; et al. MicroRNA-137 is a novel hypoxia-responsive microRNA that inhibits mitophagy via regulation of two mitophagy receptors FUNDC1 and NIX. *J. Biol. Chem.* **2014**, *289*, 10691–10701. [CrossRef] [PubMed]

86. Kim, J.; Fiesel, F.C.; Belmonte, K.C.; Hudec, R.; Wang, W.X.; Kim, C.; Nelson, P.T.; Springer, W.; Kim, J. MiR-27a and miR-27b regulate autophagic clearance of damaged mitochondria by targeting PTEN-induced putative kinase 1 (PINK1). *Mol. Neurodegener.* **2016**, *11*, 55. [CrossRef] [PubMed]

87. Cheng, M.; Liu, L.; Lao, Y.; Liao, W.; Liao, M.; Luo, X.; Wu, J.; Xie, W.; Zhang, Y.; Xu, N. MicroRNA-181a suppresses parkin-mediated mitophagy and sensitizes neuroblastoma cells to mitochondrial uncoupler-induced apoptosis. *Oncotarget* **2016**, *7*, 42274–42287. [CrossRef] [PubMed]

88. Yang, F.; Li, T. MicroRNA-410 is involved in mitophagy after cardiac ischemia/reperfusion injury by targeting high-mobility group box 1 protein. *J. Cell. Biochem.* **2018**, *119*, 2427–2439. [CrossRef] [PubMed]

89. Verjans, R.; van Bilsen, M.; Schroen, B. MiRNA deregulation in cardiac aging and associated disorders. *Int. Rev. Cell Mol. Biol.* **2017**, *334*, 207–263. [PubMed]

90. Yan, L.; Gao, S.; Ho, D.; Park, M.; Ge, H.; Wang, C.; Tian, Y.; Lai, L.; De Lorenzo, M.S.; Vatner, D.E.; et al. Calorie restriction can reverse, as well as prevent, aging cardiomyopathy. *Age* **2013**, *35*, 2177–2182. [CrossRef] [PubMed]

91. Egan, D.F.; Shackelford, D.B.; Mihaylova, M.M.; Gelino, S.; Kohnz, R.A.; Mair, W.; Vasquez, D.S.; Joshi, A.; Gwinn, D.M.; Taylor, R.; et al. Phosphorylation of ULK1 (hATG1) by AMP-activated protein kinase connects energy sensing to mitophagy. *Science* **2011**, *331*, 456–461. [CrossRef] [PubMed]

92. Kim, J.; Kim, Y.C.; Fang, C.; Russell, R.C.; Kim, J.H.; Fan, W.; Liu, R.; Zhong, Q.; Guan, K.L. Differential regulation of distinct Vps34 complexes by AMPK in nutrient stress and autophagy. *Cell* **2013**, *152*, 290–303. [CrossRef] [PubMed]

93. Russell, R.C.; Tian, Y.; Yuan, H.; Park, H.W.; Chang, Y.Y.; Kim, J.; Kim, H.; Neufeld, T.P.; Dillin, A.; Guan, K.L. ULK1 induces autophagy by phosphorylating Beclin-1 and activating VPS34 lipid kinase. *Nat. Cell Biol.* **2013**, *15*, 741–750. [CrossRef] [PubMed]

94. Wu, H.; Wang, F.; Hu, S.; Yin, C.; Li, X.; Zhao, S.; Wang, J.; Yan, X. MiR-20a and miR-106b negatively regulate autophagy induced by leucine deprivation via suppression of ULK1 expression in C2C12 myoblasts. *Cell. signal.* **2012**, *24*, 2179–2186. [CrossRef] [PubMed]

95. Wan, G.; Xie, W.; Liu, Z.; Xu, W.; Lao, Y.; Huang, N.; Cui, K.; Liao, M.; He, J.; Jiang, Y.; et al. Hypoxia-induced miR155 is a potent autophagy inducer by targeting multiple players in the mtor pathway. *Autophagy* **2014**, *10*, 70–79. [CrossRef] [PubMed]

96. Martins-Marques, T.; Ribeiro-Rodrigues, T.; Pereira, P.; Codogno, P.; Girao, H. Autophagy and ubiquitination in cardiovascular diseases. *DNA Cell Biol.* **2015**, *34*, 243–251. [CrossRef] [PubMed]

97. Pan, W.; Zhong, Y.; Cheng, C.; Liu, B.; Wang, L.; Li, A.; Xiong, L.; Liu, S. MiR-30-regulated autophagy mediates angiotensin II-induced myocardial hypertrophy. *PloS ONE* **2013**, *8*, e53950. [CrossRef] [PubMed]

98. Chen, C.; Yang, S.; Li, H.; Yin, Z.; Fan, J.; Zhao, Y.; Gong, W.; Yan, M.; Wang, D.W. MiR30c is involved in diabetic cardiomyopathy through regulation of cardiac autophagy via BECN1. *Mol. Therapy-Nucleic Acids* **2017**, *7*, 127–139. [CrossRef] [PubMed]

99. Huang, J.; Sun, W.; Huang, H.; Ye, J.; Pan, W.; Zhong, Y.; Cheng, C.; You, X.; Liu, B.; Xiong, L.; et al. MiR-34a modulates angiotensin ii-induced myocardial hypertrophy by direct inhibition of atg9a expression and autophagic activity. *PloS ONE* **2014**, *9*, e94382. [CrossRef] [PubMed]

100. Su, M.; Chen, Z.; Wang, C.; Song, L.; Zou, Y.; Zhang, L.; Hui, R.; Wang, J. Cardiac-specific overexpression of miR-222 induces heart failure and inhibits autophagy in mice. *Cell. Physiol. Biochem.* **2016**, *39*, 1503–1511. [CrossRef] [PubMed]

101. Lai, L.; Chen, J.; Wang, N.; Zhu, G.; Duan, X.; Ling, F. MiRNA-30e mediated cardioprotection of ACE2 in rats with doxorubicin-induced heart failure through inhibiting cardiomyocytes autophagy. *Life Sci.* **2017**, *169*, 69–75. [CrossRef] [PubMed]

102. Qiu, J.; Wang, A.; Xu, Y.; Qiao, S.; An, J.; Li, H.; Wang, C. Role of microRNA-1-mediated AMP-activated protein kinase pathway in cardiac fibroblasts induced by high glucose in rats. *Zhonghua wei zhong bing ji jiu yi xue* **2018**, *30*, 145–150. [PubMed]

103. D'Avenia, M.; Citro, R.; De Marco, M.; Veronese, A.; Rosati, A.; Visone, R.; Leptidis, S.; Philippen, L.; Vitale, G.; Cavallo, A.; et al. A novel miR-371a-5p-mediated pathway, leading to BAG3 upregulation in cardiomyocytes in response to epinephrine, is lost in takotsubo cardiomyopathy. *Cell Death Dis.* **2015**, *6*, e1948. [CrossRef] [PubMed]

104. Shao, H.; Yang, L.; Wang, L.; Tang, B.; Wang, J.; Li, Q. MicroRNA-34a protects myocardial cells against ischemia-reperfusion injury through inhibiting autophagy via regulating TNAα expression. *Biochem. Cell Biol.* **2018**, *96*, 349–354. [CrossRef] [PubMed]

105. Liu, X.; Deng, Y.; Xu, Y.; Jin, W.; Li, H. MicroRNA-223 protects neonatal rat cardiomyocytes and h9c2 cells from hypoxia-induced apoptosis and excessive autophagy via the akt/mtor pathway by targeting parp-1. *J. Mol. Cell. Cardiol.* **2018**, *118*, 133–146. [CrossRef] [PubMed]

106. Sun, T.; Dong, Y.H.; Du, W.; Shi, C.Y.; Wang, K.; Tariq, M.A.; Wang, J.X.; Li, P.F. The role of microRNAs in myocardial infarction: From molecular mechanism to clinical application. *Int. J. Mol. Sci.* **2017**, *18*, 745. [CrossRef] [PubMed]

107. Bo, L.; Su-Ling, D.; Fang, L.; Lu-Yu, Z.; Tao, A.; Stefan, D.; Kun, W.; Pei-Feng, L. Autophagic program is regulated by miR-325. *Cell Death Differ.* **2014**, *21*, 967–977. [CrossRef] [PubMed]

108. Zheng, J.; Li, J.; Kou, B.; Yi, Q.; Shi, T. MicroRNA-30e protects the heart against ischemia and reperfusion injury through autophagy and the notch1/Hes1/Akt signaling pathway. *Int. J. Mol. Med.* **2018**, *41*, 3221–3230. [CrossRef] [PubMed]

109. Li, Q.; Xie, J.; Li, R.; Shi, J.; Sun, J.; Gu, R.; Ding, L.; Wang, L.; Xu, B. Overexpression of microRNA-99a attenuates heart remodelling and improves cardiac performance after myocardial infarction. *J. Cell. Mol. Med.* **2014**, *18*, 919–928. [CrossRef] [PubMed]

110. Li, X.; Zeng, Z.; Li, Q.; Xu, Q.; Xie, J.; Hao, H.; Luo, G.; Liao, W.; Bin, J.; Huang, X.; et al. Inhibition of microRNA-497 ameliorates anoxia/reoxygenation injury in cardiomyocytes by suppressing cell apoptosis and enhancing autophagy. *Oncotarget* **2015**, *6*, 18829–18844. [CrossRef] [PubMed]

111. Wang, X.; Cui, T. Autophagy modulation: A potential therapeutic approach in cardiac hypertrophy. *Am. J. Physiol. Heart Circ. Physiol.* **2017**, *313*, H304–H319. [CrossRef] [PubMed]

112. Song, L.; Su, M.; Wang, S.; Zou, Y.; Wang, X.; Wang, Y.; Cui, H.; Zhao, P.; Hui, R.; Wang, J. MiR-451 is decreased in hypertrophic cardiomyopathy and regulates autophagy by targeting tsc1. *J. Cell. Mol. Med.* **2014**, *18*, 2266–2274. [CrossRef] [PubMed]

113. Zhao, X.D.; Qin, R.H.; Yang, J.J.; Xu, S.S.; Tao, H.; Ding, X.S.; Shi, K.H. DNMT3A controls miR-200b in cardiac fibroblast autophagy and cardiac fibrosis. *Inflamm. Res.* **2018**, *67*, 681–690. [CrossRef] [PubMed]

cells

MDPI

Article

miR-338-3p Is Regulated by Estrogens through GPER in Breast Cancer Cells and Cancer-Associated Fibroblasts (CAFs)

Adele Vivacqua [1,*], **Anna Sebastiani** [1], **Anna Maria Miglietta** [2], **Damiano Cosimo Rigiracciolo** [1], **Francesca Cirillo** [1], **Giulia Raffaella Galli** [1], **Marianna Talia** [1], **Maria Francesca Santolla** [1], **Rosamaria Lappano** [1], **Francesca Giordano** [1], **Maria Luisa Panno** [1] and **Marcello Maggiolini** [1,*]

[1] Department of Pharmacy, Health and Nutritional Sciences, University of Calabria, 87036 Rende, Italy; annasebastiani86@gmail.com (A.S.); damianorigiracciolo@yahoo.it (D.C.R.); francesca89cirillo@libero.it (F.C.); giulia.r.galli@gmail.com (G.R.G.); mariannatalia11@gmail.com (M.T.); m.f.s@hotmail.it (M.F.S.); lappanorosamaria@yahoo.it (R.L.); francesca.giordano@unical.it (F.G.); mamissina@yahoo.it (M.L.P.)

[2] Regional HospitalCosenza, 87100 Cosenza, Italy; annamariamiglietta@virgilio.it

* Correspondence: adele.vivacqua@unical.it (A.V.); marcellomaggiolini@yahoo.it (M.M.); Tel.: +39-0984-493-048 (A.V.); +39-0984-493-076 (M.M.)

Received: 12 October 2018; Accepted: 7 November 2018; Published: 9 November 2018

Abstract: Estrogens acting through the classic estrogen receptors (ERs) and the G protein estrogen receptor (GPER) regulate the expression of diverse miRNAs, small sequences of non-coding RNA involved in several pathophysiological conditions, including breast cancer. In order to provide novel insights on miRNAs regulation by estrogens in breast tumor, we evaluated the expression of 754 miRNAs by TaqMan Array in ER-negative and GPER-positive SkBr3 breast cancer cells and cancer-associated fibroblasts (CAFs) upon 17β-estradiol (E2) treatment. Various miRNAs were regulated by E2 in a peculiar manner in SkBr3 cancer cells and CAFs, while miR-338-3p displayed a similar regulation in both cell types. By METABRIC database analysis we ascertained that miR-338-3p positively correlates with overall survival in breast cancer patients, according to previous studies showing that miR-338-3p may suppress the growth and invasion of different cancer cells. Well-fitting with these data, a miR-338-3p mimic sequence decreased and a miR-338-3p inhibitor sequence rescued the expression of genes and the proliferative effects induced by E2 through GPER in SkBr3 cancer cells and CAFs. Altogether, our results provide novel evidence on the molecular mechanisms by which E2 may regulate miR-338-3p toward breast cancer progression.

Keywords: breast cancer; CAFs; estrogens; GPER; miR-338-3p; c-Fos; Cyclin D1

1. Introduction

Estrogens play a crucial role in diverse pathophysiological conditions, including cancer [1]. The action of estrogens are mainly mediated by the classic estrogen receptors (ERs) [2], however several data have also indicated that the G protein estrogen receptor (GPER) may trigger a network of transduction pathways toward the progression of several types of tumors [3–8]. Among numerous biological targets, estrogens may modulate the expression of diverse microRNAs (miRNAs) [6], which are small non-coding RNA molecules of 22–25 nucleotides [9]. In particular, miRNAs inhibit the expression of certain genes at both transcriptional and post-transcriptional levels binding to complementary sequences located within the 3′ untranslated region (UTR) of target mRNAs [10,11]. Therefore, miRNAs may be involved in important biological processes, including cancer development [12–20]. The involvement of ERs in miRNA regulation by estrogens has been established [6]. Likewise, it has been also reported that GPER may regulate the expression of certain miRNAs in normal and cancer cell contexts characterized by the presence or absence of ERs [21–25].

MiR-338-3p is a highly conserved gene located on the chromosome 17q25 and precisely on the 7th intron of the apoptosis-associated tyrosine kinase (AATK) [26,27]. MiR-338-3p, initially identified as a brain specifically expressed miRNA, has been involved in the formation of basolateral polarity and regulation of axonal respiration [28,29]. Various studies have also shown that miR-338-3p is downregulated in many types of malignancies, hence suggesting its potential role in tumor progression [30–34]. Nevertheless, the biological function of miR-338-3p and its prognostic significance remains to be fully understood.

In this present study we provide novel insights into the ability of estrogens to regulate miR-338-3p expression and function through GPER in ER-negative breast cancer cells and cancer associated fibroblasts (CAFs), which are main components of the tumor microenvironment [35,36]. On the basis of our findings miR-338-3p may be included among the miRNAs involved in breast tumor development.

2. Materials and Methods

2.1. Reagents

17β-estradiol (E2) was purchased from Sigma-Aldrich Corp. (Milan, Italy); rel-1-[4-(6-bromo-1,3-benzodioxol-5-yl)-3aR,4S,5,9bS-tetrahydro-3H-cyclopenta[c]quinolin-8-yl]-ethanone (G-1) was obtained from Tocris Bioscience (Space, Milan, Italy). All compounds were solubilized in dimethyl sulfoxide (DMSO).

2.2. Cell Cultures

Breast cancer cell line SkBr3 (ER-negative and GPER-positive) was obtained from ATCC (Manassas, VA, USA), used less than six months after revival and routinely tested and authenticated according to the ATCC suggestions. CAFs (ER-negative and GPER-positive) were extracted from invasive mammary ductal carcinomas obtained from mastectomies. Briefly, samples were cut into smaller pieces (1–2 mm diameter), placed in digestion solution (400 IU collagenase I, 100 IU hyaluronidase, and 10% FBS, containing antibiotic and antimycotic solution) and incubated overnight at 37 °C. The cells were then separated by differential centrifugation at $90\times g$ for 2 min. Supernatant containing fibroblasts was centrifuged at $485\times g$ for 8 min; the pellet obtained was suspended in fibroblasts growth medium (Medium 199 and Ham's F12 mixed 1:1 and supplemented with 10% FBS) and cultured at 37 °C in 5% CO_2. Primary cells cultures of breast fibroblasts were characterized by immunofluorescence. Briefly cells were incubated with human anti-vimentin (V9, sc-6260) and human anti-cytokeratin 14 (LL001 sc-53253), both from Santa Cruz Biotechnology (DBA, Milan, Italy) (data not shown). To characterize fibroblasts activation, we used anti-fibroblast activated protein α (FAPα) antibody (SS-13, sc-100528; Santa Cruz Biotechnology, DBA, Milan, Italy) (data not shown). Signed informed consent from all the patients was obtained and samples were collected, identified and used in accordance with approval by the Institutional Ethical Committee Board (Regional Hospital, Cosenza, Italy). Cell types were grown in a 37 °C incubator with 5% CO_2. SkBr3 breast cancer cells were maintained in RPMI-1640 without phenol red supplemented with 10% fetal bovine serum (FBS) and 100 μg/mL of penicillin/streptomycin (Gibco, Life Technologies, Milan, Italy). CAFs were cultured in a mixture of MEDIUM 199 and HAM'S F-12 (1:1) supplemented with 10% FBS and 100 μg/mL of penicillin/streptomycin (Gibco, Life Technologies, Milan, Italy). Cells were switched to medium without serum the day before experimental analysis.

2.3. RNA Extraction

Cells were maintained in regular growth medium and then switched to medium lacking serum before performing the indicated assays. Total RNA was extracted from cultured cells using miRVana Isolation Kit (Ambion, Life Technologies, Milan, Italy) according to the manufacturer's recommendations. The RNA concentrations were determined using Gene5 2.01 Software in Synergy H1 Hybrid Multi-Mode Microplate Reader (BioTek, AHSI, Milan, Italy).

2.4. miRNA Expression Profiling

TaqMan™ Array Human MicroRNA A+B Cards Set v3.0 was used for global miRNA profiling. The panel includes two 384-well microfluidic cards (human miRNA pool A and pool B) that contain primers and probes for 754 different miRNAs in addition to small nucleolar RNAs that function as endogenous controls for data normalization. Equal quantity (100 ng) of RNA extracted from SkBr3 breast cancer cells and CAFs treated with vehicle or 100 nM E2 for 4 h was reverse-transcribed for cDNA synthesis using the Megaplex RT Primer Pool A or B and the TaqMan MicroRNA Reverse Transcription kit (Applied Biosystems).in a final volume of 7.5 µL (Applied Biosystems, Milan, Italy). The reverse transcription reaction was incubated for 2 min at 16 °C, 1 min at 42 °C and 1 s at 50 °C for 40 cycles, followed by 5 min at 85 °C to deactivate the enzyme. The cDNA obtained was pre-amplified using Megaplex Preamp primer pool A or B and TaqMan PreAmp Master Mix 2X in a final volume of 25 µL using the same temperature conditions above described. The product was diluted 1:4 in TE 0.1X, to which were added TaqMan Universal Master Mix no UNG 2X and nuclease free water. 100 µL of the sample/master mix for each multiplex pool were loaded into fill reservoirs on the microfluidic card. The array was then centrifuged, mechanically sealed with the Applied Biosystems sealer device and run on QuantStudio 6&7 Flex Real Time PCR System (Applied Biosystems, Life Technologies, Milan, Italy). The raw array data were analysed by DataAssist™. The baseline was set automatically, while the threshold was set manually at 0.2. Samples that had Ct values>32 were removed from the analysis. Each miRNA was normalized against the mean of the four RNU6B and its expression was then assessed in the E2 treated cells against the vehicle treated cells using the $2^{-\Delta\Delta CT}$ method [37]. miRNAs showing an increased value of 2-fold expression and a 50% reduction respect to vehicle-treated cells were selected. Venn diagram was obtained by http://bioinformatics.psb.ugent.be/cgibin/liste/Venn/calculate_venn.htpl.

2.5. Analysis of Public Data Set from METABRIC and Kaplan-Meier Plotter

Prognostic values of miR-338-3p levels, using METABRIC data set, were analyzed by Kaplan–Meier survival curves of breast cancer patients, using Kaplan-Meier Plotter (www.kmplot.com/analysis) [38]. Log-rank test was used for statistical analysis.

2.6. Real Time-PCR

cDNA for miRNA expression was synthesized from 100 ng of total RNA using the TaqMan microRNA Reverse Transcription Kit (Applied Biosystems, Life Technologies, Milan, Italy). The expression levels of miR-338-3p were quantified by TaqMan microRNA Assay Kit (Applied Biosystems, Milan, Italy), using the primers for the internal control RNU6B (assay ID 001093) and miR-338-3p (assay ID 002252). In order to measure the mRNA levels of c-Fos and Cyclin D1, 3µg of total RNA were reversely transcribed using the murine leukemia virus reverse transcriptase (Life Technologies, Milan, Italy), as indicated by the manufacturer. The quantitative PCR was performed using SYBR Green PCR Master Mix (Applied Biosystems, Life Technologies, Milan, Italy). Specific primers for Actin, which was used as internal control, c-Fos and Cyclin D1 genes were designed using Primer Express version 2.0 software (Applied Biosystems Inc, Milano, Italy). The sequences were as follows: Actin Fwd: 5′-AAGCCAACCCCACTTCTCTCTAA-3′ and Rev: 5′-CACCTCCCCTGTGTGGACTT-3′; c-Fos Fwd: 5′-CGAGCCCTTTGATGACTTCCT-3′ and Rev: 5′-GGAGCGGGCTGTCTCAGA-3′; Cyclin D1 Fwd: 5′-CCGTCCATGCGGAAGATC-3′ and Rev: 5′-ATGGCCAGCGGGAAGAC-3′. All experiments were performed in triplicate using QuantStudio 6&7 Flex Real Time PCR System (Applied Biosystems, Life Technologies, Milan, Italy). Data were normalized to the geometric mean of housekeeping gene to control the variability into expression levels and fold changes were calculated by relative quantification compared to respective scrambled controls [32].

2.7. Bioinformatic Tools

The sites miRNAbase (http://www.miRNAbase.org), Targetscan (http://www.targetscan.org) and miRDip (http://ophid.utoronto.ca/mirDIP/) were used to identified miR-338-3p target genes.

2.8. Constructs and Transfections

The negative control (miR-Ctrl), the miR-338-3p mimic (miR-338-3p m) (ID MC10716) and miR-338-3p inhibitor (miR-338-3p i) (ID MH10716) sequences were purchased from Ambion (Life Technologies, Milan, Italy) and transfected into the cells 48 h before the treatments, using X-treme GENE 9 DNA Transfection Reagent (Roche Diagnostics, Sigma-Adrich, Milan, Italy). Silencing of GPER expression was obtained by using the construct previously described [39]. The plasmid DN-Fos, which encodes a c-Fos mutant that heterodimerizes with c-Fos dimerization partners but does not allow DNA binding, was a kind gift from Dr. C. Vinson (NIH, Bethesda, MD, USA).

2.9. Western Blotting

Cells were maintained in complete medium before the transfection assays, which are performed in medium without serum for 48 h and then treated as indicated. Cells were lysed in RIPA buffer containing a mixture of protease inhibitors. Equal amounts of protein extract were resolved on SDS-polyacrylamide gel, transferred to a nitrocellulose membrane (Amersham Biosciences, Italy), probed overnight at 4 °C with antibodies against: c-Fos (E-8, sc-166940) and β-Actin (AC-15, sc-69879) (Santa Cruz Biotechnology, DBA, Italy), GPER (AB137479) (Abcam, Euroclone, Milan, Italy) and Cyclin D1 (Origene, DBA, Milan, Italy). Proteins were detected by horseradish peroxidase-linked secondary antibodies (Biorad, Milan, Italy) and revealed using the chemiluminescent substrate for western blotting Westar Nova 2.0 (Cyanagen, Biogenerica, Catania, Italy).

2.10. Luciferase Assays

Cells were seeded in regular growth medium into 24-well plates. The next day the growth medium was replaced with medium lacking serum and the transfection was performed using X-tremeGene9 reagent, as recommended by the manufacturer (Roche Diagnostics), with a mixture containing Cyclin-D1-luc, the internal control pRL-TK and miR-Ctrl, miR-338-3p m, alone or in presence of miR-338-3p i, shGPER, DN-Fos as indicated. The cells were treated overnight with 100 nM of E2 or G1. Luciferase activity was measured using the Dual Luciferase kit (Promega, Milan, Italy) according to the manufacturer's instructions. Firefly luciferase values were normalized to the internal transfection control provided by the Renilla luciferase activity. The normalized relative light unit (RLU) values obtained from cells transfected with respective scrambled controls were set as 1-fold induction upon which the activity induced by the treatment was calculated.

2.11. Cell Proliferation Assays

For quantitative proliferation assay, cells (1×10^4) were seeded in 24-well plates in regular growth medium. Cells were washed, once they had attached, and then incubated in medium containing 2.5% charcoal stripped fetal bovine serum, before the transfection with 25 nM miR-338-p m and 50 nM miR-338-3p i, as indicated. Transfection was renewed every 2 day, while the cells were treated every day. Evaluation of cell growth was performed on day 6 using automatic counter (Countess™-Invitrogen).

2.12. Cell Cycle Analysis

To analyze cell cycle distribution, CAFs were cultured in regular medium and shifted in medium containing 2.5% charcoal-stripped FBS at the 70% confluence. Next, miRNA sequences as indicated were added to cells using X-treamGene9 reagent (Roche Diagnostics, Milan, Italy). After 24 h, 100 nM E2 or 100 nM G-1 were put in the medium for additional 24 h. Cells were pelleted, once washed with phosphate buffered saline and stained with a solution containing 50 µg/mL propidium iodide in 1 x

PBS (PI), 20 U/mL RNAse-A and 0.1% Triton (Sigma-Aldrich, Milan, Italy). The DNA content was measured using a FACScan flow cytometer (Becton Dickinson, Mountain View, CA, USA) and the data acquired using CellQuest software. Cell cycle profiles were determined using ModFit LT. The proportion of the cells in G0/G1, S and G2/M phases was each estimated as a percentage of the total events (10,000 cells).

2.13. Statistical Analysis

Data were analyzed by one-way ANOVA with Dunnett's multiple comparisons where applicable, using GraphPad Prism version 6.01 (GraphPad Software, Inc., San Diego, CA, USA). $p < 0.05$ (*) was considered statistically significant.

3. Results

3.1. miRNAs Expression by E2 in SkBr3 Cancer Cells and CAFs

In order to provide novel insights on the action of estrogens toward miRNAs modulation in breast cancer, the ER-negative SkBr3 breast cancer cells and CAFs were treated with 100 nM E2 for 4 h and then analyzed by TaqMan™ Array Human MicroRNA. A total amount of 754 miRNAs involved in diverse pathophysiological conditions (www.thermofisher.com/order/catalog/product/4444913) were evaluated, thereafter we focused our attention on miRNAs displaying a Ct< 32 along with at least 2 fold increase or 50% reduction upon E2 exposure respect to vehicle-treated cells. On the basis of these criteria, we identified 25 and 29 E2-regulated miRNAs in SkBr3 cancer cells (Figure 1A) and CAFs (Figure 2A), respectively. In particular, in SkBr3 cancer cells 23 miRNAs were up-regulated and 2 miRNAs were down-regulated by E2 treatment (Figure 1B). As it concerns CAFs, among the 29 E2-regulated miRNAs, 7 showed an increase and 22 a reduction upon E2 stimulation (Figure 2B). To identify unique and shared E2-regulated miRNAs in both cell types, we then calculated a Venn diagram. SkBr3s cancer cells and CAFs shared only the expression of 2 miRNAs (Figure 3A), namely miR-144 and miR-338-3p, which exhibited a similar response (Figure 3B). Considering that in our previous studies we evaluated the role of miR-144 in tumor cell growth [25], in the present investigation we aimed to determine the mechanisms leading to the estrogen regulation of miR-338-3p and its action in breast cancer. Hence, we began our study ascertaining that miR-338-3p expression correlates positively with the overall survival in 1283 breast tumor patients, as reported in the Molecular Taxonomy of Breast Cancer International Consortium (METABRIC) database [40] (Figure 3C). Nicely fitting with these findings, previous evidence has suggested that miR-338-3p may function as a tumor suppressor in certain malignancies including breast cancer [30–34].

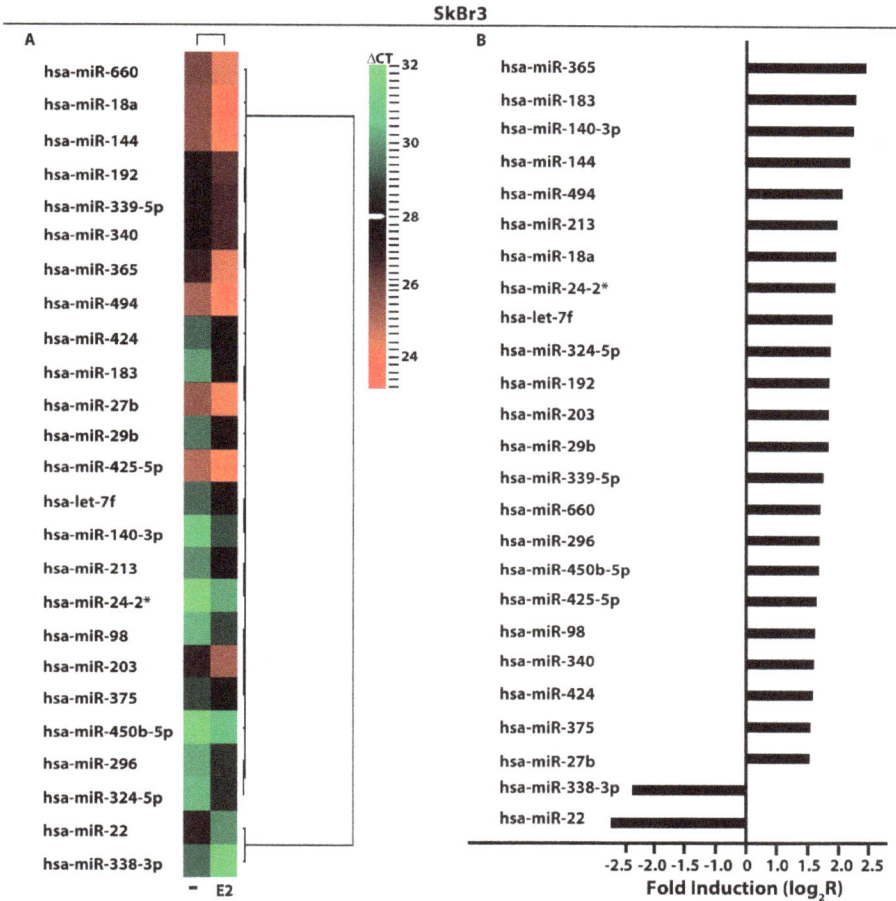

Figure 1. E2-modulated miRNAs expression in SkBr3 breast cancer cells. (**A**) Heat Map representation of E2-modulated miRNAs in SkBr3 cancer cells treated with 100 nM E2 for 4 h and analyzed by TaqMan Low-Density Array Human miRNA. Row represents a miRNA and column represents the treatment used. Each column is illustrated according to a color scale from green (low expression) to red (high expression). The distance measured is Euclidean Distance and the clustering method is complete linkage. Dendrograms of clustering analysis for miRNAs and samples are displayed on the top and right, respectively. (**B**) Up- and down-regulated miRNAs in SkBr3 breast cancer cells upon E2 stimulation. The values are indicated as log2 fold change (R) calculated respect to vehicle (-).

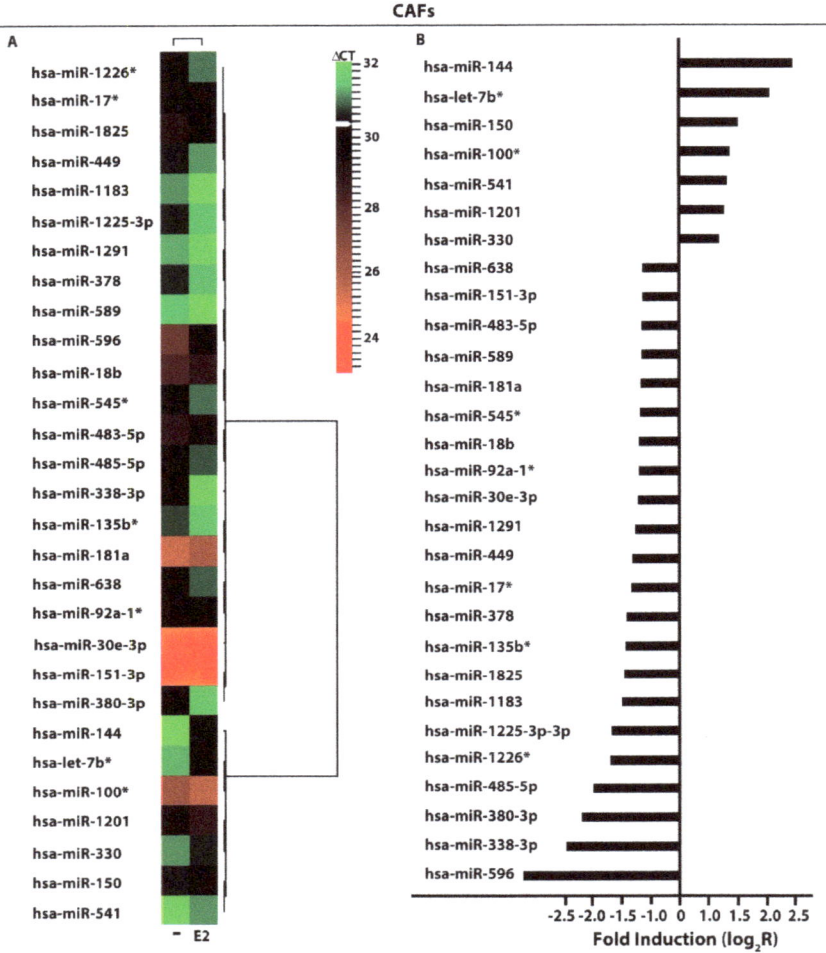

Figure 2. E2-modulated miRNAs expression in CAFs. (**A**) Heat Map representation of E2-modulated miRNAs in CAFs treated with 100 nM E2 for 4 h and analyzed by TaqMan Low-Density Array Human miRNA. Row represents a miRNA and column represents the treatment used. Each column is illustrated according to a color scale from green (low expression) to red (high expression). The distance measured is Euclidean Distance and the clustering method is complete linkage. Dendrograms of clustering analysis for miRNAs and samples are displayed on the top and right, respectively. (**B**) Up- and down-regulated miRNAs in CAFs upon E2 stimulation. The values are indicated as log2 fold change (R) calculated respect to vehicle (-).

A

SkBr3   CAFs

23   2   27

B

Cells	n° miRNA		miRNA names
SkBr3	23	22↑	hsa-miR-660; hsa-miR-18a; hsa-miR-192; hsa-miR-339-5p; hsa-miR-340; hsa-miR-365; hsa-miR-494; hsa-miR-424; hsa-miR-183; hsa-miR-27b; hsa-miR-29b-5p; hsa-miR-425-5p; hsa-let-7f; hsa-miR-140-3p; hsa-miR-213; hsa-miR-24-2*; hsa-miR-98; hsa-miR-203; hsa-miR-375; hsa-miR-450b-5p; hsa-miR-296; hsa-miR-324-5p
		1↓	hsa-miR-22
CAFs	27	6↑	hsa-let-7b*; hsa-miR-100*; hsa-miR-1201; hsa-miR-330; hsa-miR-150-5p; hsa-miR-541
		21↓	hsa-miR-1226* hsa-miR-17* hsa-miR-1825; hsa-miR-449; hsa-miR-1183; hsa-miR-1225-3p; hsa-miR-1291; hsa-miR-378; hsa-miR-589; hsa-miR-596; hsa-miR-18b; hsa-miR-545*; hsa-miR-483-5p; hsa-miR-485-5p; hsa-miR-135b*; hsa-miR-181a; hsa-miR-638; hsa-miR-92a-1; hsa-miR-30e-3p hsa-miR-151-3p; hsa-miR-380-3p
SkBr3 and CAFs	2	1↑	hsa-miR-144
		1↓	hsa-miR-338-3p

C

hsa−miR−338−3p

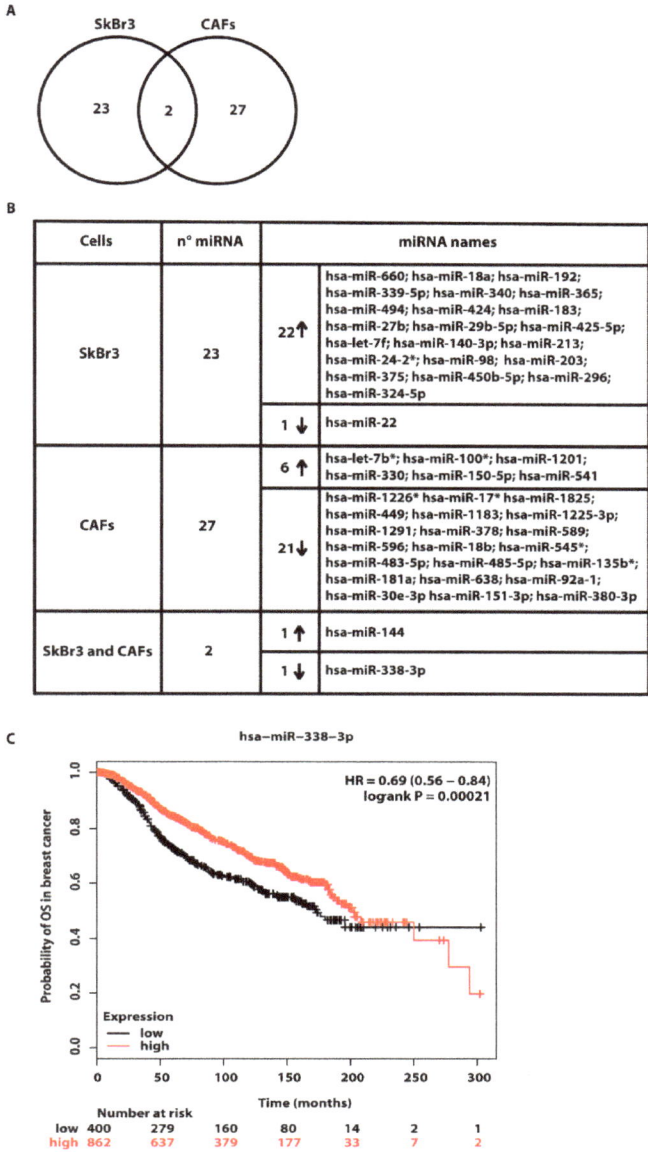

HR = 0.69 (0.56 − 0.84)
logrank P = 0.00021

Probability of OS in breast cancer

Expression
— low
— high

Time (months)

Number at risk
low   400   279   160   80   14   2   1
high 862   637   379   177   33   7   2

Figure 3. Exclusive and shared expression of miRNAs between SkBr3 and CAFs. (**A**) Venn Diagram of E2-modulated miRNAs in SkBr3 cancer cells and CAFs. (**B**) Up and down-regulated miRNAs by 100 nM E2 treatment for 4 h in SkBr3 cancer cells and CAFs. (**C**) The expression of miR-338-3p is associated with higher overall survival in breast cancer patients. The evaluation was performed by Kaplan–Meier Plotter (http://www.kmplot.com). Statistical analysis was made using the log-rank test.

3.2. GPER Is Involved in the Regulation of miR-338-3p by E2 and G-1 in SkBr3 Cancer Cells and CAFs

On the basis of the aforementioned results, we then attempted to define the molecular mechanisms involved in the estrogenic regulation of miR-338-3p performing a time-course study upon 100 nM of E2 and 100 nM of the selective GPER ligand G-1. Worthy, the inhibitory effects of E2 and G-1 on miR-338-3p expression were no longer evident silencing GPER in SkBr3 cancer cells (Figure 4A–C) and in CAFs (Figure 4D–F). Thereafter, we aimed to identify putative target genes of miR-338-3p by a bioinformatic analysis of available algorithms (http://ophid.utoronto.ca/mirDIP; http://www.microrna.org;http://www.targetscan.org). Among others, two putative target sequences of miR-338-3p located within the 3'-UTR of the oncogene c-Fos were found (Figure 5A). According to our previous studies showing that estrogens regulate c-Fos levels in diverse cancer cell types [41–44], the induction of c-Fos mRNA and protein expression upon a 4 h treatment with 100 nM E2 and 100 nM G-1 was abolished silencing GPER in SkBr3 cancer cells (Figure 5B,C) and CAFs (Figure 5D,E). Next, we found that in SkBr3 cells and CAFs transfected for 48 h with a miR-338-3p mimic sequence, the treatment for 4 h with 100 nM E2 and 100 nM G-1 is no longer able to induce c-Fos mRNA and protein levels, a response rescued transfecting the miR-338-3p mimic sequence in combination with a miR-338-3p inhibitor sequence (Figure 6A–F).

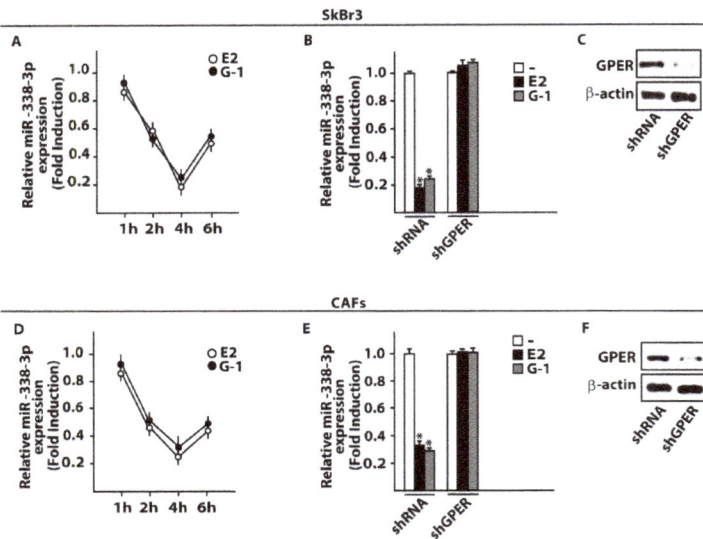

Figure 4. E2 and G-1 down-regulate miR-338-3p levels in SkBr3 cancer cells and CAFs. SkBr3 breast cancer cells (**A**) and CAFs (**D**) were stimulated with 100 nM E2 or 100 nM G-1 as indicated and analyzed by RT-PCR. Each point is plotted as fold changes of cells receiving treatments respect to cells treated with vehicle (-) and represents the mean ± SD of three independent experiments performed in triplicate. MiR-338-3p expression upon a 4 h treatment with 100 nM E2 or 100 nM G-1 in SkBr3 cells (**B**) and CAFs (**E**) previously transfected with shRNA or shGPER for 48 h. Each column represents the mean ± SD of three independent experiments performed in triplicate. Efficacy of GPER silencing in SkBr3 cells (**C**) and CAFs (**F**). β-actin serves as a loading control. (*) indicates $p < 0.05$, for cells receiving treatments vs cells treated with vehicle.

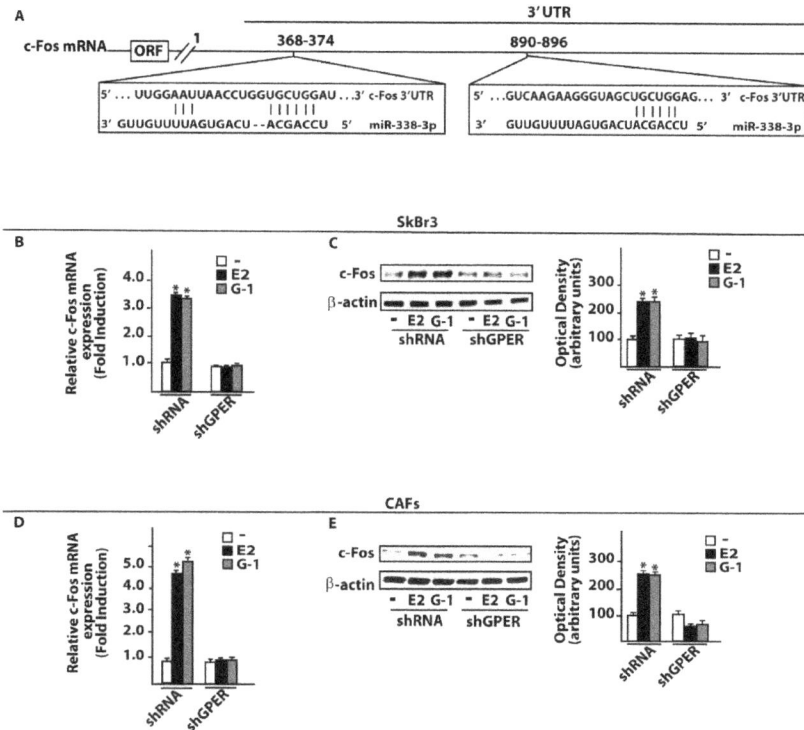

Figure 5. c-Fos is a target gene of miR-338-3p. (**A**) Schematic alignment between the miR-338-3p sequence and the 3′-UTR mRNA region of c-Fos. mRNA expression of c-Fos in SkBr3 cancer cells (**B**) and CAFs (**D**) transfected with shRNA or shGPER for 48 h and then treated for 4 h with 100 nM E2 or 100 nM G-1. Each column represents the mean ± SD of three independent experiments performed in triplicate. c-Fos protein expression in SkBr3 cancer cells (**C**) and CAFs (**E**) transfected with shRNA or shGPER for 48 h and then treated for 4 h with 100 nM E2 or 100 nM G-1. Side panels show densitometry analysis of the blots normalized to the loading control β-actin.

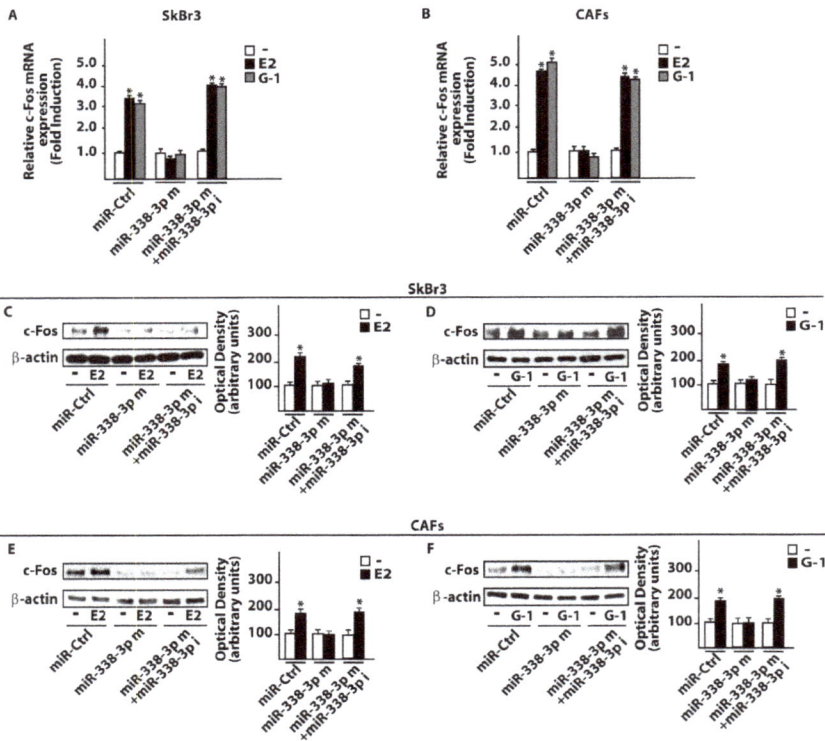

Figure 6. miR-338-3p prevents c-fos induction by E2 and G-1 in SkBr3 cancer cells and CAFs. mRNA levels of c-Fos in SkBr3 cancer cells (**A**) and CAFs (**B**) transfected for 48 h with 25 nM miR-Ctrl or miR-338-3p mimic (miR-338-3p m) in combination or not with 50 nM miR-338-3p inhibitor (miR-338-3p i) and then treated for 4 h with 100 nM E2 or 100 nM G-1. Each column represents the mean ± SD of three independent experiments performed in triplicate. c-Fos protein levels in SkBr3 cancer cells (**C, D**) and CAFs (**E, F**) transfected for 48 h with 25 nM miR-Ctrl or miR-338-3p mimic (miR-338-3p m) in combination or not with 50 nM miR-338-3p inhibitor (miR-338-3p i) and then stimulated for 4 h with 100 nM E2 or 100 nM G-1. Side panels show densitometry analysis of the blots normalized to the loading control β-actin. (*) indicates $p < 0.05$, for cells receiving treatments vs cells treated with vehicle (-).

3.3. miR-338-3p Triggers Inhibitory Effects on the Proliferation Induced by E2 and G-1

As in our previous investigations c-Fos was involved in the regulation of cyclins [43,45], we assessed that the transactivation of the Cyclin D1 promoter sequence by 100 nM E2 and 100 nM G-1 was prevented co-transfecting a dominant negative c-Fos expression construct (DN-Fos) in SkBr3 and CAFs (Figure 7A,B). Nicely recapitulating the aforementioned results, the Cyclin D1 promoter luciferase activity induced by 100 nM E2 and 100 nM G-1 was inhibited using the miR-338-3p mimic, an effect rescued by the miR-338-3p inhibitor sequence (Figure 7C,D). In addition, similar findings were observed evaluating the regulation of Cyclin D1 at both mRNA (Figure 7E,F) and protein levels (Figure 8A–D). As biological counterpart, the proliferative responses elicited by 100 nM E2 and 100 nM G-1 in SkBr3 cancer cells and CAFs were prevented silencing GPER or transfecting the DN-Fos construct (Figure 9A,B). Furthermore, the miR-338-3p mimic sequence decreased the proliferation induced by 100 nM E2 and 100 nM G-1 (Figure 9A,B), however this effect was rescued co-transfecting the miR-338-3p inhibitor (Figure 9A,B). Further supporting the aforementioned findings, the treatment for 24 h with 100 nM E2 and 100 nM G-1 triggered inhibitory effects on cell cycle progression transfecting CAFs with the miR-338-3p mimic sequence, however this response was rescued in the presence of the miR-338-3p inhibitor sequence (Figure 9C). Overall, these results suggest that estrogenic GPER signaling regulates miR-338-3p expression and function in SkBr3 cancer cells and CAFs.

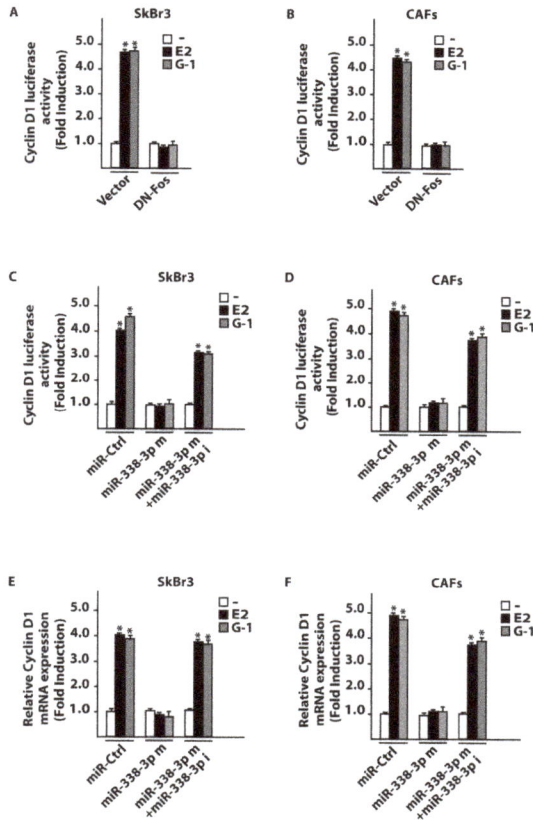

Figure 7. c-Fos and miR-338-3p are involved in Cyclin D1 regulation in SkBr3 cancer cells and CAFs. Luciferase activity of Cyclin D1 reporter gene in SkBr3 cancer cells (**A**) and CAFs (**B**) transfected for 8 h with a vector or a dominant-negative c-Fos construct (DN-Fos) before treatment with 100 nM of E2 and 100 nM G-1 for 18 h. Luciferase activity of Cyclin D1 reporter gene in SkBr3 cancer cells (**C**) and CAFs (**D**) transfected for 24 h with 25 nM miR-Ctrl or miR-338-3p mimic (miR-338-3p m) in combination or not with 50 nM miR-338-3p inhibitor (miR-338-3p i) before treatment for 18 h with 100 nM E2 or 100 nM G-1. The luciferase activity was normalized to the internal transfection control, values of cells receiving vehicle (-) were set as 1-fold induction upon which the activity obtained upon the indicated treatments was calculated. mRNA expression of Cyclin D1 in SkBr3 cells (**E**) and CAFs (**F**) transfected for 48 h with 25 nM miR-Ctrl or miR-338-3p mimic (miR-338-3p m) in combination or not with 50 nM miR-338-3p inhibitor (miR-338-3p i) before treatment for 8 h with 100 nM E2 or 100 nM G-1. Each column represents the mean ± SD of three independent experiments performed in triplicate. (*) indicates $p < 0.05$ for cells receiving treatments vs cells treated with vehicle (-).

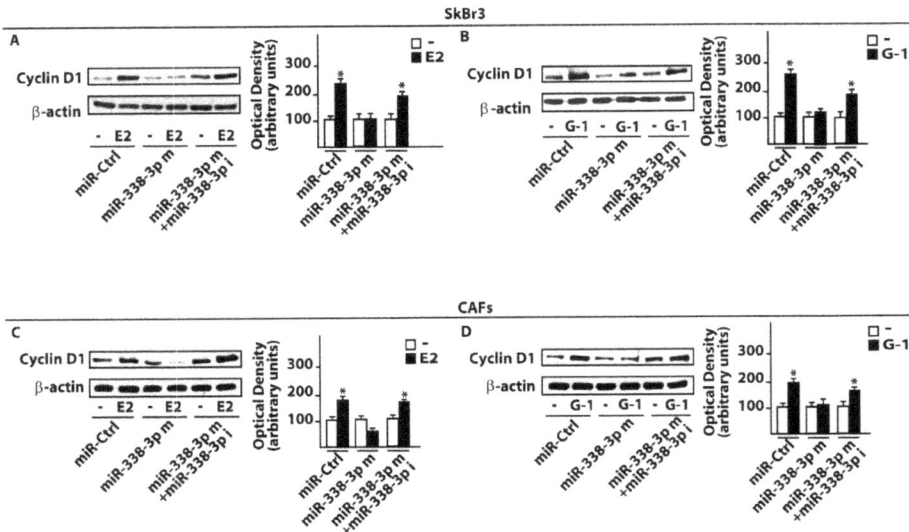

Figure 8. miR-338-3p prevents Cyclin D1 protein induction by E2 and G1 in SkBr3 cancer cells and CAFs. Cyclin D1 protein expression in SkBr3 cancer cells (**A,B**) and CAFs (**C,D**) transfected for 48 h with 25 nM miR-Ctrl or miR-338-3p mimic (miR-338-3p m) in combination or not with 50 nM miR-338-3p inhibitor (miR-338-3p i) before treatment for 12h with 100 nM E2 or 100 nM G-1. Side panels show densitometry analysis of the blots normalized to the loading control β-actin. (*) indicates $p < 0.05$ for cells receiving treatments vs cells treated with vehicle (-).

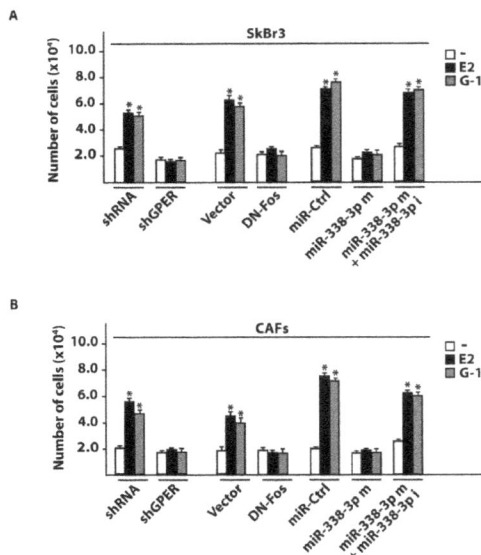

Figure 9. *Cont.*

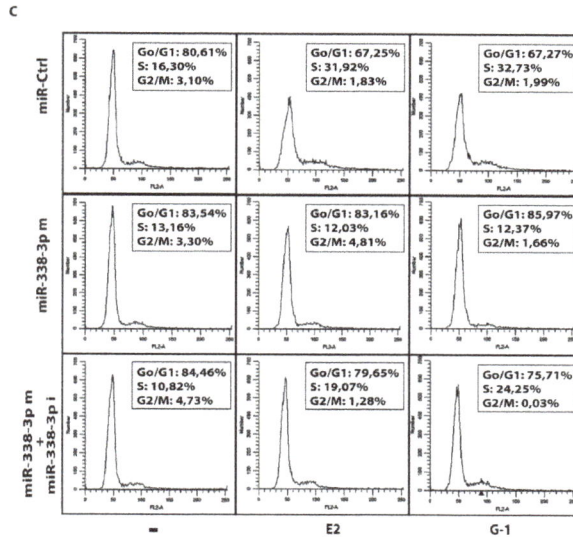

Figure 9. miR-338-3p decreases the proliferation of SkBr3 cancer cells and CAFs induced by E2 and G-1. Cell proliferation in SkBr3 cancer cells (**A**) and CAFs (**B**) transfected every 2 days with 100ng shRNA or shGPER, 100ng vector or a dominant-negative c-Fos construct (DN-Fos) and 25 nM miR-Ctrl or miR-338-3p mimic (miR-338-3p m) in combination or not with 50 nM miR-338-3p inhibitor (miR-338-3p i). Cells were treated every day with 100 nM E2 or 100 nM G-1 and counted on day 6. Each column represents the mean ± SD of three independent experiments performed in triplicate. (*) indicates $p <$ 0.05 for cells receiving treatments vs cells treated with vehicle (-). (**C**) Representative pictures of cell cycle analysis in CAFs transfected for 48 h with 25 nM miR-Ctrl or miR-338-3p mimic (miR-338-3p m) in combination or not with 50 nM miR-338-3p inhibitor (miR-338-3p i) before the treatment for 24 h with 100 nM E2 and 100 nM G-1. In each panel, the percentages of cells in G0/G1, S and G2/M phases of the cell cycle are indicated. Values represent the mean ± SD of three independent experiments.

4. Discussion

Performing a microarray analysis of 754 miRNAs involved in diverse diseases, in the present study we determined that diverse miRNAs are regulated by E2 in both SkBr3 breast cancer cells and CAFs. In particular, we assessed that E2 increases 23 miRNAs and lowers 2 miRNAs in SkBr3 cells, while E2 triggers the up-regulation of 7 miRNAs and the down-regulation of 22 miRNAs in CAFs. In addition, in both cell types E2 induced the expression of miR-144 and repressed the levels of miR-338-3p, which is known as an inhibitor of cancer progression [30–34]. Considering that miR-144 was investigated in our previous study [25], we attempted to provide novel insights into the estrogen regulation of miR-338-3p. First, we performed a METABRIC analysis that revealed a positive correlation of miR-338-3p with the overall survival in breast cancer patients. Then, we evidenced that a miR-338-3p mimic sequence prevents the expression of c-Fos, Cyclin-D1 and the growth effects induced by E2 and G-1 through GPER in SkBr3 cells and CAFs. Worthy, these effects triggered by E2 and G-1 were rescued using a miR-338-3p inhibitor sequence. Altogether, the aforementioned results provide new insights on the molecular mechanisms involved in the expression and function of certain miRNAs upon estrogen exposure in both breast cancer cells and CAFs.

Breast tumor is the most common malignancy in females and its incidence is increasing worldwide [46]. Several studies are ongoing in order to identify novels biological targets that may be considered toward innovative therapeutic approaches. To date, few markers like the estrogen receptor (ER), the progesterone receptor (PR) and the human epidermal growth factor receptor 2 (HER2), have been identified as predictors of clinical responses to breast cancer treatments [47]. None of these markers, however, well evaluates tumor invasion or provides early detection of cancer progression [48]. In this context, GPER has been suggested as a further predictor of breast cancer aggressiveness as its expression was found positively associated with clinic-pathological features of cancer progression and poor survival rates [49,50]. Moreover, GPER has been also indicated as an independent factor to predict a reduced disease-free survival in patients treated with tamoxifen [49]. The lack of GPER in the plasma membrane was also related to excellent long-term prognosis in ER-positive breast tumors treated with tamoxifen, an observation that highlighted the potential importance of GPER expression in different cancer cell types [51].

Despite the stimulatory effects elicited by GPER on the growth of diverse cancer cells [3–6], high doses of the GPER agonist G-1 (≥ 1 μM) have been shown to exert an inhibitory action on the proliferation of certain cancer cell lines [52–56]. Therefore, the different biological responses mediated by GPER in distinct tumor cell contexts may depend on the receptor expression repertoire, the signaling pathways activated and other factors that remain to be fully elucidated.

The involvement of diverse miRNAs in breast cancer progression has been well established [6]. For instance, it has been reported that let-7d, miR-210 and miR-221 are down-regulated in the breast ductal carcinoma in situ and up-regulated following the invasive transition. Moreover, miR-9, miR-10b, miR-21, miR-29a, miR-155 and miR-373-520 family were found to promote the metastatic tumor dissemination [57]. Next, member of the let-7, miR-200, miR-34 and miR-125b families, were able to regulate the epithelial-mesenchymal transition in breast cancer [57]. According to the results obtained in the present investigation, previous studies have indicated that in diverse pathophysiological conditions, including breast cancer, the regulation of certain miRNAs by E2 may involve GPER activation [21,25,58,59]. It has been shown that GPER activation by estrogens stimulates a network of transduction pathways, which triggers key factors involved in cell growth, differentiation and transformation, like c-Fos [5,44,60,61]. The proto-oncogene c-Fos represents a prototypical "immediate early" gene since its expression is rapidly induced by different extracellular stimuli through the activation of the serine-threonine kinases of mitogen-activated protein kinase (MAPK) family [62,63]. The nuclear protein encoded by c-Fos interacts with Jun family members to form the heterodimeric activating protein-1 transcription factor complex (AP-1), which binds to TGAC/GTC/AA sequences (AP-1 responsive elements) located within the promoter sequences of target genes [62,64]. Many studies focusing on the oncogenic functions of c-Fos have demonstrated its involvement in tumor growth through the modulation of Cyclin D1, which is a nuclear regulatory subunit of the cyclin-dependent kinases (CDK)-4 and CDK-6 [65–67]. Nicely fitting with these data, we determined that in SkBr3 cancer cells and CAFs E2 and G-1 induce c-Fos and Cyclin D1 expression toward cell proliferation. According to the inhibitory function of miR-338-3p in certain cancer types [30–34], we also found that miR-338-3p abrogates the abovementioned effects triggered by E2 and G-1 in SkBr3 cells and in important components of the tumor microenvironment as CAFs [35,36]. In this regard, our data highlight additional mechanisms by which tumor cells and CAFs cooperate toward worse cancer features. Well-fitting with the present findings, it has been established that cancer development involves the functional interaction of malignant cells with the tumor microenvironment [68,69]. For instance, stromal cells like CAFs generate a dynamic signaling network through the secretion of growth factors and cytokines that stimulate the proliferation and dissemination of cancer cells [70,71]. In this context, the regulation of miR-338-3p shared by breast cancer cells and CAFs may be a further mechanism linking the estrogen stimulation of both the tumor microenvironment and tumor cells.

5. Conclusions

miRNAs target numerous genes involved in the cell growth and survival of diverse types of tumors, including breast cancer [72]. Therefore, changes in miRNAs expression may have a prognostic role along with a therapeutic perspective in cancer patients. Here, we have provided novel insights on the molecular mechanisms through which estrogenic GPER signaling in both breast cancer cells and

CAFs lowers the expression of miR-338-3p, which has been reported to act as an inhibitor of cancer cell growth and invasion [30–34]. Further studies are needed to better define the functions of miR-338-3p and its usefulness in innovative therapeutic approaches in breast cancer patients.

Author Contributions: A.V., M.L.P. and M.M. conceived and designed the study. A.V., A.S., D.C.R., F.C., G.R.G., M.T., M.F.S., R.L., F.G. performed the experiments. A.M.M. provided breast tumor samples. A.V., M.L.P. and M.M. analyzed and interpreted the data. A.V. and M.M. wrote the manuscript. M.M. acquired the funding. All authors have read and approved the final manuscript.

Funding: This work has been supported by the Italian Association for Cancer Research (AIRC). MFS was supported by Fondazione Umberto Veronesi (Post-Doctoral Fellowship 2018).

Conflicts of Interest: The authors declare no conflict of interest.

References

1. Burns, K.A.; Korach, K.S. Estrogen receptors and human disease: An update. *Arch. Toxicol.* **2012**, *86*, 1491–1504. [CrossRef] [PubMed]
2. Nilsson, S.; Gustafsson, J. Estrogen receptors: Therapies targeted to receptor subtypes. *Clin. Pharmacol. Ther.* **2011**, *89*, 44–55. [CrossRef] [PubMed]
3. Cirillo, F.; Pellegrino, M.; Malivindi, R.; Rago, V.; Avino, S.; Muto, L.; Dolce, V.; Vivacqua, A.; Rigiracciolo, D.C.; De Marco, P.; et al. GPER is involved in the regulation of the estrogen-metabolizing CYP1B1 enzyme in breast cancer. *Oncotarget* **2017**, *8*, 106608–106624. [CrossRef] [PubMed]
4. Santolla, M.F.; Avino, S.; Pellegrino, M.; De Francesco, E.M.; De Marco, P.; Lappano, R.; Vivacqua, A.; Cirillo, F.; Rigiracciolo, D.C.; Scarpelli, A.; et al. SIRT1 is involved in oncogenic signaling mediated by GPER in breast cancer. *Cell Death Dis.* **2015**, *6*, e1834. [CrossRef] [PubMed]
5. Prossnitz, E.R.; Barton, M. Estrogen biology: New insights into GPER function and clinical opportunities. *Mol. Cell. Endocrinol.* **2014**, *389*, 71–83. [CrossRef] [PubMed]
6. Vrtačnik, P.; Ostanek, B.; Mencej-Bedrač, S.; Marc, J. The many faces of estrogen signaling. *Biochem. Med.* **2014**, *24*, 329–342. [CrossRef] [PubMed]
7. Santolla, M.F.; Lappano, R.; De Marco, P.; Pupo, M.; Vivacqua, A.; Sisci, D.; Abonante, S.; Iacopetta, D.; Cappello, A.R.; Dolce, V.; et al. G protein-coupled estrogen receptor mediates the up-regulation of fatty acid synthase induced by 17β-estradiol in cancer cells and cancer-associated fibroblasts. *J. Biol. Chem.* **2012**, *287*, 43234–43245. [CrossRef] [PubMed]
8. Vivacqua, A.; Romeo, E.; De Marco, P.; De Francesco, E.M.; Abonante, S.; Maggiolini, M. GPER mediates the Egr-1 expression induced by 17β-estradiol and 4-hydroxitamoxifen in breast and endometrial cancer cells. *Breast Cancer Res. Treat.* **2012**, *133*, 1025–1035. [CrossRef] [PubMed]
9. Bartel, D.P. MicroRNAs: Genomics, biogenesis, mechanism, and function. *Cell* **2004**, *116*, 281–297. [CrossRef]
10. Bartel, D.P. MicroRNAs: Target recognition and regulatory functions. *Cell* **2009**, *136*, 215–233. [CrossRef] [PubMed]
11. Fabian, M.R.; Sonenberg, N.; Filipowicz, W. Regulation of mRNA translation and stability by microRNAs. *Annu. Rev. Biochem.* **2010**, *79*, 351–379. [CrossRef] [PubMed]
12. Ben-Hamo, R.; Efroni, S. MicroRNA regulation of molecular pathways as a generic mechanism and as a core disease phenotype. *Oncotarget* **2015**, *6*, 1594–1604. [CrossRef] [PubMed]
13. Gross, N.; Kropp, J.; Khatib, H. MicroRNA signaling in embryo development. *Biology* **2017**, *6*, 34. [CrossRef] [PubMed]
14. Montagner, S.; Dehó, L.; Monticelli, S. MicroRNAs in hematopoietic development. *BMC Immunol.* **2014**, *15*, 14. [CrossRef] [PubMed]
15. Singh, R.P.; Massachi, I.; Manickavel, S.; Singh, S.; Rao, N.P.; Hasan, S.; Mc Curdy, D.K.; Sharma, S.; Wong, D.; Hahn, B.H.; et al. The role of miRNA in inflammation and autoimmunity. *Autoimmun. Rev.* **2013**, *12*, 1160–1165. [CrossRef] [PubMed]
16. Malemud, C.J. MicroRNAs and osteoarthritis. *Cells* **2018**, *7*, 92. [CrossRef] [PubMed]
17. Vannini, I.; Fanini, F.; Fabbri, M. Emerging roles of microRNAs in cancer. *Curr. Opin. Genet. Dev.* **2018**, *48*, 128–133. [CrossRef] [PubMed]

18. Santolla, M.F.; Lappano, R.; Cirillo, F.; Rigiracciolo, D.C.; Sebastiani, A.; Abonante, S.; Tassone, P.; Tagliaferri, P.; Di Martino, M.T.; Maggiolini, M.; et al. miR-221 stimulates breast cancer cells and cancer-associated fibroblasts (CAFs) through selective interference with the A20/c-Rel/CTGF signaling. *J. Exp. Clin. Cancer Res.* **2018**, *37*, 94. [CrossRef] [PubMed]

19. Inamura, K. Major tumor suppressor and oncogenic non-coding RNAs: Clinical relevance in lung cancer. *Cells* **2017**, *6*, 12. [CrossRef] [PubMed]

20. Vivacqua, A.; De Marco, P.; Belfiore, A.; Maggiolini, M. Recent advances on the role of microRNAs in both insulin resistance and cancer. *Curr. Pharm. Des.* **2017**, *23*, 3658–3666. [CrossRef] [PubMed]

21. Jacovetti, C.; Abderrahmani, A.; Parnaud, G.; Jonas, J.C.; Peyot, M.L.; Cornu, M.; Laybutt, R.; Meugnier, E.; Rome, S.; Thorens, B.; et al. MicroRNAs contribute to compensatory β cell expansion during pregnancy and obesity. *J. Clin. Investig.* **2012**, *122*, 3541–3551. [CrossRef] [PubMed]

22. Wang, Y.; Liu, Z.; Shen, J. MicroRNA-421-targeted PDCD4 regulates breast cancer cell proliferation. *Int. J. Mol. Med.* **2018**. [CrossRef] [PubMed]

23. Tao, S.; He, H.; Chen, Q.; Yue, W. GPER mediated estradiol reduces miR-148a to promote HLA-G expression in breast cancer. *Biochem. Biophys. Res. Commun.* **2014**, *451*, 74–78. [CrossRef] [PubMed]

24. Zhang, Y.; Fang, J.; Zhao, H.; Yu, Y.; Cao, X.; Zhang, B. Downregulation of microRNA-1469 promotes the development of breast cancer via targeting HOXA1 and activating PTEN/PI3K/AKT and Wnt/β-catenin pathways. *J. Cell. Biochem.* **2018**. [CrossRef] [PubMed]

25. Vivacqua, A.; De Marco, P.; Santolla, M.F.; Cirillo, F.; Pellegrino, M.; Panno, M.L.; Abonante, S.; Maggiolini, M. Estrogenic gper signaling regulates mir144 expression in cancer cells and cancer-associated fibroblasts (cafs). *Oncotarget* **2015**, *6*, 16573–16587. [CrossRef] [PubMed]

26. Rodriguez, A.; Griffiths-Jones, S.; Ashurst, J.L.; Bradley, A. Identification of mammalian microRNA host genes and transcription units. *Genome Res.* **2004**, *14*, 1902–1910. [CrossRef] [PubMed]

27. Raghunath, M.; Patti, R.; Bannerman, P.; Lee, C.M.; Baker, S.; Sutton, L.N.; Phillips, P.C.; Damodar Reddya, C. A novel kinase, AATYK induces and promotes neuronal differentiation in a human neuroblastoma (SH-SY5Y) cell line. *Mol. Brain Res.* **2000**, *77*, 151–162. [CrossRef]

28. Tsuchiya, S.; Oku, M.; Imanaka, Y.; Kunimoto, R.; Okuno, Y.; Terasawa, K.; Sato, F.; Tsujimoto, G.; Shimizu, K. MicroRNA-338-3p and microRNA-451 contribute to the formation of basolateral polarity in epithelial cells. *Nucleic Acids Res.* **2009**, *37*, 3821–3827. [CrossRef] [PubMed]

29. Aschrafi, A.; Schwechter, A.D.; Mameza, M.G.; Natera-Naranjo, O.; Gioio, A.E.; Kaplan, B.B. MicroRNA-338 regulates local cytochrome c oxidase IV mRNA levels and oxidative phosphorylation in the axons of sympathetic neurons. *J. Neurosci.* **2008**, *28*, 12581–12590. [CrossRef] [PubMed]

30. Cao, R.; Shao, J.; Hu, Y.; Wang, L.; Li, Z.; Sun, G.; Gao, X. microRNA-338-3p inhibits proliferation, migration, invasion, and EMT in osteosarcoma cells by targeting activator of 90 kDa heat shock protein ATPase homolog 1. *Cancer Cell Int.* **2018**, *18*, 49. [CrossRef] [PubMed]

31. Li, Y.; Chen, P.; Zu, L.; Liu, B.; Wang, M.; Zhou, Q. MicroRNA-338-3p suppresses metastasis of lung cancer cells by targeting the EMT regulator Sox4. *Am. J. Cancer Res.* **2016**, *6*, 127–140. [PubMed]

32. Jin, Y.; Zhao, M.; Xie, Q.; Zhang, H.; Wang, Q.; Ma, Q. MicroRNA-338-3p functions as tumor suppressor in breast cancer by targeting SOX4. *Int. J. Oncol.* **2015**, *47*, 1594–1602. [CrossRef] [PubMed]

33. Wen, C.; Liu, X.; Ma, H.; Zhang, W.; Li, H. miR-338-3p suppresses tumor growth of ovarian epithelial carcinoma by targeting Runx2. *Int. J. Oncol.* **2015**, *46*, 2277–2285. [CrossRef] [PubMed]

34. Huang, X.H.; Chen, J.S.; Wang, Q.; Chen, X.L.; Wen, L.; Chen, L.Z.; Bi, J.; Zhang, L.J.; Su, Q.; Zeng, W.T. miR-338-3p suppresses invasion of liver cancer cell by targeting smoothened. *J. Pathol.* **2011**, *225*, 463–472. [CrossRef] [PubMed]

35. Farhood, B.; Najafi, M.; Mortezaee, K. Cancer-associated fibroblasts: Secretions, interactions, and therapy. *J. Cell. Biochem.* **2018**. [CrossRef] [PubMed]

36. Kalluri, R.; Zeisberg, M. Fibroblasts in cancer. *Nat. Rev. Cancer* **2006**, *6*, 392–401. [CrossRef] [PubMed]

37. Pfaffl, M.W. A new mathematical model for relative quantification in real-time RT-PCR. *Nucleic Acids Res.* **2001**, *29*, e45. [CrossRef] [PubMed]

38. Lanczky, A.; Nagy, A.; Bottai, G.; Munkacsy, G.; Paladini, L.; Szabo, A.; Santarpia, L.; Gyorffy, B. miRpower: A web-tool to validate survival-associated miRNAs utilizing expression data from 2178 breast cancer patients. *Breast Cancer Res Treat.* **2016**, *160*, 439–446. [CrossRef] [PubMed]

39. Vivacqua, A.; Lappano, R.; De Marco, P.; Sisci, D.; Aquila, S.; De Amicis, F.; Fuqua, S.A.; Andò, S.; Maggiolini, M. G protein-coupled receptor 30 expression is up-regulated by EGF and TGF alpha in estrogen receptor alpha-positive cancer cells. *Mol. Endocrinol.* **2009**, *23*, 1815–1826. [CrossRef] [PubMed]

40. Curtis, C.; Shah, S.P.; Chin, S.F.; Turashvili, G.; Rueda, O.M.; Dunning, M.J.; Speed, D.; Lynch, A.G.; Samarajiwa, S.; Yuan, Y.; et al. The genomic and transcriptomic architecture of 2,000 breast tumours reveals novel subgroups. *Nature* **2012**, *486*, 346–352. [CrossRef] [PubMed]

41. Albanito, L.; Madeo, A.; Lappano, R.; Vivacqua, A.; Rago, V.; Carpino, A.; Oprea, T.I.; Prossnitz, E.R.; Musti, A.M.; Andò, S.; et al. G protein-coupled receptor 30 (GPR30) mediates gene expression changes and growth response to 17β-estradiol and selective GPR30 ligand G-1 in ovarian cancer cells. *Cancer Res.* **2007**, *67*, 1859–1866. [CrossRef] [PubMed]

42. Vivacqua, A.; Bonofiglio, D.; Recchia, A.G.; Musti, A.M.; Picard, D.; Andò, S.; Maggiolini, M. The G protein-coupled receptor GPR30 mediates the proliferative effects induced by 17β-estradiol and hydroxytamoxifen in endometrial cancer cells. *Mol. Endocrinol.* **2006**, *20*, 631–646. [CrossRef] [PubMed]

43. Vivacqua, A.; Bonofiglio, D.; Albanito, L.; Madeo, A.; Rago, V.; Carpino, A.; Musti, A.M.; Picard, D.; Andò, S.; Maggiolini, M. 17β-estradiol, genistein, and 4-hydroxytamoxifen induce the proliferation of thyroid cancer cells through the G protein-coupled receptor GPR30. *Mol. Pharmacol.* **2006**, *70*, 1414–1423. [CrossRef] [PubMed]

44. Maggiolini, M.; Vivacqua, A.; Fasanella, G.; Recchia, A.G.; Sisci, D.; Pezzi, V.; Montanaro, D.; Musti, A.M.; Picard, D.; Andò, S. The G protein-coupled receptor GPR30 mediates c-fos up-regulation by 17β-estradiol and phytoestrogens in breast cancer cells. *J. Biol. Chem.* **2004**, *279*, 27008–27016. [CrossRef] [PubMed]

45. Madeo, A.; Maggiolini, M. Nuclear alternate estrogen receptor GPR30 mediates 17β-estradiol-induced gene expression and migration in breast cancer-associated fibroblasts. *Cancer Res.* **2010**, *70*, 6036–6046. [CrossRef] [PubMed]

46. Siegel, R.; Naishadham, D.; Jemal, A. Cancer statistics, 2013. *Cancer J. Clin.* **2013**, *63*, 11–30. [CrossRef] [PubMed]

47. Schettini, F.; Buono, G.; Cardalesi, C.; Desideri, I.; De Placido, S.; Del Mastro, L. Hormone receptor/human epidermal growth factor receptor 2-positive breast cancer: Where we are now and where we are going. *Cancer Treat. Rev.* **2016**, *46*, 20–26. [CrossRef] [PubMed]

48. Jiang, W.G.; Sanders, A.J.; Katoh, M.; Ungefroren, H.; Gieseler, F.; Prince, M.; Thompson, S.K.; Zollo, M.; Spano, D.; Dhawan, P.; et al. Tissue invasion and metastasis: Molecular, biological and clinical perspectives. *Semin. Cancer Biol.* **2015**, *35*, S244–S275. [CrossRef] [PubMed]

49. Molina, L.; Figueroa, C.D.; Bhoola, K.D.; Ehrenfeld, P. GPER-1/GPR30 a novel estrogen receptor sited in the cell membrane: Therapeutic coupling to breast cancer. *Expert Opin. Ther. Targets* **2017**, *21*, 755–766. [CrossRef] [PubMed]

50. Filardo, E.J. A role for G-protein coupled estrogen receptor (GPER) in estrogen-induced carcinogenesis: Dysregulated glandular homeostasis, survival and metastasis. *J. Steroid Biochem. Mol. Biol.* **2018**, *176*, 38–48. [CrossRef] [PubMed]

51. Sjöström, M.; Hartman, L.; Grabau, D.; Fornander, T.; Malmström, P.; Nordenskjöld, B.; Sgroi, D.C.; Skoog, L.; Stål, O.; Fredrik Leeb-Lundberg, L.M.; et al. Lack of G protein-coupled estrogen receptor (GPER) in the plasma membrane is associated with excellent long- term prognosis in breast cancer. *Breast Cancer Res. Treat.* **2014**, *145*, 61–71. [CrossRef] [PubMed]

52. Ariazi, E.A.; Brailoiu, E.; Yerrum, S.; Shupp, H.A.; Slifker, M.J.; Cunliffe, H.E.; Black, M.A.; Donato, A.L.; Arterburn, J.B.; Oprea, T.I.; et al. The G protein-coupled receptor GPR30 inhibits proliferation of estrogen receptor-positive breast cancer cells. *Cancer Res.* **2014**, *70*, 1184–1194. [CrossRef] [PubMed]

53. Chimento, A.; Casaburi, I.; Rosano, C.; Avena, P.; De Luca, A.; Campana, C.; Martire, E.; Santolla, M.F.; Maggiolini, M.; Pezzi, V.; et al. Oleuropein and hydroxytyrosol activate GPER/GPR30-dependent pathways leading to apoptosis of ERnegative SKBR3 breast cancer cells. *Mol. Nutr. Food Res.* **2014**, *58*, 478–489. [CrossRef] [PubMed]

54. Weißenborn, C.; Ignatov, T.; Ochel, H.J.; Costa, S.D.; Zenclussen, A.C.; Ignatov, Z.; Ignatov, A. GPER functions as a tumor suppressor in triple-negative breast cancer cells. *J. Cancer Res. Clin. Oncol.* **2014**, *140*, 713–723. [CrossRef] [PubMed]

55. Weißenborn, C.; Ignatov, T.; Poehlmann, A.; Wege, A.K.; Costa, S.D.; Zenclussen, A.C.; Ignatov, A. GPER functions as a tumor suppressor in MCF-7 and SKBR-3 breast cancer cells. *J. Cancer Res. Clin. Oncol.* **2014**, *140*, 663–671. [CrossRef] [PubMed]

56. Chan, Q.K.; Lam, H.M.; Ng, C.F.; Lee, A.Y.; Chan, E.S.; Ng, H.K.; Ho, S.M.; Lau, K.M. Activation of GPR30 inhibits the growth of prostate cancer cells through sustained activation of Erk1/2, c-jun/c-fos-dependent upregulation of p21, and induction of G_2 cell-cycle arrest. *Cell Death Differ.* **2010**, *7*, 1511–1523. [CrossRef] [PubMed]

57. Volinia, S.; Galasso, M.; Sana, M.E.; Wise, T.F.; Palatini, J.; Huebner, K.; Croce, C.M. Breast cancer signatures for invasiveness and prognosis defined by deep sequencing of microRNA. *Proc. Natl. Acad. Sci. USA* **2012**, *109*, 3024–3029. [CrossRef] [PubMed]

58. Vidal-Gómez, X.; Pérez-Cremades, D.; Mompeón, A.; Dantas, A.P.; Novella, S.; Hermenegildo, C. MicroRNA as crucial regulators of gene expression in estradiol-treated human endothelial cells. *Cell. Physiol. Biochem.* **2018**, *45*, 1878–1892. [CrossRef] [PubMed]

59. Zhang, H.; Wang, X.; Chen, Z.; Wang, W. MicroRNA-424 suppresses estradiol-induced cell proliferation via targeting GPER in endometrial cancer cells. *Cell. Mol. Biol.* **2015**, *61*, 96–101. [PubMed]

60. Prossnitz, E.R.; Maggiolini, M. Mechanisms of estrogen signaling and gene expression via GPR30. *Mol. Cell. Endocrinol.* **2009**, *308*, 32–38. [CrossRef] [PubMed]

61. Pandey, D.P.; Lappano, R.; Albanito, L.; Madeo, A.; Maggiolini, M.; Picard, D. Estrogenic GPR30 signalling induces proliferation and migration of breast cancer cells through CTGF. *EMBO J.* **2009**, *28*, 523–532. [CrossRef] [PubMed]

62. Durchdewald, M.; Angel, P.; Hess, J. The transcription factor Fos: A Janus-type regulator in health and disease. *Histol. Histopathol.* **2009**, *11*, 1451–1461. [CrossRef]

63. Hess, J.; Angel, P.; Schorpp-Kistner, M. AP-1 subunits: Quarrel and harmony among siblings. *J. Cell Sci.* **2004**, *117*, 5965–5973. [CrossRef] [PubMed]

64. Milde-Langosch, K. The Fos family of transcription factors and their role in tumourigenesis. *Eur. J. Cancer* **2005**, *41*, 2449–2461. [CrossRef] [PubMed]

65. Bancroft, C.C.; Chen, Z.; Yeh, J.; Sunwoo, J.B.; Yeh, N.T.; Jackson, S.; Jackson, C.; Van Waes, C. Effects of pharmacologic antagonists of epidermal growth factor receptor, PI3K and. MEK signal kinases on NF-κB and AP-1 activation and IL-8 and VEGF expression in human head and neck squamous cell carcinoma lines. *Int. J. Cancer* **2002**, *99*, 538–548. [CrossRef] [PubMed]

66. Mishra, A.; Bharti, A.C.; Saluja, D.; Das, B.C. Transactivation and expression patterns of Jun and Fos/AP-1 super-family proteins in human oral cancer. *Int. J. Cancer* **2010**, *126*, 819–829. [CrossRef] [PubMed]

67. Qie, S.; Diehl, J.A. Cyclin D1, cancer progression, and opportunities in cancer treatment. *J. Mol. Med.* **2016**, *94*, 1313–1326. [CrossRef] [PubMed]

68. Han, Y.; Zhang, Y.; Jia, T.; Sun, Y. Molecular mechanism underlying the tumor-promoting functions of carcinoma-associated fibroblasts. *Tumor Biol.* **2015**, *36*, 1385–1394. [CrossRef] [PubMed]

69. Bhowmick, N.A.; Neilson, E.G.; Moses, H.L. Stromal fibroblasts in cancer initiation and progression. *Nature* **2004**, *432*, 332–337. [CrossRef] [PubMed]

70. Cheng, N.; Chytil, A.; Shyr, Y.; Joly, A.; Moses, H.L. Transforming growth factor-beta signaling-deficient fibroblasts enhance hepatocyte growth factor signaling in mammary carcinoma cells to promote scattering and invasion. *Mol. Cancer Res.* **2008**, *6*, 1521–1533. [CrossRef] [PubMed]

71. Zhi, K.; Shen, X.; Zhang, H.; Bi, J. Cancer-associated fibroblasts are positively correlated with metastatic potential of human gastric cancers. *J. Exp. Clin. Cancer Res.* **2010**, *29*, 66. [CrossRef] [PubMed]

72. Di Leva, G.; Cheung, D.G.; Croce, C.M. miRNA clusters as therapeutic targets for hormone-resistant breast cancer. *Expert Rev. Endocrinol. Metab.* **2015**, *10*, 607–617. [CrossRef] [PubMed]

Review

Unleashing the Full Potential of Oncolytic Adenoviruses against Cancer by Applying RNA Interference: The Force Awakens

Tereza Brachtlova and Victor W. van Beusechem *

Amsterdam UMC, Vrije Universiteit Amsterdam, Medical Oncology, Cancer Center Amsterdam,
De Boelelaan 1117, 1007 MB Amsterdam, The Netherlands; t.brachtlova@vumc.nl
* Correspondence: VW.vanBeusechem@vumc.nl; Tel.: +31(0)20-4442162

Received: 31 October 2018; Accepted: 21 November 2018; Published: 23 November 2018

Abstract: Oncolytic virus therapy of cancer is an actively pursued field of research. Viruses that were once considered as pathogens threatening the wellbeing of humans and animals alike are with every passing decade more prominently regarded as vehicles for genetic and oncolytic therapies. Oncolytic viruses kill cancer cells, sparing healthy tissues, and provoke an anticancer immune response. Among these viruses, recombinant adenoviruses are particularly attractive agents for oncolytic immunotherapy of cancer. Different approaches are currently examined to maximize their therapeutic effect. Here, knowledge of virus–host interactions may lead the way. In this regard, viral and host microRNAs are of particular interest. In addition, cellular factors inhibiting viral replication or dampening immune responses are being discovered. Therefore, applying RNA interference is an attractive approach to strengthen the anticancer efficacy of oncolytic viruses gaining attention in recent years. RNA interference can be used to fortify the virus' cancer cell-killing and immune-stimulating properties and to suppress cellular pathways to cripple the tumor. In this review, we discuss different ways of how RNA interference may be utilized to increase the efficacy of oncolytic adenoviruses, to reveal their full potential.

Keywords: RNA interference; small interfering RNA; microRNA; oncolytic virotherapy; conditionally replicating adenovirus (CRAd)

1. Oncolytic Virotherapy

Despite many different types of treatments used to combat cancer, cancer is still one of the leading causes of death worldwide. While on one hand the specific traits cancer cells acquire, such as infinite replicative potential and evading apoptosis [1], provide angles for anticancer treatment, these traits on the other hand complicate treatment and, in many cases, contribute to treatment failure. Constant replication and genetic instability allow for adaptation to medication and emerged resistance. There is, therefore, a great need for new treatments that are also effective against therapy-resistant cancers. One promising approach is oncolytic virus therapy (OVT), using viruses that selectively replicate in and kill cancer cells [2,3]. There are several reasons why OVT is considered an elegant option for cancer treatment. Firstly, oncolytic viruses exhibit specific replication in cancer cells, often dependent on the genetic changes that discriminate cancer cells from non-malignant cells [4,5]. Hence, genetic changes underlying malignant cell growth are utilized for selective therapy. Secondly, oncolytic viruses kill their host cell as an integral part of their life cycle. Replication in a host cell inevitably leads to death of the host cell. Oncolytic viruses, thus, have profound direct anticancer cell activity. Thirdly, selective replication in cancer cells produces large numbers of progeny viruses that upon their release are capable of infecting new cancer cells causing subsequent cycles of lytic replication. Thus, the anticancer effect is amplified in situ. On top of these direct anticancer effects, oncolytic viruses engage immune system responses of the host [6,7]. Oncolytic virus replication in a tumor can help override the immune suppressive conditions that exist in the tumor microenvironment [7]. Upon virus infection, an antitumor immune response is induced by the production of immune-stimulatory cytokines by virus-infected cancer cells and the concerted release of endogenous danger-associated molecular patterns, virus-derived pathogen-associated molecular patterns, and tumor-specific or tumor-associated antigens from lysed cancer cells in a pro-inflammatory context. This may induce tumor-specific immunity, also against cancer cells at distant sites not subjected to virus infection [7].

Several different viruses are considered for use in OVT. Some of them advanced to clinical trials. Of these, an oncolytic herpes simplex virus (HSV) was recently approved for treatment of melanoma by the competent authorities in the United States and Europe [8,9] and an oncolytic adenovirus was already approved in 2005 in China for treatment of head and neck cancer in combination with chemotherapy [10]. Generally, oncolytic viruses can be divided into three main groups. The first group contains viruses with a natural propensity to preferentially replicate in cancer cells, while being non-pathogenic in humans. Many of these viruses are of non-human origin. This group comprises autonomous parvovirus, myxoma virus, Newcastle disease virus, and reovirus [11–13]. Many of the second group consists of viruses that were attenuated by propagation in vitro. These attenuated strains of, e.g., measles virus or vaccinia viruses were used safely in humans as vaccine vectors and were, therefore, considered safe starting points to build oncolytic viruses. The third group consists of viruses that were genetically engineered by introducing mutations in their genome to ensure selective replication in cancer cells. The latter group includes oncolytic viruses derived from adenovirus, HSV, and vesicular stomatitis virus (VSV), as well as an attenuated and genetically modified strain of poliovirus [11–13]. Oncolytic adenoviruses, often also referred to as conditionally replicating adenoviruses (CRAds), represent one of the most extensively researched oncolytic viruses. This review focuses on oncolytic adenoviruses, but the general concepts discussed apply to other oncolytic viruses as well.

Oncolytic Adenoviruses

Adenoviruses are small, non-enveloped double-stranded DNA viruses (34–36 kb), continuously present in the population and responsible for self-limiting infections in immune-competent individuals, including mild respiratory, conjunctival, and gastrointestinal diseases, or cystitis. Infections in immunocompromised patients are much more severe, including infections in liver (hepatitis), lung (pneumonia), or heart (myocarditis) that can result in death (reviewed by Lion [14]). Adenoviral replication is strongly dependent on expression of the immediate early E1A region. To establish a successful replication in a host cell, the adenovirus seizes control over retinoblastoma protein (pRb) and p53 protein (using E1A and E1B viral proteins), which suspends the host cell cycle in the synthesis (S) phase [15].

Non-replicating adenoviruses have been used as gene delivery vectors in cancer gene therapy strategies for many years. Replication-deficient adenovirus vectors lack genes from the E1 region and may also have E3 genes deleted to accommodate newly introduced genes [16]. Although these adenoviruses showed low toxicity and promising results compared to conventional therapy, their efficacy in clinical trials remained low, which minimizes their use.

Current research focuses mainly on CRAds that are considered much more promising agents for the treatment of cancer. Replication of CRAds is controlled either by regulating mostly E1A gene expression with a tumor-specific promoter or by deletions in the viral genome that require cellular factors for compensation [17]. Examples of the first strategy include human telomerase (hTERT) promoter-driven oncolytic adenoviruses [18–20]. The first CRAd to be evaluated for cancer treatment, ONYX-015, was made following the second strategy. This adenovirus variant was designed to selectively replicate in p53-deficient cancer cells by deletion of the *E1B55K* gene [21]. Although ONYX-015 showed cancer cell-selective replication, its efficacy was disappointing [22]. Since then, newer generations CRAds with improved selectivity and potency were developed, including Ad5-Δ24 and ICOVIR-5 [23,24]. Nevertheless, despite very encouraging results from in vitro and animal studies, the anticancer efficacy of CRAds, as well as of other oncolytic viruses, as a single agent in humans is generally modest [25]. Thus, there is a clear need to increase the efficacy of OVT. This could be achieved using more effective delivery methods or by enhancing the potency of CRAds to kill cancer cells or to induce an antitumor immune response. In addition, while most efforts are on improving anticancer treatment efficacy, studies are also undertaken to more stringently control CRAd replication in healthy cells.

2. Strategies to Increase the Efficacy of Oncolytic Virus Therapy with CRAds

2.1. Achieving More Effective Delivery of Oncolytic Adenovirus to Tumors

Effective OVT with CRAds requires that viruses are delivered to tumors in the human body and that they enter cancer cells to initiate oncolysis. Notably, cancer cells are sometimes resistant to CRAd infection due to low expression of the primary receptor molecule coxsackie-adenovirus receptor (CAR) [26]. Typical neoplasms in which downregulation of CAR expression was observed include prostate, colon, and kidney cancers [27]. Retargeting strategies allow overcoming this obstacle, specifically by diversion of the virus to other cell surface receptors. Strategies that were successfully followed to accomplish this were, e.g., incorporation of a cyclic RGD4C peptide motif in the adenovirus fiber knob to allow entry via $\alpha_v\beta3$ and $\alpha_v\beta5$ integrins [28], pseudotyping the viral capsid with proteins from other serotype adenoviruses or with chimeric capsid proteins [29,30], or expressing bispecific adapter molecules from the CRAd genome targeting virus entry via an alternative cell surface receptor [31]. Generally, these modifications resulted in more effective CRAds with broader applicability in OVT.

The administration route to deliver the virus to tumor cells in the human body poses another challenge. Systemic administration of CRAds was proven quite ineffective since most injected virions are eliminated before they reach their target. Much research is put into the development of methods to chemically modify viral capsids to shield them from sequestration in the liver and inactivation by the immune system [32]. Another interesting approach is to use carrier cells as temporary virus hosts delivering oncolytic viruses, including CRAds, to tumor sites. This Trojan horse concept is very attractive, because it not only hides the virus from the immune system, but also exploits the capacity of cells to extravasate from the circulation and home to tissues [33,34]. However, several major challenges remain, including premature expression of viral proteins in the carrier cell, complicated timing of the delivery, acquired adaptive immunity to carrier cells, or the inability to pass through capillaries, which results in the accumulation in, e.g., lungs, and subsequent release of the virus before delivering it to the tumor [33,35,36]. Moreover, there is a contradiction in delivering a virus with cancer-selective replication properties using a non-malignant carrier cell. At least a single virus lifecycle should be completed in this cell to allow release of infectious progeny virus at the tumor site. This means that either the virus should not be entirely cancer-selective, or the carrier cell should have cancer cell-like properties, such as a deregulation in growth control. Both options may raise safety concerns that need to be addressed.

2.2. Improving Oncolytic Adenovirus Specificity by Employing microRNA-Dependent Replication

A novel strategy to make CRAds safer by further limiting their replication in non-malignant cells exploits microRNA (miRNA) technology. Here, a miRNA target sequence is incorporated in the 3′ untranslated region (UTR) of the transcript encoding a viral protein essential for replication, usually E1A. Infection of cells expressing the relevant miRNA results in silencing of the essential protein and, consequently, suppression of viral replication. In recognition of the fact that human adenoviruses have strong tropism for liver cells and are efficiently delivered to liver cells when administered systemically, which could lead to liver toxicity, particular efforts were made to inhibit CRAd replication in liver cells. Ylösmäki et al. [37] constructed Δ24-type CRAds carrying three target sites for the liver-specific miRNA miR-122 in the Δ24-E1A 3′ UTR. While this reduced Δ24-E1A expression in miR-122-expressing hepatoma cells by more than 90%, this did not inhibit virus replication. By introducing a second mutation in the E1A messenger RNA (mRNA) that reduces overall expression levels, this problem was successfully addressed. The resulting virus was severely attenuated, failing to induce cytopathic effects in hepatoma cells and producing more than 10,000-fold reduced titers in hepatoma cells compared to non-small lung cancer cells. The same approach was taken independently by others; CRAds with miR-122 binding sites exhibited reduced E1A expression and virus replication in liver cells and almost absent liver toxicity upon injection in mice [38,39].

Knowledge of differential expression of miRNAs in cancer cells versus non-malignant cells was used similarly to construct CRAds with reduced replication in non-cancer cells. In particular, miR-199, which is downregulated in cancer cells, was considered for this purpose [19,40]. Significantly reduced virus replication and toxicity was observed in non-malignant cell cultures and in livers of newborn mice if miR-199 target sequences were incorporated into the E1A 3′ UTR. By combining target sequences for miR-199a and liver-specific miR-122a in the 3′ UTR of an hTERT promoter-driven E1A gene, a most selective CRAd was obtained [19]. Oncolytic potency in cancer cell lines was fully retained. Others translated the general concept of miRNA-dependent suppression of CRAd replication to more specific applications, where viral toxicity might arise upon local tissue administration. MiRNA-regulated CRAds were made in a wild-type adenovirus background for selective treatment of pancreatic cancer and glioma [41,42]. The viruses were effective in animal models and exhibited strongly reduced toxicity profiles. For this, insertion of four miRNA target sites appeared sufficient, but this might depend on host miRNA expression levels and the strength of the viral gene promoter. Thus, preclinical data suggest that host cell miRNA-regulation of adenovirus replication alone can already provide sufficient CRAd selectivity. However, further development of this type of CRAd will likely be done in the background of CRAds carrying mutations with proven clinical safety.

Another strategy to restrict CRAd replication in healthy cells was taken by Gürlevik et al. [43]. They expressed multiple artificial miRNAs targeting essential adenovirus genes from a single polycistronic transcript, driven by a tumor suppressor p53-dependent promoter, inserted in the adenovirus genome. In cells with functional p53 expression, but not in p53 mutant cells, the artificial miRNAs were expressed and silenced their target genes. Although this attenuated viral replication in p53 wild-type cells, virus progeny production was not prevented. The p53-dependent attenuated phenotype was also observed in mice injected with the virus, where p53 wild-type mice carried lower virus copy numbers in their liver than did p53 knockout mice. The inhibition of virus replication that could be achieved in this study appeared inferior to that achieved when relying on host miRNAs. This can possibly be explained by the fact that the latter are already highly expressed in the host cell at the time of infection, whereas CRAd-encoded miRNAs need to be produced after cell entry and their level of expression might be replication-dependent, as copy numbers increase tremendously during replication. Hence, functional knockdown of viral genes by virus-encoded miRNAs is probably not reached before viral replication is initiated.

2.3. Strategies for Improving the Potency of Oncolytic Adenoviruses

The strong selectivity of CRAd replication in cancer cells that is obtained with currently available technology, providing excellent safety, allows exploration of methods to enhance the cancer cell-killing potency of CRAds. Apart from the tropism modifications mentioned above, investigated approaches include the introduction of potency-enhancing genome mutations and the expression of potency-enhancing transgenes. The former is usually based on identification of such mutations in virus variants that are either naturally occurring or were selected using molecular evolution methods. A good example of this approach is the genetic bioselection of a virus with a truncating mutation in the endoplasmatic retention domain of the E3/19K protein that promotes virus progeny release from infected cells [44]. Incorporation of this T1 mutation in an Ad5-Δ24 backbone with RGD4C fiber modification produced a CRAd with strong antitumor activity in vitro and in vivo, without compromising its selectivity toward cancer cells [45]. Another example is ColoAd1, a recombination product between two naturally occurring serotypes selected for rapid replication on human cancer cell lines, which exhibits increased potency in particular on colorectal carcinoma cells [46]. Transgenes that can be carried by the virus to increase its potency include genes encoding fusogenic proteins or tissue-remodeling peptides, enzymes for use in suicide gene therapy approaches, pro-apoptotic or tumor suppressor proteins, and immune-modulating proteins. Expression of a fusogenic protein induces cell–cell fusions, forming syncytia consisting of CRAd-infected cells and adjacent non-infected cells, leading to death also of the fused non-infected cells and, thus, more effective tumor eradication [47]. Another way to promote spread of CRAds through tumor tissues is to express molecules that remodel the extracellular matrix, allowing better virus dispersion in a solid tumor mass [48–50]. The suicide gene therapy concept, where an enzyme is expressed that converts a non-toxic pro-drug into a toxic compound, was also applied in the context of CRAds [51,52]. This was met with various levels of success, as premature death of the host cell inhibits virus propagation. Careful timing of virus and pro-drug administration is, thus, essential, which makes the design of clinical treatment schedules challenging. The same can be said about expression of apoptosis-inducing proteins, where premature cell death inhibits, and properly timed cell death promotes virus spread [53]. In contrast, expression of tumor suppressor protein p53 was not met with this limitation, presumably because adenovirus-encoded E1B55K protein interacts with p53 and regulates its activity during adenovirus replication [54,55]. In the late phase of the virus lifecycle, p53 promotes cell death, thereby accelerating virus progeny release [54,56]. In addition, p53 increases late adenoviral gene expression, presumably by enhancing transcription from the virus major late promoter (MLP) [57] (Figure 1). Exogenous p53 expression augmented CRAd replication in the majority of cancer cell lines, primary tumor cell cultures, and xenograft tumor models tested [54,55,58–63]. Notably, transgene expression from the adenovirus genome during virus replication in host cells can be delayed by coupling its transcription to the MLP. Apart from avoiding premature transgene expression, this strategy offers the additional advantage of confining transgene expression to cancer cells in which the CRAd replicates selectively [64]. In addition to the aforementioned ways of enhancing the primary tumor-eradicating potency of CRAds, there is currently much interest in promoting the secondary, immune-stimulating effect of these viruses, by incorporating genes encoding immune-stimulatory molecules, such as cytokines, chemokines, and checkpoint inhibitors (reviewed by de Gruijl et al. [7]). Altogether, these developments led to substantially improved CRAd variants with more potent cancer cell-killing properties than earlier generation viruses. Nevertheless, one area of potential CRAd improvement was long overlooked. While cancers may lack certain molecules needed for effective OVT that can be supplemented by transgene expression from the CRAd genome, they could also express molecules inhibiting OVT that should, thus, be depleted. Currently available gene suppression techniques allow investigating this and constructing even more powerful CRAds capable of overcoming obstacles for OVT in cancer cells. The next section focuses on this new development.

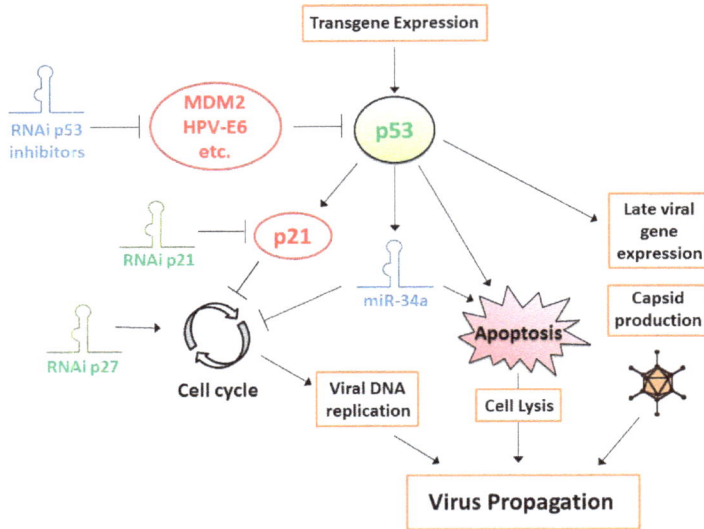

Figure 1. Tumor suppressor p53-dependent strategies to increase the potency of conditionally replicating adenoviruses (CRAds). Expression of functional p53 in cancer cells promotes CRAd propagation by stimulating late viral gene expression and accelerating cell lysis to release virus progeny. In contrast, direct p53 target p21 inhibits the cell cycle and, thereby, viral DNA replication. Silencing cyclin-dependent kinase inhibitors p21 or p27 through RNA interference (RNAi) alleviates cell cycle arrest and, thus, promotes viral DNA replication. Potential other means of promoting CRAd propagation include silencing of p53 inhibitors, which are known to inhibit CRAd replication, and silencing of p53 transcriptional target microRNA (miR)-34a. Molecules in red depict confirmed inhibitors of CRAd efficacy; molecules in green depict confirmed stimulators of CRAd efficacy; molecules in blue depict proposed, but not yet experimentally confirmed modulators of CRAd efficacy; orange delineated viral processes can be influenced to increase potency; cellular processes that are affected by viruses during infection and that can be modulated by transgenes and RNAi molecules expressed by armed CRAds are delineated in black.

3. Gene Suppression to Make Oncolytic Viruses More Effective

There are two requirements for improving OVT by suppressing inhibitors of CRAds in cancer cells, i.e., (i) identification of the inhibitory gene products in cancer cells, and (ii) design of an effective way of suppressing genes, using molecules that can be carried by CRAds. For the former, one may build on available knowledge of virus–host cell interactions or use high-throughput functional screening approaches. The latter, we know now, can be successfully achieved by using RNA interference (RNAi).

Currently, RNAi is the most widely used method to specifically downregulate functional gene expression. RNAi is a conserved cellular surveillance system in eukaryotes that activates a sequence-specific degradation of RNA species homologous to short non-coding double-stranded RNA (dsRNA) molecules [65]. These dsRNA molecules, including miRNAs and short interfering RNAs (siRNAs), are products of the RNase III enzymes Drosha and Dicer [66]. One of the strands of the miRNAs and siRNAs, the active or leading strand, is incorporated into a multiprotein complex called RNA-induced silencing complex (RISC) and directs the target RNA recognition leading to its degradation. In addition, miRNAs direct translational repression and mRNA deadenylation, further contributing to gene silencing [67]. The mechanism of this strongly conserved regulation of cellular mRNAs by miRNAs was thoroughly reviewed in several publications [65,68,69].

A means of producing exogenous siRNAs or miRNAs in a cell is to express small RNA molecules as single transcripts that form a stem-loop structure, generally referred to as short hairpin RNAs (shRNAs) and short hairpin miRNAs (shmiRNAs). Their structure is very similar to that of pre-miRNA, i.e., the Drosha-cleaved product in endogenous miRNA biogenesis. The dsRNA in the stem of these molecules is processed by Dicer to form mature siRNA and miRNA duplexes, respectively, that will then incorporate their active strand into RISC and direct RNAi. Successful RNAi induced by adenovirus-encoded exogenous shRNA was achieved initially using replication-defective vectors [70]. This opened the possibility of silencing genes in cancer cells with a therapeutic intent.

Because the efficiency of gene delivery with replication-defective vectors is generally considered inadequate for effective cancer treatment, interest rose in using CRAds as shRNA delivery vectors. However, achieving effective RNAi in cells infected with a replication-competent adenovirus was not considered trivial. RNAi is recognized as a cellular defense mechanism against virus infection and this led many viruses to evolve molecules that inhibit RNAi [71,72]. Although the importance of RNAi as an antiviral response in mammalian cells is heavily debated (e.g., References [73,74]), there was reason to assume that the process of RNAi would be hampered in cells in which a virus is replicating. In fact, several independent reports showed that adenovirus-encoded virus-associated RNAs (VA RNAs) inhibit the RNAi machinery (Figure 2). VA RNAs (I and II) were found to inhibit the host cell RNAi pathway at the level of nuclear export, by competing with miRNA precursor molecules for binding to Exportin 5; at the level of processing RNAi precursor molecules, by inhibiting the activity of Dicer; and at the effector level, by competing with siRNA and miRNA for incorporation into RISC [75,76]. In addition, it was later reported that the titration of Exportin 5 by VA RNA also reduced Dicer mRNA nuclear export and, thereby, Dicer protein levels [77]. Apart from this direct inhibition of the RNAi machinery by virus-encoded RNA, it was recently reported that the type I interferon response evoked by adenovirus vectors in cells inhibits the processing of shRNAs by Dicer, resulting in a less effective knockdown of target genes [78]. Altogether, there were reasons to expect that effective CRAd replication and RNAi could be incompatible.

Nevertheless, in a proof-of-concept study using firefly luciferase as target, Carette et al. showed that CRAd-induced shRNA can induce proper gene silencing in human cancer cells [79]. That suggested that, although interferon-responses and VA RNAs inhibit the RNAi machinery, they do not prohibit RNAi brought about by adenovirus-encoded shRNA. Still, Machitani et al. showed that RNAi mediated by adenovirus-encoded shRNA can be improved by deleting VA RNA sequences [80]. Notably, this study was done with replication-defective adenovirus vectors. Since VA RNA expression is highly replication-dependent, competitive incorporation into the RNAi machinery is expectedly even higher in a replication competent context. Thus, to achieve highly effective RNAi using CRAds it could be considered to (partly) delete VA RNA sequences. However, adenovirus lacking major VA RNA I cannot efficiently translate viral mRNA in the late phase of the infection, resulting in poor replication [81]. In addition, it was recently reported that VA RNA II-encoded viral miRNA promotes adenovirus replication, presumably via suppression of cullin 4-mediated inhibition of Jun N-terminal kinase (JNK) signaling [82]. VA RNA-depleted CRAds are, therefore, considered less attractive agents for OVT. However, observations reported by Kamel et al. suggested that VA RNA I expression is required primarily to counteract cellular anti-viral interferon responses, not so much the RNAi machinery [83]. This notion is supported by the observation that processing of VA RNAs by Dicer into viral miRNAs inhibits adenovirus replication, apparently by depleting VA RNAs capable of inhibiting double-stranded RNA-activated kinase (PKR) [84]. Thus, the RNAi machinery appears to contribute to the cellular defense against adenovirus infection indirectly, by inhibiting the capacity of the virus to suppress the major anti-viral interferon–PKR response. This implies that if the PKR and RNAi inhibitory functions of VA RNA I can be dissected by introducing specific mutations, or, alternatively, VA RNA I is deleted and the interferon response is inhibited through other means, such as an shRNA targeting PKR, more effectively replicating and gene-silencing CRAds could perhaps be made.

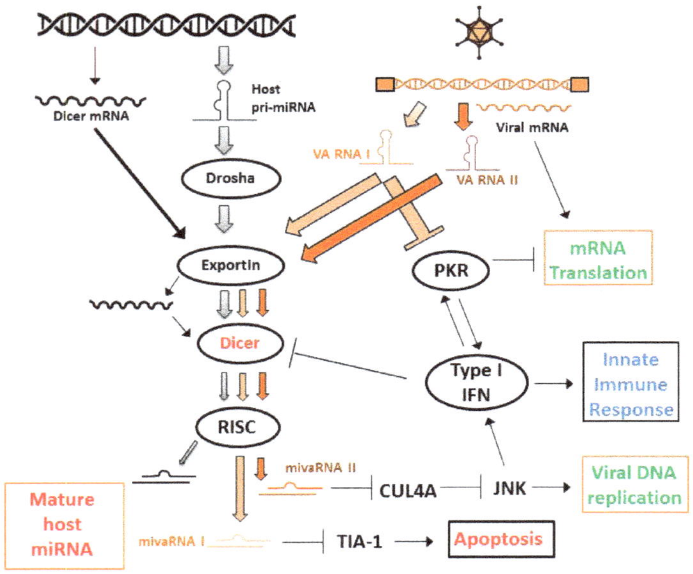

Figure 2. Interaction between adenovirus-encoded small RNA molecules and the host cell RNAi machinery and interferon–double-stranded RNA-activated kinase (PKR) response in the late phase of infection. Virus-associated (VA) RNAs I and II inhibit endogenous miRNA processing via competition at several levels in the RNAi machinery, as well as by reducing Dicer messenger RNA (mRNA) nuclear export. Upon processing of VA RNAs into viral miRNAs (mivaRNAs), they silence cellular mRNAs to inhibit apoptosis and promote viral DNA replication. VA RNA I directly inhibits PKR to promote (viral) mRNA translation and, thus, protein expression for capsid production. Regulation of type I interferons evoking innate immune responses is complex, with on one hand mivaRNA II-mediated stimulation and on the other hand VA RNA I-mediated inhibition of PKR-dependent stimulation. The type I interferon response inhibits Dicer function, contributing to inhibition of miRNA processing. Colors illustrate the different effects in cells in the late phase of infection: red, inhibited; green, stimulated; blue, complex regulation; orange, virus-encoded and virus-related; black, cellular.

Although the efficacy of CRAd-mediated RNAi can probably be improved, the findings by Carette et al. [79] fueled the development of CRAds capable of silencing a variety of target genes in cancer cells to achieve more effective anticancer treatment. In many cases, the target genes were not necessarily expected to impact the efficacy of OVT, but were chosen merely for their known or anticipated anticancer effects. Examples include, among others, silencing an oncogene to inhibit tumorigenicity, silencing a pro-angiogenic growth factor to inhibit angiogenesis, or silencing an apoptosis inhibitor to promote cell death [85–87]. These RNAi-inducing CRAds could, thus, be considered therapeutic agents providing a combination of OVT and targeted therapy. The two independent activities provided by CRAd RNAi were often additive, resulting in a more effective treatment in preclinical models than could be achieved with unarmed control CRAds or replication-defective RNAi vectors. Below, we focus on approaches to achieve more effective OVT.

3.1. Combining OVT with Suppression of CRAd-Inhibitory Target Genes in Cancer Cells

As a logical extension of their above-described observation that Dicer inhibits adenovirus replication, Machitani et al. developed a CRAd expressing an shRNA against Dicer [88]. While replicating in cancer cells, the virus silenced Dicer expression, allowing an increase in VA RNA copy numbers available for inhibiting PKR activity. Consequently, the virus produced higher genome copy numbers and functional progeny titers than did a control virus silencing luciferase. This resulted in a more effective lysis of cancer cells in vitro and a more effective inhibition of tumor growth in vivo. Apparently, in this respect, the gain of more effective VA RNA I-mediated PKR inhibition was more important than the loss of VA RNA II-mediated JNK activation. Although expression of the shRNA was driven by the H1 mammalian polymerase (Pol) III promoter, which is not expected to provide cancer cell selectivity, Dicer expression was not significantly reduced in non-malignant cells. This can best be explained by the fact that CRAds do not replicate in non-cancer cells and, consequently, genome copy numbers in these cells remain very low. Thus, while it can be argued that inhibition of the RNAi machinery in cells could interfere with endogenous miRNA processing causing unpredictable and possibly detrimental effects, the efficiency of shRNA expression in the absence of viral replication was apparently low enough to avoid such unwanted effects.

Another approach that was successfully taken to increase the potency of CRAds is to suppress cell-cycle inhibition in cancer cells. Since CRAds, by design, lost their capacity to take control over the cell cycle, to induce the S phase for efficient genome replication, their replication efficiency depends on the spontaneous progression of cancer cells through the cell cycle. Although proper regulation of the cell cycle is lost in cancer cells, partial regulation of cell-cycle progression may still exist. It could, therefore, be conceived that cell-cycle inhibition, in particular at the Gap 1 (G1) checkpoint, inhibits adenovirus replication. This notion was supported by the finding that overexpression of the major cyclin-dependent kinase (CDK) inhibitor $p21^{WAF1/CIP1}$ in cancer cells inhibited CRAd replication [89] and by the observation that CRAds replicated more efficiently, inducing stronger cytopathic effects, in $p21^{WAF1/CIP1}$ knockout cells [90]. In experiments where CRAds were combined with siRNAs targeting the CDK inhibitors $p21^{WAF1/CIP1}$ or $p27^{KIP}$, virus progeny production and cell-killing potency were increased, the latter depending on cellular p53 status [90] (Figure 1). These observations provided incentive to construct a prostate-specific CRAd expressing an shRNA targeting $p21^{WAF1/CIP1}$ [91]. As expected, the virus exhibited increased replication in and oncolysis of prostate cancer cells in vitro and provided enhanced tumor suppression and survival in a tumor xenograft animal model. Interestingly, apart from the sought effect on adenovirus replication via cell-cycle control, a less anticipated positive effect was observed as well. Silencing $p21^{WAF1/CIP1}$ increased expression of the androgen receptor in prostate cancer cells, thereby stimulating expression of the viral E1A gene driven by the used androgen receptor-dependent prostate-specific promoter. While the relative contributions of both effects are difficult to dissect, the total effect of expressing a $p21^{WAF1/CIP1}$-silencing shRNA from the CRAd genome on the efficacy of OVT was evident.

A foreseeable further improvement of CRAd potency could be to arm the CRAd with both a p53 gene and a $p21^{WAF1/CIP1}$-silencing shRNA. As discussed above, the utility of expressing p53 from the CRAd genome to increase its potency was shown in a variety of cancer models. Since $p21^{WAF1/CIP1}$ is a direct transcriptional target of p53, an unwanted side effect of expressing p53 could be induction of $p21^{WAF1/CIP1}$ expression, limiting CRAd replication. Combined expression of p53 and an shRNA targeting $p21^{WAF1/CIP1}$ could then overcome this limitation. A replication-defective adenovirus vector with cocistronic expression of p53 and a $p21^{WAF1/CIP1}$-targeting artificial miRNA was already tested for its anticancer properties [92]. This viral vector caused more effective apoptosis induction in cancer cells and decreased xenograft tumor growth compared to control vectors expressing only p53 or $p21^{WAF1/CIP1}$-targeting miRNA. A similar favorable combination effect of expressing p53 and silencing $p21^{WAF1/CIP1}$ is to be expected in the context of OVT with a CRAd and, therefore, deserves to be explored.

Along the same line of reasoning, CRAds expressing p53 could probably be made more effective by silencing p53 inhibitors (Figure 1). Many p53 inhibitors are known and confirmed active in cancer cells. Paradigm examples are human papillomavirus (HPV)-E6 protein in cervical cancer cells and amplified mouse double minute 2 homolog (MDM2) in sarcomas. We already showed that CRAds expressing p53 variants incapable of binding to HPV-E6 or MDM2 are more effective in these cancers [61,62]. A limitation of this approach could be that the introduction of mutations in the p53 protein to abrogate binding to its inhibitor might compromise its transcriptional activity. Therefore, combining the expression of wild-type p53 with an shRNA targeting the inhibitor could perhaps be more effective. In addition, cancer cells often express multiple p53 inhibitors. This can be deduced from functional screens for p53 modulators done on cancer cell lines [93–95]. Preventing binding to a single inhibitor might, therefore, not be sufficient in many cases. It will be difficult, if not impossible, to prevent binding of p53 to multiple inhibitors by mutating its amino-acid sequence. In contrast, since their small size allows for multiple shRNAs to be inserted into a single CRAd genome, it should not be too difficult to simultaneously silence multiple p53 inhibitors, maximizing the activity of p53-expressing CRAds.

3.2. Combining OVT with Targeting Immune Suppression

As mentioned above, stimulation of the immune system to evoke an anti-tumor response is currently considered a crucial property of oncolytic viruses. This inherent property can be further enhanced by arming the virus with, e.g., immunostimulatory cytokines [7]. While this resulted in much improved OVT systems, active immune suppression in the tumor microenvironment was not fully addressed. Many immune-suppressive molecules, expressed by cancer cells or by suppressive cells of the immune system, were already identified in the tumor microenvironment. Surprisingly, with the technology to combine CRAds with RNAi at hand, arming CRAds with shRNAs targeting such molecules is still quite an unexplored approach. The exception is a CRAd encoding a combination of MART1 melanoma antigen, granulocyte-macrophage colony-stimulating factor (GM-CSF) cytokine, and an shRNA silencing transforming growth factor β2 (TGF-β2) [96]. The rationale for the combination was that, while GM-CSF has many immune-stimulating properties, such as inducing maturation of dendritic cells (DCs), expansion and differentiation of lymphocytes, and recruitment of natural killer (NK) cells, it is also known to enhance the expansion of myeloid-derived suppressor cells (MDSC). TGF-β2 antagonizes GM-CSF-induced DC maturation, inhibits tumor antigen-specific T-cell activation, and stimulates MDSC. Concomitant expression of GM-CSF and suppression of TGF-β2 was, thus, expected to provide a more potent induction of antitumor immunity than expressing GM-CSF alone. This CRAd was tested in a prime-boost therapeutic vaccination strategy, using immune-competent mice and mouse melanoma cells engineered to allow human adenovirus replication. A plasmid expressing MART1 was injected intramuscularly to generate MART1-specific memory T cells and the CRAd was injected intratumorally to boost the immune response. The strategy significantly increased tumor infiltration with cytotoxic T cells, NK cells, NKT cells, (mature) DCs, and macrophages, and decreased regulatory T cells. Immune cell activity was evident from high levels of interferon-γ (IFN-γ) expressed. Together with the oncolytic activity of the virus, this resulted in delayed tumor growth. The contribution of silencing TGF-β2 to the total effect was, however, not entirely clear from the reported results.

Another CRAd that could reactivate antitumor immunity through suppressing an immune suppressive target gene in cancer cells was made, but not tested as such. This CRAd expressing antisense RNA against signal transducer and activator of transcription 3 (STAT3) was shown capable of depleting STAT3 protein in infected cancer cells, as well as its downstream targets [97]. The virus exhibited increased replication and more potent antitumor activity, presumably through counteracting the effects of STAT3 on cell survival, cell migration, and angiogenesis. STAT3 is, however, also involved in immune evasion. It would, therefore, be very interesting to investigate the effects of the STAT3-suppressing virus on the antitumor immune response. The host specificity of human adenoviruses is, however, a major challenge in the design of such experiments. It is very difficult to develop an immune-competent animal model with available tools for immunological studies, that allows replication of human adenoviruses. Although Syrian hamsters are semi-permissive for human adenovirus replication and, thus, provide a useful immune-competent animal model for CRAd efficacy studies [98], and CRAds armed with immune-stimulating transgenes were successfully tested in these animals (e.g., References [99,100]), only few antibodies are available to study hamster immune cell subsets, precluding in-depth analysis of immune responses. For this, ex vivo cultures of human whole tumor-derived single-cell suspensions or fresh tissue slides, or novel mouse models carrying patient-derived xenograft tumors and reconstituted with immune cells from the same donor could prove highly valuable.

3.3. Exploiting Virus–Host Interactions via MicroRNAs

Although, as mentioned above, a physiological role for RNAi in response to virus infection in mammals is being challenged, miRNA-mediated virus–host interactions were described for a number of mammalian viruses (reviewed in References [101–103]). This includes anti-viral or pro-viral effects of host miRNAs, as well as virus-encoded miRNAs that modify host cell biology to establish an environment conducive to completion of the virus lifecycle or that exploit the host RNAi machinery to regulate viral gene expression. Paradigm examples of human miRNAs that promote virus replication include miR-122 and miR-132 that promote hepatitis C virus replication and human immunodeficiency virus 1 (HIV-1) replication, respectively [104,105]. Inhibitory effects of human miRNAs on viruses are, however, more common. They could be part of the host innate antiviral response to recover from an infection, or alternatively, be part of a strategy of the virus to establish a persistent infection [106].

In the case of adenovirus, the VA RNAs constitute the virus-encoded miRNAs (Figure 2). They are processed by the RNAi machinery and are functional in directing silencing of target sequences [76,107,108]. As mentioned above, VA RNA II-encoded viral miRNA was shown to promote adenovirus replication [82]. The functional importance of VA RNA I acting as miRNAs is, however, questioned. While specific inhibition of miRNA target site binding using chemically modified oligonucleotides (i.e., antagomirs) reduced adenovirus replication [76], introduction of mutations in the miRNA seed sequence did not appear to influence adenovirus propagation [83]. Nevertheless, specific cellular target genes for VA RNA I-encoded miRNA, including T-cell-restricted intracellular antigen 1 (TIA-1), were identified and their silencing was confirmed in virus-infected cells [109]. The relevance of silencing these genes for the virus lifecycle is so far mostly unresolved. However, it seems unlikely that adenoviruses evolved specific silencing of target genes in host cells without any functional importance.

Adenoviruses also modulate the expression of host miRNAs. We found that, on top of a general suppression of miRNA levels in adenovirus-infected cancer cells, which is in agreement with the described interferon- and VA RNA-dependent effects on the RNAi machinery, adenovirus serotype 5 significantly reduced expression of a subset of human miRNAs, in particular miR-27, miR-100, miR-155, miR-181, and miR-222 in the late phase of replication [110]. Interestingly, others reported that these miRNAs were also downregulated in cells infected with other viruses, suggesting that these miRNAs could be involved in general antiviral responses, and that viruses evolved specific countermeasures against them. In contrast, only very few miRNAs were induced upon adenovirus infection [110]. Interestingly, among these was miR-132, which was also reported overexpressed in human monocytes infected with several other viruses, suppressing innate antiviral immunity and promoting viral replication [111]. Hence, different viruses exhibit striking similarities in their communication with host cells via miRNAs. As miRNAs usually have many targets, modulation of a single host miRNA by a virus can have complex biological effects, silencing multiple genes that together impact virus replication. In addition, not all target genes predicted on the basis of sequence complementarity are validated as genuine miRNA targets. They may not or only under certain circumstances be silenced. This makes it very difficult to dissect virus–host interactions via miRNAs and identify the crucial host genes contributing to virus biology.

Since miRNA expression is deregulated in cancer cells, CRAds that are replicating in cancer cells might encounter miRNA expression patterns that differ from the ones that human adenoviruses evolved to cope with in their natural differentiated host cells. Consequently, CRAds might not exploit host miRNA expression to their full benefit, and host antiviral miRNA responses to viral infection might not always be effectively counteracted. It can be envisioned, therefore, that overexpression or inhibition of certain miRNAs in cancer cells increases CRAd replication and, thus, oncolytic potency. The identification of miRNAs or miRNA inhibitors for this purpose could perhaps be based on known virus–host interactions and deregulated miRNA expression patterns. For example, tumor suppressor miR-34a expression is often absent or low in cancer cells as a consequence of chromosomal deletion or p53 deficiency. As a direct transcriptional target of p53 and silencer of genes required for cell proliferation and survival, miR-34 mimics part of the p53 activities [112]. In view of the established effects of expressing p53 on CRAd efficacy, it could, thus, be envisioned that expressing miR-34a from the CRAd genome similarly promotes CRAd propagation and cancer cell killing (see Figure 1). MiR-34a-expressing CRAds were presented in two studies. In the first study, hTERT promoter-driven E1A-Δ24-type CRAds were armed with miR-34a and/or interleukin-24 (IL-24) and tested in hepatocellular carcinoma models [20]. The cytokine IL-24 was chosen for its pro-apoptotic and anti-angiogenic properties. Knockdown of miR-34a target genes was confirmed in infected cells. The effects of miR-34a on cytotoxicity in vitro and tumor growth inhibition in vivo were minimal. In contrast, the effects of expressing IL-24 were significant and most prominent anti-tumor effects, including complete tumor regression, were achieved with the virus carrying both miR-34a and IL-24. This suggests that the augmented efficacy of the CRAds depended primarily on the anti-tumor activities of IL-24. In the second study, carcinoembryonic antigen promoter-driven E1A-type CRAds were armed with miR-34a and/or miR-126 and tested in pancreatic cancer models [113]. These two tumor suppressor miRNAs were chosen because of their low expression in pancreatic cancer and confirmed anti-tumor effects. Also in this study the effect of expressing miR-34a was very small. The significance of miR-34a expression for tumor growth control in comparison to a control CRAd was not demonstrated. Interestingly, the miR-34a-expressing CRAd appeared to produce somewhat lower virus titers in pancreatic cancer cell lines than the control virus. In this respect, the virus, thus, did not reproduce previous observations made with p53-expressing CRAds.

Apart from rationally choosing miRNAs to empower CRAds, one may also take an unbiased approach to identify useful miRNAs through functional screening. The availability of genome-wide miRNA mimic and inhibitor libraries allows global analysis of miRNA effects on the lifecycle of viruses [110,114]. Screening for miRNA effects on the replication of a number of human and mouse herpes viruses, including human and mouse cytomegalovirus (CMV), as well as validation on an evolutionary unrelated RNA virus, identified several miRNAs with broad proviral or antiviral properties [114]. In particular, miR-30 broadly stimulated virus replication, and miR-199a and miR-214 inhibited virus replication. Since these observations were made on a diversity of viruses, the miRNAs are not likely to act on viral elements. Instead, most probably, they regulate host cell pathways that support or protect against virus replication. Interestingly, miR-199a and miR-214 are expressed from the same intronic cluster [115] and were found downregulated in CMV-infected cells. Thus, these viruses appear to counteract two antiviral miRNAs by suppressing a single transcript. While miR-30 was not reported differentially expressed in this study, several other viruses were shown to induce the expression of miR-30 variants [116,117]. Intriguingly, on one hand, this upregulation appears IFN-dependent [116], whereas, on the other hand, miR-30 inhibited type-I IFN signaling, allowing increased virus replication [117]. In any event, these findings suggest that certain virus–host interactions via miRNAs are broadly conserved. Hence, observations made in miRNA screens with one virus might be applicable to other viruses as well.

We performed screens for adenovirus replication and cytotoxicity on prostate cancer cells using the same miRNA libraries as were used in the aforementioned study [110]. We identified several miRNAs that stimulate adenovirus-induced cell death. Some of these inhibited adenovirus replication, possibly as a consequence of premature cell death occurring before adenovirus replication was completed. In general, strong induction of cell death was associated with reduced virus production. The exception was miR-181a. Expression of this miRNA strongly induced cell death and slightly increased virus production. The most striking results were obtained with miR-26b. Transfection of miR-26b mimics increased adenovirus-induced cell death, as well as infectious virus production and release. Consequently, adenovirus propagated more rapidly in human prostate cancer cell lines, producing larger plaque sizes. This makes miR-26 a promising candidate for incorporation in the genome of CRAds, to augment OVT of prostate cancer and perhaps other cancers. Construction and characterization of such a novel CRAd is, therefore, warranted. An important question that needs to be answered then is if the timing and levels of miRNA expression obtained during CRAd replication in cancer cells will reproduce the functional effects obtained by combining miRNA mimic transfection and virus infection.

3.4. RNAi Screening for Inhibitors of Oncolytic Virus Efficacy in Cancer Cells

Clearly, virus–host interactions are complex and our current knowledge of the cellular pathways involved is incomplete. There are, therefore, probably many more opportunities to improve the potency of oncolytic viruses than we can rationally hypothesize. The advent of high-throughput screening technology with whole-genome RNAi libraries offered the opportunity to functionally interrogate every single gene, one by one, for its biological role. Many investigations used this technology to identify cellular factors involved in susceptibility to virus infection (reviewed in Reference [118]). These screens helped identify genes responsible for viral entry, uncoating of the virus, replication of the genome, particle assembly, and spread of progeny viruses. Not surprisingly, these studies focused primarily on viruses that constitute a threat to public health, where the knowledge obtained can be used to develop novel antiviral treatments. However, genome-wide RNAi screening can also be used to identify determinants of oncolytic virus anticancer efficacy. With the ultimate goal of increasing the efficacy of OVT by arming the virus with an shRNA in mind, the most important factors of the virus–host interaction to be studied are those involved in virus genome replication and infectious progeny production, host cell killing, and modulation of the immune response. Early steps in the virus lifecycle, such as entry, uncoating, and early gene expression are of less relevance, since these processes cannot be influenced by virus-encoded shRNAs.

Functional genomic screens usually have a single readout of biological effect. Although high-content (image) analyses can measure more complex phenotypes, such as multiple descriptors of cellular morphology, most screens measure a single parameter, such as cell viability, reporter gene expression, or binding/uptake/secretion of a (fluorescent) molecule. Hence, the most appropriate readout for oncolytic virus efficacy needs to be carefully chosen. In addition, it is crucial that the screening effort is followed by thorough examination of the effects of hits on other aspects of the virus lifecycle that are not measured in the discovery screen. For example, in our miRNA library screens described above [110], we identified a number of miRNAs that strongly promoted adenovirus-induced cell death, but this was at the expense of reducing infectious progeny virus production. Thus, interference with host cell gene expression can have opposing effects on different steps in the virus lifecycle, negating the overall effect of gene silencing on OVT.

Until now, there are a few examples reported to show that genome-wide RNAi screening can be used effectively to increase our understanding of the interactions between oncolytic viruses and their host cancer cells and to identify determinants of their anticancer efficacy [119–121].

In the first example, an arrayed siRNA screen was performed, measuring effects on the killing of three cancer cell lines by Maraba virus [119]. Although Maraba is not a widely used oncolytic virus, it is related to the more commonly used VSV, and the major findings in this study could be validated for both viruses. Among the hits identified in the screens, genes within endoplasmic reticulum (ER) stress response pathways were enriched. Knocking down a key gene in the ER stress response pathway activated caspase-dependent apoptosis in virus-infected cells. This increased oncolytic virus potency in many cancer cell lines, reaching up to four orders of magnitude more effective killing. Importantly, augmented killing was much less pronounced on non-malignant cells, providing a high therapeutic index. Notably, in the screen design used, siRNAs were transfected three days before virus infection. Thus, ER stress responses were already inhibited at the time of infection. In follow-up time scheduling experiments with chemical compounds targeting key proteins in the ER stress response pathway, it was found that virus-induced cell killing was augmented when ER stress responses were inhibited prior to infection, but not when they were inhibited during infection. This pointed at a preconditioning process, where inhibition of ER stress response rewired cellular signaling, leading to increased cell death upon subsequent virus infection. Hence, while the findings of this study can be used to develop combination treatments consisting of preconditioning with a chemical compound followed by virus administration, they do not provide footholds for arming viruses with transgenes or shRNAs. For this purpose, it is probably better to delay gene silencing relative to virus infection in the screen design.

In the second example, a pooled shRNA cell depletion screen was done on breast cancer cells that were infected with an oncolytic HSV [120]. This identified a component of the RNA splicing machinery, arginine-rich splicing factor 2 (SRSF2). Although the effects of silencing SRSF2 were not very strong, they could be reproduced on different cell lines and with different reagents. Silencing SRSF2 changed the abundance of pro- and anti-apoptotic mRNAs, and increased apoptosis. In contrast, it did not stimulate virus replication. In fact, a trend toward inhibition of infectious virus production was observed. SRSF2 depletion appeared to dampen HSV-induced activation of mammalian target of rapamycin (mTOR) signaling, thereby probably inhibiting multiple processes, including cell cycle, virus replication, and cellular antiviral responses. Thus, the effects of targeting SRSF2 could promote, as well as inhibit, the HSV oncolytic potency. Apparently, under the applied experimental conditions the overall effect was more potent cytotoxicity. Importantly, also in this study, gene knockdown or knockout was already established before cells were subjected to the virus. It is difficult to predict if delayed silencing of SRSF2 after virus entry will also tip the scales toward increased oncolysis. Interestingly, treatment of cancer cells with an mTOR inhibitor increased the oncolytic potency of CRAds [122,123]. This suggests a similarity in the virus–host interactions of different oncolytic viruses that could allow translation of the findings to broader applications.

The third example is the systematic analysis of human cell factors affecting replication of Myxoma virus [121]. Genome-wide arrayed siRNA screens were done on a breast cancer cell line that was infected with virus three days after siRNA transfection. A recombinant virus was used that expresses a marker gene in the late phase of replication. Marker gene expression served as a surrogate marker for virus replication. The screens yielded a large database of genes that either enhance or inhibit Myxoma virus replication. Follow-up experiments zooming in on certain cellular pathways provided support for their role in virus replication, but interpretation was difficult and some observations seemed contradictory. This is illustrated, e.g., by the observations made when silencing the Raf/mitogen-activated protein kinase kinase (MEK)/extracellular signal-regulated kinase (ERK) pathway. Whereas silencing of many genes upstream of MEK/ERK enhanced virus replication, silencing more downstream pathway genes, including ERK1, inhibited virus replication. Effects of targeting single genes were usually modest. This is in agreement with the concept that viruses modulate cellular pathways by converting multiple switches in a concerted manner, and that host cells respond to virus infection by activating many systems that collectively create antiviral defenses. Finally, based on the known involvement of RNA helicases in permissiveness of cells for replication of a variety of viruses, Rahman et al. [124] performed a focused siRNA library screen on cervical cancer cells using marker gene-expressing Myxoma viruses, followed by confirmation experiments in other cell lines. This effort identified several RNA helicases that consistently increased or decreased Myxoma virus production. An intriguing finding was that knockdown of retinoic acid-induced gene 1 (RIG-1) reduced Myxoma replication. This was highly unexpected, because of the known role of RIG-1 in antiviral innate immune responses. Knockdown of five RNA helicases stimulated Myxoma virus production, without an apparent effect on viral gene expression or replication. Although their mechanism of action remains to be resolved, these genes could be appropriate targets to construct more potent oncolytic Myxoma viruses.

Similar screens, as far as we are aware, are yet to be reported for CRAds. In addition, despite the recognized crucial role of the immune system in the anticancer effects of oncolytic viruses, RNAi screens with oncolytic viruses have not studied this important aspect. While the design of relevant high-throughput assays for modulation of the immune system will be a challenge, such efforts are warranted. They are expected to provide highly valuable information to design next-generation CRAds for more effective OVT.

4. Conclusions

For the past decades, molecular cancer and virology research advanced forward with incredible speed and efficiency. This brought OVT from the bench to the bedside. Novel molecular targets were identified, oncolytic virus constructs designed, and combined therapies proposed or tested in clinical trials. So far, only a few OVTs obtained market authorization, but others are likely to follow. Oncolytic virus therapy moved from a mere concept to a functioning approach. RNA interference became one of the successfully applied methods to make OVT more effective. Here, knowledge of miRNA expression in cancer and healthy cells, as well as of miRNA-mediated virus–host interactions, is implemented. Clearly, a great amount of work needs to be done before OVT will reach its full potential. The contribution of RNAi technology to reaching this goal is difficult to predict. However, although the battle is not over, new research results are constantly made that bring us closer to victory.

Funding: This project received funding from the European Union's Horizon 2020 research and innovation program under grant agreement No. 643130.

Acknowledgments: The authors would like to thank Tanja D. de Gruijl for critical reading of the manuscript and helpful suggestions.

Conflicts of Interest: The authors declare no conflicts of interest.

References

1. Hanahan, D.; Weinberg, R.A. The hallmarks of cancer. *Cell* **2000**, *100*, 57–70. [CrossRef]
2. Parato, K.A.; Senger, D.; Forsyth, P.A.J.; Bell, J.C. Recent progress in the battle between oncolytic viruses and tumours. *Nat. Rev. Cancer* **2005**, *5*, 965–976. [CrossRef] [PubMed]
3. Vähä-Koskela, M.J.V.; Heikkilä, J.E.; Hinkkanen, A.E. Oncolytic viruses in cancer therapy. *Cancer Lett.* **2007**, *254*, 178–216. [CrossRef] [PubMed]
4. Sinkovics, J.G.; Horvath, J.C. Natural and genetically engineered viral agents for oncolysis and gene therapy of human cancers. *Arch. Immunol. Ther. Exp. (Warsz.)* **2008**, *56*, 1–59. [CrossRef] [PubMed]
5. Nguyen, T.L.A.; Tumilasci, V.F.; Singhroy, D.; Arguello, M.; Hiscott, J. The emergence of combinatorial strategies in the development of RNA oncolytic virus therapies. *Cell. Microbiol.* **2009**, *11*, 889–897. [CrossRef] [PubMed]
6. Diaconu, I.; Cerullo, V.; Hirvinen, M.L.M.; Escutenaire, S.; Ugolini, M.; Pesonen, S.K.; Bramante, S.; Parviainen, S.; Kanerva, A.; Loskog, A.S.I.; et al. Immune response is an important aspect of the antitumor effect produced by a CD40L-encoding oncolytic adenovirus. *Cancer Res.* **2012**, *72*, 2327–2338. [CrossRef] [PubMed]
7. de Gruijl, T.D.; Janssen, A.B.; van Beusechem, V.W. Arming oncolytic viruses to leverage antitumor immunity. *Expert Opin. Biol. Ther.* **2015**, *15*, 959–971. [CrossRef] [PubMed]
8. Pol, J.; Buqué, A.; Aranda, F.; Bloy, N.; Cremer, I.; Eggermont, A.; Erbs, P.; Fucikova, J.; Galon, J.; Limacher, J.-M.; et al. Trial Watch—Oncolytic viruses and cancer therapy. *Oncoimmunology* **2016**, *5*, e1117740. [CrossRef] [PubMed]
9. Andtbacka, R.H.I.; Kaufman, H.L.; Collichio, F.; Amatruda, T.; Senzer, N.; Chesney, J.; Delman, K.A.; Spitler, L.E.; Puzanov, I.; Agarwala, S.S.; et al. Talimogene Laherparepvec Improves Durable Response Rate in Patients With Advanced Melanoma. *J. Clin. Oncol.* **2015**, *33*, 2780–2788. [CrossRef] [PubMed]
10. Yu, W.; Fang, H. Clinical Trials with Oncolytic Adenovirus in China. *Curr. Cancer Drug Targets* **2007**, *7*, 141–148. [CrossRef] [PubMed]
11. Chiocca, E.A.; Rabkin, S.D. Oncolytic viruses and their application to cancer immunotherapy. *Cancer Immunol. Res.* **2014**, *2*, 295–300. [CrossRef] [PubMed]
12. Russell, S.J.; Peng, K.-W.; Bell, J.C. Oncolytic virotherapy. *Nat. Biotechnol.* **2012**, *30*, 658–670. [CrossRef] [PubMed]

13. Cattaneo, R.; Miest, T.; Shashkova, E.V.; Barry, M.A. Reprogrammed viruses as cancer therapeutics: Targeted, armed and shielded. *Nat. Rev. Microbiol.* **2008**, *6*, 529–540. [CrossRef] [PubMed]

14. Lion, T. Adenovirus Infections in Immunocompetent and Immunocompromised Patients. *Clin. Microbiol. Rev.* **2014**, *27*, 441–462. [CrossRef] [PubMed]

15. Ben-Israel, H.; Kleinberger, T. Adenovirus and cell cycle control. *Front. Biosci.* **2002**, *7*, 1369–1395. [CrossRef]

16. Robert-Guroff, M. Replicating and non-replicating viral vectors for vaccine development. *Curr. Opin. Biotechnol.* **2007**, *18*, 546–556. [CrossRef] [PubMed]

17. Bauerschmitz, G.J.; Barker, S.D.; Hemminki, A. Adenoviral gene therapy for cancer: From vectors to targeted and replication competent agents (review). *Int. J. Oncol.* **2002**, *21*, 1161–1174. [CrossRef] [PubMed]

18. Doloff, J.C.; Waxman, D.J. Dual E1A oncolytic adenovirus: Targeting tumor heterogeneity with two independent cancer-specific promoter elements, DF3/MUC1 and hTERT. *Cancer Gene Ther.* **2011**, *18*, 153–166. [CrossRef] [PubMed]

19. Sugio, K.; Sakurai, F.; Katayama, K.; Tashiro, K.; Matsui, H.; Kawabata, K.; Kawase, A.; Iwaki, M.; Hayakawa, T.; Fujiwara, T.; et al. Enhanced Safety Profiles of the Telomerase-Specific Replication-Competent Adenovirus by Incorporation of Normal Cell-Specific microRNA-Targeted Sequences. *Clin. Cancer Res.* **2011**, *17*, 2807–2818. [CrossRef] [PubMed]

20. Lou, W.; Chen, Q.; Ma, L.; Liu, J.; Yang, Z.; Shen, J.; Cui, Y.; Bian, X.; Qian, C. Oncolytic adenovirus co-expressing miRNA-34a and IL-24 induces superior antitumor activity in experimental tumor model. *J. Mol. Med.* **2013**, *91*, 715–725. [CrossRef] [PubMed]

21. Bischoff, J.R.; Kirn, D.H.; Williams, A.; Heise, C.; Horn, S.; Muna, M.; Ng, L.; Nye, J.A.; Sampson-Johannes, A.; Fattaey, A.; et al. An adenovirus mutant that replicates selectively in p53-deficient human tumor cells. *Science* **1996**, *274*, 373–376. [CrossRef] [PubMed]

22. Nemunaitis, J.; Khuri, F.; Ganly, I.; Arseneau, J.; Posner, M.; Vokes, E.; Kuhn, J.; McCarty, T.; Landers, S.; Blackburn, A.; et al. Phase II trial of intratumoral administration of ONYX-015, a replication-selective adenovirus, in patients with refractory head and neck cancer. *J. Clin. Oncol.* **2001**, *19*, 289–298. [CrossRef] [PubMed]

23. Cascallo, M.; Alonso, M.M.; Rojas, J.J.; Perez-Gimenez, A.; Fueyo, J.; Alemany, R. Systemic Toxicity–Efficacy Profile of ICOVIR-5, a Potent and Selective Oncolytic Adenovirus Based on the pRB Pathway. *Mol. Ther.* **2007**, *15*, 1607–1615. [CrossRef] [PubMed]

24. Fueyo, J.; Gomez-Manzano, C.; Alemany, R.; Lee, P.S.Y.; McDonnell, T.J.; Mitlianga, P.; Shi, Y.-X.; Levin, V.A.; Yung, W.K.A.; Kyritsis, A.P. A mutant oncolytic adenovirus targeting the Rb pathway produces anti-glioma effect in vivo. *Oncogene* **2000**, *19*, 2–12. [CrossRef] [PubMed]

25. Eager, R.M.; Nemunaitis, J. Clinical development directions in oncolytic viral therapy. *Cancer Gene Ther.* **2011**, *18*, 305–317. [CrossRef] [PubMed]

26. Li, Y.; Pong, R.C.; Bergelson, J.M.; Hall, M.C.; Sagalowsky, A.I.; Tseng, C.P.; Wang, Z.; Hsieh, J.T. Loss of adenoviral receptor expression in human bladder cancer cells: A potential impact on the efficacy of gene therapy. *Cancer Res.* **1999**, *59*, 325–330. [PubMed]

27. Reeh, M.; Bockhorn, M.; Görgens, D.; Vieth, M.; Hoffmann, T.; Simon, R.; Izbicki, J.R.; Sauter, G.; Schumacher, U.; Anders, M. Presence of the Coxsackievirus and Adenovirus Receptor (CAR) in human neoplasms: A multitumour array analysis. *Br. J. Cancer* **2013**, *109*, 1848–1858. [CrossRef] [PubMed]

28. Suzuki, K.; Fueyo, J.; Krasnykh, V.; Reynolds, P.N.; Curiel, D.T.; Alemany, R. A conditionally replicative adenovirus with enhanced infectivity shows improved oncolytic potency. *Clin. Cancer Res.* **2001**, *7*, 120–126. [CrossRef] [PubMed]

29. Guse, K.; Ranki, T.; Ala-Opas, M.; Bono, P.; Särkioja, M.; Rajecki, M.; Kanerva, A.; Hakkarainen, T.; Hemminki, A. Treatment of metastatic renal cancer with capsid-modified oncolytic adenoviruses. *Mol. Cancer Ther.* **2007**, *6*, 2728–2736. [CrossRef] [PubMed]

30. Kanerva, A.; Mikheeva, G.V.; Krasnykh, V.; Coolidge, C.J.; Lam, J.T.; Mahasreshti, P.J.; Barker, S.D.; Straughn, M.; Barnes, M.N.; Alvarez, R.D.; et al. Targeting adenovirus to the serotype 3 receptor increases gene transfer efficiency to ovarian cancer cells. *Clin. Cancer Res.* **2002**, *8*, 275–280. [PubMed]

31. van Beusechem, V.W.; Mastenbroek, D.C.J.; van den Doel, P.B.; Lamfers, M.L.M.; Grill, J.; Würdinger, T.; Haisma, H.J.; Pinedo, H.M.; Gerritsen, W.R. Conditionally replicative adenovirus expressing a targeting adapter molecule exhibits enhanced oncolytic potency on CAR-deficient tumors. *Gene Ther.* **2003**, *10*, 1982–1991. [CrossRef] [PubMed]

32. Choi, J.-W.; Lee, Y.S.; Yun, C.-O.; Kim, S.W. Polymeric oncolytic adenovirus for cancer gene therapy. *J. Control. Release* **2015**, *219*, 181–191. [CrossRef] [PubMed]

33. Power, A.T.; Bell, J.C. Taming the Trojan horse: Optimizing dynamic carrier cell/oncolytic virus systems for cancer biotherapy. *Gene Ther.* **2008**, *15*, 772–779. [CrossRef] [PubMed]

34. Yong, R.L.; Shinojima, N.; Fueyo, J.; Gumin, J.; Vecil, G.G.; Marini, F.C.; Bogler, O.; Andreeff, M.; Lang, F.F. Human Bone Marrow-Derived Mesenchymal Stem Cells for Intravascular Delivery of Oncolytic Adenovirus Δ24-RGD to Human Gliomas. *Cancer Res.* **2009**, *69*, 8932–8940. [CrossRef] [PubMed]

35. Ong, H.T.; Hasegawa, K.; Dietz, A.B.; Russell, S.J.; Peng, K.-W. Evaluation of T cells as carriers for systemic measles virotherapy in the presence of antiviral antibodies. *Gene Ther.* **2007**, *14*, 324–333. [CrossRef] [PubMed]

36. Power, A.T.; Wang, J.; Falls, T.J.; Paterson, J.M.; Parato, K.A.; Lichty, B.D.; Stojdl, D.F.; Forsyth, P.A.J.; Atkins, H.; Bell, J.C. Carrier cell-based delivery of an oncolytic virus circumvents antiviral immunity. *Mol. Ther.* **2007**, *15*, 123–130. [CrossRef] [PubMed]

37. Ylosmaki, E.; Hakkarainen, T.; Hemminki, A.; Visakorpi, T.; Andino, R.; Saksela, K. Generation of a Conditionally Replicating Adenovirus Based on Targeted Destruction of E1A mRNA by a Cell Type-Specific MicroRNA. *J. Virol.* **2008**, *82*, 11009–11015. [CrossRef] [PubMed]

38. Cawood, R.; Chen, H.H.; Carroll, F.; Bazan-Peregrino, M.; van Rooijen, N.; Seymour, L.W. Use of Tissue-Specific MicroRNA to Control Pathology of Wild-Type Adenovirus without Attenuation of Its Ability to Kill Cancer Cells. *PLoS Pathog.* **2009**, *5*, e1000440. [CrossRef] [PubMed]

39. Leja, J.; Nilsson, B.; Yu, D.; Gustafson, E.; Åkerström, G.; Öberg, K.; Giandomenico, V.; Essand, M. Double-Detargeted Oncolytic Adenovirus Shows Replication Arrest in Liver Cells and Retains Neuroendocrine Cell Killing Ability. *PLoS ONE* **2010**, *5*, e8916. [CrossRef] [PubMed]

40. Callegari, E.; Elamin, B.K.; D'Abundo, L.; Falzoni, S.; Donvito, G.; Moshiri, F.; Milazzo, M.; Altavilla, G.; Giacomelli, L.; Fornari, F.; et al. Anti-Tumor Activity of a miR-199-dependent Oncolytic Adenovirus. *PLoS ONE* **2013**, *8*, e73964. [CrossRef] [PubMed]

41. Bofill-De Ros, X.; Gironella, M.; Fillat, C. MiR-148a- and miR-216a-regulated Oncolytic Adenoviruses Targeting Pancreatic Tumors Attenuate Tissue Damage Without Perturbation of miRNA Activity. *Mol. Ther.* **2014**, *22*, 1665–1677. [CrossRef] [PubMed]

42. Yao, W.; Guo, G.; Zhang, Q.; Fan, L.; Wu, N.; Bo, Y. The application of multiple miRNA response elements enables oncolytic adenoviruses to possess specificity to glioma cells. *Virology* **2014**, *458*, 69–82. [CrossRef] [PubMed]

43. Gürlevik, E.; Woller, N.; Schache, P.; Malek, N.P.; Wirth, T.C.; Zender, L.; Manns, M.P.; Kubicka, S.; Kühnel, F. p53-dependent antiviral RNA-interference facilitates tumor-selective viral replication. *Nucleic. Acids Res.* **2009**, *37*, e84. [CrossRef] [PubMed]

44. Gros, A.; Martínez-Quintanilla, J.; Puig, C.; Guedan, S.; Molleví, D.G.; Alemany, R.; Cascallo, M. Bioselection of a gain of function mutation that enhances adenovirus 5 release and improves its antitumoral potency. *Cancer Res.* **2008**, *68*, 8928–8937. [CrossRef] [PubMed]

45. Dong, W.; van Ginkel, J.-W.H.; Au, K.Y.; Alemany, R.; Meulenberg, J.J.M.; van Beusechem, V.W. ORCA-010, a Novel Potency-Enhanced Oncolytic Adenovirus, Exerts Strong Antitumor Activity in Preclinical Models. *Hum. Gene Ther.* **2014**, *25*, 897–904. [CrossRef] [PubMed]

46. Kuhn, I.; Harden, P.; Bauzon, M.; Chartier, C.; Nye, J.; Thorne, S.; Reid, T.; Ni, S.; Lieber, A.; Fisher, K.; et al. Directed Evolution Generates a Novel Oncolytic Virus for the Treatment of Colon Cancer. *PLoS ONE* **2008**, *3*, e2409. [CrossRef] [PubMed]

47. Guedan, S.; Grases, D.; Rojas, J.J.; Gros, A.; Vilardell, F.; Vile, R.; Mercade, E.; Cascallo, M.; Alemany, R. GALV expression enhances the therapeutic efficacy of an oncolytic adenovirus by inducing cell fusion and enhancing virus distribution. *Gene Ther.* **2012**, *19*, 1048–1057. [CrossRef] [PubMed]

48. Guedan, S.; Rojas, J.J.; Gros, A.; Mercade, E.; Cascallo, M.; Alemany, R. Hyaluronidase expression by an oncolytic adenovirus enhances its intratumoral spread and suppresses tumor growth. *Mol. Ther.* **2010**, *18*, 1275–1283. [CrossRef] [PubMed]

49. Kim, J.H.; Lee, Y.S.; Kim, H.; Huang, J.H.; Yoon, A.R.; Yun, C.O. Relaxin expression from tumor-targeting adenoviruses and its intratumoral spread, apoptosis induction, and efficacy. *J. Natl. Cancer Inst.* **2006**, *98*, 1482–1493. [CrossRef] [PubMed]

50. Choi, I.-K.; Lee, Y.-S.; Yoo, J.Y.; Yoon, A.-R.; Kim, H.; Kim, D.-S.; Seidler, D.G.; Kim, J.-H.; Yun, C.-O. Effect of decorin on overcoming the extracellular matrix barrier for oncolytic virotherapy. *Gene Ther.* **2010**, *17*, 190–201. [CrossRef] [PubMed]

51. Morris, J.C.; Wildner, O. Therapy of Head and Neck Squamous Cell Carcinoma with an Oncolytic Adenovirus Expressing HSV-tk. *Mol. Ther.* **2000**, *1*, 56–62. [CrossRef] [PubMed]

52. Oosterhoff, D.; Pinedo, H.M.; Witlox, M.A.; Carette, J.E.; Gerritsen, W.R.; van Beusechem, V.W. Gene-directed enzyme prodrug therapy with carboxylesterase enhances the anticancer efficacy of the conditionally replicating adenovirus AdΔ24. *Gene Ther.* **2005**, *12*, 1011–1018. [CrossRef] [PubMed]

53. Mi, J.; Li, Z.-Y.; Ni, S.; Steinwaerder, D.; Lieber, A. Induced Apoptosis Supports Spread of Adenovirus Vectors in Tumors. *Hum. Gene Ther.* **2001**, *12*, 1343–1352. [CrossRef] [PubMed]

54. Van Beusechem, V.W.; Van den Doel, P.B.; Grill, J.; Pinedo, H.M.; Gerritsen, W.R. Conditionally replicative adenovirus expressing p53 exhibits enhanced oncolytic potency. *Cancer Res.* **2002**, *62*, 6165–6171. [PubMed]

55. Sauthoff, H.; Pipiya, T.; Heitner, S.; Chen, S.; Norman, R.G.; Rom, W.N.; Hay, J.G. Late Expression of p53 from a Replicating Adenovirus Improves Tumor Cell Killing and Is More Tumor Cell Specific than Expression of the Adenoviral Death Protein. *Hum. Gene Ther.* **2002**, *13*, 1859–1871. [CrossRef] [PubMed]

56. Hall, A.R.; Dix, B.R.; O'Carroll, S.J.; Braithwaite, A.W. p53-dependent cell death/apoptosis is required for a productive adenovirus infection. *Nat. Med.* **1998**, *4*, 1068–1072. [CrossRef] [PubMed]

57. Royds, J.A.; Hibma, M.; Dix, B.R.; Hananeia, L.; Russell, I.A.; Wiles, A.; Wynford-Thomas, D.; Braithwaite, A.W. p53 promotes adenoviral replication and increases late viral gene expression. *Oncogene* **2006**, *25*, 1509–1520. [CrossRef] [PubMed]

58. Wang, X.; Su, C.; Cao, H.; Li, K.; Chen, J.; Jiang, L.; Zhang, Q.; Wu, X.; Jia, X.; Liu, Y.; et al. A novel triple-regulated oncolytic adenovirus carrying p53 gene exerts potent antitumor efficacy on common human solid cancers. *Mol. Cancer Ther.* **2008**, *7*, 1598–1603. [CrossRef] [PubMed]

59. Geoerger, B.; Vassal, G.; Opolon, P.; Dirven, C.M.F.; Morizet, J.; Laudani, L.; Grill, J.; Giaccone, G.; Vandertop, W.P.; Gerritsen, W.R.; et al. Oncolytic Activity of p53-Expressing Conditionally Replicative Adenovirus AdΔ24-p53 against Human Malignant Glioma. *Cancer Res.* **2004**, *64*, 5753–5759. [CrossRef] [PubMed]

60. Geoerger, B.; van Beusechem, V.W.; Opolon, P.; Morizet, J.; Laudani, L.; Lecluse, Y.; Barrois, M.; Idema, S.; Grill, J.; Gerritsen, W.R.; et al. Expression of p53, or targeting towards EGFR, enhances the oncolytic potency of conditionally replicative adenovirus against neuroblastoma. *J. Gene Med.* **2005**, *7*, 584–594. [CrossRef] [PubMed]

61. van Beusechem, V.W.; van den Doel, P.B.; Gerritsen, W.R. Conditionally replicative adenovirus expressing degradation-resistant p53 for enhanced oncolysis of human cancer cells overexpressing murine double minute 2. *Mol. Cancer Ther.* **2005**, *4*, 1013–1018. [CrossRef] [PubMed]

62. Heideman, D.A.M.; Steenbergen, R.D.M.; van der Torre, J.; Scheffner, M.; Alemany, R.; Gerritsen, W.R.; Meijer, C.J.L.M.; Snijders, P.J.F.; van Beusechem, V.W. Oncolytic Adenovirus Expressing a p53 Variant Resistant to Degradation by HPV E6 Protein Exhibits Potent and Selective Replication in Cervical Cancer. *Mol. Ther.* **2005**, *12*, 1083–1090. [CrossRef] [PubMed]

63. Graat, H.C.A.; Carette, J.E.; Schagen, F.H.E.; Vassilev, L.T.; Gerritsen, W.R.; Kaspers, G.J.L.; Wuisman, P.I.J.M.; van Beusechem, V.W. Enhanced tumor cell kill by combined treatment with a small-molecule antagonist of mouse double minute 2 and adenoviruses encoding p53. *Mol. Cancer Ther.* **2007**, *6*, 1552–1561. [CrossRef] [PubMed]

64. Carette, J.E.; Graat, H.C.A.; Schagen, F.H.E.; Abou El Hassan, M.A.I.; Gerritsen, W.R.; van Beusechem, V.W. Replication-dependent transgene expression from a conditionally replicating adenovirus via alternative splicing to a heterologous splice-acceptor site. *J. Gene Med.* **2005**, *7*, 1053–1062. [CrossRef] [PubMed]

65. Hannon, G.J. RNA interference. *Nature* **2002**, *418*, 244–251. [CrossRef] [PubMed]

66. Carmell, M.A.; Hannon, G.J. RNase III enzymes and the initiation of gene silencing. *Nat. Struct. Mol. Biol.* **2004**, *11*, 214–218. [CrossRef] [PubMed]

67. Huntzinger, E.; Izaurralde, E. Gene silencing by microRNAs: Contributions of translational repression and mRNA decay. *Nat. Rev. Genet.* **2011**, *12*, 99–110. [CrossRef] [PubMed]

68. Doench, J.G.; Sharp, P.A. Specificity of microRNA target selection in translational repression. *Genes Dev.* **2004**, *18*, 504–511. [CrossRef] [PubMed]

69. Bushati, N.; Cohen, S.M. microRNA Functions. *Annu. Rev. Cell Dev. Biol.* **2007**, *23*, 175–205. [CrossRef] [PubMed]

70. Xia, H.; Mao, Q.; Paulson, H.L.; Davidson, B.L. siRNA-mediated gene silencing in vitro and in vivo. *Nat. Biotechnol.* **2002**, *20*, 1006–1010. [CrossRef] [PubMed]

71. Cullen, B.R. RNA interference: Antiviral defense and genetic tool. *Nat. Immunol.* **2002**, *3*, 597–599. [CrossRef] [PubMed]

72. Roth, B.M.; Pruss, G.J.; Vance, V.B. Plant viral suppressors of RNA silencing. *Virus Res.* **2004**, *102*, 97–108. [CrossRef] [PubMed]

73. Backes, S.; Langlois, R.A.; Schmid, S.; Varble, A.; Shim, J.V.; Sachs, D.; TenOever, B.R. The Mammalian response to virus infection is independent of small RNA silencing. *Cell Rep.* **2014**, *8*, 114–125. [CrossRef] [PubMed]

74. tenOever, B.R. Questioning antiviral RNAi in mammals. *Nat. Microbiol.* **2017**, *2*, 17052. [CrossRef] [PubMed]

75. Lu, S.; Cullen, B.R. Adenovirus VA1 noncoding RNA can inhibit small interfering RNA and MicroRNA biogenesis. *J. Virol.* **2004**, *78*, 12868–12876. [CrossRef] [PubMed]

76. Aparicio, O.; Razquin, N.; Zaratiegui, M.; Narvaiza, I.; Fortes, P. Adenovirus Virus-Associated RNA Is Processed to Functional Interfering RNAs Involved in Virus Production. *J. Virol.* **2006**, *80*, 1376–1384. [CrossRef] [PubMed]

77. Bennasser, Y.; Chable-Bessia, C.; Triboulet, R.; Gibbings, D.; Gwizdek, C.; Dargemont, C.; Kremer, E.J.; Voinnet, O.; Benkirane, M. Competition for XPO5 binding between Dicer mRNA, pre-miRNA and viral RNA regulates human Dicer levels. *Nat. Struct. Mol. Biol.* **2011**, *18*, 323–327. [CrossRef] [PubMed]

78. Machitani, M.; Sakurai, F.; Wakabayashi, K.; Takayama, K.; Tachibana, M.; Mizuguchi, H. Type I Interferons Impede Short Hairpin RNA-Mediated RNAi via Inhibition of Dicer-Mediated Processing to Small Interfering RNA. *Mol. Ther.-Nucleic Acids* **2017**, *6*, 173–182. [CrossRef] [PubMed]

79. Carette, J.E.; Overmeer, R.M.; Schagen, F.H.E.; Alemany, R.; Barski, O.A.; Gerritsen, W.R.; Van Beusechem, V.W. Conditionally Replicating Adenoviruses Expressing Short Hairpin RNAs Silence the Expression of a Target Gene in Cancer Cells. *Cancer Res.* **2004**, *64*, 2663–2667. [CrossRef] [PubMed]

80. Machitani, M.; Sakurai, F.; Katayama, K.; Tachibana, M.; Suzuki, T.; Matsui, H.; Yamaguchi, T.; Mizuguchi, H. Improving adenovirus vector-mediated RNAi efficiency by lacking the expression of virus-associated RNAs. *Virus Res.* **2013**, *178*, 357–363. [CrossRef] [PubMed]

81. Thimmappaya, B.; Weinberger, C.; Schneider, R.J.; Shenk, T. Adenovirus VAI RNA is required for efficient translation of viral mRNAs at late times after infection. *Cell* **1982**, *31*, 543–551. [CrossRef]

82. Wakabayashi, K.; Machitani, M.; Tachibana, M.; Sakurai, F.; Mizuguchi, H. A microRNA derived from adenovirus virus-associated RNAII promotes virus infection *via* post-transcriptional gene silencing. *J. Virol.* **2018**, Epub ahead of print. [CrossRef] [PubMed]

83. Kamel, W.; Segerman, B.; Öberg, D.; Punga, T.; Akusjärvi, G. The adenovirus VA RNA-derived miRNAs are not essential for lytic virus growth in tissue culture cells. *Nucleic Acids Res.* **2013**, *41*, 4802–4812. [CrossRef] [PubMed]

84. Machitani, M.; Sakurai, F.; Wakabayashi, K.; Tomita, K.; Tachibana, M.; Mizuguchi, H. Dicer functions as an antiviral system against human adenoviruses via cleavage of adenovirus-encoded noncoding RNA. *Sci. Rep.* **2016**, *6*, 27598. [CrossRef] [PubMed]

85. Zhang, Y.-A.; Nemunaitis, J.; Samuel, S.K.; Chen, P.; Shen, Y.; Tong, A.W. Antitumor Activity of an Oncolytic Adenovirus-Delivered Oncogene Small Interfering RNA. *Cancer Res.* **2006**, *66*, 9736–9743. [CrossRef] [PubMed]

86. Chu, L.; Gu, J.; Sun, L.; Qian, Q.; Qian, C.; Liu, X. Oncolytic adenovirus-mediated shRNA against Apollon inhibits tumor cell growth and enhances antitumor effect of 5-fluorouracil. *Gene Ther.* **2008**, *15*, 484–494. [CrossRef] [PubMed]

87. Yoo, J.Y.; Kim, J.-H.; Kwon, Y.-G.; Kim, E.-C.; Kim, N.K.; Choi, H.J.; Yun, C.-O. VEGF-specific short hairpin RNA-expressing oncolytic adenovirus elicits potent inhibition of angiogenesis and tumor growth. *Mol. Ther.* **2007**, *15*, 295–302. [CrossRef] [PubMed]

88. Machitani, M.; Sakurai, F.; Wakabayashi, K.; Tachibana, M.; Fujiwara, T.; Mizuguchi, H. Enhanced Oncolytic Activities of the Telomerase-Specific Replication-Competent Adenovirus Expressing Short-Hairpin RNA against Dicer. *Mol. Cancer Ther.* **2017**, *16*, 251–259. [CrossRef] [PubMed]

89. Höti, N.; Chowdhury, W.; Hsieh, J.-T.; Sachs, M.D.; Lupold, S.E.; Rodriguez, R. Valproic Acid, a Histone Deacetylase Inhibitor, Is an Antagonist for Oncolytic Adenoviral Gene Therapy. *Mol. Ther.* **2006**, *14*, 768–778. [CrossRef] [PubMed]

90. Shiina, M.; Lacher, M.D.; Christian, C.; Korn, W.M. RNA interference-mediated knockdown of p21WAF1 enhances anti-tumor cell activity of oncolytic adenoviruses. *Cancer Gene Ther.* **2009**, *16*, 810–819. [CrossRef] [PubMed]

91. Höti, N.; Chowdhury, W.H.; Mustafa, S.; Ribas, J.; Castanares, M.; Johnson, T.; Liu, M.; Lupold, S.E.; Rodriguez, R. Armoring CRAds with p21/Waf-1 shRNAs: The next generation of oncolytic adenoviruses. *Cancer Gene Ther.* **2010**, *17*, 585–597. [CrossRef] [PubMed]

92. Idogawa, M.; Sasaki, Y.; Suzuki, H.; Mita, H.; Imai, K.; Shinomura, Y.; Tokino, T. A Single Recombinant Adenovirus Expressing p53 and p21-targeting Artificial microRNAs Efficiently Induces Apoptosis in Human Cancer Cells. *Clin. Cancer Res.* **2009**, *15*, 3725–3732. [CrossRef] [PubMed]

93. Huang, Q.; Raya, A.; DeJesus, P.; Chao, S.-H.; Quon, K.C.; Caldwell, J.S.; Chanda, S.K.; Izpisua-Belmonte, J.C.; Schultz, P.G. Identification of p53 regulators by genome-wide functional analysis. *Proc. Natl. Acad. Sci. USA* **2004**, *101*, 3456–3461. [CrossRef] [PubMed]

94. Llanos, S.; Efeyan, A.; Monsech, J.; Dominguez, O.; Serrano, M. A High-Throughput Loss-of-Function Screening Identifies Novel p53 Regulators. *Cell Cycle* **2006**, *5*, 1880–1885. [CrossRef] [PubMed]

95. Siebring-van Olst, E.; Blijlevens, M.; de Menezes, R.X.; van der Meulen-Muileman, I.H.; Smit, E.F.; van Beusechem, V.W. A genome-wide siRNA screen for regulators of tumor suppressor p53 activity in human non-small cell lung cancer cells identifies components of the RNA splicing machinery as targets for anticancer treatment. *Mol. Oncol.* **2017**, *11*, 534–551. [CrossRef] [PubMed]

96. Kim, S.Y.; Kang, D.; Choi, H.J.; Joo, Y.; Kim, J.-H.; Song, J.J. Prime-boost immunization by both DNA vaccine and oncolytic adenovirus expressing GM-CSF and shRNA of TGF-β2 induces anti-tumor immune activation. *Oncotarget* **2017**, *8*, 15858–15877. [CrossRef] [PubMed]

97. Han, Z.; Hong, Z.; Chen, C.; Gao, Q.; Luo, D.; Fang, Y.; Cao, Y.; Zhu, T.; Jiang, X.; Ma, Q.; et al. A novel oncolytic adenovirus selectively silences the expression of tumor-associated STAT3 and exhibits potent antitumoral activity. *Carcinogenesis* **2009**, *30*, 2014–2022. [CrossRef] [PubMed]

98. Thomas, M.A.; Spencer, J.F.; La Regina, M.C.; Dhar, D.; Tollefson, A.E.; Toth, K.; Wold, W.S.M. Syrian Hamster as a Permissive Immunocompetent Animal Model for the Study of Oncolytic Adenovirus Vectors. *Cancer Res.* **2006**, *66*, 1270–1276. [CrossRef] [PubMed]

99. Cerullo, V.; Pesonen, S.; Diaconu, I.; Escutenaire, S.; Arstila, P.T.; Ugolini, M.; Nokisalmi, P.; Raki, M.; Laasonen, L.; Särkioja, M.; et al. Oncolytic Adenovirus Coding for Granulocyte Macrophage Colony-Stimulating Factor Induces Antitumoral Immunity in Cancer Patients. *Cancer Res.* **2010**, *70*, 4297–4309. [CrossRef] [PubMed]

100. Wang, P.; Li, X.; Wang, J.; Gao, D.; Li, Y.; Li, H.; Chu, Y.; Zhang, Z.; Liu, H.; Jiang, G.; et al. Re-designing Interleukin-12 to enhance its safety and potential as an anti-tumor immunotherapeutic agent. *Nat. Commun.* **2017**, *8*, 1395. [CrossRef] [PubMed]

101. Schütz, S.; Sarnow, P. Interaction of viruses with the mammalian RNA interference pathway. *Virology* **2006**, *344*, 151–157. [CrossRef] [PubMed]

102. Skalsky, R.L.; Cullen, B.R. Viruses, microRNAs, and Host Interactions. *Annu. Rev. Microbiol.* **2010**, *64*, 123–141. [CrossRef] [PubMed]

103. Grundhoff, A.; Sullivan, C.S. Virus-encoded microRNAs. *Virology* **2011**, *411*, 325–343. [CrossRef] [PubMed]

104. Jopling, C.L. Modulation of Hepatitis C Virus RNA Abundance by a Liver-Specific MicroRNA. *Science* **2005**, *309*, 1577–1581. [CrossRef] [PubMed]

105. Chiang, K.; Liu, H.; Rice, A.P. miR-132 enhances HIV-1 replication. *Virology* **2013**, *438*, 1–4. [CrossRef] [PubMed]

106. Russo, A.; Potenza, N. Antiviral effects of human microRNAs and conservation of their target sites. *FEBS Lett.* **2011**, *585*, 2551–2555. [CrossRef] [PubMed]

107. Andersson, M.G.; Haasnoot, P.C.J.; Xu, N.; Berenjian, S.; Berkhout, B.; Akusjarvi, G. Suppression of RNA Interference by Adenovirus Virus-Associated RNA. *J. Virol.* **2005**, *79*, 9556–9565. [CrossRef] [PubMed]

108. Sano, M.; Kato, Y.; Taira, K. Sequence-specific interference by small RNAs derived from adenovirus VAI RNA. *FEBS Lett.* **2006**, *580*, 1553–1564. [CrossRef] [PubMed]

109. Aparicio, O.; Carnero, E.; Abad, X.; Razquin, N.; Guruceaga, E.; Segura, V.; Fortes, P. Adenovirus VA RNA-derived miRNAs target cellular genes involved in cell growth, gene expression and DNA repair. *Nucleic Acids Res.* **2010**, *38*, 750–763. [CrossRef] [PubMed]

110. Hodzic, J.; Sie, D.; Vermeulen, A.; van Beusechem, V.W. Functional screening identifies human miRNAs that modulate adenovirus propagation in prostate cancer cells. *Hum. Gene Ther.* **2017**, *2016*, 143. [CrossRef] [PubMed]

111. Lagos, D.; Pollara, G.; Henderson, S.; Gratrix, F.; Fabani, M.; Milne, R.S.B.; Gotch, F.; Boshoff, C. miR-132 regulates antiviral innate immunity through suppression of the p300 transcriptional co-activator. *Nat. Cell Biol.* **2010**, *12*, 513–519. [CrossRef] [PubMed]

112. He, L.; He, X.; Lowe, S.W.; Hannon, G.J. microRNAs join the p53 network—Another piece in the tumour-suppression puzzle. *Nat. Rev. Cancer* **2007**, *7*, 819–822. [CrossRef] [PubMed]

113. Feng, S.-D.; Mao, Z.; Liu, C.; Nie, Y.-S.; Sun, B.; Guo, M.; Su, C. Simultaneous overexpression of miR-126 and miR-34a induces a superior antitumor efficacy in pancreatic adenocarcinoma. *OncoTargets Ther.* **2017**, *10*, 5591–5604. [CrossRef] [PubMed]

114. Santhakumar, D.; Forster, T.; Laqtom, N.N.; Fragkoudis, R.; Dickinson, P.; Abreu-Goodger, C.; Manakov, S.A.; Choudhury, N.R.; Griffiths, S.J.; Vermeulen, A.; et al. Combined agonist-antagonist genome-wide functional screening identifies broadly active antiviral microRNAs. *Proc. Natl. Acad. Sci. USA* **2010**, *107*, 13830–13835. [CrossRef] [PubMed]

115. Lee, Y.-B.; Bantounas, I.; Lee, D.-Y.; Phylactou, L.; Caldwell, M.A.; Uney, J.B. Twist-1 regulates the miR-199a/214 cluster during development. *Nucleic Acids Res.* **2009**, *37*, 123–128. [CrossRef] [PubMed]

116. Thornburg, N.J.; Hayward, S.L.; Crowe, J.E. Respiratory syncytial virus regulates human microRNAs by using mechanisms involving beta interferon and NF-κB. *mBio* **2012**, *3*, e00220-12. [CrossRef] [PubMed]

117. Zhang, Q.; Huang, C.; Yang, Q.; Gao, L.; Liu, H.-C.; Tang, J.; Feng, W. MicroRNA-30c Modulates Type I IFN Responses To Facilitate Porcine Reproductive and Respiratory Syndrome Virus Infection by Targeting JAK1. *J. Immunol.* **2016**, *196*, 2272–2282. [CrossRef] [PubMed]

118. Panda, D.; Cherry, S. Cell-based genomic screening: Elucidating virus–host interactions. *Curr. Opin. Virol.* **2012**, *2*, 784–792. [CrossRef] [PubMed]

119. Mahoney, D.J.; Lefebvre, C.; Allan, K.; Brun, J.; Sanaei, C.A.; Baird, S.; Pearce, N.; Grönberg, S.; Wilson, B.; Prakesh, M.; et al. Virus-Tumor Interactome Screen Reveals ER Stress Response Can Reprogram Resistant Cancers for Oncolytic Virus-Triggered Caspase-2 Cell Death. *Cancer Cell* **2011**, *20*, 443–456. [CrossRef] [PubMed]

120. Workenhe, S.T.; Ketela, T.; Moffat, J.; Cuddington, B.P.; Mossman, K.L. Genome-wide lentiviral shRNA screen identifies serine/arginine-rich splicing factor 2 as a determinant of oncolytic virus activity in breast cancer cells. *Oncogene* **2016**, *35*, 2465–2474. [CrossRef] [PubMed]

121. Teferi, W.M.; Dodd, K.; Maranchuk, R.; Favis, N.; Evans, D.H. A Whole-Genome RNA Interference Screen for Human Cell Factors Affecting Myxoma Virus Replication. *J. Virol.* **2013**, *87*, 4623–4641. [CrossRef] [PubMed]

122. Homicsko, K.; Lukashev, A.; Iggo, R.D. RAD001 (Everolimus) Improves the Efficacy of Replicating Adenoviruses that Target Colon Cancer. *Cancer Res.* **2005**, *65*, 6882–6890. [CrossRef] [PubMed]

123. Alonso, M.M.; Jiang, H.; Yokoyama, T.; Xu, J.; Bekele, N.B.; Lang, F.F.; Kondo, S.; Gomez-Manzano, C.; Fueyo, J. Delta-24-RGD in Combination With RAD001 Induces Enhanced Anti-glioma Effect via Autophagic Cell Death. *Mol. Ther.* **2008**, *16*, 487–493. [CrossRef] [PubMed]

124. Rahman, M.M.; Bagdassarian, E.; Ali, M.A.M.; McFadden, G. Identification of host DEAD-box RNA helicases that regulate cellular tropism of oncolytic Myxoma virus in human cancer cells. *Sci. Rep.* **2017**, *7*, 15710. [CrossRef] [PubMed]

Review

The Role of Extracellular Vesicles in Cancer: Cargo, Function, and Therapeutic Implications

James Jabalee [1], Rebecca Towle [1] and Cathie Garnis [1,2,*]

1   Department of Integrative Oncology, British Columbia Cancer Research Center,
    Vancouver V5Z 1L3, BC, Canada; jjabalee@bccrc.ca (J.J.); rtowle@bccrc.ca (R.T.)
2   Division of Otolaryngology, Department of Surgery, University of British Columbia,
    Vancouver V6T 1Z4, BC, Canada
*   Correspondence: cgarnis@bccrc.ca; Tel.: +1-604-675-8041

Received: 9 July 2018; Accepted: 29 July 2018; Published: 1 August 2018

Abstract: Extracellular vesicles (EVs) are a heterogeneous collection of membrane-bound structures that play key roles in intercellular communication. EVs are potent regulators of tumorigenesis and function largely via the shuttling of cargo molecules (RNA, DNA, protein, etc.) among cancer cells and the cells of the tumor stroma. EV-based crosstalk can promote proliferation, shape the tumor microenvironment, enhance metastasis, and allow tumor cells to evade immune destruction. In many cases these functions have been linked to the presence of specific cargo molecules. Herein we will review various types of EV cargo molecule and their functional impacts in the context of oncology.

Keywords: extracellular vesicles; cancer; therapeutics

---

1. Introduction

Extracellular vesicles (EVs) are a collection of lipid-bilayer enclosed vesicles secreted by virtually all cell types including cancer cells. EVs can be divided into subtypes based on their biogenesis, size and morphology, and collection method [1]. While other subtypes certainly exist, we use the term "EVs" to refer primarily to exosomes, ectosomes, and apoptotic bodies. Exosomes, the most heavily studied subtype, are ≈50–150 nm EVs formed by invagination of the multivesicular body (MVB) membrane. Once formed, exosomes are released into the extracellular space via fusion of the MVB with the plasma membrane. Ectosomes (sometimes called microvesicles or shedding microvesicles) are more heterogeneous, ranging in size from ≈100–1000 nm, and are formed through an outward blebbing of the plasma membrane. Apoptotic bodies are vesicles secreted by cells undergoing apoptotic cell death and range in size from ≈1000–5000 nm. Much of the work referenced herein is done on populations of EVs isolated via differential ultracentrifugation, which are likely enriched for exosomes compared to other subtypes. However, we use the term EVs to reflect the heterogeneous nature of these vesicles and the imperfect methods used to isolate them (which can lead to a mixture of EV subtypes). We occasionally use the term "exosomes" to refer to EVs pelleted by centrifugation at ≈100,000× g, "ectosomes" for EVs pelleted at ≈10,000 g, and apoptotic bodies for EVs pelleted at ≈2000 g.

Although initially thought to function exclusively in the removal of unwanted molecules from cells, EVs are now recognized as important mediators of cell–cell communication. EVs play key roles in both normal and disease processes and are important regulators of cancer progression. EVs are known to contain cell-type specific cargo, including RNA, DNA, and protein, which are selectively sorted into EVs [2]. Once released, EVs can interact with cells in the immediate vicinity or at distant locations via transfer through the circulation. EVs interact with recipient cells in a number of ways, including ligand–receptor interaction [3], release of vesicle contents in the extracellular space by bursting [4], direct fusion with the plasma membrane [5], and endocytosis into the cell [6]. The latter mechanisms are of specific interest here as they result in a transfer of molecular cargo from EVs to the recipient cells [2,7]. In cancer, tumor cells both release and receive EVs. This crosstalk between tumor and stromal cells regulates numerous aspects of tumorigenesis, including growth of tumor vasculature [8], recruitment of cancer-associated fibroblasts [9], metastatic potential [8], and evasion of immune destruction [10]. Herein we describe each of the major cargo types associated with EVs, assess the functional impact of EVs on cancer biology, and address the potential clinical uses of EVs relating to their roles as biomarkers and therapeutics.

2. EV Cargo

2.1. EV Isolation and Cargo Profiling

The identification and accurate functional characterization of EV cargo requires appropriate isolation and purification methods. Currently, the most popular method of EV isolation remains differential ultracentrifugation (DU) [11]. DU involves removal of contaminating material through a series of low-speed centrifugations followed by pelleting of EVs at higher speeds. DU is a low-cost, high-throughput method, making it an ideal means of EV isolation in many labs. However, co-purification of non-vesicular proteins and other contaminants is an issue [12]. Combination of DU with other purification methods, such as ultrafiltration or density gradient centrifugation, can improve the purity of the collected vesicles at the expense of particle yield [12]. Furthermore, density gradient centrifugation can be laborious and is unable to separate EVs from contaminants of similar density. The use of alternative techniques, such as size-exclusion chromatography, immunoaffinity capture, microfluidics, and precipitation-based methods are on the rise, each with their own advantages and disadvantages that have been thoroughly examined elsewhere [13]. Briefly, size-exclusion chromatography results in excellent purity, but dilutes samples and therefore requires EVs be re-concentrated following isolation. In contrast, precipitation methods tend to result in high yield but relatively low purity. Immunoaffinity capture and microfluidics can be used to isolate EVs sharing a specific characteristic; for example, all EVs expressing CD63 on their surface [14]. However, markers capable of distinguishing between exosomes, ectosomes, apoptotic bodies, and other EVs have not yet been identified, and markers such as CD63 can be found on various EV subtypes [13]. It is worth noting that because these techniques isolate EVs according to different characteristics (e.g., size, density, presence of specific surface markers, etc.), they may enrich different vesicle subpopulations, potentially generating misleading results [15].

Once EVs have been isolated and purified, consideration must be given to how the cargo of interest (i.e., RNA, protein, DNA, etc.) is to be extracted and profiled. Highlighting the importance of extraction technique, EV RNA yield and size distribution were found to differ greatly depending on the method used, with column-based methods resulting in both the highest yield and broadest size range [16]. In some cases, additional treatment may be required to remove protein and nucleic acids from the outside surface of EVs. Indeed, the International Society of Extracellular Vesicles recommends investigators quantify EV RNA before and after vesicles are treated with proteinase/RNase to determine the contribution of surface-bound RNAs to the total RNA content collected [15]. For additional detail, we refer the reader to excellent recent reviews [13,15]. These studies serve to highlight the need for consistent and detailed reporting of experimental methods for the accurate interpretation of EV studies.

2.2. MicroRNA

Next-generation sequencing (NGS) of EV RNA cargo has revealed the presence of various classes of RNA in EVs derived from normal and cancer cells, including messenger RNA, transfer RNA, ribosomal RNA, microRNA, and more [17–22]. Among these, microRNAs (miRNAs), $\approx$22-nucleotide non-coding RNA molecules, are perhaps the most intriguing and heavily studied. A single miRNA species can regulate the translation and degradation of numerous mRNA targets, which makes miRNAs powerful regulators of cell phenotype. Although EV-mediated miRNA transfer has been strongly linked to cancer progression, the mechanisms underlying miRNA sorting into EVs are poorly understood.

MiRNA abundance in EVs is highly variable and depends on the cell line under study. MiRNAs were reported to comprise from 5–30% of EV small RNA content in colorectal and breast cancer cell lines [18,19], ≈50% in non-tumorigenic murine hepatocyte cells [23], and <1% in HEK293T cells [22]. The miRNA content of EVs is distinct from that of the parental cell, indicating that specific miRNA are selectively sorted into or excluded from EVs [7]. Although the mechanisms by which sorting occurs remain unclear, recent work has suggested that the recognition of specific miRNA motifs by RNA binding proteins (RBPs) may play a role. Specifically, the RBP hnRNPA2B1 was found to selectively sort miRNAs containing the GGAG motif in the 3′ half of their sequence [24], and the RBP synaptotagmin binding cytoplasmic RNA interacting protein (SYNCRIP) was found to selectively sort miRNAs containing the guanine-guanine-cytosine-uracil (GGCU) motif [23]. Furthermore, hnRNPA2B1 must be attached to small ubiquitin-like modifiers (SUMOylated) in order for sorting to occur, thus adding an additional layer of regulation [24]. How the RBP-miRNA complex is then selected for sorting remains an open question. In some cases, such as for hnRNPA2B1, the protein can be detected within EVs, suggesting the entire complex is sorted [24]. However, this may not be the case for all RBPs. Ubiquitination of the RBP HuR causes the protein to release its bound miRNA; ubiquitinated HuR was found to associate primarily with the MVB, where sorting into exosomes occurs, and had low affinity for its target miRNA compared to the non-ubiquitinated protein [25]. Interestingly, silencing of the RBP hnRNPH1 increased the total RNA in EVs, suggesting that RBPs may also exclude specific miRNAs from EVs [26]. In many cases, RBP-mediated changes in EV miRNA sorting have been linked to tumorigenesis. For example, major vault protein (MVP) regulates sorting of tumor suppressive miR-193a into EVs, effectively removing it from cells and leading to more aggressive disease [27]. The tumor suppressor VPS4A was found to have the opposite effect; in this case, overexpression of VPS4A was found to cause can accumulation of tumor suppressive miRNAs in cells and oncogenic miRNAs in their EVs, thus decreasing the growth, migration, and invasion of the cancer cells [28]. Additional examples of RBPs with known oncogenic or tumor suppressive functions that have been linked to EV miRNA sorting include Annexin A2 (ANXA2) [29], Kirsten rat sarcoma (KRAS) [18], Y-box binding protein 1 (YB-1) [22,30], MEX3C [31], and Argonaute 2 (AGO2) [32–34].

Interestingly, AGO2 is not the only component of the miRNA processing machinery found in EVs. The major components of the RNA-induced silencing complex (RISC)-loading complex machinery, including AGO2, Dicer, and trans-activation responsive RNA-binding protein (TRBP), were found in EVs of breast cancer cells where they actively processed pre-miRNA to mature miRNA, and Dicer sorting was found to be CD43-dependent [33]. In contrast, EVs of cultured monocytes were found to contain single-stranded, mature miRNAs but only low levels of AGO2 [35], suggesting that miRNA sorting occurs independently of RISC. Thus, two independent pathways, one involving the sorting of pre-miRNAs along with the RISC machinery and one involving the sorting of mature miRNAs, may exist [36].

In addition to RBPs, 3′-end nucleotide additions (NTAs) also regulate miRNA sorting. MiRNAs with 3′-end adenylation tend to be overrepresented in cells whereas those with 3′-end uridylation are overrepresented in EVs, particularly exosomes [37]. However, the underlying mechanisms remain unclear. It is unknown if uridylated miRNAs are specifically sorted into EVs and, if so, how sorting occurs. NTAs change the stability and activity of miRNAs, which may in turn affect their availability for sorting. Adenylation stabilizes miRNAs, allowing them to interact with their mRNA targets in the cell, whereas uridylation achieves the opposite [38]. The poor activity of uridylated miRNAs may decrease miRNA-mRNA interaction, allowing for the sorting of the free miRNA. Indeed, altering the expression level of a miRNA or its target mRNA can alter the quantity of that miRNA in EVs [39].

Finally, the biogenesis of EVs and the sorting of miRNA contents is regulated by the membrane lipid ceramide. In the case of exosomes, ceramide is produced via the breakdown of sphingomyelin by neutral sphingomyelinases (nSMases) at the MVB membrane. Inhibition of ceramide generation by the nSMase inhibitor GW4869 greatly reduces the small RNA content of EVs [40] and decreases the quantity of EVs released by cells in vitro [2]. Further, treatment of breast cancer cells with GW4869 decreased EV sorting of pro-angiogenic miR-210, thereby inhibiting tumor angiogenesis [8]. Ceramide is thought to play a role in sorting via the formation of ceramide-rich lipid microdomains. Sorting may occur directly via interaction of miRNA with ceramide-rich microdomains in the MVB membrane or indirectly via microdomain-dependent recruitment of proteins to the site of sorting [41]. Specific miRNA sequences show greater affinity for ceramide than others, thus providing a potential mechanism for sequence-based miRNA sorting [41]. Interestingly, ceramide appears to be required for miRNA sorting by some RBPs. Ceramide has been shown to co-localize with hnRNPA2B1 in the cytoplasm, and GW4869 inhibited HuR-mediated sorting of miR-122 and MEX3C-mediated sorting of miR-451a into exosomes [24,25,31]. Sorting of miR-451a was independent of the endosomal sorting complexes required for transport (ESCRT) pathway [31], discussed below.

EV miRNA content can also be modified by external factors, including hypoxia [42], carcinogens such as asbestos [43] or toluene [44], and infection with oncogenic viruses [45–47]. As with RBPs, these factors can alter EV miRNA content in ways that may promote or inhibit tumorigenesis. Viral infection provides an especially intriguing example of the power of EV miRNAs to promote tumorigenesis by influencing the tumor microenvironment. The genome of Kaposi's sarcoma herpesvirus (KSHV) encodes a set of 12 latency-associated miRNAs, all of which have been found in the EVs of infected host cells [47]. Transfer of the viral miRNAs to non-infected cells via EVs results in a shift toward aerobic glycolysis which supports the growth of cancer cells by providing them with energy-rich metabolites [47]. In this way, virally-encoded EV miRNAs are used to reprogram the tumor microenvironment to enhance growth of KSHV-infected cancer cells.

The sensitivity of EV miRNA content to genetic and environmental stimuli suggests their use as biomarkers. EV miRNA biomarkers show great promise; among their most exciting aspects are stability in the face of non-ideal collection methods, ability to be collected non-invasively, and specificity to certain disease states. With regard to non-invasive collection, potential EV miRNA biomarkers have been identified in numerous body fluids, including blood [43,48–58], urine [59–62], pleural effusion [63,64], and saliva [65]. Intriguingly, EV miRNA biomarkers appear exquisitely sensitive to specific disease states, and have been shown to discriminate among closely-related diseases, such as metastatic versus non-metastatic tumors [50], recurrent versus non-recurrent tumors [60], and high-grade versus low-grade tumors [56,60]. In addition, numerous studies have shown that EV miRNA biomarkers return to normal levels upon surgical resection of the tumor [51,56,61], suggesting their use in monitoring disease progression. While these results are encouraging, large-scale validation studies are required before EV miRNA biomarkers can be approved for clinical use. Along these lines, it is worth noting a few particularly promising biomarkers which have been found numerous times in separate studies. The oncomiR miR-21 has been proposed as a biomarker for various cancer types, including breast [49], prostate [59], bladder [61], brain [56], larynx [66], and liver [67]. MiR-375 [59,60,68] and the miR-200 family [21,51,63] show similar promise.

2.3. mRNA and Other RNA Types

Like miRNA, mRNA appears to make up a minority of EV RNA. In EVs derived from glioma stem-like cultures, mRNA accounted for <10% of the total RNA reads as assayed by NGS, and it is unclear what proportion of mRNA reads can be attributed to full-length mRNAs as opposed to mRNA fragments [18,69]. This may reflect a difference among EV subtypes, since mRNA appears to be more enriched in ectosomes compared to exosomes [69]. Despite uncertainty regarding mRNA abundance, numerous studies have shown that functional mRNAs can be transferred via EVs to recipient cells where they are translated and alter recipient cell phenotype [70–73]. For example, hTERT mRNA, which encodes the catalytic subunit of the telomerase enzyme, is transferred via EVs into nearby fibroblasts, thus increasing their proliferation, extending their life span, postponing senescence, and protecting from DNA damage [72]. Furthermore, hTERT mRNA was found in serum-derived EVs from 67.5% of

133 individuals with various cancers, but none of the 45 healthy controls [74], suggesting its use as a biomarker for detection of multiple cancer types.

Recent studies suggest that the presence of specific sequence motifs play a key role in mRNA sorting into EVs. Interestingly, the YB-1 protein, which has also been linked to miRNA sorting [22,30], has been found to interact specifically with mRNAs whose 3′ untranslated region (UTR) contains any of three motifs, while the methyltransferase NSUN interacts with one motif [75,76]. YB-1 is overexpressed in numerous cancer types and drives cell proliferation [77], thus providing a link between cancer progression and EV packaging. Similarly, a 25-nucleotide motif was found to be enriched in the 3′ UTR of exosomal mRNAs from glioblastoma cells [78], whereas mRNAs harboring the signal peptide sequence were excluded from EVs [79].

Additional classes of RNA found in EVs include ribosomal RNA, transfer RNA, mitochondrial RNA, long non-coding RNA, piwi-interacting RNA, small nucleolar RNAs, and circular RNA [20,22,80–82]. Recent results have linked EV long non-coding RNAs to chemosensitivity [83] and cancer progression [84]. While outside the scope of the current manuscript, we refer the reader to an excellent recent review on this topic [85].

2.4. DNA

Extracellular vesicles carry DNA, which may be genomic (gDNA) [86,87] or mitochondrial (mtDNA) [88–90] in origin. Depending on the cell line and context, DNA may be single- or double-stranded, and may reside within the lumen or on the surface of EVs [88,91–95]. Surface-bound DNA can alter the ability of EVs to adhere to fibronectin [89], suggesting it may help determine how EVs interact with extracellular matrix molecules, such as those found in the tumor microenvironment or pre-metastatic niche. Luminal DNA, like other cargo types, can be transferred from donor to recipient cells resulting in increased mRNA and protein production, and oncogenes can be distributed among different cell types via this mechanism [92,94]. Intriguingly, EV-mediated spread of oncogenes has been shown to promote disease progression in mice. EVs derived from chronic myeloid leukemia (CML) cells transfer DNA encoding the breakpoint cluster region/Abelson murine leukemia viral oncogene homolog (BCR/ABL) fusion oncogene to the neutrophils of Sprague-Dawley rats and non-obese diabetic/severe combined immunodeficient (NOD/SCID) mice in vivo resulting in increased BCR/ABL mRNA and protein in the recipient murine cells and the eventual onset of CML-like characteristics [96]. Mitochondrial DNA in EVs also appears to be functional in recipient cells, and cancer-associated fibroblast-derived mtDNA has been found to play a role in the resistance of breast cancer cells to hormone therapy [90].

While current work has shed light on the functions of EV DNA, little is known regarding how it is sorted into EVs. Numerous studies have reported that cancer cell-derived EVs contain gDNA from all chromosomes [86,93,95,97], and at least one study has provided evidence for the packaging of the entire mitochondrial genome within EVs [90]. These results suggest that selective sorting of specific DNA sequences may not occur. Interestingly, knockdown of EV release in human fibroblasts and various cancer cell lines results in the accumulation of damaged DNA in the cytoplasm [98]. Such cytoplasmic DNA can be recognized by DNA sensing proteins, the activation of which results in genomic DNA damage, senescence, or apoptosis [98]. Considering this evidence, it is possible that EVs play a role in maintaining cellular homeostasis through non-selective removal of cytoplasmic DNA [98]. In contrast to the idea of non-specific gDNA packaging, apoptotic bodies, ectosomes, and exosomes were found to contain shared and unique DNA sequences [86], suggesting that independent DNA packaging mechanisms for each of the vesicle subtypes may exist. More work is required to clarify the mechanisms underlying EV DNA packaging and its functional relevance to normal and cancer cells.

Intriguingly, cells with specific mutations in their gDNA release EVs containing DNA that harbor identical mutations. Indeed, EV DNA containing mutations identical to the gDNA has been found in cell culture supernatants [86,93,95], the plasma of tumor-bearing mice [95], and the blood (serum and plasma) of human cancer patients [93,97,99–102] (Table 1). Blood-derived EV DNA provides a rich source of clinically relevant information. NGS was used to detect at least 10 potentially clinically actionable mutations in the EV DNA of patients with pancreatic cancer [97], and a similar approach was used to detect specific mutations in three well-known oncogenes in patients with different cancer types with an overall sensitivity of 95% [100]. A major advantage of mutational analysis of EV DNA compared to other techniques, such as standard biopsies, is the ability to collect samples throughout the course of treatment, thus decreasing the need for invasive biopsies. This flexibility allows physicians to monitor genomic changes in the tumor over time.

Table 1. Summary of recent publications using EV DNA for detection of specific mutations in cancer-related genes.

Study	Fluid	Cancer Type	Patients	Technique	Genes Analyzed	Results
Kahlert et al., 2014 [93]	Blood (serum)	Pancreatic ductal adenocarcinoma	Human; 2 cancer (no stage given), 2 healthy	PCR, Sequencing (BigDye terminator kit)	KRAS, TP53	Detected two different KRAS mutations and one TP53 mutation.
Lázaro-Ibáñez et al., 2014 [96]	Blood (plasma)	Prostate	Human; 4 cancer (T stages 1–3), 4 healthy	PCR, sequencing (BigDye terminator kit)	MLH1, PTEN, TP53	Unable to detect specific mutations.
Thakur et al., 2014 [95]	Blood (plasma)	Melanoma	SK-MEL-28 cells xenografted into NOD/SCID mice; EVs collected when tumors reached max allowable size	Allele-specific PCR	BRAF (V600E)	Mutation detected.
San Lucas et al., 2016 [97]	Blood and pleural fluid	Pancreatic ductal adenocarcinoma (PDAC) and ampullary adenocarcinoma	Human; 2 PDAC and 1 ampullary adenocarcinoma	Next-generation sequencing	Whole genome	At least 10 potentially clinically actionable mutations identified in each patient.
Allenson et al., 2017 [99]	Blood (plasma)	Pancreatic ductal adenocarcinoma (PDAC)	Human; 68 PDAC (all stages), 20 PDAC patients whose blood was drawn after resection with curative intent, and 54 healthy controls	Droplet digital PCR	KRAS	Mutations detected in 7.4%, 66.7%, 80%, and 85% of controls, localized, locally advanced, and metastatic PDAC patients.
Möhrmann et al., 2017 [100]	Blood (plasma)	46.5% colorectal, 18.6% melanoma, 14.0% non-small cell lung cancer, 20.9% other	Human; 43 progressing advanced cancers	Next-generation sequencing	BRAF^{V600}, KRAS$^{G12/G13}$, EGFR^{exon19del,L858R}	Mutations in EV DNA which correspond to those in tissue found in 95% of cases. EV DNA did not contain mutations not present in the parental tumor cells.
Yang et al., 2017 [101]	Blood (serum)	Pancreatic ductal adenocarcinoma (PDAC), chronic pancreatitis (CP), intraductal papillary mucinous neoplasm (IPMN)	Human; 48 PDAC, 9 CP, 7 IPMN, 114 healthy controls	Digital PCR	KRAS^{G12D}, TP53^{R273H}	KRAS mutation detected in 39.6% PDAC, 28.6% IPMN, 55.6% CP, 2.6% healthy controls. TP53 mutation detected in 4.2% PDAC, 14.2% IPMN, 0% CP, 0% healthy controls.
Castellanos-Rizaldos et al., 2018 [102]	Blood (serum)	Non-small cell lung cancer	Human; Training and test cohorts each with 51 mutation positive and 54 mutation negative samples	Allele-specific PCR	EGFR^{T790M}	Training: 81% sensitivity, 95% specificity. Test: 92% sensitivity, 89% specificity

Interestingly, recent reports suggest that cell-free circulating tumor DNA (cfDNA), which was previously thought to derive primarily from apoptotic and necrotic tumor cells, is comprised largely of EV DNA [103]. Many companies offer panels for the comprehensive detection of cfDNA mutations in cancer-related genes, and analysis of mutations in cfDNA is currently being used to guide patient management [104]. By simultaneously collecting and analyzing cfDNA, exosomal DNA, and exosomal RNA from the serum of non-small cell lung cancer patients, Castellanos-Rizaldo and colleagues detected the EGFR T790M mutation with 92% specificity and 89% sensitivity when compared to tissue biopsy [102]. These results highlight the exciting clinical applications of EV DNA. A deeper understanding of EV biology-such as how specific molecular targets are selected for packaging and the identification of markers for the separation of tumor and non-tumor-derived EVs-will serve to further refine the utility of EV-based biomarkers in the clinic.

2.5. Protein

In addition to being transferred to recipient cells and influencing cell phenotype, proteins also regulate the sorting of other EV components, determine which cell types are able to receive EVs (i.e., determine tropism), provide markers for the separation of EV subtypes, bind to and activate receptors on recipient cells, and carry out cell-independent reactions inside of EVs after their release. Here we will discuss the sorting of specific proteins followed by a brief discussion of a few of their various other functions in relation to cancer biology.

A key pathway of EV protein sorting involves the endosomal sorting complexes required for transport (ESCRT), a series of four protein complexes (ESCRT-0, I, II, and III), and accessory proteins. ESCRTs recognize and bind ubiquitinated proteins, facilitating their sorting into EVs [105]. In addition to ubiquitination, other post-translation modifications appear to play an important role in EV protein sorting through both ESCRT-dependent and ESCRT-independent pathways, at least for some proteins. Phosphorylation may either promote or inhibit EV sorting, as evidenced by EPHA2 and AGO2, respectively [34,106]. As mentioned above, SUMOylation of the RBP hnRNPA2B1 regulates sorting into exosomes [24]. Other mechanisms of protein sorting into EVs involve dimerization [107], and recruitment via other proteins, such as tetraspanins [108].

The oncogenic activity of EVs is dependent not only upon their intraluminal cargo, but also on the array of proteins that span the EV membrane. One intriguing function of such transmembrane proteins involves determining EV tropism; i.e., which cell types are most likely to take up, and thus be influenced by, EVs. EV tropism has been observed in vitro among different cell types. In vivo, EVs from metastatic cell lines are more likely to be taken up by resident cells at sites to which those lines commonly metastasize [109,110]. Especially important in this process are integrins (ITGs), a class of proteins known to facilitate cell-extracellular matrix interactions. $ITG\alpha_6$, and its partners $ITG\beta_4$ and $ITG\beta_1$, are abundant in EVs that distribute mainly to the lung where they are taken up primarily by S100A4-positive fibroblasts, $ITG\beta_5$ and $ITG\alpha_V$ are abundant in EVs that distribute mainly to the liver where they are taken up primarily by Kupffer cells, and $ITG\beta_3$ is abundant in EVs that distribute mainly to the brain where they are primarily taken up by endothelial cells [110]. The uptake of EVs at specific locations within the body appears to play a key role in determining the location of metasteses, as evidenced by the observation that injection of lung-tropic EVs into mice increased the lung metastatic capacity of breast cancer cells which normally metastasize preferentially to bone [110]. Moreover, EVs can alter gene expression in recipient cells of the pre-metastatic niche, which may include cancer and stromal cells [109–112]. For example, astrocyte-derived EV miRNAs inhibit the tumor suppressor PTEN in cells that metastasize to the brain, thus priming them for metastatic outgrowth [111]. In addition, miR-122-containing breast cancer cell-derived EVs prime the premetastatic niche by decreasing expression of the glucose metabolizing enzyme pyruvate kinase in nearby stromal cells, thus increasing nutrient availability for metastasizing cancer cells [112]. In contrast to integrins, which direct EVs to their preferred targets, additional proteins have been identified which function to modify or block EV uptake. Along these lines, REG3β interferes with the uptake of EVs into target cells by binding to glycoproteins on the EV surface [113]. Similarly, CD47, an anti-phagocytic signal, was found to block uptake of EVs by immune cells, thus prolonging their time in circulation [114]. These results suggest that manipulation of the surface proteins of EVs could be used to alter their tropism and block their pro-tumor effects.

In addition to tropism-determining proteins, the presence of which is dependent upon cell type, other proteins have been shown to be more ubiquitously found in EVs and can provide information on their cellular origin. In an intriguing example, immunoaffinity capture was used to separate A33-positive and EpCAM-positive exosomes secreted from colorectal cancer organoids, and each population was found to be enriched for distinct proteins [115]. These results suggest that even within the EV subpopulation of exosomes, additional subpopulations, which may differ in aspects of their biogenesis and cargo, are likely to exist, a result supported by others [116]. Tetraspanins, a family of membrane-spanning proteins which includes CD63, CD9, and CD81, are among the most commonly cited "exosome markers". Unfortunately, the assumption that such markers are found on all exosomes is an over-simplification; truly specific markers that are found on all exosomes do not exist. CD63 and CD9 were also found, to differing degrees, to be present on other EV subtypes including ectosomes and apoptotic bodies [116], and particles isolated by ultracentrifugation (which are assumed to be enriched for exosomes) can be separated into CD63+, CD9+, CD81+, and non-tetraspanin-bearing subpopulations, although overlap between markers on a single EV is common [116]. The distinction among exosome subtypes may be an important one due to the many key roles tetraspanins play in EV biology, including regulation of biogenesis [117], cargo sorting [117], and tropism [118]. Thus, subpopulations of exosomes that differ in their tetraspanin content may also differ in their biological function. Further detailed investigation into how best to obtain and purify EVs based on protein markers, and whether such distinctions are truly biologically relevant, is warranted.

The separation of EV subtypes based on surface proteins may prove clinically useful. By studying the surface proteins of EVs collected from cancer and non-cancer cell lines, Melo and colleagues found dozens of cancer cell-specific markers [119]. One of these, GPC1, was found to increase in the blood of patients with breast and pancreatic cancer compared to controls, suggesting its use as a disease detection biomarker [119]. Using similar methodology, Castillo and colleagues identified proteins specific to the EVs of pancreatic cancer cells but not normal controls [120]. Unfortunately, GPC1 was not found in pancreatic cancer EVs, and appeared instead to be selectively expressed in non-cancerous tissues [120]. Once refined, separation of cancer and normal EVs will increase the yield of tumor-specific material and decrease unwanted background in downstream analyses.

3. Extracellular Vesicle Function in Cancer

3.1. Impact of EVs on Fibroblasts

Release of EVs by tumor cells is believed to play a major role in intercellular communication, facilitating signaling to surrounding tumor cells and to distant sites via blood or other biological fluid transportation (Figure 1). A primary focus of inquiry has been the impact of EVs on the tumor stroma, including fibroblasts, endothelial cells, and immune cells.

Fibroblasts comprise a major component of the tumor stroma. Under tumorigenic conditions, fibroblasts can undergo morphological changes that confer a phenotype similar to myofibroblasts, which are activated, mobile fibroblasts. Interestingly, tumor-derived EVs are able to induce the transformation of normal stromal fibroblasts into activated cancer-associated fibroblasts (CAFs) [121–125]. For instance, TGFβ-containing prostate cancer-derived EVs are sufficient to induce fibroblast transformation to a CAF-like phenotype [123]. Further, the resultant increases in wound healing and endothelial cell growth were more pronounced in EV-exposed fibroblasts as compared to fibroblasts transformed by soluble TGFβ alone [123]. The ability of EVs to promote CAF activation was found to correlate with the aggressiveness of the tumor cells, with EVs from a more aggressive cell line prompting higher CAF marker expression, proliferation rate, and enzyme release by treated fibroblasts than EVs from a less aggressive cell line [126].

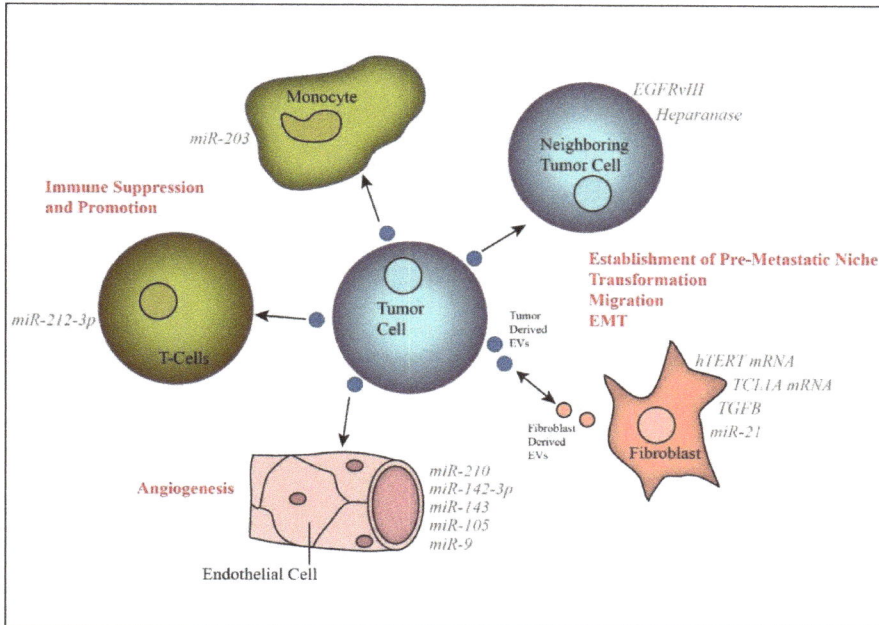

Figure 1. Extracellular vesicle-mediated transfer of specific cargo molecules alters the phenotype of recipient cells, including neighboring tumor cells, fibroblasts, endothelial cells, and immune cells.

Once activated, CAFs secrete EVs that promote tumorigenesis by increasing proliferation, motility, epithelial-mesenchymal transition, migration, and metabolic changes in tumor, endothelial cells, and other fibroblasts [126–129]. Especially intriguing is the role of EVs in mediating resistance to chemotherapy [127,130]. For instance, exposing pancreatic ductal carcinoma cells to conditioned pancreatic fibroblast media was sufficient to confer resistance to gemcitabine, potentially due to up-regulation of snail family transcriptional repressor 1 (SNAIL) and miR-146a in the recipient cells [127]. Gemcitabine treatment also led to an increase in CAF EV secretion, indicating a potential mechanism of drug resistance in pancreatic cancer [127].

3.2. EVs Induce Angiogenesis in Endothelial Cells

The ability to induce angiogenesis is a hallmark of cancer, and recent evidence suggests that EVs are key regulators of tumor vascularization via transfer of pro-angiogenic molecules from tumor to endothelial cells. Indeed, EVs have been shown to increase tube formation, migration, cell–cell adhesion, and proliferation in endothelial cells in a variety of cancer types [8,122,131–138]. For example, activated EGFR found in EVs is sufficient to induce EGFR and VEGFR signaling in recipient endothelial cells, and blocking EV-mediated EGFR transfer decreased tumor growth and angiogenesis [131,139]. Furthermore, EVs produced by hypoxic tumor cells have been shown to have a more pronounced effect on endothelial cells in promoting angiogenesis than those derived from normoxic cells [134,137]. Hypoxia increases the production of tumor and stromal cell-derived EVs and alters their cargo [42,137,140–142]. For example, miR-23a is found in the EVs of hypoxic, but not normoxic, lung cancer cells and promotes angiogenesis through the inhibition of prolyl hydroxylase in recipient endothelial cells [141]. Increased EV production by hypoxic endothelial cells was abrogated by siRNA targeting hypoxia inducible factor 1α, thus providing a clear link between cell response to hypoxia and EV production [142]. Other EV-derived molecules that have been shown to play a role in promoting angiogenesis include miR-9, miR-105, miR-142-3p, miR-210, and H19 lncRNA [132,133,135,138,141,143,144].

Several groups have looked at the impact of sub-populations with cell markers indicative of tumor initiating cells. In renal cell carcinoma cell lines, CD105-positive cells were found to release EVs that increase proliferation, vessel formation, and invasion in HUVEC endothelial cells, whereas CD105-negative cells did not [134]. Similarly, in liver cancer cells, CD90-positive cells were found to secrete EVs that promote tube formation and cell–cell adhesion via transfer of H19 lncRNA [132]. These results highlight the heterogeneity found within tumors and suggest that subsets of tumor cells secrete EVs carrying a unique set of cargo capable of altering stromal cell phenotypes in specific ways.

3.3. Extracellular Vesicles in Immunomodulation

EVs are an important mode of communication among cells of the immune system and are key regulators of the anti-cancer immune response. Initial reports showed that EVs secreted by dendritic cells induce an antitumor immune response, suggesting the use of immune cell-derived EVs as an anti-cancer vaccine as discussed below [145]. This work was strengthened by the observation that EVs contain proteins involved in antigen presentation and immune stimulation, including tumor antigens and major histocompatibility complex (MHC) proteins [146,147]. However, additional work indicated that tumor-derived EVs often promote immune suppression.

Among the most heavily studied immune cell recipients of tumor-derived EVs are dendritic cells and T cells. In many instances, tumor-derived EVs have been found to have an inhibitory effect on dendritic cell function [146,148]. For instance, pancreatic cancer-derived EVs containing miR-203 were found to impair dendritic cell function via reduction of toll-like receptor 4 expression [146]. Furthermore, pancreatic cancer cell-derived EVs were found to alter the transcriptional profile of recipient dendritic cells via transfer of miRNA, specifically miR-212-3p [148]. This led to a decrease in MHCII expression and suppressed immune function. There are conflicting reports on the role of tumor-derived EVs in dendritic cell maturation, with different studies citing either inhibitory or stimulatory effects [149–152].

EVs can also impact T-cell function either directly or via inhibition of other immune cell types, such as dendritic cells [152]. Several studies found that tumor-derived EVs can affect T-cell function, specifically by increasing proliferation, differentiation, and induction of T regulatory cells that function to blunt the immune response [151,153]. Interestingly, direct inhibition of T cell function via EV-associated PD-L1 has also been reported [154]. Furthermore, prostate cancer cell-derived EVs have been found to down-regulate NKG2 in natural T-killer cells and this could contribute to immune suppression [150]. These examples serve to underscore the variety of ways in which tumor-derived EVs can inhibit the immune system.

EVs may also act on cells of the immune system to promote tumor-supportive inflammation. For example, tumor cell-derived EVs can stimulate macrophages to release pro-inflammatory cytokines via activation of the NFκB pathway [155,156]. In breast cancer cells, this pathway activation led to an increase in the secretion of pro-inflammatory cytokines including IL-6, TNFα, GCSF, and CCL2 [156]. Furthermore, tumor EV-associated miR-21 and miR-29a can trigger a pro-inflammatory response in immune cells [157]. Interestingly, these miRs appear to function by acting as EV-associated ligands for toll-like receptors, rather than through their internalization into the cell [157].

3.4. Tumor Promoting Effects of Other Extracellular Vesicles

As noted above, the majority of studies have focused on the functional impact of EV populations collected via ultracentrifugation at 100,000× *g*, which are assumed to be enriched for exosomes. Several papers, however, have also assessed larger EV species, including ectosomes and large oncosomes, which arise from non-apoptotic blebbing of the plasma membrane [124,158,159].

This subset of EVs is obtained by collecting pellets from cell culture media supernatant centrifuged at 10,000× *g* and further purified by density gradient centrifugation. Minciacchi and colleagues found that large oncosomes were able to reprogram prostate fibroblasts via alterations in MYC/AKT1 pathways, rather than via TGFβ as was observed by exosomes [123,158]. Thus, different subtypes of EV reprogram fibroblasts using different mechanisms. Interestingly, by comparing the tumorigenic capabilities of the 10,000 *g* and 100,000 *g* EV fractions from the same cell line, Lindoso and colleagues found that EVs collected after centrifugation at 10,000× *g* were more effective at stimulating angiogenesis, whereas EVs collected after centrifugation at 100,000× *g* were more effective at increasing migration of endothelial cells [124]. These results strengthen the conclusion that different EV subtypes perform unique functions within the tumor niche.

4. Therapeutic Implications of Extracellular Vesicles

In addition to EV biomarkers derived from serum or other biological fluids, a topic that has been thoroughly reviewed above and by others [160,161], EVs have significant potential for use in anti-cancer therapy. Strategies include using EVs as potential cancer vaccines or drug delivery systems, developing interventions to sequester tumor-derived EVs in patients, and developing drugs that target factors involved in EV release.

One of the first indications that EVs may have utility as cancer therapeutics was the observation that dendritic cells secrete antigen-presenting vesicles and that tumor peptide-pulsed dendritic cell-derived EVs decrease tumor growth in mice [145]. This finding drove interest in using dendritic cell-derived EVs as tumor vaccines and spurred multiple clinical trials. Three Phase I trials confirmed the safety of use of dendritic cell-derived EVs in anti-cancer treatments; however, the injected EVs exhibited poor potential in stimulating a T-cell response in the patients [162–164]. More recently, a Phase II trial was completed using dendritic cell-derived EVs as a vaccine. This involved EVs derived from IFN-γ-matured dendritic cells rather than immature dendritic cells [165]. Unfortunately, the endpoint goal (4 months of disease-free survival in 50% of patients) was not reached. A major hurdle in the use of EVs as therapeutics involves the standardization of techniques used to collect and analyze EVs and their molecular cargo, as discussed in Section 2.1 [13]. Interestingly, Tkach and colleagues found that EVs derived from immature dendritic cells are functionally heterogeneous, with large (2000 *g*) and small (100,000 *g*) EVs resulting in different cytokine expression profiles in recipient cells [166]. However, no such heterogeneity was observed for EVs derived from mature dendritic cells [166]. While promising, further work is required to develop a suitable strategy for use of dendritic cell-derived EVs as a form of anti-cancer therapy.

EVs display characteristics that make them ideal options for drug delivery. They are well tolerated in the body, easily taken up by cells, and can be targeted for uptake by specific tissues [167]. A general strategy involves engineering EVs to contain a specific cargo, such as pro-apoptotic proteins, miRNAs, or siRNAs, chemotherapeutic drugs, or molecules targeting specific oncogenes [114,168–173]. Indeed, several papers have described the successful insertion of siRNAs into exosomes [114,169–171]. For example, Alvarez-Erviti and colleagues successfully used self-derived dendritic cell EVs loaded with siRNA to target the brains of mice, finding that this approach did indeed facilitate knockdown of target mRNAs [169]. Further, injection of mice with EVs engineered to contain siRNA targeting mutant KRAS suppressed pancreatic cancer growth and improved overall survival [114].

An additional approach involves counteracting the pro-tumorigenic effects of EVs. One approach is to directly remove EVs from circulation. For example, Marleau and colleagues describe an extracorporeal hemofiltration system which filters blood for components under 200 nm and removes them using affinity agents for target molecules [174]. However, additional studies are required to test the clinical utility of this device. As another example, Nishida-Aoki and colleagues found that treatment of mice with anti-CD9 or anti-CD63 antibodies stimulated EV removal by macrophages, thus greatly decreasing EV concentration in the blood [118]. Although this treatment had no effect on the primary tumor, the authors observed a significant reduction in metastasis [118]. Blocking EV biogenesis in tumor cells by silencing genes encoding EV-related machinery is another potential avenue for inhibiting tumorigenesis. For example, knockdown of *SMPD3* and *RAB27A* resulted in reduced EV secretion and decreased tumorigenesis in mouse models [127,133,143]. However, such strategies may interfere with the normal process of EV-mediated communication; thus, a strategy which serves to minimize off-target effects is required.

5. Summary

In the past few years we have learned a great deal regarding the myriad of cargo molecules contained within EVs and the complex roles EVs play in the tumor microenvironment. The pace of research on this topic has vastly increased in the past couple of years, and we will no doubt make great strides in the years ahead in understanding the complexities underlying the role of EVs in cancer. Though much of this research is still in its infancy, there no doubt there lies many exciting therapeutic and biomarker opportunities ahead.

Author Contributions: Conceptualization, J.J., R.T., and C.G.; Writing-Original Draft Preparation, J.J. and R.T.; Writing-Review & Editing, J.J., R.T., and C.G.

Funding: This research received no external funding.

Acknowledgments: We thank Timon Buys for helpful discussion.

Conflicts of Interest: The authors declare no conflict of interest.

References

1. Gould, S.J.; Raposo, G. As we wait: Coping with an imperfect nomenclature for extracellular vesicles. *J. Extracell. Vesicles* **2013**, *2*, 20389. [CrossRef] [PubMed]
2. Trajkovic, K.; Hsu, C.; Chiantia, S.; Rajendran, L.; Wenzel, D.; Wieland, F.; Schwille, P.; Brügger, B.; Simons, M. Ceramide triggers budding of exosome vesicles into multivesicular endosomes. *Science* **2008**, *319*, 1244–1247. [CrossRef] [PubMed]
3. Abusamra, A.J.; Zhong, Z.; Zheng, X.; Li, M.; Ichim, T.E.; Chin, J.L.; Min, W.P. Tumor exosomes expressing fas ligand mediate CD8+ T-cell apoptosis. *Blood Cells Mol. Dis.* **2005**, *35*, 169–173. [CrossRef] [PubMed]
4. Taraboletti, G.; D'Ascenzo, S.; Giusti, I.; Marchetti, D.; Borsotti, P.; Millimaggi, D.; Giavazzi, R.; Pavan, A.; Dolo, V. Bioavailability of VEGF in tumor-shed vesicles depends on vesicle burst induced by acidic pH. *Neoplasia* **2006**, *8*, 96–103. [CrossRef] [PubMed]
5. Montecalvo, A.; Larregina, A.T.; Shufesky, W.J.; Stolz, D.B.; Sullivan, M.L.; Karlsson, J.M.; Baty, C.J.; Gibson, G.A.; Erdos, G.; Wang, Z. Mechanism of transfer of functional microRNAs between mouse dendritic cells via exosomes. *Blood* **2011**. [CrossRef] [PubMed]
6. Tian, T.; Zhu, Y.L.; Zhou, Y.Y.; Liang, G.F.; Wang, Y.Y.; Hu, F.H.; Xiao, Z.D. Exosome uptake through clathrin-mediated endocytosis and macropinocytosis and mediating miR-21 delivery. *J. Biol. Chem.* **2014**, *289*, 22258–22267. [CrossRef] [PubMed]
7. Valadi, H.; Ekström, K.; Bossios, A.; Sjöstrand, M.; Lee, J.J.; Lötvall, J.O. Exosome-mediated transfer of mrnas and microRNAs is a novel mechanism of genetic exchange between cells. *Nat. Cell Biol.* **2007**, *9*, 654–659. [CrossRef] [PubMed]
8. Kosaka, N.; Iguchi, H.; Hagiwara, K.; Yoshioka, Y.; Takeshita, F.; Ochiya, T. Neutral sphingomyelinase 2 (nSMase2)-dependent exosomal transfer of angiogenic microRNAs regulate cancer cell metastasis. *J. Biol. Chem.* **2013**, *288*, 10849–10859. [CrossRef] [PubMed]
9. Webber, J.; Steadman, R.; Mason, M.D.; Tabi, Z.; Clayton, A. Cancer exosomes trigger fibroblast to myofibroblast differentiation. *Cancer Res.* **2010**, *70*, 9621–9630. [CrossRef] [PubMed]
10. Wieckowski, E.U.; Visus, C.; Szajnik, M.; Szczepanski, M.J.; Storkus, W.J.; Whiteside, T.L. Tumor-derived microvesicles promote regulatory T cell expansion and induce apoptosis in tumor-reactive activated CD8+ T lymphocytes. *J. Immunol.* **2009**, *183*, 3720–3730. [CrossRef] [PubMed]
11. Gardiner, C.; Di Vizio, D.; Sahoo, S.; Thery, C.; Witwer, K.W.; Wauben, M.; Hill, A.F. Techniques used for the isolation and characterization of extracellular vesicles: Results of a worldwide survey. *J. Extracell. Vesicles* **2016**, *5*, 32945. [CrossRef] [PubMed]

12. Webber, J.; Clayton, A. How pure are your vesicles? *J. Extracell. Vesicles* **2013**, *2*, 19861. [CrossRef] [PubMed]
13. Ramirez, M.I.; Amorim, M.G.; Gadelha, C.; Milic, I.; Welsh, J.A.; Freitas, V.M.; Nawaz, M.; Akbar, N.; Couch, Y.; Makin, L.; et al. Technical challenges of working with extracellular vesicles. *Nanoscale* **2018**, *10*, 881–906. [CrossRef] [PubMed]
14. Kanwar, S.S.; Dunlay, C.J.; Simeone, D.M.; Nagrath, S. Microfluidic device (exochip) for on-chip isolation, quantification and characterization of circulating exosomes. *Lab Chip* **2014**, *14*, 1891–1900. [CrossRef] [PubMed]
15. Mateescu, B.; Kowal, E.J.; van Balkom, B.W.; Bartel, S.; Bhattacharyya, S.N.; Buzas, E.I.; Buck, A.H.; de Candia, P.; Chow, F.W.; Das, S.; et al. Obstacles and opportunities in the functional analysis of extracellular vesicle RNA—An isev position paper. *J. Extracell. Vesicles* **2017**, *6*, 1286095. [CrossRef] [PubMed]
16. Eldh, M.; Lotvall, J.; Malmhall, C.; Ekstrom, K. Importance of RNA isolation methods for analysis of exosomal RNA: Evaluation of different methods. *Mol. Immunol.* **2012**, *50*, 278–286. [CrossRef] [PubMed]
17. Ji, H.; Chen, M.; Greening, D.W.; He, W.; Rai, A.; Zhang, W.; Simpson, R.J. Deep sequencing of RNA from three different extracellular vesicle (ev) subtypes released from the human lim1863 colon cancer cell line uncovers distinct miRNA-enrichment signatures. *PLoS ONE* **2014**, *9*, e110314. [CrossRef] [PubMed]
18. Cha, D.J.; Franklin, J.L.; Dou, Y.; Liu, Q.; Higginbotham, J.N.; Beckler, M.D.; Weaver, A.M.; Vickers, K.; Prasad, N.; Levy, S. Kras-dependent sorting of miRNA to exosomes. *eLife* **2015**, *4*, e07197. [CrossRef] [PubMed]
19. Fiskaa, T.; Knutsen, E.; Nikolaisen, M.A.; Jørgensen, T.E.; Johansen, S.D.; Perander, M.; Seternes, O.M. Distinct small RNA signatures in extracellular vesicles derived from breast cancer cell lines. *PLoS ONE* **2016**, *11*, e0161824. [CrossRef] [PubMed]
20. Amorim, M.G.; Valieris, R.; Drummond, R.D.; Pizzi, M.P.; Freitas, V.M.; Sinigaglia-Coimbra, R.; Calin, G.A.; Pasqualini, R.; Arap, W.; Silva, I.T. A total transcriptome profiling method for plasma-derived extracellular vesicles: Applications for liquid biopsies. *Sci. Rep.* **2017**, *7*, 14395. [CrossRef] [PubMed]
21. Endzeliņš, E.; Berger, A.; Melne, V.; Bajo-Santos, C.; Soboļevska, K.; Ābols, A.; Rodriguez, M.; Šantare, D.; Rudņickiha, A.; Lietuvietis, V. Detection of circulating miRNAs: Comparative analysis of extracellular vesicle-incorporated miRNAs and cell-free miRNAs in whole plasma of prostate cancer patients. *BMC Cancer* **2017**, *17*, 730. [CrossRef] [PubMed]
22. Shurtleff, M.J.; Yao, J.; Qin, Y.; Nottingham, R.M.; Temoche-Diaz, M.M.; Schekman, R.; Lambowitz, A.M. Broad role for YBX1 in defining the small noncoding RNA composition of exosomes. *Proc. Natl. Acad. Sci. USA* **2017**, *114*, E8987–E8995. [CrossRef] [PubMed]
23. Santangelo, L.; Giurato, G.; Cicchini, C.; Montaldo, C.; Mancone, C.; Tarallo, R.; Battistelli, C.; Alonzi, T.; Weisz, A.; Tripodi, M. The RNA-binding protein SYNCRIP is a component of the hepatocyte exosomal machinery controlling microRNA sorting. *Cell Rep.* **2016**, *17*, 799–808. [CrossRef] [PubMed]
24. Villarroya-Beltri, C.; Gutiérrez-Vázquez, C.; Sánchez-Cabo, F.; Pérez-Hernández, D.; Vázquez, J.; Martin-Cofreces, N.; Martinez-Herrera, D.J.; Pascual-Montano, A.; Mittelbrunn, M.; Sánchez-Madrid, F. Sumoylated HNRNPA2B1 controls the sorting of miRNAs into exosomes through binding to specific motifs. *Nat. Commun.* **2013**, *4*, 2980. [CrossRef] [PubMed]
25. Mukherjee, K.; Ghoshal, B.; Ghosh, S.; Chakrabarty, Y.; Shwetha, S.; Das, S.; Bhattacharyya, S.N. Reversible hur-microRNA binding controls extracellular export of miR-122 and augments stress response. *EMBO Rep.* **2016**, *17*, 1184–1203. [CrossRef] [PubMed]
26. Statello, L.; Maugeri, M.; Garre, E.; Nawaz, M.; Wahlgren, J.; Papadimitriou, A.; Lundqvist, C.; Lindfors, L.; Collen, A.; Sunnerhagen, P.; et al. Identification of RNA-binding proteins in exosomes capable of interacting with different types of RNA: RBP-facilitated transport of RNAs into exosomes. *PLoS ONE* **2018**, *13*, e0195969. [CrossRef] [PubMed]
27. Teng, Y.; Ren, Y.; Hu, X.; Mu, J.; Samykutty, A.; Zhuang, X.; Deng, Z.; Kumar, A.; Zhang, L.; Merchant, M.L. MVP-mediated exosomal sorting of miR-193a promotes colon cancer progression. *Nat. Commun.* **2017**, *8*, 14448. [CrossRef] [PubMed]
28. Wei, J.x.; Lv, L.h.; Wan, Y.l.; Cao, Y.; Li, G.l.; Lin, H.m.; Zhou, R.; Shang, C.z.; Cao, J.; He, H. Vps4A functions as a tumor suppressor by regulating the secretion and uptake of exosomal microRNAs in human hepatoma cells. *Hepatology* **2015**, *61*, 1284–1294. [CrossRef] [PubMed]

29. Hagiwara, K.; Katsuda, T.; Gailhouste, L.; Kosaka, N.; Ochiya, T. Commitment of annexin a2 in recruitment of microRNAs into extracellular vesicles. *FEBS Lett.* **2015**, *589*, 4071–4078. [CrossRef] [PubMed]

30. Shurtleff, M.J.; Temoche-Diaz, M.M.; Karfilis, K.V.; Ri, S.; Schekman, R. Y-box protein 1 is required to sort microRNAs into exosomes in cells and in a cell-free reaction. *eLife* **2016**, *5*, e19276. [CrossRef] [PubMed]

31. Lu, P.; Li, H.; Li, N.; Singh, R.N.; Bishop, C.E.; Chen, X.; Lu, B. Mex3c interacts with adaptor-related protein complex 2 and involves in miR-451a exosomal sorting. *PLoS ONE* **2017**, *12*, e0185992. [CrossRef] [PubMed]

32. Le, M.T.; Hamar, P.; Guo, C.; Basar, E.; Perdigão-Henriques, R.; Balaj, L.; Lieberman, J. MiR-200–containing extracellular vesicles promote breast cancer cell metastasis. *J. Clin. Investig.* **2014**, *124*, 5109–5128. [CrossRef] [PubMed]

33. Melo, S.A.; Sugimoto, H.; O'Connell, J.T.; Kato, N.; Villanueva, A.; Vidal, A.; Qiu, L.; Vitkin, E.; Perelman, L.T.; Melo, C.A.; et al. Cancer exosomes perform cell-independent microRNA biogenesis and promote tumorigenesis. *Cancer Cell* **2014**, *26*, 707–721. [CrossRef] [PubMed]

34. McKenzie, A.J.; Hoshino, D.; Hong, N.H.; Cha, D.J.; Franklin, J.L.; Coffey, R.J.; Patton, J.G.; Weaver, A.M. KRAS-MEK signaling controls Ago2 sorting into exosomes. *Cell Rep.* **2016**, *15*, 978–987. [CrossRef] [PubMed]

35. Gibbings, D.J.; Ciaudo, C.; Erhardt, M.; Voinnet, O. Multivesicular bodies associate with components of miRNA effector complexes and modulate miRNA activity. *Nat. Cell Biol.* **2009**, *11*, 1143–1149. [PubMed]

36. Fatima, F.; Nawaz, M. Long distance metabolic regulation through adipose-derived circulating exosomal miRNAs: A trail for RNA-based therapies? *Front. Physiol.* **2017**, *8*, 545. [CrossRef] [PubMed]

37. Koppers-Lalic, D.; Hackenberg, M.; Bijnsdorp, I.V.; van Eijndhoven, M.A.J.; Sadek, P.; Sie, D.; Zini, N.; Middeldorp, J.M.; Ylstra, B.; de Menezes, R.X.; et al. Nontemplated nucleotide additions distinguish the small RNA composition in cells from exosomes. *Cell Rep.* **2014**, *8*, 1649–1658. [CrossRef] [PubMed]

38. Song, J.; Song, J.; Mo, B.; Chen, X. Uridylation and adenylation of RNAs. *Sci. China Life Sci.* **2015**, *58*, 1057–1066. [CrossRef] [PubMed]

39. Squadrito, M.L.; Baer, C.; Burdet, F.; Maderna, C.; Gilfillan, G.D.; Lyle, R.; Ibberson, M.; De Palma, M. Endogenous RNAs modulate microRNA sorting to exosomes and transfer to acceptor cells. *Cell Rep.* **2014**, *8*, 1432–1446. [CrossRef] [PubMed]

40. Kubota, S.; Chiba, M.; Watanabe, M.; Sakamoto, M.; Watanabe, N. Secretion of small/microRNAs including mir-638 into extracellular spaces by sphingomyelin phosphodiesterase 3. *Oncol. Rep.* **2015**, *33*, 67–73. [CrossRef] [PubMed]

41. Janas, T.; Janas, M.M.; Sapon, K.; Janas, T. Mechanisms of RNA loading into exosomes. *FEBS Lett.* **2015**, *589*, 1391–1398. [CrossRef] [PubMed]

42. Zhang, G.; Zhang, Y.; Cheng, S.; Wu, Z.; Liu, F.; Zhang, J. CD133 positive U87 glioblastoma cells-derived exosomal microRNAs in hypoxia-versus normoxia-microenviroment. *J. Neuro-Oncol.* **2017**, *135*, 37–46. [CrossRef] [PubMed]

43. Cavalleri, T.; Angelici, L.; Favero, C.; Dioni, L.; Mensi, C.; Bareggi, C.; Palleschi, A.; Rimessi, A.; Consonni, D.; Bordini, L. Plasmatic extracellular vesicle microRNAs in malignant pleural mesothelioma and asbestos-exposed subjects suggest a 2-miRNA signature as potential biomarker of disease. *PLoS ONE* **2017**, *12*, e0176680. [CrossRef] [PubMed]

44. Lim, J.-H.; Song, M.-K.; Cho, Y.; Kim, W.; Han, S.O.; Ryu, J.-C. Comparative analysis of microRNA and mRNA expression profiles in cells and exosomes under toluene exposure. *Toxicol. In Vitro* **2017**, *41*, 92–101. [CrossRef] [PubMed]

45. Harden, M.E.; Munger, K. Human papillomavirus 16 e6 and e7 oncoprotein expression alters microRNA expression in extracellular vesicles. *Virology* **2017**, *508*, 63–69. [CrossRef] [PubMed]

46. Xiong, L.; Zhen, S.; Yu, Q.; Gong, Z. HCV-E2 inhibits hepatocellular carcinoma metastasis by stimulating mast cells to secrete exosomal shuttle microRNAs. *Oncol. Lett.* **2017**, *14*, 2141–2146. [CrossRef] [PubMed]

47. Yogev, O.; Henderson, S.; Hayes, M.J.; Marelli, S.S.; Ofir-Birin, Y.; Regev-Rudzki, N.; Herrero, J.; Enver, T. Herpesviruses shape tumour microenvironment through exosomal transfer of viral microRNAs. *PLoS Pathog.* **2017**, *13*, e1006524. [CrossRef] [PubMed]

48. Rabinowits, G.; Gerçel-Taylor, C.; Day, J.M.; Taylor, D.D.; Kloecker, G.H. Exosomal microRNA: A diagnostic marker for lung cancer. *Clin. Lung Cancer* **2009**, *10*, 42–46. [CrossRef] [PubMed]

49. Hannafon, B.N.; Trigoso, Y.D.; Calloway, C.L.; Zhao, Y.D.; Lum, D.H.; Welm, A.L.; Zhao, Z.J.; Blick, K.E.; Dooley, W.C.; Ding, W. Plasma exosome microRNAs are indicative of breast cancer. *Breast Cancer Res.* **2016**, *18*, 90. [CrossRef] [PubMed]

50. Li, Z.; Ma, Y.-Y.; Wang, J.; Zeng, X.-F.; Li, R.; Kang, W.; Hao, X.-K. Exosomal microRNA-141 is upregulated in the serum of prostate cancer patients. *OncoTargets Ther.* **2016**, *9*, 139–148.

51. Meng, X.; Müller, V.; Milde-Langosch, K.; Trillsch, F.; Pantel, K.; Schwarzenbach, H. Diagnostic and prognostic relevance of circulating exosomal miR-373, miR-200a, miR-200b and miR-200c in patients with epithelial ovarian cancer. *Oncotarget* **2016**, *7*, 16923–16935. [CrossRef] [PubMed]

52. Ostenfeld, M.S.; Jensen, S.G.; Jeppesen, D.K.; Christensen, L.-L.; Thorsen, S.B.; Stenvang, J.; Hvam, M.L.; Thomsen, A.; Mouritzen, P.; Rasmussen, M.H. MiRNA profiling of circulating epcam+ extracellular vesicles: Promising biomarkers of colorectal cancer. *J. Extracell. Vesicles* **2016**, *5*, 31488. [CrossRef] [PubMed]

53. Caivano, A.; La Rocca, F.; Simeon, V.; Girasole, M.; Dinarelli, S.; Laurenzana, I.; De Stradis, A.; De Luca, L.; Trino, S.; Traficante, A. MicroRNA-155 in serum-derived extracellular vesicles as a potential biomarker for hematologic malignancies-a short report. *Cell. Oncol.* **2017**, *40*, 97–103. [CrossRef] [PubMed]

54. Lan, F.; Qing, Q.; Pan, Q.; Hu, M.; Yu, H.; Yue, X. Serum exosomal miR-301a as a potential diagnostic and prognostic biomarker for human glioma. *Cell. Oncol.* **2018**, *41*, 25–33. [CrossRef] [PubMed]

55. Qu, Z.; Wu, J.; Wu, J.; Ji, A.; Qiang, G.; Jiang, Y.; Jiang, C.; Ding, Y. Exosomal miR-665 as a novel minimally invasive biomarker for hepatocellular carcinoma diagnosis and prognosis. *Oncotarget* **2017**, *8*, 80666–80678. [CrossRef] [PubMed]

56. Santangelo, A.; Imbrucè, P.; Gardenghi, B.; Belli, L.; Agushi, R.; Tamanini, A.; Munari, S.; Bossi, A.M.; Scambi, I.; Benati, D. A microRNA signature from serum exosomes of patients with glioma as complementary diagnostic biomarker. *J. Neuro-Oncol.* **2018**, *136*, 51–62. [CrossRef] [PubMed]

57. Xu, J.-F.; Wang, Y.-P.; Zhang, S.-J.; Chen, Y.; Gu, H.-F.; Dou, X.-F.; Xia, B.; Bi, Q.; Fan, S.-W. Exosomes containing differential expression of microRNA and mRNA in osteosarcoma that can predict response to chemotherapy. *Oncotarget* **2017**, *8*, 75968–75978. [CrossRef] [PubMed]

58. Yan, S.; Han, B.; Gao, S.; Wang, X.; Wang, Z.; Wang, F.; Zhang, J.; Xu, D.; Sun, B. Exosome-encapsulated microRNAs as circulating biomarkers for colorectal cancer. *Oncotarget* **2017**, *8*, 60149–60158. [CrossRef] [PubMed]

59. Koppers-Lalic, D.; Hackenberg, M.; de Menezes, R.; Misovic, B.; Wachalska, M.; Geldof, A.; Zini, N.; de Reijke, T.; Wurdinger, T.; Vis, A. Non-invasive prostate cancer detection by measuring miRNA variants (isomiRs) in urine extracellular vesicles. *Oncotarget* **2016**, *7*, 22566–22578. [CrossRef] [PubMed]

60. Andreu, Z.; Oshiro, R.O.; Redruello, A.; López-Martín, S.; Gutiérrez-Vázquez, C.; Morato, E.; Marina, A.I.; Gómez, C.O.; Yáñez-Mó, M. Extracellular vesicles as a source for non-invasive biomarkers in bladder cancer progression. *Eur. J. Pharm. Sci.* **2017**, *98*, 70–79. [CrossRef] [PubMed]

61. Matsuzaki, K.; Fujita, K.; Jingushi, K.; Kawashima, A.; Ujike, T.; Nagahara, A.; Ueda, Y.; Tanigawa, G.; Yoshioka, I.; Ueda, K. MiR-21-5p in urinary extracellular vesicles is a novel biomarker of urothelial carcinoma. *Oncotarget* **2017**, *8*, 24668–24678. [CrossRef] [PubMed]

62. Xu, Y.; Qin, S.; An, T.; Tang, Y.; Huang, Y.; Zheng, L. MiR-145 detection in urinary extracellular vesicles increase diagnostic efficiency of prostate cancer based on hydrostatic filtration dialysis method. *Prostate* **2017**, *77*, 1167–1175. [CrossRef] [PubMed]

63. Lin, J.; Wang, Y.; Zou, Y.-Q.; Chen, X.; Huang, B.; Liu, J.; Xu, Y.-M.; Li, J.; Zhang, J.; Yang, W.-M. Differential miRNA expression in pleural effusions derived from extracellular vesicles of patients with lung cancer, pulmonary tuberculosis, or pneumonia. *Tumor Biol.* **2016**, *37*, 15835–15845. [CrossRef] [PubMed]

64. Wang, Y.; Xu, Y.-M.; Zou, Y.-Q.; Lin, J.; Huang, B.; Liu, J.; Li, J.; Zhang, J.; Yang, W.-M.; Min, Q.-H. Identification of differential expressed PE exosomal miRNA in lung adenocarcinoma, tuberculosis, and other benign lesions. *Medicine* **2017**, *96*, e8361. [CrossRef] [PubMed]

65. Langevin, S.; Kuhnell, D.; Parry, T.; Biesiada, J.; Huang, S.; Wise-Draper, T.; Casper, K.; Zhang, X.; Medvedovic, M.; Kasper, S. Comprehensive microRNA-sequencing of exosomes derived from head and neck carcinoma cells in vitro reveals common secretion profiles and potential utility as salivary biomarkers. *Oncotarget* **2017**, *8*, 82459–82474. [CrossRef] [PubMed]

66. Wang, J.; Zhou, Y.; Lu, J.; Sun, Y.; Xiao, H.; Liu, M.; Tian, L. Combined detection of serum exosomal miR-21 and HOTAIR as diagnostic and prognostic biomarkers for laryngeal squamous cell carcinoma. *Med. Oncol.* **2014**, *31*, 148. [CrossRef] [PubMed]

67. Wang, H.; Hou, L.; Li, A.; Duan, Y.; Gao, H.; Song, X. Expression of serum exosomal microRNA-21 in human hepatocellular carcinoma. *BioMed Res. Int.* **2014**, *2014*, 864894. [CrossRef] [PubMed]

68. Huang, X.; Yuan, T.; Liang, M.; Du, M.; Xia, S.; Dittmar, R.; Wang, D.; See, W.; Costello, B.A.; Quevedo, F. Exosomal miR-1290 and miR-375 as prognostic markers in castration-resistant prostate cancer. *Eur. Urol.* **2015**, *67*, 33–41. [CrossRef] [PubMed]

69. Wei, Z.; Batagov, A.O.; Schinelli, S.; Wang, J.; Wang, Y.; El Fatimy, R.; Rabinovsky, R.; Balaj, L.; Chen, C.C.; Hochberg, F. Coding and noncoding landscape of extracellular RNA released by human glioma stem cells. *Nat. Commun.* **2017**, *8*, 1145. [CrossRef] [PubMed]

70. Lai, C.P.; Kim, E.Y.; Badr, C.E.; Weissleder, R.; Mempel, T.R.; Tannous, B.A.; Breakefield, X.O. Visualization and tracking of tumour extracellular vesicle delivery and RNA translation using multiplexed reporters. *Nat. Commun.* **2015**, *6*, 7029. [CrossRef] [PubMed]

71. Zomer, A.; Maynard, C.; Verweij, F.J.; Kamermans, A.; Schäfer, R.; Beerling, E.; Schiffelers, R.M.; de Wit, E.; Berenguer, J.; Ellenbroek, S.I.J. In vivo imaging reveals extracellular vesicle-mediated phenocopying of metastatic behavior. *Cell* **2015**, *161*, 1046–1057. [CrossRef] [PubMed]

72. Gutkin, A.; Uziel, O.; Beery, E.; Nordenberg, J.; Pinchasi, M.; Goldvaser, H.; Henick, S.; Goldberg, M.; Lahav, M. Tumor cells derived exosomes contain hTERT mRNA and transform nonmalignant fibroblasts into telomerase positive cells. *Oncotarget* **2016**, *7*, 59173–59188. [CrossRef] [PubMed]

73. Yokoi, A.; Yoshioka, Y.; Yamamoto, Y.; Ishikawa, M.; Ikeda, S.-I.; Kato, T.; Kiyono, T.; Takeshita, F.; Kajiyama, H.; Kikkawa, F. Malignant extracellular vesicles carrying mmp1 mRNA facilitate peritoneal dissemination in ovarian cancer. *Nat. Commun.* **2017**, *8*, 14470. [CrossRef] [PubMed]

74. Goldvaser, H.; Gutkin, A.; Beery, E.; Edel, Y.; Nordenberg, J.; Wolach, O.; Rabizadeh, E.; Uziel, O.; Lahav, M. Characterisation of blood-derived exosomal hTERT mRNA secretion in cancer patients: A potential pan-cancer marker. *Br. J. Cancer* **2017**, *117*, 353. [CrossRef] [PubMed]

75. Kossinova, O.A.; Gopanenko, A.V.; Tamkovich, S.N.; Krasheninina, O.A.; Tupikin, A.E.; Kiseleva, E.; Yanshina, D.D.; Malygin, A.A.; Ven'yaminova, A.G.; Kabilov, M.R. Cytosolic YB-1 and NSUN2 are the only proteins recognizing specific motifs present in mRNAs enriched in exosomes. *Biochim. Biophys. Acta (BBA)-Proteins Proteom.* **2017**, *1865*, 664–673. [CrossRef] [PubMed]

76. Yanshina, D.D.; Kossinova, O.A.; Gopanenko, A.V.; Krasheninina, O.A.; Malygin, A.A.; Venyaminova, A.G.; Karpova, G.G. Structural features of the interaction of the 3'-untranslated region of mRNA containing exosomal RNA-specific motifs with YB-1, a potential mediator of mRNA sorting. *Biochimie* **2018**, *144*, 134–143. [CrossRef] [PubMed]

77. Maurya, P.K.; Mishra, A.; Yadav, B.S.; Singh, S.; Kumar, P.; Chaudhary, A.; Srivastava, S.; Murugesan, S.N.; Mani, A. Role of y box protein-1 in cancer: As potential biomarker and novel therapeutic target. *J. Cancer* **2017**, *8*, 1900–1907. [CrossRef] [PubMed]

78. Bolukbasi, M.F.; Mizrak, A.; Ozdener, G.B.; Madlener, S.; Ströbel, T.; Erkan, E.P.; Fan, J.-B.; Breakefield, X.O.; Saydam, O. MiR-1289 and "zipcode"-like sequence enrich mrnas in microvesicles. *Mol. Ther. Nucleic Acids* **2012**, *1*, e10. [CrossRef] [PubMed]

79. Conley, A.; Minciacchi, V.R.; Lee, D.H.; Knudsen, B.S.; Karlan, B.Y.; Citrigno, L.; Viglietto, G.; Tewari, M.; Freeman, M.R.; Demichelis, F. High-throughput sequencing of two populations of extracellular vesicles provides an mRNA signature that can be detected in the circulation of breast cancer patients. *RNA Biol.* **2017**, *14*, 305–316. [CrossRef] [PubMed]

80. Gezer, U.; Özgür, E.; Cetinkaya, M.; Isin, M.; Dalay, N. Long non-coding RNAs with low expression levels in cells are enriched in secreted exosomes. *Cell Biol. Int.* **2014**, *38*, 1076–1079. [CrossRef] [PubMed]

81. Freedman, J.E.; Gerstein, M.; Mick, E.; Rozowsky, J.; Levy, D.; Kitchen, R.; Das, S.; Shah, R.; Danielson, K.; Beaulieu, L. Diverse human extracellular RNAs are widely detected in human plasma. *Nat. Commun.* **2016**, *7*, 11106. [CrossRef] [PubMed]

82. Li, Y.; Zheng, Q.; Bao, C.; Li, S.; Guo, W.; Zhao, J.; Chen, D.; Gu, J.; He, X.; Huang, S. Circular RNA is enriched and stable in exosomes: A promising biomarker for cancer diagnosis. *Cell Res.* **2015**, *25*, 981–984. [CrossRef] [PubMed]

83. Takahashi, K.; Yan, I.K.; Kogure, T.; Haga, H.; Patel, T. Extracellular vesicle-mediated transfer of long non-coding RNA ror modulates chemosensitivity in human hepatocellular cancer. *FEBS Open Bio* **2014**, *4*, 458–467. [CrossRef] [PubMed]

84. Pan, L.; Liang, W.; Fu, M.; Huang, Z.-H.; Li, X.; Zhang, W.; Zhang, P.; Qian, H.; Jiang, P.-C.; Xu, W.-R. Exosomes-mediated transfer of long noncoding RNA ZFAS1 promotes gastric cancer progression. *J. Cancer Res. Clin. Oncol.* **2017**, *143*, 991–1004. [CrossRef] [PubMed]

85. Fatima, F.; Nawaz, M. Vesiculated long non-coding RNAs: Offshore packages deciphering trans-regulation between cells, cancer progression and resistance to therapies. *Noncoding RNA* **2017**, *3*, 10. [CrossRef] [PubMed]

86. Lázaro-Ibáñez, E.; Sanz-Garcia, A.; Visakorpi, T.; Escobedo-Lucea, C.; Siljander, P.; Ayuso-Sacido, Á.; Yliperttula, M. Different gdna content in the subpopulations of prostate cancer extracellular vesicles: Apoptotic bodies, microvesicles, and exosomes. *Prostate* **2014**, *74*, 1379–1390. [CrossRef] [PubMed]

87. García-Romero, N.; Carrión-Navarro, J.; Esteban-Rubio, S.; Lázaro-Ibáñez, E.; Peris-Celda, M.; Alonso, M.M.; Guzmán-De-Villoria, J.; Fernández-Carballal, C.; de Mendivil, A.O.; García-Duque, S. DNA sequences within glioma-derived extracellular vesicles can cross the intact blood-brain barrier and be detected in peripheral blood of patients. *Oncotarget* **2017**, *8*, 1416–1428. [CrossRef] [PubMed]

88. Guescini, M.; Genedani, S.; Stocchi, V.; Agnati, L.F. Astrocytes and glioblastoma cells release exosomes carrying mtDNA. *J. Neural Transm.* **2010**, *117*, 1. [CrossRef] [PubMed]

89. Németh, A.; Orgovan, N.; Sódar, B.W.; Osteikoetxea, X.; Pálóczi, K.; Szabó-Taylor, K.É.; Vukman, K.V.; Kittel, Á.; Turiák, L.; Wiener, Z. Antibiotic-induced release of small extracellular vesicles (exosomes) with surface-associated DNA. *Sci. Rep.* **2017**, *7*, 8202. [CrossRef] [PubMed]

90. Sansone, P.; Savini, C.; Kurelac, I.; Chang, Q.; Amato, L.B.; Strillacci, A.; Stepanova, A.; Iommarini, L.; Mastroleo, C.; Daly, L. Packaging and transfer of mitochondrial DNA via exosomes regulate escape from dormancy in hormonal therapy-resistant breast cancer. *Proc. Natl. Acad. Sci. USA* **2017**. [CrossRef] [PubMed]

91. Balaj, L.; Lessard, R.; Dai, L.; Cho, Y.-J.; Pomeroy, S.L.; Breakefield, X.O.; Skog, J. Tumour microvesicles contain retrotransposon elements and amplified oncogene sequences. *Nat. Commun.* **2011**, *2*, 180. [CrossRef] [PubMed]

92. Cai, J.; Han, Y.; Ren, H.; Chen, C.; He, D.; Zhou, L.; Eisner, G.M.; Asico, L.D.; Jose, P.A.; Zeng, C. Extracellular vesicle-mediated transfer of donor genomic DNA to recipient cells is a novel mechanism for genetic influence between cells. *J. Mol. Cell Biol.* **2013**, *5*, 227–238. [CrossRef] [PubMed]

93. Kahlert, C.; Melo, S.A.; Protopopov, A.; Tang, J.; Seth, S.; Koch, M.; Zhang, J.; Weitz, J.; Chin, L.; Futreal, A. Identification of double-stranded genomic DNA spanning all chromosomes with mutated kras and p53 DNA in the serum exosomes of patients with pancreatic cancer. *J. Biol. Chem.* **2014**, *289*, 3869–3875. [CrossRef] [PubMed]

94. Lee, T.H.; Chennakrishnaiah, S.; Audemard, E.; Montermini, L.; Meehan, B.; Rak, J. Oncogenic ras-driven cancer cell vesiculation leads to emission of double-stranded DNA capable of interacting with target cells. *Biochem. Biophys. Res. Commun.* **2014**, *451*, 295–301. [CrossRef] [PubMed]

95. Thakur, B.K.; Zhang, H.; Becker, A.; Matei, I.; Huang, Y.; Costa-Silva, B.; Zheng, Y.; Hoshino, A.; Brazier, H.; Xiang, J. Double-stranded DNA in exosomes: A novel biomarker in cancer detection. *Cell Res.* **2014**, *24*, 766–769. [CrossRef] [PubMed]

96. Cai, J.; Wu, G.; Tan, X.; Han, Y.; Chen, C.; Li, C.; Wang, N.; Zou, X.; Chen, X.; Zhou, F. Transferred bcr/abl DNA from k562 extracellular vesicles causes chronic myeloid leukemia in immunodeficient mice. *PLoS ONE* **2014**, *9*, e105200. [CrossRef] [PubMed]

97. San Lucas, F.; Allenson, K.; Bernard, V.; Castillo, J.; Kim, D.; Ellis, K.; Ehli, E.; Davies, G.; Petersen, J.; Li, D. Minimally invasive genomic and transcriptomic profiling of visceral cancers by next-generation sequencing of circulating exosomes. *Ann. Oncol.* **2015**, *27*, 635–641. [CrossRef] [PubMed]

98. Takahashi, A.; Okada, R.; Nagao, K.; Kawamata, Y.; Hanyu, A.; Yoshimoto, S.; Takasugi, M.; Watanabe, S.; Kanemaki, M.T.; Obuse, C. Exosomes maintain cellular homeostasis by excreting harmful DNA from cells. *Nat. Commun.* **2017**, *8*, 15827. [CrossRef] [PubMed]

99. Allenson, K.; Castillo, J.; San Lucas, F.; Scelo, G.; Kim, D.; Bernard, V.; Davis, G.; Kumar, T.; Katz, M.; Overman, M. High prevalence of mutant KRAS in circulating exosome-derived DNA from early-stage pancreatic cancer patients. *Ann. Oncol.* **2017**, *28*, 741–747. [CrossRef] [PubMed]

100. Möhrmann, L.; Huang, H.; Hong, D.S.; Tsimberidou, A.M.; Fu, S.; Piha-Paul, S.; Subbiah, V.; Karp, D.D.; Naing, A.; Krug, A.K. Liquid biopsies using plasma exosomal nucleic acids and plasma cell-free DNA compared with clinical outcomes of patients with advanced cancers. *Clin. Cancer Res.* **2017**. [CrossRef] [PubMed]

101. Yang, S.; Che, S.P.; Kurywchak, P.; Tavormina, J.L.; Gansmo, L.B.; Correa de Sampaio, P.; Tachezy, M.; Bockhorn, M.; Gebauer, F.; Haltom, A.R. Detection of mutant KRAS and TP53 DNA in circulating exosomes from healthy individuals and patients with pancreatic cancer. *Cancer Biol. Ther.* **2017**, *18*, 158–165. [CrossRef] [PubMed]

102. Castellanos-Rizaldos, E.; Grimm, D.G.; Tadigotla, V.; Hurley, J.; Healy, J.; Neal, P.L.; Sher, M.; Venkatesan, R.; Karlovich, C.; Raponi, M.; et al. Exosome-based detection of EGFR T790m in plasma from non-small cell lung cancer patients. *Clin. Cancer Res.* **2018**, *24*, 2944–2950. [CrossRef] [PubMed]

103. Fernando, M.R.; Jiang, C.; Krzyzanowski, G.D.; Ryan, W.L. New evidence that a large proportion of human blood plasma cell-free DNA is localized in exosomes. *PLoS ONE* **2017**, *12*, e0183915. [CrossRef] [PubMed]

104. Lanman, R.B.; Mortimer, S.A.; Zill, O.A.; Sebisanovic, D.; Lopez, R.; Blau, S.; Collisson, E.A.; Divers, S.G.; Hoon, D.S.; Kopetz, E.S. Analytical and clinical validation of a digital sequencing panel for quantitative, highly accurate evaluation of cell-free circulating tumor DNA. *PLoS ONE* **2015**, *10*, e0140712. [CrossRef] [PubMed]

105. Williams, R.L.; Urbe, S. The emerging shape of the escrt machinery. *Nat. Rev. Mol. Cell Biol.* **2007**, *8*, 355–368. [CrossRef] [PubMed]

106. Takasugi, M.; Okada, R.; Takahashi, A.; Virya Chen, D.; Watanabe, S.; Hara, E. Small extracellular vesicles secreted from senescent cells promote cancer cell proliferation through EPHA2. *Nat. Commun.* **2017**, *8*, 15729. [CrossRef] [PubMed]

107. Itoh, S.; Mizuno, K.; Aikawa, M.; Aikawa, E. Dimerization of sortilin regulates its trafficking to extracellular vesicles. *J. Biol. Chem.* **2018**, *293*, 4532–4544. [CrossRef] [PubMed]

108. Perez-Hernandez, D.; Gutierrez-Vazquez, C.; Jorge, I.; Lopez-Martin, S.; Ursa, A.; Sanchez-Madrid, F.; Vazquez, J.; Yanez-Mo, M. The intracellular interactome of tetraspanin-enriched microdomains reveals their function as sorting machineries toward exosomes. *J. Biol. Chem.* **2013**, *288*, 11649–11661. [CrossRef] [PubMed]

109. Peinado, H.; Alečković, M.; Lavotshkin, S.; Matei, I.; Costa-Silva, B.; Moreno-Bueno, G.; Hergueta-Redondo, M.; Williams, C.; García-Santos, G.; Ghajar, C.M. Melanoma exosomes educate bone marrow progenitor cells toward a pro-metastatic phenotype through met. *Nat. Med.* **2012**, *18*, 883–891. [CrossRef] [PubMed]

110. Hoshino, A.; Costa-Silva, B.; Shen, T.-L.; Rodrigues, G.; Hashimoto, A.; Mark, M.T.; Molina, H.; Kohsaka, S.; Di Giannatale, A.; Ceder, S. Tumour exosome integrins determine organotropic metastasis. *Nature* **2015**, *527*, 329–335. [CrossRef] [PubMed]

111. Zhang, L.; Zhang, S.; Yao, J.; Lowery, F.J.; Zhang, Q.; Huang, W.C.; Li, P.; Li, M.; Wang, X.; Zhang, C.; et al. Microenvironment-induced pten loss by exosomal microRNA primes brain metastasis outgrowth. *Nature* **2015**, *527*, 100–104. [CrossRef] [PubMed]

112. Fong, M.Y.; Zhou, W.; Liu, L.; Alontaga, A.Y.; Chandra, M.; Ashby, J.; Chow, A.; O'Connor, S.T.; Li, S.; Chin, A.R.; et al. Breast-cancer-secreted miR-122 reprograms glucose metabolism in premetastatic niche to promote metastasis. *Nat. Cell Biol.* **2015**, *17*, 183–194. [CrossRef] [PubMed]

113. Bonjoch, L.; Gironella, M.; Iovanna, J.L.; Closa, D. REG3β modifies cell tumor function by impairing extracellular vesicle uptake. *Sci. Rep.* **2017**, *7*, 3143. [CrossRef] [PubMed]

114. Kamerkar, S.; LeBleu, V.S.; Sugimoto, H.; Yang, S.; Ruivo, C.F.; Melo, S.A.; Lee, J.J.; Kalluri, R. Exosomes facilitate therapeutic targeting of oncogenic KRAS in pancreatic cancer. *Nature* **2017**, *546*, 498–503. [CrossRef] [PubMed]

115. Tauro, B.J.; Greening, D.W.; Mathias, R.A.; Mathivanan, S.; Ji, H.; Simpson, R.J. Two distinct populations of exosomes are released from lim1863 colon carcinoma cell-derived organoids. *Mol. Cell. Proteom.* **2013**, *12*, 587–598. [CrossRef] [PubMed]

116. Kowal, J.; Arras, G.; Colombo, M.; Jouve, M.; Morath, J.P.; Primdal-Bengtson, B.; Dingli, F.; Loew, D.; Tkach, M.; Théry, C. Proteomic comparison defines novel markers to characterize heterogeneous populations of extracellular vesicle subtypes. *Proc. Natl. Acad. Sci. USA* **2016**, *113*, E968–E977. [CrossRef] [PubMed]

117. Hurwitz, S.N.; Nkosi, D.; Conlon, M.M.; York, S.B.; Liu, X.; Tremblay, D.C.; Meckes, D.G. CD63 regulates epstein-barr virus LMP1 exosomal packaging, enhancement of vesicle production, and noncanonical NF-κB signaling. *J. Virol.* **2017**, *91*, e02251-16. [CrossRef] [PubMed]

118. Nishida-Aoki, N.; Tominaga, N.; Takeshita, F.; Sonoda, H.; Yoshioka, Y.; Ochiya, T. Disruption of circulating extracellular vesicles as a novel therapeutic strategy against cancer metastasis. *Mol. Ther.* **2017**, *25*, 181–191. [CrossRef] [PubMed]

119. Melo, S.A.; Luecke, L.B.; Kahlert, C.; Fernandez, A.F.; Gammon, S.T.; Kaye, J.; LeBleu, V.S.; Mittendorf, E.A.; Weitz, J.; Rahbari, N. Glypican-1 identifies cancer exosomes and detects early pancreatic cancer. *Nature* **2015**, *523*, 177–182. [CrossRef] [PubMed]

120. Castillo, J.; Bernard, V.; San Lucas, F.; Allenson, K.; Capello, M.; Kim, D.; Gascoyne, P.; Mulu, F.; Stephens, B.; Huang, J. Surfaceome profiling enables isolation of cancer-specific exosomal cargo in liquid biopsies from pancreatic cancer patients. *Ann. Oncol.* **2017**, *29*, 223–229. [CrossRef] [PubMed]

121. Paggetti, J.; Haderk, F.; Seiffert, M.; Janji, B.; Distler, U.; Ammerlaan, W.; Kim, Y.J.; Adam, J.; Lichter, P.; Solary, E. Exosomes released by chronic lymphocytic leukemia cells induce the transition of stromal cells into cancer-associated fibroblasts. *Blood* **2015**. [CrossRef] [PubMed]

122. Song, Y.H.; Warncke, C.; Choi, S.J.; Choi, S.; Chiou, A.E.; Ling, L.; Liu, H.Y.; Daniel, S.; Antonyak, M.A.; Cerione, R.A.; et al. Breast cancer-derived extracellular vesicles stimulate myofibroblast differentiation and pro-angiogenic behavior of adipose stem cells. *Matrix Biol.* **2017**, *60–61*, 190–205. [CrossRef] [PubMed]

123. Webber, J.P.; Spary, L.K.; Sanders, A.J.; Chowdhury, R.; Jiang, W.G.; Steadman, R.; Wymant, J.; Jones, A.T.; Kynaston, H.; Mason, M.D.; et al. Differentiation of tumour-promoting stromal myofibroblasts by cancer exosomes. *Oncogene* **2015**, *34*, 290–302. [CrossRef] [PubMed]

124. Lindoso, R.S.; Collino, F.; Camussi, G. Extracellular vesicles derived from renal cancer stem cells induce a pro-tumorigenic phenotype in mesenchymal stromal cells. *Oncotarget* **2015**, *6*, 7959–7969. [CrossRef] [PubMed]

125. Baroni, S.; Romero-Cordoba, S.; Plantamura, I.; Dugo, M.; D'Ippolito, E.; Cataldo, A.; Cosentino, G.; Angeloni, V.; Rossini, A.; Daidone, M.G.; et al. Exosome-mediated delivery of miR-9 induces cancer-associated fibroblast-like properties in human breast fibroblasts. *Cell Death Dis.* **2016**, *7*, e2312. [CrossRef] [PubMed]

126. Giusti, I.; Di Francesco, M.; D'Ascenzo, S.; Palmerini, M.G.; Macchiarelli, G.; Carta, G.; Dolo, V. Ovarian cancer-derived extracellular vesicles affect normal human fibroblast behavior. *Cancer Biol. Ther.* **2018**, *19*, 722–734. [CrossRef] [PubMed]

127. Richards, K.E.; Zeleniak, A.E.; Fishel, M.L.; Wu, J.; Littlepage, L.E.; Hill, R. Cancer-associated fibroblast exosomes regulate survival and proliferation of pancreatic cancer cells. *Oncogene* **2017**, *36*, 1770–1778. [CrossRef] [PubMed]

128. Donnarumma, E.; Fiore, D.; Nappa, M.; Roscigno, G.; Adamo, A.; Iaboni, M.; Russo, V.; Affinito, A.; Puoti, I.; Quintavalle, C. Cancer-associated fibroblasts release exosomal microRNAs that dictate an aggressive phenotype in breast cancer. *Oncotarget* **2017**, *8*, 19592–19608. [CrossRef] [PubMed]

129. Zhao, H.; Yang, L.; Baddour, J.; Achreja, A.; Bernard, V.; Moss, T.; Marini, J.C.; Tudawe, T.; Seviour, E.G.; San Lucas, F.A.; et al. Tumor microenvironment derived exosomes pleiotropically modulate cancer cell metabolism. *eLife* **2016**, *5*, e10250. [CrossRef] [PubMed]

130. Au Yeung, C.L.; Co, N.N.; Tsuruga, T.; Yeung, T.L.; Kwan, S.Y.; Leung, C.S.; Li, Y.; Lu, E.S.; Kwan, K.; Wong, K.K.; et al. Exosomal transfer of stroma-derived miR21 confers paclitaxel resistance in ovarian cancer cells through targeting apaf1. *Nat. Commun* **2016**, *7*, 11150. [CrossRef] [PubMed]

131. Al-Nedawi, K.; Meehan, B.; Kerbel, R.S.; Allison, A.C.; Rak, J. Endothelial expression of autocrine VEGF upon the uptake of tumor-derived microvesicles containing oncogenic EGFR. *Proc. Natl. Acad. Sci. USA* **2009**, *106*, 3794–3799. [CrossRef] [PubMed]

132. Conigliaro, A.; Costa, V.; Lo Dico, A.; Saieva, L.; Buccheri, S.; Dieli, F.; Manno, M.; Raccosta, S.; Mancone, C.; Tripodi, M.; et al. CD90+ liver cancer cells modulate endothelial cell phenotype through the release of exosomes containing H19 LncRNA. *Mol. Cancer* **2015**, *14*, 155. [CrossRef] [PubMed]

133. Dickman, C.T.; Lawson, J.; Jabalee, J.; MacLellan, S.A.; LePard, N.E.; Bennewith, K.L.; Garnis, C. Selective extracellular vesicle exclusion of miR-142-3p by oral cancer cells promotes both internal and extracellular malignant phenotypes. *Oncotarget* **2017**, *8*, 15252–15266. [CrossRef] [PubMed]

134. Grange, C.; Tapparo, M.; Collino, F.; Vitillo, L.; Damasco, C.; Deregibus, M.C.; Tetta, C.; Bussolati, B.; Camussi, G. Microvesicles released from human renal cancer stem cells stimulate angiogenesis and formation of lung premetastatic niche. *Cancer Res.* **2011**, *71*, 5346–5356. [CrossRef] [PubMed]

135. Lawson, J.; Dickman, C.; MacLellan, S.; Towle, R.; Jabalee, J.; Lam, S.; Garnis, C. Selective secretion of microRNAs from lung cancer cells via extracellular vesicles promotes CAMK1D-mediated tube formation in endothelial cells. *Oncotarget* **2017**, *8*, 83913–83924. [CrossRef] [PubMed]

136. Schillaci, O.; Fontana, S.; Monteleone, F.; Taverna, S.; Di Bella, M.A.; Di Vizio, D.; Alessandro, R. Exosomes from metastatic cancer cells transfer amoeboid phenotype to non-metastatic cells and increase endothelial permeability: Their emerging role in tumor heterogeneity. *Sci. Rep.* **2017**, *7*, 4711. [CrossRef] [PubMed]

137. Umezu, T.; Tadokoro, H.; Azuma, K.; Yoshizawa, S.; Ohyashiki, K.; Ohyashiki, J.H. Exosomal miR-135b shed from hypoxic multiple myeloma cells enhances angiogenesis by targeting factor-inhibiting HIF-1. *Blood* **2014**, *124*, 3748–3757. [CrossRef] [PubMed]

138. Zhuang, G.; Wu, X.; Jiang, Z.; Kasman, I.; Yao, J.; Guan, Y.; Oeh, J.; Modrusan, Z.; Bais, C.; Sampath, D.; et al. Tumour-secreted miR-9 promotes endothelial cell migration and angiogenesis by activating the Jak-stat pathway. *EMBO J.* **2012**, *31*, 3513–3523. [CrossRef] [PubMed]

139. Al-Nedawi, K.; Meehan, B.; Micallef, J.; Lhotak, V.; May, L.; Guha, A.; Rak, J. Intercellular transfer of the oncogenic receptor EGFRVIII by microvesicles derived from tumour cells. *Nat. Cell Biol.* **2008**, *10*, 619–624. [CrossRef] [PubMed]

140. King, H.W.; Michael, M.Z.; Gleadle, J.M. Hypoxic enhancement of exosome release by breast cancer cells. *BMC Cancer* **2012**, *12*, 421. [CrossRef] [PubMed]

141. Hsu, Y.L.; Hung, J.Y.; Chang, W.A.; Lin, Y.S.; Pan, Y.C.; Tsai, P.H.; Wu, C.Y.; Kuo, P.L. Hypoxic lung cancer-secreted exosomal miR-23a increased angiogenesis and vascular permeability by targeting prolyl hydroxylase and tight junction protein zo-1. *Oncogene* **2017**, *36*, 4929–4942. [CrossRef] [PubMed]

142. Burnley-Hall, N.; Willis, G.; Davis, J.; Rees, D.A.; James, P.E. Nitrite-derived nitric oxide reduces hypoxia-inducible factor 1alpha-mediated extracellular vesicle production by endothelial cells. *Nitric Oxide* **2017**, *63*, 1–12. [CrossRef] [PubMed]

143. Zhou, W.; Fong, M.Y.; Min, Y.; Somlo, G.; Liu, L.; Palomares, M.R.; Yu, Y.; Chow, A.; O'Connor, S.T.; Chin, A.R.; et al. Cancer-secreted mir-105 destroys vascular endothelial barriers to promote metastasis. *Cancer Cell* **2014**, *25*, 501–515. [CrossRef] [PubMed]

144. Zitvogel, L.; Regnault, A.; Lozier, A.; Wolfers, J.; Flament, C.; Tenza, D.; Ricciardi-Castagnoli, P.; Raposo, G.; Amigorena, S. Eradication of established murine tumors using a novel cell-free vaccine: Dendritic cell-derived exosomes. *Nat. Med.* **1998**, *4*, 594–600. [CrossRef] [PubMed]

145. Zhou, M.; Chen, J.; Zhou, L.; Chen, W.; Ding, G.; Cao, L. Pancreatic cancer derived exosomes regulate the expression of TLR4 in dendritic cells via miR-203. *Cell. Immunol.* **2014**, *292*, 65–69. [CrossRef] [PubMed]

146. Wolfers, J.; Lozier, A.; Raposo, G.; Regnault, A.; Thery, C.; Masurier, C.; Flament, C.; Pouzieux, S.; Faure, F.; Tursz, T.; et al. Tumor-derived exosomes are a source of shared tumor rejection antigens for CTL cross-priming. *Nat. Med.* **2001**, *7*, 297–303. [CrossRef] [PubMed]

147. Ding, G.; Zhou, L.; Qian, Y.; Fu, M.; Chen, J.; Chen, J.; Xiang, J.; Wu, Z.; Jiang, G.; Cao, L. Pancreatic cancer-derived exosomes transfer miRNAs to dendritic cells and inhibit rfxap expression via miR-212-3p. *Oncotarget* **2015**, *6*, 29877–29888. [CrossRef] [PubMed]

148. Marton, A.; Vizler, C.; Kusz, E.; Temesfoi, V.; Szathmary, Z.; Nagy, K.; Szegletes, Z.; Varo, G.; Siklos, L.; Katona, R.L.; et al. Melanoma cell-derived exosomes alter macrophage and dendritic cell functions in vitro. *Immunol. Lett.* **2012**, *148*, 34–38. [CrossRef] [PubMed]

149. Lundholm, M.; Schroder, M.; Nagaeva, O.; Baranov, V.; Widmark, A.; Mincheva-Nilsson, L.; Wikstrom, P. Prostate tumor-derived exosomes down-regulate NKG2D expression on natural killer cells and CD8+ T cells: Mechanism of immune evasion. *PLoS ONE* **2014**, *9*, e108925. [CrossRef] [PubMed]

150. Szajnik, M.; Czystowska, M.; Szczepanski, M.J.; Mandapathil, M.; Whiteside, T.L. Tumor-derived microvesicles induce, expand and up-regulate biological activities of human regulatory T cells (TREG). *PLoS ONE* **2010**, *5*, e11469. [CrossRef] [PubMed]

151. Ning, Y.; Shen, K.; Wu, Q.; Sun, X.; Bai, Y.; Xie, Y.; Pan, J.; Qi, C. Tumor exosomes block dendritic cells maturation to decrease the T cell immune response. *Immunol. Lett.* **2018**, *199*, 36–43. [CrossRef] [PubMed]

152. Ye, S.B.; Li, Z.L.; Luo, D.H.; Huang, B.J.; Chen, Y.S.; Zhang, X.S.; Cui, J.; Zeng, Y.X.; Li, J. Tumor-derived exosomes promote tumor progression and T-cell dysfunction through the regulation of enriched exosomal microRNAs in human nasopharyngeal carcinoma. *Oncotarget* **2014**, *5*, 5439–5452. [CrossRef] [PubMed]

153. Yang, Y.; Li, C.W.; Chan, L.C.; Wei, Y.; Hsu, J.M.; Xia, W.; Cha, J.H.; Hou, J.; Hsu, J.L.; Sun, L.; et al. Exosomal PD-L1 harbors active defense function to suppress T cell killing of breast cancer cells and promote tumor growth. *Cell Res.* **2018**. [CrossRef] [PubMed]

154. Wu, L.; Zhang, X.; Zhang, B.; Shi, H.; Yuan, X.; Sun, Y.; Pan, Z.; Qian, H.; Xu, W. Exosomes derived from gastric cancer cells activate NF-kappaB pathway in macrophages to promote cancer progression. *Tumour Biol.* **2016**, *37*, 12169–12180. [CrossRef] [PubMed]

155. Chow, A.; Zhou, W.; Liu, L.; Fong, M.Y.; Champer, J.; Van Haute, D.; Chin, A.R.; Ren, X.; Gugiu, B.G.; Meng, Z.; et al. Macrophage immunomodulation by breast cancer-derived exosomes requires toll-like receptor 2-mediated activation of NF-kappaB. *Sci. Rep.* **2014**, *4*, 5750. [CrossRef] [PubMed]

156. Fabbri, M.; Paone, A.; Calore, F.; Galli, R.; Gaudio, E.; Santhanam, R.; Lovat, F.; Fadda, P.; Mao, C.; Nuovo, G.J.; et al. MicroRNAs bind to toll-like receptors to induce prometastatic inflammatory response. *Proc. Natl. Acad. Sci. USA* **2012**, *109*, E2110–E2116. [CrossRef] [PubMed]

157. Minciacchi, V.R.; Spinelli, C.; Reis-Sobreiro, M.; Cavallini, L.; You, S.; Zandian, M.; Li, X.; Mishra, R.; Chiarugi, P.; Adam, R.M.; et al. Myc mediates large oncosome-induced fibroblast reprogramming in prostate cancer. *Cancer Res.* **2017**, *77*, 2306–2317. [CrossRef] [PubMed]

158. Minciacchi, V.R.; You, S.; Spinelli, C.; Morley, S.; Zandian, M.; Aspuria, P.-J.; Cavallini, L.; Ciardiello, C.; Sobreiro, M.R.; Morello, M. Large oncosomes contain distinct protein cargo and represent a separate functional class of tumor-derived extracellular vesicles. *Oncotarget* **2015**, *6*, 11327–11341. [CrossRef] [PubMed]

159. Nawaz, M.; Camussi, G.; Valadi, H.; Nazarenko, I.; Ekstrom, K.; Wang, X.; Principe, S.; Shah, N.; Ashraf, N.M.; Fatima, F.; et al. The emerging role of extracellular vesicles as biomarkers for urogenital cancers. *Nat. Rev. Urol.* **2014**, *11*, 688–701. [CrossRef] [PubMed]

160. Nawaz, M.; Fatima, F.; Nazarenko, I.; Ekström, K.; Murtaza, I.; Anees, M.; Sultan, A.; Neder, L.; Camussi, G.; Valadi, H. Extracellular vesicles in ovarian cancer: Applications to tumor biology, immunotherapy and biomarker discovery. *Expert Rev. Proteom.* **2016**, *13*, 395–409. [CrossRef] [PubMed]

161. Escudier, B.; Dorval, T.; Chaput, N.; Andre, F.; Caby, M.P.; Novault, S.; Flament, C.; Leboulaire, C.; Borg, C.; Amigorena, S.; et al. Vaccination of metastatic melanoma patients with autologous dendritic cell (DC) derived-exosomes: Results of the first phase I clinical trial. *J. Transl. Med.* **2005**, *3*, 10. [CrossRef] [PubMed]

162. Morse, M.A.; Garst, J.; Osada, T.; Khan, S.; Hobeika, A.; Clay, T.M.; Valente, N.; Shreeniwas, R.; Sutton, M.A.; Delcayre, A.; et al. A phase I study of dexosome immunotherapy in patients with advanced non-small cell lung cancer. *J. Transl. Med.* **2005**, *3*, 9. [CrossRef] [PubMed]

163. Dai, S.; Wei, D.; Wu, Z.; Zhou, X.; Wei, X.; Huang, H.; Li, G. Phase I clinical trial of autologous ascites-derived exosomes combined with GM-CSF for colorectal cancer. *Mol. Ther.* **2008**, *16*, 782–790. [CrossRef] [PubMed]

164. Besse, B.; Charrier, M.; Lapierre, V.; Dansin, E.; Lantz, O.; Planchard, D.; Le Chevalier, T.; Livartoski, A.; Barlesi, F.; Laplanche, A.; et al. Dendritic cell-derived exosomes as maintenance immunotherapy after first line chemotherapy in NSCLC. *Oncoimmunology* **2016**, *5*, e1071008. [CrossRef] [PubMed]

165. Tkach, M.; Kowal, J.; Zucchetti, A.E.; Enserink, L.; Jouve, M.; Lankar, D.; Saitakis, M.; Martin-Jaular, L.; Thery, C. Qualitative differences in t-cell activation by dendritic cell-derived extracellular vesicle subtypes. *EMBO J.* **2017**, *36*, 3012–3028. [CrossRef] [PubMed]

166. Vader, P.; Mol, E.A.; Pasterkamp, G.; Schiffelers, R.M. Extracellular vesicles for drug delivery. *Adv. Drug Deliv. Rev.* **2016**, *106*, 148–156. [CrossRef] [PubMed]

167. Ohno, S.; Takanashi, M.; Sudo, K.; Ueda, S.; Ishikawa, A.; Matsuyama, N.; Fujita, K.; Mizutani, T.; Ohgi, T.; Ochiya, T.; et al. Systemically injected exosomes targeted to EGFR deliver antitumor microRNA to breast cancer cells. *Mol. Ther.* **2013**, *21*, 185–191. [CrossRef] [PubMed]

168. Alvarez-Erviti, L.; Seow, Y.; Yin, H.; Betts, C.; Lakhal, S.; Wood, M.J. Delivery of siRNA to the mouse brain by systemic injection of targeted exosomes. *Nat. Biotechnol.* **2011**, *29*, 341–345. [CrossRef] [PubMed]

169. Greco, K.A.; Franzen, C.A.; Foreman, K.E.; Flanigan, R.C.; Kuo, P.C.; Gupta, G.N. PLK-1 silencing in bladder cancer by siRNA delivered with exosomes. *Urology* **2016**, *91*, e241–e247. [CrossRef] [PubMed]

170. Wahlgren, J.; De, L.K.T.; Brisslert, M.; Vaziri Sani, F.; Telemo, E.; Sunnerhagen, P.; Valadi, H. Plasma exosomes can deliver exogenous short interfering RNA to monocytes and lymphocytes. *Nucleic Acids Res.* **2012**, *40*, e130. [CrossRef] [PubMed]

171. Mizrak, A.; Bolukbasi, M.F.; Ozdener, G.B.; Brenner, G.J.; Madlener, S.; Erkan, E.P.; Strobel, T.; Breakefield, X.O.; Saydam, O. Genetically engineered microvesicles carrying suicide mRNA/protein inhibit schwannoma tumor growth. *Mol. Ther.* **2013**, *21*, 101–108. [CrossRef] [PubMed]

172. Tian, Y.; Li, S.; Song, J.; Ji, T.; Zhu, M.; Anderson, G.J.; Wei, J.; Nie, G. A doxorubicin delivery platform using engineered natural membrane vesicle exosomes for targeted tumor therapy. *Biomaterials* **2014**, *35*, 2383–2390. [CrossRef] [PubMed]

173. Marleau, A.M.; Chen, C.S.; Joyce, J.A.; Tullis, R.H. Exosome removal as a therapeutic adjuvant in cancer. *J. Transl. Med.* **2012**, *10*, 134. [CrossRef] [PubMed]

174. Bobrie, A.; Krumeich, S.; Reyal, F.; Recchi, C.; Moita, L.F.; Seabra, M.C.; Ostrowski, M.; Thery, C. Rab27a supports exosome-dependent and -independent mechanisms that modify the tumor microenvironment and can promote tumor progression. *Cancer Res.* **2012**, *72*, 4920–4930. [CrossRef] [PubMed]

cells

MDPI

Article

Expanding the miRNA Repertoire in Atlantic Salmon; Discovery of IsomiRs and miRNAs Highly Expressed in Different Tissues and Developmental Stages

Nardos Tesfaye Woldemariam [1], Oleg Agafonov [2], Bjørn Høyheim [3], Ross D. Houston [4], John B. Taggart [5] and Rune Andreassen [1,*]

[1]   Department of Life Sciences and Health, Faculty of Health Sciences, OsloMet–Oslo Metropolitan University, Pilestredet 50, N-0130 Oslo, Norway; nate@oslomet.no
[2]   Bioinformatics Core Facility, Department of Core Facilities, Institute of Cancer Research, Radium Hospital, Oslo University Hospital, 0379 Oslo, Norway; olegag@ifi.uio.no
[3]   Department of Basic Sciences and Aquatic Medicine, Faculty of Veterinary Medicine, Norwegian University of Life Sciences, 0454 Oslo, Norway; bjorn.hoyheim@nmbu.no
[4]   Division of Genetics and Genomics, The Roslin Institute and Royal (Dick) School of Veterinary Studies, University of Edinburgh, Midlothian EH25 9RG, UK; ross.houston@roslin.ed.ac.uk
[5]   Institute of Aquaculture, University of Stirling, Stirling, FK9 4LA, UK; j.b.taggart@stir.ac.uk
*   Correspondence: rune.andreassen@oslomet.no; Tel.: +47-6723-6274

Received: 19 November 2018; Accepted: 18 December 2018; Published: 11 January 2019

Abstract: MicroRNAs (miRNAs) are important post-transcriptional gene expression regulators. Here, 448 different miRNA genes, including 17 novel miRNAs, encoding for 589 mature Atlantic salmon miRNAs were identified after sequencing 111 samples (fry, pathogen challenged fry, various developmental and adult tissues). This increased the reference miRNAome with almost one hundred genes. Prior to isomiR characterization (mature miRNA variants), the proportion of erroneous sequence variants (ESVs) arising in the analysis pipeline was assessed. The ESVs were biased towards 5′ and 3′ end of reads in unexpectedly high proportions indicating that measurements of ESVs rather than Phred score should be used to avoid misinterpreting ESVs as isomiRs. Forty-three isomiRs were subsequently discovered. The biological effect of the isomiRs measured as increases in target diversity was small (<3%). Five miRNA genes showed allelic variation that had a large impact on target gene diversity if present in the seed. Twenty-one miRNAs were ubiquitously expressed while 31 miRNAs showed predominant expression in one or few tissues, indicating housekeeping or tissue specific functions, respectively. The miR-10 family, known to target *Hox* genes, were highly expressed in the developmental stages. The proportion of miR-430 family members, participating in maternal RNA clearance, was high at the earliest developmental stage.

Keywords: Teleostei; embryogenesis; tissue-enriched miRNAs; post-transcriptional gene regulation

1. Introduction

MicroRNAs (miRNAs) are short non-coding RNAs that play an important role in post-transcriptional regulation of gene expression [1]. After transcription, the large primary miRNA transcripts (pri-miRNAs) are cleaved into shorter miRNA precursors (pre-miRNAs) that are exported out of the nucleus by Exportin 5. Here they are processed further by Dicer to produce the mature 5p and 3p miRNAs that are about 20–24 nts long. The mature miRNAs are then loaded onto Argonaute proteins and incorporated in the RNA induced silencing complex (miRISC). As part of the miRISC they form partially complementary bindings with their target mRNAs which subsequently leads to degradation or translational repression of the target transcripts [1,2]. The characteristics of these miRNA precursors and the mature miRNAs produced from Dicer cleavage of the miRNA precursors may be utilized to design bioinformatics tools that identify the miRNA genes and their mature miRNAs. High-throughput sequencing of small RNAs that are analyzed with such dedicated bioinformatics tools against a reference genome allows for massive parallel identification of a large number of miRNA genes [3,4].

Several studies have demonstrated that miRNAs are involved in many biological processes such as development, growth, tissue differentiation and apoptosis [5,6]. There is also evidence that some miRNAs are important regulators of immune function and immune responses [7,8]. As their biological function is defined by the sequence of the mature miRNAs, characterization of the miRNA repertoire is a first step to understand how miRNAs participate in the regulation of a species gene networks.

Genomic research in Atlantic salmon (*Salmo salar*) has been carried out due to the cultural and recreational importance of this species, and more importantly, its economic importance as an aquaculture species (www.fao.org/fishery/affris/species-profiles/atlantic-salmon/atlantic-salmon-home/en/). Atlantic salmon miRNAs were first characterized in 2013 by Andreassen et al. [9] and Bekaert et al. [10] utilizing the first assembly of the Atlantic salmon genome sequence. These miRNA reference sequences (miRNAome) has been used to develop RT-qPCR methods to measure miRNA expression, to investigate the role of miRNAs in host-virus responses, miRNAs that may affect sea louse infestation and miRNAs that may affect testis development [11–14].

The development of sequencing technology and the increased sensitivity of deep sequencing has led to the discovery of isomiRs [15–17]. IsomiRs are mature miRNA sequence variants with different 5′ and/or 3′ ends compared to their corresponding canonical mature miRNAs. IsomiRs are assumed to be products of imprecise cleavage of pre-miRNAs, RNA editing and non-templated nucleotide additions at the 3′ end of miRNAs [17,18]. Knowledge about the function of isomiRs is limited, but it has been suggested that isomiRs cooperate with their corresponding canonical miRNAs to target common biological pathways. Any post-transcriptional modification of the 5′ end of a mature miRNA (5′ isomiRs) is of particular importance as such changes affect the "seed" sequences. Any change of "seed" would affect target specificity, and by this, potentially increase the number of transcripts targeted [18,19].

Small-RNA sequencing datasets consist of a huge number of reads. Some differences in length or sequence among reads are due to errors arising in the sequencing pipeline like RNA degradation, sequence errors introduced in cDNA synthesis or during sequencing [20,21]. Such artefacts may be misinterpreted as isomiRs. Therefore, to characterize isomiRs that are true products of biological post-transcriptional processing one need to distinguish isomiRs from such artefacts arising in the deep sequencing pipeline.

So far, there have been no studies to characterize isomiRs in Atlantic salmon. Although the first miRNA discovery studies provided an important species-specific reference [9,10], the materials sequenced were from relative few samples. No samples from different developmental stages or tissues from fish challenged with pathogens were investigated.

In the present study, about ten times more samples have been included compared to the initial study [9]. One hundred and eleven small RNA libraries from different tissues sampled from adult fish, from different developmental stages, from normal fry, as well as from fry challenged with pathogens were sequenced. Thus, the materials investigated has allowed us to identify miRNAs likely to have tissue and developmental stage specific functions. The proportion of isomiR-like artifacts generated in the small-RNA sequencing pipeline was also revealed. On this background, the first characterization of isomiRs in Atlantic salmon was carried out. Nearly one hundred new miRNA genes, both miRNAs conserved in teleosts, as well as novel miRNAs, were discovered. The new reference miRNAome, where miRNA gene locations are assigned to the present genome sequence [22], forms an important updated resource for miRNA expression studies, as well as for comparative studies of miRNA gene evolution.

2. Materials and Methods

2.1. Materials

A total of 111 samples were sequenced in this study. Samples 1–96 in Table S1 were from fry, while samples 97–111 were tissue samples from particular organs from fully developed adults (intestine, gills, gonads, head-kidney or mid-kidney) and from early developmental stages. The samples from different developmental stages comprised of embryos sampled at 4, 19, 39, and 50 days post fertilization (dpf), an eyed-egg 63dpf, and an alevin one-day post hatching (alevin 1dph). These 111 samples were used in the miRdeep2 analysis (version 0.0.7) [4], the analysis of tissue enriched miRNAs and for isomiR characterization (Table S1). Results from the eleven small-RNA sequenced samples in Andreassen et al. [9] were also included in the datasets to identify miRNAs enriched in particular tissues. An additional 24 tissue samples (Table S2) from brain (n = 4), gills (n = 3), heart (n = 4), intestine (n = 5), liver (n = 5) and white muscle (n = 3) were included to investigate miRNAs enriched in those particular tissues. These samples were used for RT-qPCR analysis as described in Section 2.9.

Fry materials, sampled in Scotland were euthanized using a procedure specifically listed on the appropriate Home Office (UK) license, and all experiments were performed under the approval of Cefas ethical review committee and complied with the Animals Scientific Procedures Act. Some of the pathogens challenged fry were part of a challenge study using infectious pancreatic necrosis virus (IPNV) described in in Robledo et al. [23]. Sacrifice procedure of the other fish in the materials sampled in Norway was approved by the official ethics board FOTS (forsøksdyrutvalgets tilsyns-og søknadssystem). Dissection and sampling of materials were performed in agreement with the provisions enforced by the Norwegian Animal Research Authority.

2.2. Small RNA Extraction

Sampling and extraction of RNA from the fry materials (samples 1–96, Table S1) were carried out, as described in Robledo et al. [23]. Total RNA was isolated using TRI reagent (Sigma–Aldrich®, St. Louis, MO, USA) following the manufacturer's instructions. The RNA quantity and quality were determined using spectrophotometry (NanoDrop ND-1000, Thermo Scientific, Wilmington, DE, USA) and agarose gel electrophoresis, respectively. Total RNA from the remaining materials was extracted by using the mirVana Isolation Kit (Ambion, Life Technologies, Carlsbad, CA, USA) according to the manufacturer's protocol. The RNA concentration and purity were determined using spectrophotometry (NanoDrop ND-1000, Thermo Scientific, Wilmington, DE, USA).

2.3. Library Preparation and Small-RNA Sequencing

Small-RNA library construction and the small-RNA sequencing was performed at the Norwegian Genomics Consortium's genomic core facility (NGC). The Illumina NEBNext small RNA Library Prep Set (New England Biolabs, Inc. Ipswich, MA, USA) was used to prepare the 96 libraries from the 96 fry samples, while Illumina® TruSeq Small RNA sample preparation kit (Illumina, San Diego, CA, USA) was used to construct 15 libraries from the tissue and developmental stage specific samples. 1 μg of total RNA was used per sample as input in the library preparation in accordance with the manufacturer's protocols. The sequencing was carried out on the Illumina Genome Analyzer IIx sequencing platform.

2.4. Pre-Processing and Quality Assesment of Small-RNA Sequence Reads

The raw sequencing reads (fastq-files) from each sample were pre-processed to ensure that the raw data used in downstream analysis were of good quality and desired sizes. An assessment of the raw sequence reads was first carried out using FASTQC (v.0.11.5) [24], to ensure that the small RNA read quality was satisfactory before adaptor sequences were removed using cutadapt (v.1.13) [25]. Additional size filtering was carried out to discard reads that were outside the expected size range of mature miRNAs (18–25 nts). The quality of the trimmed and size filtered reads were analyzed by a second FASTQC analysis. Reads that passed this post-trim QC were used in the downstream analysis of the small RNA sequenced samples.

2.5. Identification of Atlantic Salmon miRNA Precursors, Their Mature miRNAs and miRNA Gene Locations

High quality reads from 108 samples (samples 1–97, 99, 102–111, Table S1) were used for identification of miRNAs applying the miRDeep2 software package. Default settings were used [3,4] to predict miRNA precursors along with their 5p and 3p mature miRNAs. Each sample was independently analyzed with miRDeep2 to allow for detection of miRNAs expressed in particular tissues or developmental stages. The present version of the Atlantic salmon genome assembly, ICSASG_v2, GenBank accession number: GCA_000233375.4 [22], and a genome index consisting of the existing reference Atlantic salmon miRNAs [9,10] was also used in the miRdeep2 analysis. The characterization pipeline is illustrated in Figure 1. The reads were mapped to the Atlantic salmon genome sequence using the miRDeep2 mapper module. Guided by the mapped reads genomic sequences that showed features expected from precursor sequences (e.g., ability to form hairpins) and with aligned reads showing expected characteristics of mature miRNAs processed by Dicer/Drosha were identified. These were assigned a log-odds score (the miRDeep2 score) based on an algorithm that integrates the statistics of the read positions, the frequencies of reads within hairpins, and the posterior probability that the hairpin was derived from a true miRNA gene [3]. To prevent false positive detection of miRNA precursors, a miRDeep2 score of ≥ 2 was used as a cut-off. All predicted precursors with miRDeep2 scores equal to or above the threshold were included in the downstream analysis. They were further analyzed by Basic Local Alignment Search Tool (BLAST) searches, using the putative precursor sequences as input, against miRNAs in the miRNA sequence database (miRBase) (version 22) (http://www.mirbase.org/search.shtml) [26]. A significant hit was defined as matches with an e-value of $\leq 1 \times 10^{-7}$ against any hairpin precursor in the database. Those identified as Atlantic salmon miRNA genes in the first study [9] were annotated with new genome locations in the present Atlantic salmon genome. Other matches were identified as Atlantic salmon orthologs of miRNA genes discovered in other teleost species. These were annotated according to the miRBase nomenclature guidelines (ssa-prefix and same number as in other teleosts) [27,28]. The precursors that did not provide significant matches with any of the miRNAs in miRBase were potentially novel miRNAs. These were further analyzed by blastn (http://blast.ncbi.nlm.nih.gov/Blast) searches against RNA databases in GenBank (http://blast.ncbi.nlm.nih.gov/Blast), the small RNA databases Rfam v.13.0. (http://rfam.xfam.org/search) [29], and the functional RNA database fRNAdb v.3.0. (https://dbarchive.biosciencedbc.jp/blast) [30]. Candidates that matched other kinds of small RNAs in these databases were removed (e-value threshold of $\leq 1 \times 10^{-7}$). The remaining precursors were used as queries in blastn analysis against the Atlantic salmon genome sequence. Any putative precursor that provided a significant hit (e-value threshold of $\leq 1 \times 10^{-7}$) against more than 15 loci in the genome reference sequence were annotated as interspersed repeats and removed. RNA secondary structure of the remaining miRNA precursors and their aligned reads were manually inspected. Novel miRNA genes were identified based on passing the following miRBase criteria [26]; (1) detected in at least two independent samples, (2) at least 10 sequence reads of mature and star miRNAs mapped (with no mismatches) to the hairpin precursor, (3) the mature microRNAs were paired with the precursor hairpin with 0–4 nt overhang at their 3' ends, (4) the reads mapped supported a consistent pre-processing of 5' end (5' homogeneity) and (5) more than 60% of the bases of the mature sequences paired in the hairpin structure. Those that passed all these criteria were annotated as novel Atlantic salmon miRNAs. These were designated "ssa-miR-novel-x" with identifying numbers (x; e.g., ssa-mir-novel-1). All miRNAs, including the novel ones discovered in this study have been submitted to miRBase. When assigned unique miRNA identities by miRBase these will be added as Addendum to this paper.

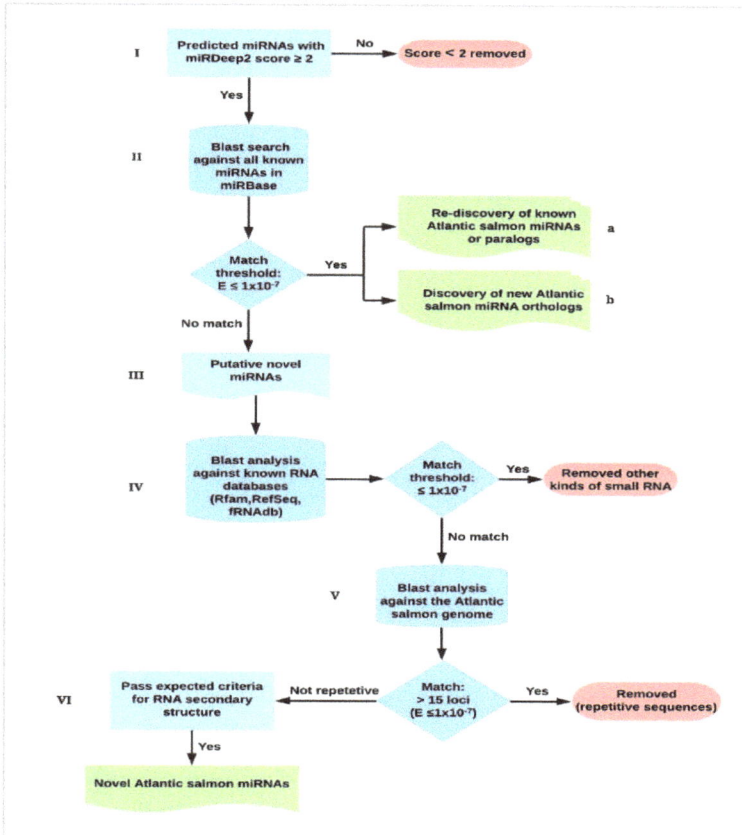

Figure 1. Identification and characterization of Atlantic salmon microRNAs (miRNAs). (I) Candidate miRNAs with miRdeep2 score ≥ 2 were regarded as putative miRNAs. (II) miRNA precursors were used as input for BLAST search against miRBase to identify; (**a**) Atlantic salmon miRNAs already in the database and new paralogs, (**b**) Orthologues of other teleost miRNAs not discovered in prior studies of Atlantic salmon. (III) Candidate miRNAs with no significant matches (e $\leq 1 \times 10^{-7}$) to miRBase were considered as putative novel Atlantic salmon miRNAs. (IV) Putative novel miRNA precursor sequences were BLAST analyzed against other RNA databases (Rfam, NCBI RefSeq and fRNAdb) to remove those that were other kinds of small RNA. (V) Sequences with no matches in these databases were further used for BLAST analysis against the Atlantic salmon reference genome sequence. Significant matches (e $\leq 1 \times 10^{-7}$) to more than 15 loci in the Atlantic salmon genome were annotated as interspersed repeats and removed. (VI) The remaining putative novel miRNAs were compared against expected features of precursors and their mature miRNAs. The candidates that passed all these criteria were regarded as novel Atlantic salmon miRNAs.

2.6. Annotation of Clustered miRNA Genes

The genome location of all Atlantic salmon miRNA genes was identified using the second and improved version of the Atlantic salmon genome assembly. The amount of clustered miRNA genes in the Atlantic salmon genome was examined by comparing their locations in ICSASG_v2. Any miRNA genes located in the same direction in a contig, within a distance of 10 kb or less were annotated as clustered miRNA genes (same definition as used by miRBase). We also compared the miRNA clusters from our study to miRNA clusters in zebrafish (*Danio rerio*) and Atlantic cod (*Gadus morhua*), as given in miRBase to reveal evolutionarily conserved miRNA gene clusters.

2.7. Sequence Errors Arising in the Sequencing Pipeline and IsomiR Detection

Prior to characterizing the abundance of isomiRs the amount and type of reads with erroneous sequences (erroneous sequence variants, ESVs) that arise in the sequencing pipeline (extraction, cDNA synthesis, library prep, small RNA sequencing) must be assessed. High quality reads from 48 fry samples were pooled (441,320,712 reads with Phred quality score above 32). This dataset was collapsed into a total of 7,004,792 unique reads annotated with their read count number. The phiX Control is commonly used as an internal control to measure the proportion of ESVs when larger fragments are sequenced [31,32]. This control can however, not be applied in small RNA sequencing. Instead, we used two highly abundant and ubiquitously expressed Atlantic salmon RNAs; (18S rRNA (Genbank accession: FJ710886.1), and 60S ribosomal protein L37 (GenBank accession: BT058368.1)) for this purpose. The two RNAs, the rRNA (1750 bp) and the short mRNA (540 bp), were used as references as reads that are derived from anywhere out of these larger sequences are not expected to be RNA edited. All reads were aligned against these two references. The aligned reads with bp differences compared to the references would be ESVs (or polymorphism). By aligning the unique sequence reads to those reference sequences in Sequencher software version 5.4.6 (Gene Codes Corporation, Ann Arbor, MI, USA), the ESVs were visualized. Reads that aligned perfectly to the references and reads that represented reads with ESVs were counted. Any bias in error rate within reads (e.g., higher error rate in 3' end) was also assessed. The ratio between perfectly aligned reads and ESVs were used to set an error threshold. Reads present in lower amount than this threshold and with non-templated sequence variation compared to the canonical reference miRNAs are likely to be ESVs, not isomiR variants that are products of RNA editing.

Detection of isomiRs was carried out by use of isomiR Seed Extension Aligner (isomiR-SEA) software (version 1.60) [33]. The tool detects mature miRNAs and mature miRNA isoforms by a seed-based alignment procedure. The unique reads were aligned to reference miRNAs (mature canonical miRNAs identified in this study) using default settings and commands. The tool classifies miRNA variations into four categories with respect to the reference miRNA sequence. The reads matching perfectly to the reference miRNA (canonical miRNAs), reads with any nucleotide variation in the 5' end (5' isomiRs), reads with any nucleotide variation in the 3' end (3' isomiRs and 3' length isomiRs), and reads with mismatches anywhere in their sequences. Count number of the reads identical to canonical miRNA sequences was used to set a lower threshold based on the error ratios revealed in alignments to the reference sequences. If below this threshold, they were removed assuming they were erroneous sequence variants (ESVs) that had arisen in the pipeline, not isomiRs. All reads variants passing the error rate cut-off filtering were aligned to their reference miRNA sequences using Sequencher software. Degraded miRNAs would be identical, but shorter than reference miRNAs. Shorter 5' and 3' isomiR variants could therefore not be reliably identified. The total RNA extracted also consists of precursor miRNAs, occasionally present in a larger amount than the mature miRNAs derived from the precursor [34]. Any templated 3' length isomiRs larger in size than the reference miRNAs could therefore not be reliably detected as such reads could arise from degradation of precursor molecules. These shorter and longer template sequences were excluded from further analysis. The isomiR variants that could be reliably detected in small RNA sequencing data sets were therefore non-templated 5' isomiRs, non-templated 3' isomiRs and non-templated 3' length isomiRs. The identification of miRNAs (Section 2.4) is based on the alignment of reads to the genome sequence. Only reads that aligned perfectly would identify a miRNA gene. As ESVs would be approximately one fifth of all reads aligning, and those not aligning perfectly, they would not interfere with detection of a miRNA gene and it's templated, canonical mature miRNAs (reads aligning perfectly to the template).

2.8. Examining the Biological Effect on Target Gene Specificity from IsomiR Variation

Target gene predictions of isomiRs and their corresponding canonical mature miRNAs was carried out using the target gene prediction software RNAhybrid version 2.2 [35]. The mature miRNAs (canonical and isomiR variants) were tested against 3'UTRs from all Atlantic salmon mRNA transcripts in the Refseq database of Genbank (https://www.ncbi.nlm.nih.gov/refseq/). The analysis was performed applying the following conditions: Helix constraint 2–8 and no G:U in seed. This allowed only target genes with perfect seed complementarity to be detected. Minimum free energy threshold for RNA hybrids was set to -18 kcal/mol to retrieve results (target site matches) from RNA hybrids with high stability.

2.9. Identification of Tissue Enriched miRNAs

The high-quality reads from 23 tissue samples (samples 6, 7, 97, 98, 100, 102, 104, 106–111 (Table S1) and the 10 samples sequenced in a previous study [9]) were used to provide rough estimations of the expression of individual miRNAs across the different tissues. Each mature miRNA was counted using the reference miRNAome and STAR aligner (version 2.5.2b) [36]. The number of miRNAs in each sample was normalized by reads per million scaling factor (RPM) [37]. In cases where there were biological replicates of same tissue (liver, spleen, kidney, head-kidney and intestine) we used the average RPM values for these tissues. The normalized read counts of individual miRNAs within each tissue were compared to identify miRNAs that were highly expressed. The fifteen most abundant miRNAs in each of the tissues, a total of 43 miRNAs, were identified in this manner. The RPM values from these 43 miRNAs were then compared to identify those miRNAs highly expressed across all tissues (three-fold or less difference) and those highly expressed in one or a few particular tissues (more than ten-fold increase in one or a few tissues). Additional tissue enriched miRNAs that were not among the fifteen most abundant miRNAs from a tissue, were identified in the same manner (RPM comparison). These were also analyzed by RT-qPCR to validate they were enriched in particular tissues (see Section 2.9). Those that showed more than ten times higher expression in a particular tissue (measured by RPM and RT-qPCR or RPM only) was termed highly expressed or tissue enriched.

2.10. RT-qPCR Analysis of Tissue Enriched miRNAs

Nineteen miRNAs (ssa-miR-9a/b-5p, ssa-miR-153a-3p, ssa-miR-122-5p, ssa-miR-499a-5p, ssa-miR-194b-5p and ssa-miR-192a-5p, ssa-miR-96, ssa-miR-129, ssa-miR-132, ssa-miR-135c, ssa-miR-212, ssa-miR-219, ssa-miR-723, ssa-miR-734a, ssa-miR-8163, ssa-miR-736, ssa-miR-459 and ssa-miR-140) were analyzed by RT-qPCR as their RPM values indicated high enrichment in one or few particular tissues. A summary of the samples used for verification of tissue enriched miRNA expression by RT-qPCR, a total of 24 samples, is given in Table S2.

First strand cDNA synthesis was performed using the miScript II RT kit (Qiagen, Hilden, Germany) according to the manufacturer's instructions. miRNAs were then detected using the miScript SYBR® Green PCR kit (Qiagen, Hilden, Germany) as described by the manufacturer. The qPCR assays were carried out by means of custom designed forward primers together with a universal reverse primer provided with the miScript qPCR kit. An overview of the primers specific to each mature miRNA is given in Table S3. The qPCR reactions were performed on Mx3000p qPCR system (Stratagene, Agilent Technologies, LA Jolla, CA, USA) using 96-well plates, with thermocycling conditions of 95 °C for 15 min, followed by 40 cycles of 94 °C for 15 s, 55 °C for 30 s and 70 °C for 30 s. Cq values were obtained using the MX3000p software package (Stratagene, Agilent Technologies, USA). The relative expression of each miRNA was normalized against miR-25-3p and miR-107-3p, based on their good stability across tissues [10]. The mean normalized Cq values for each of the different tissues was calculated. The relative difference in expression between tissues was calculated using the ΔΔCt-method [38]. Statistical significance of the observed relative difference in expression was tested using student's t-test and significance levels corresponding to P-values ≤ 0.05 that were set after Bonferroni-correction based on number of tests.

3. Results

3.1. Small RNA Sequencing and Identification of Atlantic Salmon miRNAs

3.1.1. Generation of RNA Libraries and Results from Small RNA Sequencing

Small RNA libraries were successfully generated for 111 samples from salmon fry (n = 24), pathogen challenged fry (n = 72), tissue samples (intestine, gills, gonads, head-kidney and mid-kidney) from fully developed adults (n = 9), and samples from different developmental stages; embryos sampled at 4, 19, 39, 50 dpf, eyed-egg 63dpf, and an alevin 1dph. After pre-processing the raw reads (see methods), there were a total of 656,748,326 high quality, adapter trimmed and size filtered reads from fry samples and 46,047,362 reads from the different tissues and developmental stage samples. A detailed overview of sample origin, RNA concentration, quality and read numbers are given in Table S1.

3.1.2. Results from Discovery and Characterization of Atlantic Salmon miRNAs

The miRdeep2 algorithm has been shown to be a sensitive and reliable method for identifying miRNAs in different species [3,4]. Here it was successfully used to analyze the processed high-quality reads from each of the 111 samples separately. The results from the miRdeep2 analysis was 941 predicted miRNA precursors with their corresponding 5p and 3p mature reads. These were further processed as described in Figure 1 (see Section 2.4). Out of the 371 different Atlantic salmon miRNA genes already annotated and deposited in the database [9], all but one of each of the miRNA paralogs ssa-mir-210-1 and ssa-mir-29b-4 were re-discovered. In addition, new identical paralogs (i.e., identical precursor miRNA sequences at different genome locations) and new miRNA paralogs with small sequence differences, altogether a total of 533, were discovered. The re-discovery of these miRNAs in this new material provided additional confidence that they are true Atlantic salmon miRNAs. Another 20 miRNAs were identified as orthologues of miRNA genes in other teleosts and annotated as new Atlantic salmon miRNA genes in accordance with miRBase guidelines. Three of these miRNA genes had two copies (paralogs) in Atlantic salmon. We thus identified 556 evolutionary conserved miRNA genes with their corresponding mature 5p and 3p miRNAs (Table S4).

The remaining precursor candidates retrieved from the miRdeep2 analysis showed very low sequence similarity to any of the miRNA genes in miRBase. Seventy-nine of these candidates provided significant matches (e-value $\leq 1 \times 10^{-7}$) to other kinds of small RNA (e.g., rRNA, snoRNAs), and were removed. Hundred and twenty sequences provided multiple hits against the salmon genome sequence indicating they were different kinds of repetitive sequences rather than miRNA genes, and thus removed.

Finally, we applied the miRBase guidelines to identify high confidence novel miRNA genes among the remaining precursors retrieved from the miRdeep2 analysis [26]. Following these guidelines (see Section 2.4), 17 novel miRNA genes were identified. The precursor sequences from all novel miRNAs with their corresponding mature 5p and 3p sequences, annotation of arm dominance and the genome location of each precursor is given in Table 1. One of the novel miRNA genes revealed a perfect match in miRBase, (e-value $\leq 1 \times 10^{-18}$) to a salmon fluke (*Gyrodactylus salaris*) miRNA (gsa-mir-9404). However, this precursor sequence aligned perfectly (100% identity) to the Atlantic salmon genome, while there were no matches to the *Gyrodactylus salaris* genome (GenBank accession: GCA_000715275.1) (e-value > 0.5). This one was also identified as an Atlantic salmon miRNA in Bekaert et al. (miR-new156-5p) [10]. This strongly indicates that this is an Atlantic salmon miRNA rather than originating from *G. salaris*. This miRNA was annotated as ssa-miR-novel-17 in our new reference miRNAome. One novel precursor (ssa-miR-novel-15) had several identical copies clustered at two unique genomic locations, and two novels (ssa-miR-novel-6 and ssa-miR-novel-7) had two and three identical copies respectively, while the other 14 novel miRNA genes were present as single copies.

Table 1. Novel Atlantic salmon miRNAs identified in this study.

miRNA ID [1]	Mature-5p (5′-3′) [2]	Mature-3p (5′-3′) [2]	Precursor Sequence (5′-3′)
ssa-mir-novel-1	UCAGUGAUGUGU ACGCCAAAGGU	UCGGCAUACACA UCACUGACA	UCAGUGAUGUGUACGCCAAAGGUGUAAAGCU UCAAGUUCCUCGGCAUACACAUCACUGACA
ssa-mir-novel-2	AGUUUCCCGGAC ACAGAUUAAGCC	UUUUGUCUGUC UGGGAAACCGG	AGUUUCCCGGACACAGAUUAAGCCUAGUCAU AAUUAUUAUGUUUUGUCUGUCUGGGAAACCGG
ssa-mir-novel-3	UGACGAUACCUU UGGAACAAGA	UUGUACCAAUAGU AAAGUCUGA	UGACGAUACCUUUGGAACAAGAGGUGAAUUAC GUCUUAUGCUCUUGUACCAAUAGUAAAGUCUGA
ssa-mir-novel-4	CGGAUCGCUGCG UUCACCAUU	AUGGUGAAUGCAAC GAUAAGGC	CGGAUCGCUGCGUUCACCAUUAUAUUUAACU UCAACAGAAUGGUGAAUGCAACGAUAAGGC
ssa-mir-novel-5	UACGGUAUGUA CUGUAGGCUAC	UAGGCUACGGUA UGUACUGAAG	UACGGUAUGUACUGUAGGCUACGGUAUGUUA UGUACUGUAGGCUACGGUAUGUACUGAAG
ssa-mir-novel-6	UGAGCCUUGUC CUGGACUAAGA	UCAGUCCAUGAC UAGGCUUAAC	UGAGCCUUGUCCUGGACUAAGAAGUACUUCCA AUGGCUAUUUUCAGUCCAUGACUAGGCUUAAC
ssa-mir-novel-7	UUGCUGGUGA CACUGUCUGUGA	AAGGCACACUUC ACCAGUAUGG	UUGCUGGUGACACUGUCUGUGAUUUAUUUAG AAUUCAAGGCACACUUCACCAGUAUGG
ssa-mir-novel-8	AGACACCUGA CACAGCCCCCAUU	UGGGUCUGUGU CUAUUGUCUCU	AGACACCUGACACAGCCCCCAUUCUAUCUCA UAAAAGUGGGUCUGUGUCUAUUGUCUCU
ssa-mir-novel-9	UAGGCGUGUC ACUGCGUGUCACA	UGCGCACGGGG CCACGCUCUGC	UAGGCGUGUCACUGCGUGUCACAGUCACUG CUUGCGCACGGGGCCACGCUCUGC
ssa-mir-novel-10	AGGUCUGUUU GUGCUGUCUUCC	GUGACUGCACA AACGGAUCUGG	AGGUCUGUUUGUGCUGUCUUCCCAUGGCUUU GGUGACUGCACAAACGGAUCUGG
ssa-mir-novel-11	AUUGUUCAG GGCAUUCAUUUCU	UAAGUGAACC CUUGAGACAAUU	AUUGUUCAGGGCAUUCAUUUCUUGUGAACC AAUCAAUAAGUGAACCCUUGAGACAAUU
ssa-mir-novel-12	UUCGCCCCU GAGGACACACGGU	CCGAAUCCACA GAAGUGAUGC	UUCGCCCCUGAGGACACACGGUUUUUCUU UUAAUAGCACCGAAUCCACAGAAGUGAUGC
ssa-mir-novel-13	CCUUGACCA CGUAACCUGACCA	UUAGGUCAGAU GUGGUCAGGAGA	CCUUGACCACGUAACCUGACCAUAGUUUUC UUGGUUAGGUCAGAUGUGGUCAGGAGA
ssa-mir-novel-14	GGGAAUAUA CAUGACUGUGAUU	UCACAGUCGUG UAUAUUCCCUC	GGGAAUAUACAUGACUGUGAUUAUGAUUGA AGAGAAAAUACACAGUCGUGUAUAUUCCCUC
ssa-mir-novel-15	CAGAGCUCU GCUAUCUGCUGUCU	AAGGAGAAAA CAGAGCUCUGCU	CAGAGCUCUGCUAUCUGCUGUCUGUAUCUU GUUAAAGGGGAAGGAGAAAACAGAGCUCUGCU
ssa-mir-novel-16	UUGCUGUUG ACACUGUCUGUG	UCAAGGCACACU UAACCAGCAUGG	UUGCUGUUGACACUGUCUGUGAUUUAUUUA AGGCACACUUCAAGGCACACUUAACCAGCAUGG
ssa-mir-novel-17	GCGUCUCAG AGGUCAAACACAGU	UGUGUUAGGCC UCCGAGUCUGA	GCGUCUCAGAGGUCAAACACAGUAAGUCA UAUUAAGCUGUGUUAGGCCUCCGAGUCUGA

[1] Temporary annotation for the novel miRNAs identified. These will be renamed by miRBase when uploaded in the miRNA database. [2] The dominant mature miRNA is given in bold. Genomic location of the novel precursors is given in Table S4.

In summary, there were 448 different miRNA genes, including 17 novel miRNAs discovered. Adding the 102 identical paralogs there were 577 miRNA genes. Their locations in the present Atlantic salmon genome is given in supplemental material (Table S4). The new mature Atlantic salmon miRNAome reference sequences originating from these genes were a total of 589 unique 5p and 3p mature miRNAs. These are given in FASTA format to be used as a reference miRNAome in miRNA expression studies in supplemental file S5.

miRNA genes are often located in clusters [9,39–41] that may be co-transcribed as a single pri-miRNA [2]. Furthermore, miRNA genes located in the same cluster have been shown to work together to control the same gene pathways when regulating various cellular processes [42–44]. With the discovery of more than one hundred new miRNAs and new genome annotation of all miRNA genes, the amount of clustered miRNA genes was re-examined. miRNA gene clusters were defined as suggested by miRBase as two or more miRNA genes located less than 10 kb from each other in the same direction (on the same genomic strand) [45]. Applying this definition, we identified 235 miRNA genes that were grouped into 93 distinct clusters. These clusters account for 40% of all Atlantic salmon miRNA genes. All gene clusters, the miRNA genes in the cluster and their genomic locations are listed in Table S6.

Only six gene clusters were present in single copies (1–6, Table S6). The remaining clusters could be further subdivided into groups where each gene cluster within a group were paralogous copies. There were 23 such groups, including one miRNA gene cluster with a novel miRNA gene (ssa-mir-novel-15, group XXIII, Table S6). In the previous discovery of Atlantic salmon miRNA genes, Andreassen et al. reported a total of 84 miRNA gene clusters [9]. All these, as well as nine additional clusters were revealed in the new genome sequence. A simple comparison of all groups of Atlantic salmon miRNA gene clusters to the orthologue miRNA gene clusters discovered in zebrafish and Atlantic cod (miRBase release 22) showed that 86% of these gene clusters were present in Atlantic cod with at least one copy, while 90% of these were present in zebrafish with at least one copy. Those present in zebrafish and Atlantic cod are also shown in Table S6. The large percentage of orthologous gene clusters in these three species indicates that the majority of gene clusters are evolutionarily conserved in teleosts. In general, most of the evolutionarily conserved miRNA gene clusters showed a higher copy number in Atlantic salmon compared to zebrafish and Atlantic cod. This is in agreement with findings that miRNA genes have been retained in the Atlantic salmon genome in the process of re-diploidization from the evolutionary recent salmonid specific genome duplication [9,46].

3.2. Characterization of IsomiRs and Polymorphics miRNAs in Atlantic Salmon

3.2.1. Sequence Errors Arising in the Sequencing Pipeline

The characterization of isomiRs requires that the proportion of reads with ESVs that arise in the sequencing pipeline is measured. A threshold that assures that ESVs are not incorrectly reported as isomiRs can then be set. Alignment of reads to the references (see Section 2.6) allowed us to examine the number of sequence errors arising in all parts of the sequencing pipeline. Figure 2 illustrates the results from the alignment of reads against a small part (bp 150–170) of the reference 18S rRNA.

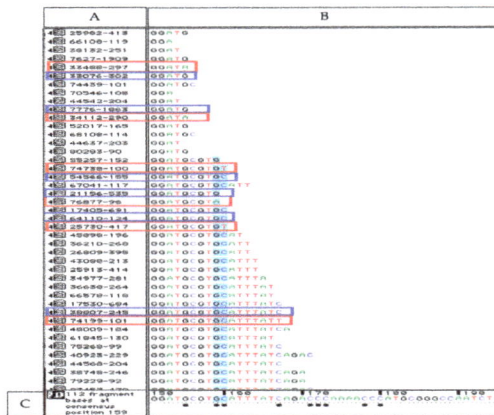

Figure 2. Illustration of aligned reads to the reference (18S rRNA) applying Sequencher software. (**A**) This window shows each collapsed read identified by a unique number followed by the count numbers of this particular read. (**B**) The sequence of each collapsed read is given in this window (**C**) The sequence at the bottom is the reference sequence (18S rRNA). Bullets below bases of the reference indicate that some of the reads differ from the reference at these basepair positions (erroneous sequence variants, ESVs). All collapsed reads shown in red boxes have 3′ ESVs. The collapsed reads shown in blue boxes are reads with correct bp in their 3′ end position. The ratio of reads with 3′ ESVs compared to reads with correct 3′ end bp's in this short part of the sequence (bp 150–bp 170) was 0.10. The other reads below 3′ ESV were all identical to reference at these base positions (e.g., bases in blue color) verifying that the misaligned bases are 3′ ESVs not polymorphic variants.

This revealed the amount of ESVs, as well as the position of the ESVs within reads. The ratio of ESVs distributed across any position within the reads were 0.0004. This was approximately as expected from the Phred quality of reads as a Phred score of 32 means there are less than 0.00063 bp errors due to sequencing alone. However, the alignments revealed that there was a strong bias in ESV position across reads. Almost all ESVs were positioned in the 3' end (independent of read length). The average ratio of ESVs in these positions (last bases at 3' end) was 0.21. The most 5' bases also revealed a higher ratio of ESVs (0.02), but far from the ESV frequency observed at the 3' end. This showed that other steps in the sequencing pipeline than the sequencing itself, e.g., the cDNA synthesis [19], generated a large amount of ESVs. If not accounted for, these ESVs could be misinterpreted as 3' or 5' isomiR variants. Alignment of reads to the two controls also showed that all different sizes of reads (18–25 nts) identical to the references were present in large numbers in the data sets. This showed, as expected, that any characterization of templated size variants (e.g., from imprecise Dicer processing) of canonical miRNAs could not be carried out as they could not be distinguished from those reads with small size variations that were just products of the pipeline itself (e.g., degradation).

3.2.2. IsomiR Characterization

After a series of filtering steps and accounting for the ratio of ESVs (see Section 3.2.1), we identified 41 isomiR variants derived from 37 mature miRNAs, including four isomiR variants of novel miRNAs. Thirty-two of these were non-templated 3' isomiRs, while eight were non-templated isomiR length variants (3' length isomiRs). One non-templated 5' isomiR was also identified (Table S7). The most common variants observed were, thus, the 3' end variants (98%). Analysis of the ratio of isomiR variants vs. their canonical forms showed, in general, the predominance of the canonical mature miRNAs (Table S7). However, ten isomiR variants that were more abundant than their respective canonical forms were also observed. The most prominent change of non-templated nucleotides was from cytosine (C) to uracil (U). All eight of the 3' length isomiRs had uracil (U) added in their 3' ends. Such uridylation is a common RNA modification found for different RNA species, including miRNAs [47,48]. These findings indicated that the editing of mature miRNAs was not random.

The biological significance of these isomiRs was further investigated by predicting the targets of canonical miRNAs and their isomiR forms using RNAhybrid [31]. All the thirty-seven mature miRNAs and their isomiRs were analyzed against the 3'UTRs of Atlantic salmon mRNAs (Refseq, Genbank). The results (Figure 3 and Table S8) showed that the 37 canonical miRNAs together could putatively target 3916 transcripts.

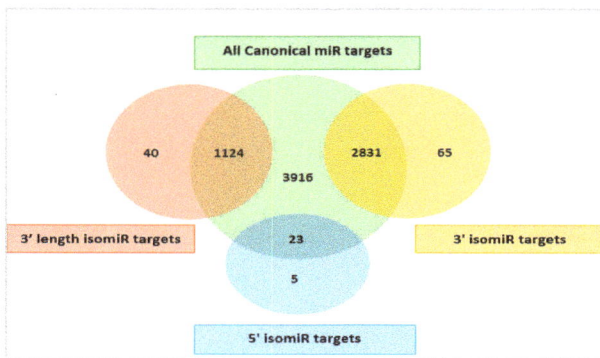

Figure 3. A Venn diagram depicting the overlap and difference between the predicted target genes of canonical miRNAs (green color), 3' isomiRs (yellow color), 3' length isomiRs (red color), and 5' isomiRs (blue color).

The 3′ isomiRs (yellow color, Figure 3) were predicted to share 2831 common targets with their canonical miRNAs. The 3′ length isomiRs (red color, Figure 3) were predicted to share 1124 targets with their canonical miRNAs, while the 5′ isomiRs (blue color, Figure 3) were predicted to share 23 targets with their canonical miRNAs. Furthermore, the 3′ isomiRs were also predicted to target additional 65 transcripts, the 3′ length isomiRs could target 40 additional transcripts, while the 5′ isomiRs could target 5 additional transcripts not predicted as targets of their canonical forms. These comparisons showed that the isomiR variants did not lead to a large increase in putative target transcripts. Only 110 new transcripts (<3%) were added as targets when including the additional transcripts targeted by all isomiR variants. Thus, the biological effect, measured as the increase in the number of targets was small.

3.2.3. Polymorphic Mature miRNAs

Five polymorphic miRNAs (allelic variants) were also identified (Table 2). In all cases, the proportion of reads with the new variant was larger than 40%. In one of the variants (ssa-miR-100a-2-3p), the polymorphism (T to C transition) was in the seed sequence. The biological significance of the variants was, similarly as with isomiRs, investigated by prediction of putative targets of both the reference and the new variants. Figure 4 and Table S9 show the results from this analysis.

Table 2. Polymorphic variants of canonical mature miRNAs.

miRNA	Reference Sequence [1]	Variant Sequence [2]
ssa-miR-16a-1-3p	CCAGTATTGTTCGTGCTGCTGA	CCAGTATTG**C**TCGTGCTGCTGA
ssa-miR-100a-2-3p	ACAAGCTTGTGTCTATAGGTATG	ACAAGCT**C**GTGTCTATAGGTATG
ssa-miR-2188-3p	GCTGTGTGAGGTCAGACCTATC	GCTGTGTGAGGTC**G**GACCTATC
ssa-miR-29b-1-5p	ACTGATTTCTTCTGGTGTTTAGA	ACTGATTTC**C**TCTGGTGTTTAGA
ssa-let-7a-2-3p	CTATACAACTTACTGTCTTTCC	CTATACAAC**A**TACTGTCTTTCC

[1] The reference sequence of mature miRNA. [2] Variant sequence with the polymorphic base given in bold. The most common variant is underlined.

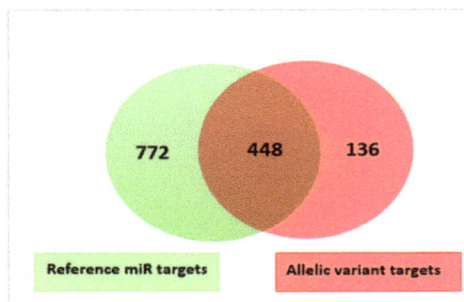

Figure 4. A Venn diagram illustrating the overlap and difference between the predicted target genes of the five reference mature miRNAs (green color) and the new mature miRNA variants identified in this study (light red color).

As shown in Figure 4, the reference mature miRNAs were predicted to target 722 transcripts. The new variants and their reference miRNAs were predicted to share 488 common targets. One hundred-thirty-six transcripts were predicted to be targeted by the new variants (increase of more than 15%) (Table S9a). The effect of allelic variation on target diversity was, thus, larger than the effect observed for isomiRs. However, this was mostly caused by the one variant where the polymorphic site was within the seed. This ssa-miR-100a-2-3p seed variant contributed 81 (60%) of all the additional targets (Table S9b). The reference mature and the new seed variant of the ssa-miR-100a-2-3p were in this case predicted to share only six common target transcripts, but mostly at different target sites on the same transcript. This indicated that a single base change in seed can significantly affect target gene specificity.

3.3. Characterization of miRNA Expression Profiles in Different Tissues and Developmental Stages

3.3.1. Housekeeping miRNAs vs. miRNAs Predominantly Expressed in Particular Tissues

Several studies have demonstrated that miRNAs play an important role in tissue development and/or in the maintenance of tissue specific functions [5,40,49–54]. Such miRNAs are often highly expressed in one or a few tissues. On the other hand, there are miRNAs with ubiquitous high expression across most tissues assumed to maintain housekeeping functions [54]. To identify miRNAs that are likely to have such housekeeping functions the fifteen most abundant miRNAs in each of the tissues brain, liver, heart, head-kidney, muscle, intestine, kidney, spleen, gonads and gills were revealed. Together, there were 43 such miRNAs. Twenty-one of these did not reveal any large expression differences when comparing across tissues (three-fold or less). Further examining the enrichment patterns of these 21 miRNAs showed that seven of these miRNAs were among the top ten most abundant miRNAs in all tissues. These were miR-143-3p, miR-181a-3p, miR-21b-5p, mir-26a-5p, miR-10b-5p, mir-10d-5p and mir-10a-5p (Table S10). Together, these seven miRNAs accounted for more than 30% of all miRNAs expressed in any tissue. Their ubiquitous nature and high expression in many or all tissues suggest that these miRNAs are constitutively expressed and have housekeeping functions.

Other miRNAs showed a predominant expression in particular tissues. Although they were among the 15 most highly expressed miRNAs in one tissue, they showed very little or no expression in the other tissues. All of these showed 10–100 times higher expression in one or few particular tissues. These are shown in the heat map in Figure 5.

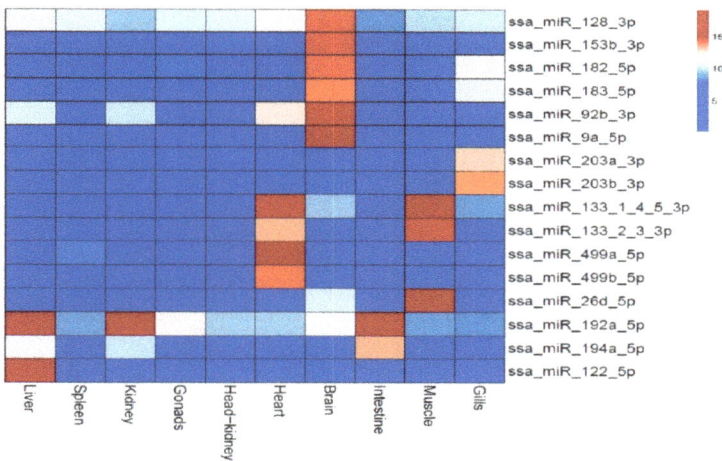

Figure 5. The heatmap illustrates the predominant expression of 16 mature miRNAs in particular tissues. The tissues compared are given at the bottom of the columns, while the miRNAs are given in the horizontal rows to the right. Four of these miRNAs (ssa-miR-9a-5p, ssa-miR-499a-5p, ssa-miR-192a-5p and ssa-miR-122-5p) were also analyzed by RT-qPCR (see Table 2). The expression values (log2) of the miRNAs are depicted in the color scale. Very little or no expression is shown in dark blue, while other colors indicate increased expression (10–100 times) with dark red color as those with the higher expression.

Six of these miRNAs (ssa-miR-128-3p, ssa-miR-153b-3p, ssa-miR-182-5p, ssa-miR-183-5p, ssa-miR-92b-3p and ssa-miR-9a-5p), were predominantly expressed in brain. Two miRNAs from the miR-499-family (ssa-miR-499a/b-5p) showed about 70 times higher expression in cardiac tissue, while all 3p mature miRNAs in the miR-133 family (ssa-miR-133-1-4-5/2-3-3p) showed higher expression in both muscle and cardiac tissue. Another miRNA that was predominantly expressed in muscle tissue was ssa-miR-26d-5p (about 20 times increase). Two miRNAs, ssa-miR-203a/b-3p, were highly expressed in gills. Ssa-miR-192a-5p was predominantly expressed in intestine, but this miRNA also showed an eight to twenty times higher expression in kidney and liver than in all other tissues indicating it was serving a function common to a cell type present in these three tissues. One more miRNA, ssa-miR-122-5p also showed a predominant expression in the liver.

Another 15 miRNAs showed lower abundance (i.e., not among the 15 most abundant miRNAs in one or more tissues), but were still enriched in specific tissues. Additional measurements by RT-qPCR were used to show that there was a significantly higher expression ($>10\times$ higher) of these in the particular tissues (p-adjusted ≤ 0.05). The results from RT-qPCR analysis of all these miRNAs (Table 3) agreed with the patterns revealed by the RPM comparisons.

Table 3. Results from RT-qPCR analysis.

miRNA	Tissue [1]	$\Delta\Delta$CT [2]	Enrichment [3]
ssa-miR-9a-5p [4]	B	−11.13	2241
ssa-miR-9b-3p	B	−10.21	1184
ssa-miR-96-5p	B	−5.18	36
ssa-miR-129-5p	B	−3.56	12
ssa-miR-132-5p	B	−7.46	176
ssa-miR-135c-5p	B	−7.05	133
ssa-miR-153a-3p	B	−10.08	1082
ssa-miR-212ab-3p	B	−7.29	156
ssa-miR-219a-3p	B	−7.66	202
ssa-miR-723-5p	B	−6.83	114
ssa-miR-734a-3p	B	−4.68	26
ssa-miR-122-5p [4]	L	−12.3	5043
ssa-miR-8163-3p	L	−8.6	388
ssa-miR-192a-5p [4]	L	−7.96	249
ssa-miR-499a-5p [4]	H	−9.7	832
ssa-miR-736-3p	H	−14.7	26616
ssa-miR-192a-5p [4]	I	−11.9	3822
ssa-miR-459-5p	I	−12.9	7643
ssa-miR-194b-5p	I	−10.8	1783
ssa-miR-140-3p	G	−3.7	13

[1] Tissue where the particular mature miRNA is highly expressed (B = brain, L = liver, H = heart, I = intestine, G = gills, M = muscle). [2] Logfold difference in highly expressed tissue relative to mean of all other. [3] Times increase in the highly enriched tissue. [4] These four miRNAs are also among those shown in Figure 4. All differences were significant at adjusted p-values ≤ 0.05.

A complete overview of all 31 miRNAs that showed a tissue specific expression pattern is given in Table 4. Brain tissue was the one tissue showing largest number of miRNAs expressed in a tissue specific manner, both among those miRNAs with highly enriched expression (Figure 4), as well as those investigated by RT-qPCR (16 miRNAs, Table 4).

Table 4. miRNAs that show tissue enriched expression patterns.

Gills	Muscle	Intestine	Brain	Heart	Kidney	Liver
ssa-miR-140-3p	ssa-miR-133-1/2-3p	ssa-miR-192a-5p	ssa-miR-128-3p	ssa-miR-133-1/2-3p	ssa-miR-192a-5p	ssa-miR-122-5p
ssa-miR-203a/b-3p	ssa-miR-26d-5p	ssa-miR-194a/b-5p	ssa-miR-129-5p	ssa-miR-499a/b-5p		ssa-miR-192a-5p
		ssa-miR-459-5p	ssa-miR-132-5p	ssa-miR-736-3p		ssa-miR-8163-3p
			ssa-miR-135c-5p			
			ssa-miR-153a/b-3p			
			ssa-miR-182-5p			
			ssa-miR-183-5p			
			ssa-miR-212ab-3p			
			ssa-miR-219a-3p			
			ssa-miR-723-5p			
			ssa-miR-734a-3p			
			ssa-miR-92b-3p			
			ssa-miR-96-5p			
			ssa-miR-9a-5p			
			ssa-miR-9b-3p			

Overview of all mature miRNAs enriched in particular tissues. Different family members that are enriched in the same tissue are indicated by a slash (e.g., ssa-miR-449a/b-5p). Identical family members are shown together as given in the reference miRNAome (Supplementary file S5) (e.g., ssa-miR-212a-3p and ssa-miR-212b-3p are shown as ssa-miR-212ab-3p).

3.3.2. Expression Patterns of miRNAs at Development Specific Stages

It has been reported that miRNAs show developmental stage-specific abundance during the embryonic development of teleosts [32,37]. The miRNA diversity showed, in general, an increase from less than one hundred to close to 600 different miRNAs during development. Embryo 4dpf, corresponding to the earliest developmental stage in our materials, showed the smallest diversity with only 63 different miRNAs expressed (Figure 6). There was a very large increase in the number of miRNAs that were expressed during the next 15 days as the embryo 19dpf sample revealed there were 318 different miRNAs expressed. The number of miRNAs detected increased by another hundred the next 20 days (embryo 39dpf), and at the eyed-egg stage (63dpf) it was at its maximum (585 different miRNAs).

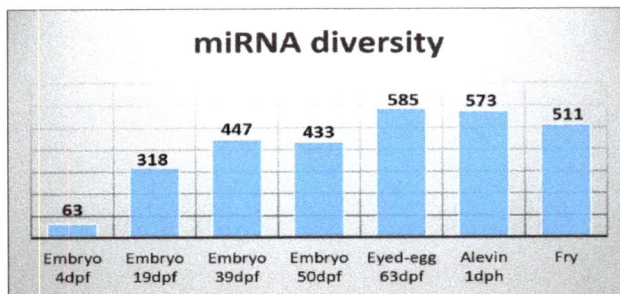

Figure 6. miRNA diversity measured as a number of different miRNAs detected across developmental stages.

Due to the limited number of samples and lack of proper normalization, we could not compare expression differences between the developmental stages. We could however compare the expression differences of miRNAs within each of the developmental stages. A few miRNAs showed high expression at specific stages, while others exhibited a ubiquitous expression pattern and were highly enriched in all the developmental stages. This was the case for the three members of the mir-10 family (ssa-miR-10b-5p, ssa-miR-10d-5p and ssa-miR-10a-5p) that constituted a large proportion (>50%) of all miRNAs at most developmental stages (Figure 7A–F), while their abundance was very low in fry (1% of all miRNAs) (Figure 8). The proportion of miRNAs from the ssa-miR-430 cluster was relatively large at the earliest stages of development (embryo 4dpf, Figure 7A), but appeared to decline rapidly as the proportion was 3% at 19dpf and less than 1% at all other stages (Figure 7B–F and Figure 8).

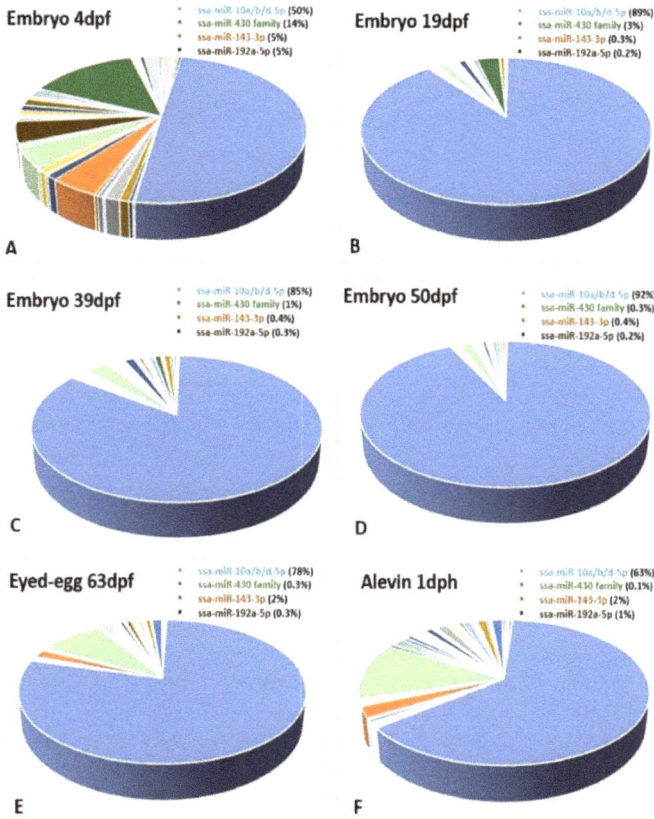

Figure 7. Abundance as a proportion of the total number of miRNAs within each of the developmental stages. The ssa-miR-10a/b/d-5p family, the miR-430 family, ssa-miR-143-3p and ssa-miR-192a-5p are shown in blue, green, orange and brown colors, respectively, in the figures (**A–F**).

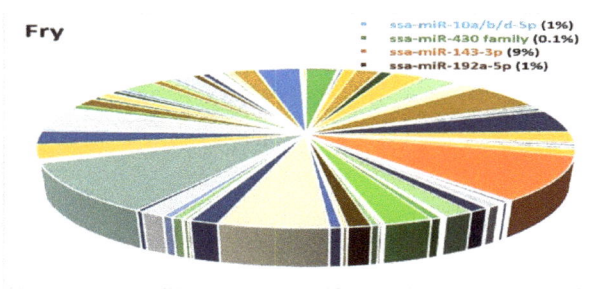

Figure 8. A pie chart demonstrating the miRNA expression diversity (511 miRNAs) in fry. The ssa-miR-10a/b/d-5p family, the miR-430 family, ssa-miR-143-3p and ssa-miR-192a-5p are shown in blue, green, orange and brown colors, respectively, in Figure 8.

The proportion of ssa-miR-143-3p and ssa-miR-192 was about 5% in embryo 4dpf (Figure 7A), but the proportion of these miRNAs declined about a tenfold in the following stages (19dpf to eyed-egg 63dpf, Figure 7B–E). These two miRNAs seemed to increase their expression post hatching (Figure 8).

4. Discussion

4.1. Small RNA Sequencing and Identification of Atlantic Salmon miRNAs

In this study, we performed a comprehensive analysis for miRNA characterization and identified 448 different miRNA genes in Atlantic salmon, including 17 novel miRNAs. Although a high number of Atlantic salmon miRNAs had been identified in 2013, nearly one hundred new miRNA genes, both miRNAs conserved in teleosts, as well as novel miRNAs, were discovered in this study. The increase in number of miRNA genes discovered could largely be attributed to the large number of materials included (111 samples from different tissues, developmental stages, fry and pathogen challenged fry). This miRNAome will be the new, improved reference to apply when investigating differential miRNA expression in Atlantic salmon. Furthermore, the genomic locations of all miRNA genes and their clustering patterns annotated in the updated Atlantic salmon genome sequence will facilitate further studies of the comparative evolution of miRNA genes.

4.2. Characterization of IsomiRs and Polymorphics miRNAs in Atlantic Salmon

With the advancement of high throughput sequencing techniques, many recent studies have reported the presence of a number of mature miRNA sequence variants with different 5′ and/or 3′ ends compared to their corresponding canonical mature miRNAs termed as isomiRs. Small-RNA sequencing projects generate datasets consisting of millions of reads differing in length and quality. Before characterizing isomiRs one need to control size variations and bp errors that arise in the pipeline. The read alignments to the larger sized quality control references indicated that most of the size variations (variation in read length) were artificial and generated in the RNA extraction or the first steps of the sequencing pipeline. There may be some templated isomiR length variants in Atlantic salmon, but due to the platform related high proportion of read length variation, they could not be detected. While there are some studies that report that they have accounted for sequence errors when characterizing isomiRs, many either have used Phred quality score estimates as an error threshold or spike-in controls to measure the sequence errors [15,55]. Despite the fact that Phred score is a good measure of sequence quality, such estimates will only account for the errors caused by the sequencing itself, not other sources of sequence errors that could be generated at other steps of the pipeline (e.g., cDNA synthesis). A high frequency of site-specific bias (especially at the 3′ ends), that would otherwise not be identified by a Phred quality score, was revealed when we estimated the average ratio of ESVs in our data (see Section 3.2.1). As isomiRs are mature miRNA sequence variants with non-templated nucleotide differences in the 5′ or 3′ ends, a large proportion of the erroneous sequence variants revealed could potentially be mistaken for isomiRs. This illustrates the importance of incorporating measurements of error beyond Phred quality to distinguish ESVs from isomiRs. The proportion of ESVs may, however, differ between library preparation methods and sequencing platforms. Nevertheless, controlling the error rate seems crucial.

Most isomiRs identified in our study were the non-templated 3′ isomiRs in the form of nucleotide substitution and/or nucleotide addition. This result is not surprising, as the 3′ isomiR variants are the most common isomiR variants observed in animals and plants [19]. As shown in Figure 3, the isomiRs variants were predicted to cause only very small changes in the number of targeted transcripts. Our findings are consistent with those reported in other studies, as targeting is mainly mediated through complementary binding of the 5′ seed (2–8 nts of mature miRNA sequence) [17,19]. Moreover, the total amount of isomiR variants was small (43 isomiRs). Together, these findings suggest that these types of modifications may have less biological impact than anticipated [56]. Allelic variation in the seed did however have a major effect on target gene diversity. The target gene analysis of the two allelic variants of miR-100a-2-3p (Table 2) showed that the new variant practically acted as a "new miRNA", with a completely new set of target transcripts. The negative selective pressure against such variation, that could allow miRNAs to adapt new regulatory functions, may be less strict when miRNAs are from families where the other members maintain regulation of their original target genes. As the partially tetraploid Atlantic salmon miRNAome has a much larger number of very similar miRNA genes (paralogs) than other diploid fish (e.g., zebrafish, Atlantic cod), this would allow for WGD-derived paralogs to change in seed and develop new functions.

Although the biological significance of 3′ isomiRs and 3′ length isomiRs seems small, these may still cause methodological issues when analyzing miRNAs by quantitative real time PCR (RT-qPCR) [12,16]. This miRNA detection methodology mainly depends on amplification that is initiated with a miRNA-specific primer. A nucleotide difference in the 3′ end (3′ isomiRs and 3′ length variants) could lead to the detection of products with different melting temperatures and thus, affect precise measurements of specific miRNA levels as both the canonical miRNA and the isomiR variant may be cross detected [57]. It is therefore important to be aware of such variants, as they may explain some of the methodological challenges one may come across when measuring miRNA expression by RT-qPCR [16].

4.3. Characterization of miRNA Expression Profiles in Different Tissues and Developmental Stages

Identifying expression patterns of Atlantic salmon miRNAs in different tissues and developmental stages provides important insight into the function of individual miRNAs. The miRNAs discovered in our study showed a wide range of expression profiles in the different tissues and developmental stages. The miRNAs ssa-miR-143-3p, ssa-miR-181a-3p, ssa-miR-21b-5p, ssa-mir-26a-5p, ssa-miR-10b-5p, ssa-mir-10d-5p and ssa-mir-10a-5p were ubiquitously expressed in all tissues tested. The high abundance and ubiquitous expression profile of most of these miRNAs have been reported in several other species [58–61]. The evolutionarily conserved high expression profiles of these miRNAs suggest they are associated with common signaling pathways in vertebrates and have the same housekeeping functions in Atlantic salmon. Other miRNAs showed a tissue specific expression pattern suggesting a specialized role for these miRNAs in tissue differentiation or maintenance of tissue specific functions. The brain-enriched miRNAs, such as the miR-9 family are e.g., known to have important roles in neurogenesis and brain development in other fish and mammals [43,53,54,58]. Also consistent with previous findings in salmon and other teleosts was the tissue specific high enrichment pattern of miR-122-5p in liver [39,51,52,62]. The high expression of miR-192 shown in liver, kidney and intestine tissues have also been reported in other fish species [39,63]. Finally, there was a high expression of the miR-133 family in both muscle and cardiac tissue, whereas the miR-499 family members were only enriched in cardiac tissue. The miR-133 family are amongst others known to regulate cardiomyocyte differentiation and proliferation, cardiac morphogenesis and stress responsive cardiac remodeling process in other species [64,65]. As demonstrated in other species, a majority of the orthologous miRNAs that showed tissue specific expression have specialized functions in different tissues. When revealing same tissue enriched expression in Atlantic salmon it is likely that they also have similar specialized functions in this species.

Several studies have suggested that miRNAs have essential roles in the developmental progression in vertebrates [66,67]. The increase in miRNA diversity along with the different stages of development (Figure 6) indicates that the developmental processes are under miRNA regulation during Atlantic salmon development. The largest change in the proportion of miRNAs within the different developmental stages was observed for the miR-430 family members. This miRNA family were highly abundant during the earliest stages of development, while their abundance decreased rapidly throughout the later stages (Figures 7 and 8). This pattern of expression agreed with those of miR-430 reported in zebrafish. It is assumed that this family of miRNAs are involved in maternal RNA clearance during early embryogenesis [49,68,69]. We also found three members of the miR-10 family that were highly abundant throughout all developmental stages up to one-day post-hatching (Figure 7). Previous studies in other teleosts have identified members of the miR-10 family members as key regulators of *Hox* genes, which are important regulators of embryonic development in vertebrates [50]. The expression of miR-10 family members is, among others, also shown to be associated with mediating cell proliferation and differentiation [70]. The sequences of these Atlantic salmon miRNA family members are highly conserved across species [70]. Together, this implies that they may have similar functions in Atlantic salmon development. The proportion of ssa-miR-143-3p and ssa-miR-192 was higher in embryo 4dpf (Figure 7A) and fry (Figure 8) (Section 3.3) compared to all the other developmental stages. The investigation of tissue specific expression showed that they have very high expression levels in all tissues (ssa-miR-143-3p) or in particular tissues (ssa-miR-192a-5p). In addition, miR-143 accounted for a large proportion (10%) in female gonads, which suggests that these two miRNAs have important functions in adults. Thus, the high abundance of these miRNAs in embryo 4dpf could reflect the maternal contribution rather than a particular function at the earliest stage of development. The rapid disappearance in the later stages could be essential to allow the expression of genes important in development. Further experimental studies are necessary to reveal the particular roles of the miRNAs that showed a specific expression in some tissues and developmental stages.

5. Conclusions

We discovered nearly one hundred new miRNA genes, both miRNAs conserved in teleosts, as well as novel miRNAs, thus, contributing to a major expansion in the number of different miRNAs characterized in Atlantic salmon. The resulting new reference miRNAome provides an important updated resource for miRNA expression studies. Further, a subset of miRNA genes highly abundant in one or more tissues and developmental stages were revealed, suggesting important biological functions of particular miRNAs in the maintenance of tissue specific functions and in the regulation of embryonic development. Together, results from this study provide insight on miRNA regulation that includes those biological processes, and that may be of economic importance to the aquaculture industry.

Supplementary Materials: The supplementary materials are available online at http://www.mdpi.com/2073-4409/8/1/42/s1.

Author Contributions: Conceived and coordinated the study, R.A. and B.H.; Methodology, R.A.; Software, O.A.; Validation, R.A. and N.T.W.; Formal analysis, R.A., O.A and N.T.W.; Investigation, R.A. and N.T.W.; Resources, R.A., R.D.H. and J.B.T.; Data curation, R.A. and N.T.W.; Writing—original draft preparation, N.T.W.; Writing—review and editing, N.T.W., R.A. and B.H.; Visualization, R.A. and N.T.W.; Supervision, R.A. and B.H.; Project administration, R.A.; Funding acquisition, R.A. All authors revised and approved the final draft.

Funding: This research was supported by funding from the Norwegian Research council, grant number 254849/E40.

Conflicts of Interest: The authors declare no conflict of interest.

References

1. Bartel, D.P. MicroRNAs: Genomics, biogenesis, mechanism, and function. *Cell* **2004**, *116*, 281–297. [CrossRef]
2. Ha, M.; Kim, V.N. Regulation of microRNA biogenesis. *Nat. Rev. Mol. Cell Biol.* **2014**, *15*, 509–524. [CrossRef]
3. Friedlander, M.R.; Chen, W.; Adamidi, C.; Maaskola, J.; Einspanier, R.; Knespel, S.; Rajewsky, N. Discovering microRNAs from deep sequencing data using miRDeep. *Nat. Biotechnol.* **2008**, *26*, 407–415. [CrossRef]
4. Friedlander, M.R.; Mackowiak, S.D.; Li, N.; Chen, W.; Rajewsky, N. miRDeep2 accurately identifies known and hundreds of novel microRNA genes in seven animal clades. *Nucleic Acids Res.* **2012**, *40*, 37–52. [CrossRef]
5. Bushati, N.; Cohen, S.M. microRNA functions. *Annu. Rev. Cell Dev. Biol.* **2007**, *23*, 175–205. [CrossRef]
6. Lynam-Lennon, N.; Maher, S.G.; Reynolds, J.V. The roles of microRNA in cancer and apoptosis. *Biol. Rev.* **2009**, *84*, 55–71. [CrossRef]
7. Forster, S.C.; Tate, M.D.; Hertzog, P.J. MicroRNA as Type I Interferon-Regulated Transcripts and Modulators of the Innate Immune Response. *Front. Immunol.* **2015**, *6*, 334. [CrossRef]
8. Sonkoly, E.; Stahle, M.; Pivarcsi, A. MicroRNAs and immunity: Novel players in the regulation of normal immune function and inflammation. *Semin. Cancer Biol.* **2008**, *18*, 131–140. [CrossRef]

9. Andreassen, R.; Worren, M.M.; Hoyheim, B. Discovery and characterization of miRNA genes in Atlantic salmon (*Salmo salar*) by use of a deep sequencing approach. *BMC Genom.* **2013**, *14*, 482. [CrossRef]

10. Bekaert, M.; Lowe, N.R.; Bishop, S.C.; Bron, J.E.; Taggart, J.B.; Houston, R.D. Sequencing and characterisation of an extensive Atlantic salmon (*Salmo salar* L.) microRNA repertoire. *PLoS ONE* **2013**, *8*, e70136. [CrossRef]

11. Andreassen, R.; Woldemariam, N.T.; Egeland, I.O.; Agafonov, O.; Sindre, H.; Hoyheim, B. Identification of differentially expressed Atlantic salmon miRNAs responding to salmonid alphavirus (SAV) infection. *BMC Genom.* **2017**, *18*, 349. [CrossRef]

12. Johansen, I.; Andreassen, R. Validation of miRNA genes suitable as reference genes in qPCR analyses of miRNA gene expression in Atlantic salmon (*Salmo salar*). *BMC Res. Notes* **2014**, *8*, 945. [CrossRef]

13. Valenzuela-Munoz, V.; Novoa, B.; Figueras, A.; Gallardo-Escarate, C. Modulation of Atlantic salmon miRNome response to sea louse infestation. *Dev. Comp. Immunol.* **2017**, *76*, 380–391. [CrossRef]

14. Skaftnesmo, K.O.; Edvardsen, R.B.; Furmanek, T.; Crespo, D.; Andersson, E.; Kleppe, L.; Taranger, G.L.; Bogerd, J.; Schulz, R.W.; Wargelius, A. Integrative testis transcriptome analysis reveals differentially expressed miRNAs and their mRNA targets during early puberty in Atlantic salmon. *BMC Genom.* **2017**, *18*, 801. [CrossRef]

15. Ebhardt, H.A.; Tsang, H.H.; Dai, D.C.; Liu, Y.; Bostan, B.; Fahlman, R.P. Meta-analysis of small RNA-sequencing errors reveals ubiquitous post-transcriptional RNA modifications. *Nucleic Acids Res.* **2009**, *37*, 2461–2470. [CrossRef]

16. Lee, L.W.; Zhang, S.; Etheridge, A.; Ma, L.; Martin, D.; Galas, D.; Wang, K. Complexity of the microRNA repertoire revealed by next-generation sequencing. *RNA* **2010**, *16*, 2170–2180. [CrossRef]

17. Guo, L.; Chen, F. A challenge for miRNA: Multiple isomiRs in miRNAomics. *Gene* **2014**, *544*, 1–7. [CrossRef]

18. Cloonan, N.; Wani, S.; Xu, Q.; Gu, J.; Lea, K.; Heater, S.; Barbacioru, C.; Steptoe, A.L.; Martin, H.C.; Nourbakhsh, E.; et al. MicroRNAs and their isomiRs function cooperatively to target common biological pathways. *Genome Biol.* **2011**, *12*, R126. [CrossRef]

19. Neilsen, C.T.; Goodall, G.J.; Bracken, C.P. IsomiRs-the overlooked repertoire in the dynamic microRNAome. *Trends. Genet.* **2012**, *28*, 544–549. [CrossRef]

20. Lin, W.; Piskol, R.; Tan, M.H.; Li, J.B. Comment on "Widespread RNA and DNA sequence differences in the human transcriptome". *Science* **2012**, *335*, 1302, author reply 1302. [CrossRef]

21. Robasky, K.; Lewis, N.E.; Church, G.M. The role of replicates for error mitigation in next-generation sequencing. *Nat. Rev. Genet.* **2014**, *15*, 56–62. [CrossRef]

22. Lien, S.; Koop, B.F.; Sandve, S.R.; Miller, J.R.; Kent, M.P.; Nome, T.; Hvidsten, T.R.; Leong, J.S.; Minkley, D.R.; Zimin, A.; et al. The Atlantic salmon genome provides insights into rediploidization. *Nature* **2016**, *533*, 200–205. [CrossRef]

23. Robledo, D.; Taggart, J.B.; Ireland, J.H.; McAndrew, B.J.; Starkey, W.G.; Haley, C.S.; Hamilton, A.; Guy, D.R.; Mota-Velasco, J.C.; Gheyas, A.A.; et al. Gene expression comparison of resistant and susceptible Atlantic salmon fry challenged with Infectious Pancreatic Necrosis virus reveals a marked contrast in immune response. *BMC Genom.* **2016**, *17*, 279. [CrossRef]

24. Andrews, S.; Krueger, F.; Seconds-Pichon, A.; Biggins, F.; Wingett, S. *FastQC: A Quality Control Tool for High Throughput Sequence Data*; Babraham Institute: Cambridge, UK, 2012.

25. Martin, M. Cutadapt removes adapter sequences from high-throughput sequencing reads. *EMBnet. J.* **2011**, *17*, 10–12. [CrossRef]

26. Kozomara, A.; Griffiths-Jones, S. miRBase: Annotating high confidence microRNAs using deep sequencing data. *Nucleic Acids Res.* **2014**, *42*, D68–D73. [CrossRef]

27. Ambros, V.; Bartel, B.; Bartel, D.P.; Burge, C.B.; Carrington, J.C.; Chen, X.; Dreyfuss, G.; Eddy, S.R.; Griffiths-Jones, S.; Marshall, M.; et al. A Uniform System for microRNA Annotation. *RNA* **2003**, *9*, 277–279. [CrossRef]

28. Griffiths-Jones, S.; Grocock, R.J.; van Dongen, S.; Bateman, A.; Enright, A.J. miRBase: MicroRNA sequences, targets and gene nomenclature. *Nucleic Acids Res.* **2006**, *34*, D140–D144. [CrossRef]

29. Kalvari, I.; Argasinska, J.; Quinones-Olvera, N.; Nawrocki, E.P.; Rivas, E.; Eddy, S.R.; Bateman, A.; Finn, R.D.; Petrov, A.I. Rfam 13.0: Shifting to a genome-centric resource for non-coding RNA families. *Nucleic Acids Res.* **2018**, *46*, D335–D342. [CrossRef]

30. Mituyama, T.; Yamada, K.; Hattori, E.; Okida, H.; Ono, Y.; Terai, G.; Yoshizawa, A.; Komori, T.; Asai, K. The Functional RNA Database 3.0: Databases to support mining and annotation of functional RNAs. *Nucleic Acids Res.* **2009**, *37*, D89–D92. [CrossRef]

31. Sanger, F.; Air, G.M.; Barrell, B.G.; Brown, N.L.; Coulson, A.R.; Fiddes, J.C.; Hutchison Iii, C.A.; Slocombe, P.M.; Smith, M. Nucleotide sequence of bacteriophage φX174 DNA. *Nature* **1977**, *265*, 687. [CrossRef]

32. Manley, L.J.; Ma, D.; Levine, S.S. Monitoring Error Rates In Illumina Sequencing. *J. Biomol. Tech.* **2016**, *27*, 125–128. [CrossRef]

33. Urgese, G.; Paciello, G.; Acquaviva, A.; Ficarra, E. isomiR-SEA: An RNA-Seq analysis tool for miRNAs/isomiRs expression level profiling and miRNA-mRNA interaction sites evaluation. *BMC Bioinform.* **2016**, *17*, 148. [CrossRef]

34. Schmittgen, T.D.; Lee, E.J.; Jiang, J.; Sarkar, A.; Yang, L.; Elton, T.S.; Chen, C. Real-time PCR quantification of precursor and mature microRNA. *Methods* **2008**, *44*, 31–38. [CrossRef]

35. Rehmsmeier, M.; Steffen, P.; Hochsmann, M.; Giegerich, R. Fast and effective prediction of microRNA/target duplexes. *RNA* **2004**, *10*, 1507–1517. [CrossRef]

36. Dobin, A.; Davis, C.A.; Schlesinger, F.; Drenkow, J.; Zaleski, C.; Jha, S.; Batut, P.; Chaisson, M.; Gingeras, T.R. STAR: Ultrafast universal RNA-seq aligner. *Bioinformatics* **2013**, *29*, 15–21. [CrossRef]

37. Stokowy, T.; Eszlinger, M.; Swierniak, M.; Fujarewicz, K.; Jarzab, B.; Paschke, R.; Krohn, K. Analysis options for high-throughput sequencing in miRNA expression profiling. *BMC Res. Notes* **2014**, *7*, 144. [CrossRef]

38. Schmittgen, T.D.; Livak, K.J. Analyzing real-time PCR data by the comparative C(T) method. *Nat. Protoc* **2008**, *3*, 1101–1108. [CrossRef]

39. Andreassen, R.; Rangnes, F.; Sivertsen, M.; Chiang, M.; Tran, M.; Worren, M.M. Discovery of miRNAs and their corresponding miRNA genes in Atlantic Cod (*Gadus morhua*): Use of stable miRNAs as reference genes reveals subgroups of miRNAs that are highly expressed in particular organs. *PLoS ONE* **2016**, *11*, e0153324. [CrossRef]

40. Chen, P.Y.; Manninga, H.; Slanchev, K.; Chien, M.; Russo, J.J.; Ju, J.; Sheridan, R.; John, B.; Marks, D.S.; Gaidatzis, D.; et al. The developmental miRNA profiles of zebrafish as determined by small RNA cloning. *Genes Dev.* **2005**, *19*, 1288–1293. [CrossRef]

41. Li, S.C.; Chan, W.C.; Ho, M.R.; Tsai, K.W.; Hu, L.Y.; Lai, C.H.; Hsu, C.N.; Hwang, P.P.; Lin, W.C. Discovery and characterization of medaka miRNA genes by next generation sequencing platform. *BMC Genom.* **2010**, *11* (Suppl. S4), S8. [CrossRef]

42. Baskerville, S.; Bartel, D.P. Microarray profiling of microRNAs reveals frequent coexpression with neighboring miRNAs and host genes. *RNA* **2005**, *11*, 241–247. [CrossRef]

43. Giraldez, A.J.; Cinalli, R.M.; Glasner, M.E.; Enright, A.J.; Thomson, J.M.; Baskerville, S.; Hammond, S.M.; Bartel, D.P.; Schier, A.F. MicroRNAs regulate brain morphogenesis in zebrafish. *Science* **2005**, *308*, 833–838. [CrossRef]

44. Sokol, N.S. The role of microRNAs in muscle development. *Curr. Top. Dev. Biol.* **2012**, *99*, 59–78. [CrossRef]

45. Kozomara, A.; Griffiths-Jones, S. miRBase: Integrating microRNA annotation and deep-sequencing data. *Nucleic Acids Res.* **2011**, *39*, D152–D157. [CrossRef]

46. Berthelot, C.; Brunet, F.; Chalopin, D.; Juanchich, A.; Bernard, M.; Noël, B.; Bento, P.; Da Silva, C.; Labadie, K.; Alberti, A.; et al. The rainbow trout genome provides novel insights into evolution after whole-genome duplication in vertebrates. *Nat. Commun.* **2014**, *5*, 3657. [CrossRef]

47. Song, J.B.; Song, J.; Mo, B.X.; Chen, X.M. Uridylation and adenylation of RNAs. *Sci. China Life Sci.* **2015**, *58*, 1057–1066. [CrossRef]

48. Wyman, S.K.; Knouf, E.C.; Parkin, R.K.; Fritz, B.R.; Lin, D.W.; Dennis, L.M.; Krouse, M.A.; Webster, P.J.; Tewari, M. Post-transcriptional generation of miRNA variants by multiple nucleotidyl transferases contributes to miRNA transcriptome complexity. *Genome Res.* **2011**, *21*, 1450–1461. [CrossRef]

49. Giraldez, A.J.; Mishima, Y.; Rihel, J.; Grocock, R.J.; Van Dongen, S.; Inoue, K.; Enright, A.J.; Schier, A.F. Zebrafish MiR-430 promotes deadenylation and clearance of maternal mRNAs. *Science* **2006**, *312*, 75–79. [CrossRef]

50. Giusti, J.; Pinhal, D.; Moxon, S.; Campos, C.L.; Münsterberg, A.; Martins, C. MicroRNA-10 modulates Hox genes expression during Nile tilapia embryonic development. *Mech. Dev.* **2016**, *140*, 12–18. [CrossRef]

51. Mennigen, J.A.; Martyniuk, C.J.; Seiliez, I.; Panserat, S.; Skiba-Cassy, S. Metabolic consequences of microRNA-122 inhibition in rainbow trout, Oncorhynchus mykiss. *BMC Genom.* **2014**, *15*, 70. [CrossRef]
52. Mennigen, J.A.; Plagnes-Juan, E.; Figueredo-Silva, C.A.; Seiliez, I.; Panserat, S.; Skiba-Cassy, S. Acute endocrine and nutritional co-regulation of the hepatic omy-miRNA-122b and the lipogenic gene fas in rainbow trout, Oncorhynchus mykiss. *Comp. Biochem. Physiol. B Biochem. Mol. Biol.* **2014**, *169*, 16–24. [CrossRef] [PubMed]
53. Radhakrishnan, B.; Alwin Prem Anand, A. Role of miRNA-9 in Brain Development. *J. Exp. Neurosci.* **2016**, *10*, 101–120. [CrossRef] [PubMed]
54. Ludwig, N.; Leidinger, P.; Becker, K.; Backes, C.; Fehlmann, T.; Pallasch, C.; Rheinheimer, S.; Meder, B.; Stahler, C.; Meese, E.; et al. Distribution of miRNA expression across human tissues. *Nucleic Acids Res.* **2016**, *44*, 3865–3877. [CrossRef] [PubMed]
55. Luo, G.Z.; Hafner, M.; Shi, Z.; Brown, M.; Feng, G.H.; Tuschl, T.; Wang, X.J.; Li, X. Genome-wide annotation and analysis of zebra finch microRNA repertoire reveal sex-biased expression. *BMC Genom.* **2012**, *13*, 727. [CrossRef] [PubMed]
56. Fernandez-Valverde, S.L.; Taft, R.J.; Mattick, J.S. Dynamic isomiR regulation in Drosophila development. *RNA* **2010**, *16*, 1881–1888. [CrossRef]
57. Schamberger, A.; Orban, T.I. 3′ IsomiR species and DNA contamination influence reliable quantification of microRNAs by stem-loop quantitative PCR. *PLoS ONE* **2014**, *9*, e106315. [CrossRef] [PubMed]
58. Bizuayehu, T.T.; Babiak, I. MicroRNA in teleost fish. *Genome Biol. Evol.* **2014**, *6*, 1911–1937. [CrossRef]
59. Lagos-Quintana, M.; Rauhut, R.; Yalcin, A.; Meyer, J.; Lendeckel, W.; Tuschl, T. Identification of tissue-specific microRNAs from mouse. *Curr. Biol.* **2002**, *12*, 735–739. [CrossRef]
60. Herkenhoff, M.E.; Oliveira, A.C.; Nachtigall, P.G.; Costa, J.M.; Campos, V.F.; Hilsdorf, A.W.S.; Pinhal, D. Fishing into the MicroRNA transcriptome. *Front. Genet.* **2018**, *9*, 88. [CrossRef] [PubMed]
61. Juanchich, A.; Bardou, P.; Rue, O.; Gabillard, J.C.; Gaspin, C.; Bobe, J.; Guiguen, Y. Characterization of an extensive rainbow trout miRNA transcriptome by next generation sequencing. *BMC Genom.* **2016**, *17*, 164. [CrossRef]
62. Trattner, S.; Vestergren, A.S. Tissue distribution of selected microRNA in Atlantic salmon. *Eur. J. Lipid Sci. Tech.* **2013**, *115*, 1348–1356. [CrossRef]
63. Xia, J.H.; He, X.P.; Bai, Z.Y.; Yue, G.H. Identification and characterization of 63 MicroRNAs in the Asian seabass Lates calcarifer. *PLoS ONE* **2011**, *6*, e17537. [CrossRef] [PubMed]
64. Mitchelson, K.R.; Qin, W.Y. Roles of the canonical myomiRs miR-1, -133 and -206 in cell development and disease. *World J. Biol. Chem.* **2015**, *6*, 162–208. [CrossRef] [PubMed]
65. Wu, G.; Huang, Z.P.; Wang, D.Z. MicroRNAs in cardiac regeneration and cardiovascular disease. *Sci. China Life Sci.* **2013**, *56*, 907–913. [CrossRef] [PubMed]
66. Alberti, C.; Cochella, L. A framework for understanding the roles of miRNAs in animal development. *Development* **2017**, *144*, 2548–2559. [CrossRef]
67. Rosa, A.; Brivanlou, A.H. MicroRNAs in early vertebrate development. *Cell Cycle* **2009**, *8*, 3513–3520. [CrossRef] [PubMed]
68. Takacs, C.M.; Giraldez, A.J. miR-430 regulates oriented cell division during neural tube development in zebrafish. *Dev. Biol.* **2016**, *409*, 442–450. [CrossRef]
69. Wienholds, E.; Kloosterman, W.P.; Miska, E.; Alvarez-Saavedra, E.; Berezikov, E.; de Bruijn, E.; Horvitz, H.R.; Kauppinen, S.; Plasterk, R.H. MicroRNA expression in zebrafish embryonic development. *Science* **2005**, *309*, 310–311. [CrossRef]
70. Tehler, D.; Hoyland-Kroghsbo, N.M.; Lund, A.H. The miR-10 microRNA precursor family. *RNA Biol.* **2011**, *8*, 728–734. [CrossRef]

Article

Predicting MicroRNA Mediated Gene Regulation between Human and Viruses

Xin Shu †, Xinyuan Zang †, Xiaoshuang Liu, Jie Yang * and Jin Wang *

The State Key Laboratory of Pharmaceutical Biotechnology and Jiangsu Engineering Research Center for MicroRNA Biology and Biotechnology, NJU Advanced Institute for Life Sciences (NAILS), School of Life Science, Nanjing University, Nanjing 210023, China; cpu_shuxin@126.com (X.S.); hsinring@foxmail.com (X.Z.); xsliunju@foxmail.com (X.L.)

* Correspondence: yangjie@nju.edu.cn (J.Y.); jwang@nju.edu.cn (J.W.)
† These authors contributed equally to this work as first authors.

Received: 26 June 2018; Accepted: 6 August 2018; Published: 8 August 2018

Abstract: MicroRNAs (miRNAs) mediate various biological processes by actively fine-tuning gene expression at the post-transcriptional level. With the identification of numerous human and viral miRNAs, growing evidence has indicated a common role of miRNAs in mediating the interactions between humans and viruses. However, there is only limited information about Cross-Kingdom miRNA target sites from studies. To facilitate an extensive investigation on the interplay among the gene regulatory networks of humans and viruses, we designed a prediction pipeline, mirTarP, that is suitable for miRNA target screening on the genome scale. By applying mirTarP, we constructed the database mirTar, which is a comprehensive miRNA target repository of bidirectional interspecies regulation between viruses and humans. To provide convenient downloading for users from both the molecular biology field and medical field, mirTar classifies viruses according to "ICTV viral category" and the "medical microbiology classification" on the web page. The mirTar database and mirTarP tool are freely available online.

Keywords: miRNA; virus; host; Cross-Kingdom; target prediction

---

1. Introduction

MicroRNAs (miRNAs) are a class of small (~24 nt), non-coding RNA molecules that play a critical role in fundamental cellular processes and many types of diseases. They negatively regulate gene expression by binding to the 3′-untranslated regions (3′UTR) of the target mRNAs in cells [1]. Recent studies have found that they are involved in viral infections and play a key role in the host–virus interaction network. Host miRNAs modulate the expression of viral genes by targeting on virus transcripts, while viruses encode miRNAs that protect them from the host's antiviral response by acting on cellular mRNAs [2–5]. Skalsky et al. [5] reported a comprehensive survey of viral and cellular miRNA targetome in Epstein-Barr virus (EBV)-infected lymphoblastoid cell lines using photoactivatable ribonucleoside-enhanced crosslinking and immunoprecipitation (PAR-CLIP) and deep sequencing technique combined with bioinformatics. In this survey, over 500 target sites of EBV miRNAs on cellular transcripts were detected in addition to the cellular miRNA targets on virus. This result may imply that viral miRNAs have a similar mode of multiple targeting as cellular miRNAs. Although the detection of miRNA targets by high throughput techniques remains a big challenge, there has been growing interest in the role of miRNAs in host–virus interactions.

The virus miRNAs can target both the host genes and viral genes in order to contribute to the creation of a propagating environment in the host cell [2]. EBV-encoded miRNA miR-BHRF1-2-5p blocks Interleukin-1 (IL-1) signaling by directly targeting the IL-1 Receptor 1 (IL1R1) [6]. Hancock et al. [7] found that human Cytomegalovirus (HCMV) also uses its own miRNAs, miR-US5-1 and miR-UL112-3p, which bind to IkB kinase (IKK) complex components IKKα and IKKβ, in order to avoid the immune response of the host. Some viral miRNAs show sequence similarity with host miRNAs and thus, may take part in the conserved cellular gene regulation network [8]. In Kaposi's sarcoma-associated herpesvirus (KSHV)-infected human cell line, Manzano et al. [9] identified that KSHV miR-K3+1 and miR-K3 share perfect and offset 5′ homology with cellular miR-23, respectively. KSHV miR-k12-11 is an ortholog of miR-155, which can inhibit the 3′UTR region of BACH-1 [10].

Host miRNAs were found to target the viral RNA transcripts to inhibit viral pathogenesis, which essentially involves being a defense against viral infections [3–5]. It was reported that human miRNA effectively restricts the accumulation of the retrovirus primate foamy virus type 1 (PFV-1) in human cells, which involves hsa-miR-32 inhibiting the proliferation of PFV-1 by targeting PFV-1 F11 sequence. However, PFV-1 also encodes a protein named Tas, which suppresses miRNA-directed functions in mammalian cells and displays Cross-Kingdom anti-silencing activities [4]. This new report focused on an EBV-encoded protein EBNA2, which subverts immune surveillance by downregulating miR-34a that targets an important immune checkpoint PD-L1 in lymphoma B cells [11]. Human liver-specific miRNA hsa-miR-122 can induce hepatitis C virus (HCV) replication by targeting the 5′-non-coding region (NCR) of the viral genome [3]. The human miRNAs let-7b and mir-199a target the 5′UTR of HCV to decrease viral replication [12,13]. Pedersen et al. [14] also found that the overexpression of miR-196 and miR-448 significantly reduced the replication of HCV as they target the NS5A coding region and core of the HCV genome, respectively.

These findings indicate a common role of miRNAs in mediating the diversified interactions between humans and viruses. A total of 2588 mature human miRNAs and 181 mature miRNAs of human-related viruses have been recruited in mirBase so far (release 21) [15]. To facilitate the extensive investigation on the interplay among the gene regulatory network of humans and viruses, computational tools and comprehensive miRNA target repositories pertaining to human–virus interactions is necessary. These resources could provide the researchers with an efficient approach and potential miRNA targets to facilitate the investigation of miRNA function and regulation mechanisms. In particular, in the era of omics when it is possible to obtain a complete set of molecular data of gene expression, prediction tools and database are essential for genome-scale or microbiome-scale data analysis and help to decipher the panorama of the gene regulation network of human–virus interplay. This will ultimately facilitate the discovery of new drug targets for viruses, including HIV [16], HCMV [7] and HCV [3].

MiRNAs suppress interspecies gene expressions by targeting the 3′-UTRs of mRNAs during the infection or antiviral processes. Although many algorithms [17–21] are available for miRNA target prediction, only a few of them can be directly used to predict the interspecies regulation between viruses and hosts [20]. Most of the tools were designed for intra-species application by predicting the miRNA targets on their own genome, such as TargetScan, PicTar, miRanda and DIANA-microT [18,22–24]. In this situation, the databases of Cross-Kingdom miRNA target sites were produced by using the multiple intra-species target prediction tools mentioned above, which may possibly create concerns regarding the methodology and thus, the accuracy.

ViTa [25] provides predicted targets of host miRNAs from humans, mice, rats and chickens (mirBase release 8.2), which are located on 2108 virus species from 23 families. VHoT [26] houses predictions of 271 viral miRNAs on six hosts, which are namely humans, mice, rats, rhesus monkeys and cows. VmiReg [27] contains predicted targets of 169 viral miRNAs (from 10 types of viruses) on humans. VIRmiRNA [28] provides experimentally validated viral miRNAs and their targets on human and other species. All of these databases provide information of interspecies miRNA targeting in one direction only, which include either the target of viral miRNAs on host genes or the target of host miRNAs on viruses. To investigate the complex and dynamic interactions between the gene regulatory network of humans and infectious viruses that are mediated by miRNAs, the database mirTar was constructed that provides a comprehensive miRNA target repository pertaining to 2588 human miRNAs (mirBase release 21) that target 386 genomes of human-related viruses as well as

181 viral miRNAs that target the human genome. The new computational pipeline that was specially designed for human–virus interspecies miRNA target prediction was presented.

2. The prediction Tool and Results

2.1. Data Collection

A total of 2588 mature human miRNAs and 181 mature viral miRNAs were downloaded from mirBase (Release 21). Human genome and virus genomes as well as their classification information and taxonomy annotation were obtained and organized from NCBI [29,30]. Meanwhile, the annotation of gene name and protein name pertaining to the mRNA transcripts were acquired from Ensemble [31]. A total of 386 human-related viral species were collected that are belong to 34 families and fall under the following 7 genome types: (1) Deltavirus; (2) dsDNA viruses, no RNA stage; (3) dsRNA viruses; (4) Retro-transcribing viruses; (5) ssDNA viruses; (6) ssRNA viruses; and (7) unclassified viruses [32]. These viruses are all human infections. Some of them (79/386) are common and medically important viral species as categorized in medical microbiology, which mostly cause diseases of the respiratory tract, gastrointestinal tract and liver.

2.2. The Prediction Tool

Mainstream miRNA target prediction tools were limited to intra-species applications as they were only capable of predicting the miRNA targets on their own genome. Thus, the databases of interspecies miRNA targets were produced by using a combination of these methods as an approach to improve the reliability of the prediction results. For example, ViTa applied miRanda and TargetScan to identify the host miRNA target sites in virus genomes [18,23]. VHot combined five miRNA target prediction tools, which were namely TargetScan, miRanda, RNAhybrid, DIANA-microT and PITA, to form its prediction engine [18,23,24,33,34]. VmiReg predicted targets of viral miRNAs by four established prediction programs, which were namely miRanda, TargetScan, RNAhybrid and PITA [18,23,33,34]. This approach may create problems in inter-species target prediction as the sequence specificity of intra-species miRNA-target interaction are included. In addition, most of these calculations are quite time-consuming and require huge processing resources for a genome-scale prediction. To find miRNA targets across different kingdoms, we designed a prediction pipeline, mirTarP, that directly seeks the potential miRNA target. This can produce results quickly and thus, is very suitable for miRNA target screening using large-scale calculations.

MirTarP was designed by integrating two classical algorithms of sequence analysis, which were Blast [35] and RNAhybrid [34]. They work as the cores of two modules included in mirTarP, which are quick match and duplex assessment. Blast uses heuristics to accelerate searches for similar segments of a sequence. A window of consecutive perfect match can be set when running the algorithm. To improve the calculation efficiency, mirTarP introduced the sequence similarity tool Blast to produce preliminary matches between the miRNA and its target mRNA sequences. The results from the quick match module were subsequently delivered to duplex assessment module, which uses the RNAhybrid program for the calculation of minimum free energy (mfe) of miRNA–mRNA hybridization duplexes based on the principles of thermodynamics. The mfe value stands for the stability of miRNA binding. To assess the influence of local secondary structures on the target accessibility, RNAfold [36] was used to calculate the minimum folding energy around the target sites. The results were listed as the supplementary data of predicted targets. The default parameters set in mirTarP include the 7-consecutive base matches as the seed of targeting and the cutoff of mfe of −25 kcal /mol for local dimer formation. The flow chart of mirTarP is illustrated in Figure 1. The advantage of mirTarP over the current prediction tools is that it operates independent of conservation and thus, can be used to find miRNA targets on virus genomes or obtain other interspecies miRNA target predictions. This tool runs quickly and is easy to use with only 2 parameters to be set. Therefore, it will be helpful to wet-lab researchers dealing with new viruses. A comparison of mirTarP to TargetScan and PITA on a dataset of 221 experimentally validated miRNA-target pairs is included in the website along with the tool mirTarP.

The prediction tool mirTarP is free for downloading in the web page.

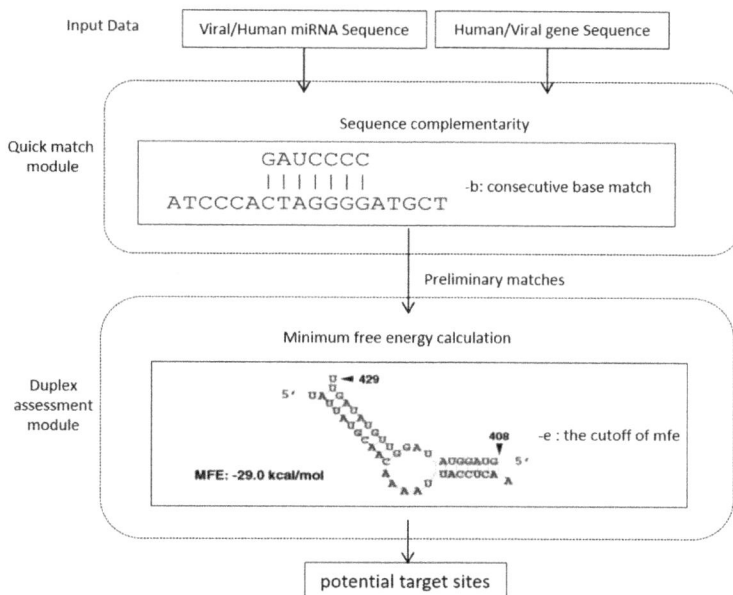

Figure 1. Flow chart of mirTarP for human–virus interspecies miRNA target prediction. The parameter '-b' represents the number of consecutive base matches between miRNA and target sequence, while '-e' represents the cutoff of minimum free energy of miRNA-target binding.

2.3. Prediction of miRNA Targets

By applying mirTarP, 2557 human mature miRNAs were found to have targets in 3133 viral genes, which corresponds to 3376 viral proteins. A total of 181 miRNA records from 13 viral species of 3 families were used for the prediction of targets on human genome. The calculation results showed that these viral miRNAs had potential target sites in 16,439 human genes.

A total of 2,680,194 entries about the miRNAs target sites within human and viral genomes were produced.

3. MirTar Database

3.1. Web Interface Development

MirTar is designed to adapt a wide variety of screen formats and devices (PCs, tablets, smartphones, etc.). All data were organized by MySQL and the website is implemented in PHP, JavaScript and HTML.

3.2. Data Download

The web page provides two ways of data downloads, i.e., customized download and the complete download. The customized download is associated with the items or viruses selected by the user. To provide easy downloading for users from both the molecular biology field and medical field, mirTar database classified the viruses in the following two ways: (1) according to the definition by medical microbiology; and (2) according to ICTV virus category [32]. Currently, the International Committee on the Taxonomy of Viruses (ICTV) provides the most comprehensive, fully annotated compendium of information on virus taxa and taxonomy. Thus, the web page provides a search function for convenient categories when retrieving an input virus. In addition, a python script of the prediction tool mirTarP is available on the web page to facilitate a quick screening of miRNA targets on new viruses. The mirTar database and mirTarP tool are freely available at http://mcube.nju.edu.cn/jwang/mirTar/docs/mirTar/ or http://118.89.139.70/mirTar/docs/mirTar/. The interface of mirTar is shown in Figure 2.

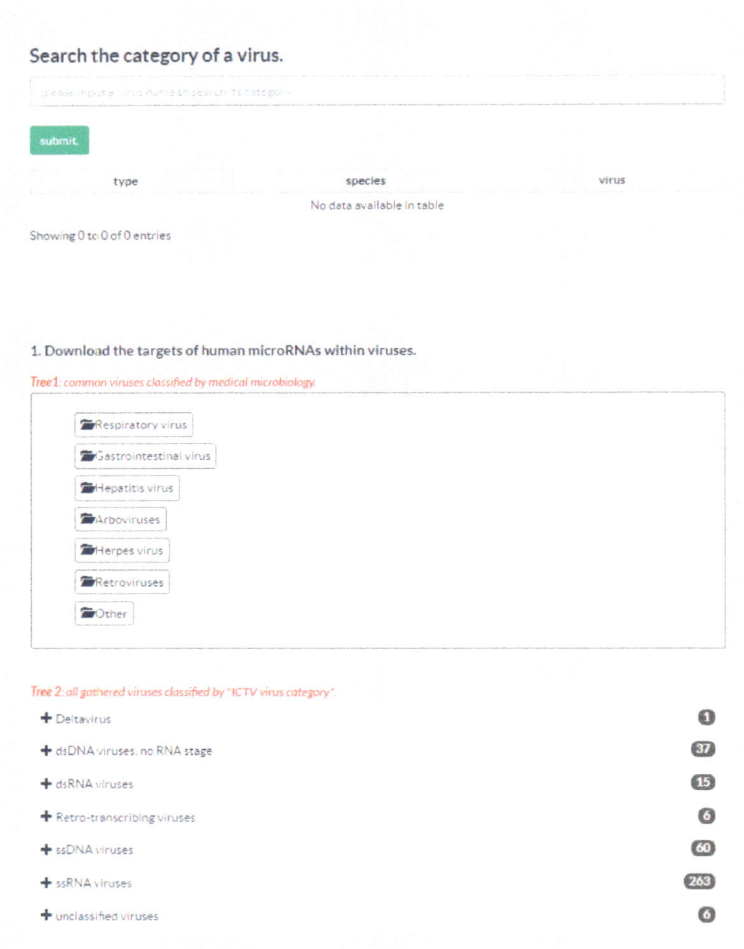

Figure 2. The interface of mirTar database.

4. Conclusions

In this paper, we provide a comprehensive miRNA target database that includes the bidirectional interspecies actions between human and the infectious viruses along with a fast miRNA target prediction program to facilitate a quick screening of miRNA targets on new viruses. The database mirTar contains 2,200,076 candidate target sites on 386 viral genomes for 2577 human mature miRNAs and 480,118 targets of 181 viral mature miRNAs on human genome. The web page of the database was designed for convenient data querying and downloading by classifying the virus species by the two categories of molecular biology and medicine. The database will benefit investigations on the crosstalk between the host and virus gene regulations and the new role of miRNAs in infections and diseases caused by latent viruses, including many cancers.

Author Contributions: Formal analysis, X.Z.; Funding acquisition, J.W.; Methodology, X.S., X.Z. and X.L.; Resources, J.Y., J.W.; Visualization, X.S.; Writing—original draft, X.S.; Writing—review & editing, X.S., X.Z. and J.W.

Funding: This research was funded by National Science Foundation of China [81250044, 31500674]; NJU-Yangzhou Institute of Optoelectronics.

Acknowledgments: The authors would acknowledge the Center of High Performance Computation of Nanjing University for the support of computational resources.

Conflicts of Interest: The authors declare no conflict of interest.

References

1. Bartel, D.P. Micrornas: Genomics, biogenesis, mechanism, and function. *Cell* **2004**, *116*, 281–297. [CrossRef]
2. Bryan, C.R.; Cullen; Bryan; Vogt, P.K. *Intrinsic Immunity*; Springer: Berlin, Germany, 2013.
3. Jopling, C.L.; Yi, M.; Lancaster, A.M.; Lemon, S.M.; Sarnow, P. Modulation of hepatitis C virus RNA abundance by a liver-specific microRNA. *Science* **2005**, *309*, 1577–1581. [CrossRef] [PubMed]
4. Lecellier, C.-H.; Dunoyer, P.; Arar, K.; Lehmann-Che, J.; Eyquem, S.; Himber, C.; Saïb, A.; Voinnet, O. A cellular microRNA mediates antiviral defense in human cells. *Science* **2005**, *308*, 557–560. [CrossRef] [PubMed]
5. Skalsky, R.L.; Corcoran, D.L.; Gottwein, E.; Frank, C.L.; Kang, D.; Hafner, M.; Nusbaum, J.D.; Feederle, R.; Delecluse, H.-J.; Luftig, M.A. The viral and cellular microRNA targetome in lymphoblastoid cell lines. *PLoS Pathog.* **2012**, *8*, e1002484. [CrossRef] [PubMed]
6. Skinner, C.M.; Ivanov, N.S.; Barr, S.A.; Chen, Y.; Skalsky, R.L. An Epstein-Barr virus microRNA blocks interleukin-1 (IL-1) signaling by targeting IL-1 receptor 1. *J. Virol.* **2017**, *91*. [CrossRef] [PubMed]
7. Hancock, M.H.; Hook, L.M.; Mitchell, J.; Nelson, J.A. Human cytomegalovirus microRNAs miR-US5-1 and miR-UL112-3p block proinflammatory cytokine production in response to NF-κB-activating factors through direct downregulation of IKKα and IKKβ. *mBio* **2017**, *8*, e00109-17. [CrossRef] [PubMed]
8. Ghosh, Z.; Mallick, B.; Chakrabarti, J. Cellular versus viral microRNAs in host-virus interaction. *Nucleic Acids Res.* **2009**, *37*, 1035–1048. [CrossRef] [PubMed]
9. Manzano, M.; Shamulailatpam, P.; Raja, A.N.; Gottwein, E. Kaposi's sarcoma-associated herpesvirus encodes a mimic of cellular miR-23. *J. Virol.* **2013**, *87*, 11821–11830. [CrossRef] [PubMed]
10. Skalsky, R.L.; Samols, M.A.; Plaisance, K.B.; Boss, I.W.; Riva, A.; Lopez, M.C.; Baker, H.V.; Renne, R. Kaposi's sarcoma-associated herpesvirus encodes an ortholog of miR-155. *J. Virol.* **2007**, *81*, 12836–12845. [CrossRef] [PubMed]
11. Anastasiadou, E.; Stroopinsky, D.; Alimperti, S.; Jiao, A.L.; Pyzer, A.R.; Cippitelli, C.; Pepe, G.; Severa, M.; Rosenblatt, J.; Etna, M.P.; et al. Epstein-Barr virus-encoded EBNA2 alters immune checkpoint PD-l1 expression by downregulating miR-34a in B-cell lymphomas. *Leukemia* **2018**. [CrossRef] [PubMed]
12. Cheng, J.C.; Yeh, Y.J.; Tseng, C.P.; Hsu, S.D.; Chang, Y.L.; Sakamoto, N.; Huang, H.D. Let-7b is a novel regulator of hepatitis C virus replication. *Cell. Mol. Life Sci. CMLS* **2012**, *69*, 2621–2633. [CrossRef] [PubMed]
13. Murakami, Y.; Aly, H.H.; Tajima, A.; Inoue, I.; Shimotohno, K. Regulation of the hepatitis C virus genome replication by miR-199a. *J. Hepatol.* **2009**, *50*, 453–460. [CrossRef] [PubMed]
14. Pedersen, I.M.; Cheng, G.; Wieland, S.; Volinia, S.; Croce, C.M.; Chisari, F.V.; David, M. Interferon modulation of cellular microRNAs as an antiviral mechanism. *Nature* **2007**, *449*, 919–922. [CrossRef] [PubMed]
15. Kozomara, A.; Griffiths-Jones, S. miRBase: Annotating high confidence microRNAs using deep sequencing data. *Nucleic Acids Res.* **2013**, *42*, D68–D73. [CrossRef] [PubMed]
16. Balasubramaniam, M.; Pandhare, J.; Dash, C. Are microRNAs important players in HIV-1 infection? An update. *Viruses* **2018**, *10*, 110. [CrossRef] [PubMed]

17. Grün, D.; Wang, Y.-L.; Langenberger, D.; Gunsalus, K.C.; Rajewsky, N. microRNA target predictions across seven *Drosophila* species and comparison to mammalian targets. *PLoS Comput. Biol.* **2005**, *1*, e13. [CrossRef] [PubMed]

18. Agarwal, V.; Bell, G.W.; Nam, J.-W.; Bartel, D.P. Predicting effective microRNA target sites in mammalian mRNAs. *eLife* **2015**, *4*, e05005. [CrossRef] [PubMed]

19. M. Witkos, T.; Koscianska, E.; J. Krzyzosiak, W. Practical aspects of microRNA target prediction. *Curr. Mol. Med.* **2011**, *11*, 93–109. [CrossRef]

20. Laganà, A.; Forte, S.; Russo, F.; Giugno, R.; Pulvirenti, A.; Ferro, A. Prediction of human targets for viral-encoded microRNAs by thermodynamics and empirical constraints. *J. RNAi Gene Silenc.* **2010**, *6*, 379.

21. Cheng, S.; Guo, M.; Wang, C.; Liu, X.; Liu, Y.; Wu, X. MiRTDL: A deep learning approach for miRNA target prediction. *IEEE/ACM Trans. Comput. Biol. Bioinform.* **2016**, *13*, 1161–1169. [CrossRef] [PubMed]

22. Krek, A.; Grün, D.; Poy, M.N.; Wolf, R.; Rosenberg, L.; Epstein, E.J.; MacMenamin, P.; Da Piedade, I.; Gunsalus, K.C.; Stoffel, M. Combinatorial microRNA target predictions. *Nat. Genet.* **2005**, *37*, 495. [CrossRef] [PubMed]

23. John, B.; Enright, A.J.; Aravin, A.; Tuschl, T.; Sander, C.; Marks, D.S. Human microRNA targets. *PLoS Biol.* **2004**, *2*, e363. [CrossRef] [PubMed]

24. Paraskevopoulou, M.D.; Georgakilas, G.; Kostoulas, N.; Vlachos, I.S.; Vergoulis, T.; Reczko, M.; Filippidis, C.; Dalamagas, T.; Hatzigeorgiou, A.G. Diana-microT web server v5.0: Service integration into miRNA functional analysis workflows. *Nucleic Acids Res.* **2013**, *41*, W169–W173. [CrossRef] [PubMed]

25. Hsu, P.W.-C.; Lin, L.-Z.; Hsu, S.-D.; Hsu, J.B.-K.; Huang, H.-D. Vita: Prediction of host microRNAs targets on viruses. *Nucleic Acids Res.* **2006**, *35*, D381–D385. [CrossRef] [PubMed]

26. Kim, H.; Park, S.; Min, H.; Yoon, S. vHoT: A database for predicting interspecies interactions between viral microRNA and host genomes. *Arch. Virol.* **2012**, *157*, 497–501. [CrossRef] [PubMed]

27. Shao, T.; Zhao, Z.; Wu, A.; Bai, J.; Li, Y.; Chen, H.; Jiang, C.; Wang, Y.; Li, S.; Wang, L. Functional dissection of virus–human crosstalk mediated by miRNAs based on the VmiReg database. *Mol. BioSyst.* **2015**, *11*, 1319–1328. [CrossRef] [PubMed]

28. Qureshi, A.; Thakur, N.; Monga, I.; Thakur, A.; Kumar, M. VIRmiRNA: A comprehensive resource for experimentally validated viral miRNAs and their targets. *Database* **2014**, *2014*. [CrossRef] [PubMed]

29. Brister, J.R.; Ako-Adjei, D.; Bao, Y.; Blinkova, O. NCBI viral genomes resource. *Nucleic Acids Res.* **2014**, *43*, D571–D577. [CrossRef] [PubMed]

30. Federhen, S. The NCBI taxonomy database. *Nucleic Acids Res.* **2011**, *40*, D136–D143. [CrossRef] [PubMed]

31. Aken, B.L.; Ayling, S.; Barrell, D.; Clarke, L.; Curwen, V.; Fairley, S.; Fernandez Banet, J.; Billis, K.; García Girón, C.; Hourlier, T. The Ensembl gene annotation system. *Database* **2016**, *2016*. [CrossRef] [PubMed]

32. Lefkowitz, E.J.; Dempsey, D.M.; Hendrickson, R.C.; Orton, R.J.; Siddell, S.G.; Smith, D.B. Virus taxonomy: The database of the International Committee on Taxonomy of Viruses (ICTV). *Nucleic Acids Res.* **2017**, *46*, D708–D717. [CrossRef] [PubMed]

33. Kertesz, M.; Iovino, N.; Unnerstall, U.; Gaul, U.; Segal, E. The role of site accessibility in microRNA target recognition. *Nat. Genet.* **2007**, *39*, 1278. [CrossRef] [PubMed]

34. Krüger, J.; Rehmsmeier, M. RNAhybrid: MicroRNA target prediction easy, fast and flexible. *Nucleic Acids'Res.* **2006**, *34*, W451–W454. [CrossRef] [PubMed]

35. Altschul, S.F.; Gish, W.; Miller, W.; Myers, E.W.; Lipman, D.J. Basic local alignment search tool. *J. Mol. Boil.* **1990**, *215*, 403–410. [CrossRef]

36. Lorenz, R.; Bernhart, S.H.; Zu Siederdissen, C.H.; Tafer, H.; Flamm, C.; Stadler, P.F.; Hofacker, I.L. ViennaRNA package 2.0. *Algorithms Mol. Biol.* **2011**, *6*, 26. [CrossRef] [PubMed]

cells

MDPI

Article

Tensor Decomposition-Based Unsupervised Feature Extraction Can Identify the Universal Nature of Sequence-Nonspecific Off-Target Regulation of mRNA Mediated by MicroRNA Transfection

Y.-H. Taguchi

Department of Physics, Chuo University, Tokyo 112-8551, Japan; tag@granular.com; Tel.: +81-3-3817-1791

Received: 3 May 2018; Accepted: 31 May 2018; Published: 4 June 2018

Abstract: MicroRNA (miRNA) transfection is known to degrade target mRNAs and to decrease mRNA expression. In contrast to the notion that most of the gene expression alterations caused by miRNA transfection involve downregulation, they often involve both up- and downregulation; this phenomenon is thought to be, at least partially, mediated by sequence-nonspecific off-target effects. In this study, I used tensor decomposition-based unsupervised feature extraction to identify genes whose expression is likely to be altered by miRNA transfection. These gene sets turned out to largely overlap with one another regardless of the type of miRNA or cell lines used in the experiments. These gene sets also overlap with the gene set associated with altered expression induced by a Dicer knockout. This result suggests that the off-target effect is at least as important as the canonical function of miRNAs that suppress translation. The off-target effect is also suggested to consist of competition for the protein machinery between transfected miRNAs and miRNAs in the cell. Because the identified genes are enriched in various biological terms, these genes are likely to play critical roles in diverse biological processes.

Keywords: tensor decomposition; miRNA transfection; sequence-nonspecific off-target regulation

1. Introduction

MicroRNA (miRNA) is short noncoding (functional) RNA whose primary function is mRNA degradation and disruption of translation [1]. Thus, it is generally expected that the primary effect of miRNA transfection (or overexpression) on mRNA expression is suppression. Based on this assumption, numerous miRNA transfection and/or overexpression experiments have been conducted to identify genes that are directly targeted by miRNAs [2]; during these analyses, only genes with expression levels inversely related to those of miRNA have been sought. Nevertheless, it was found that many mRNAs whose expression was likely to be altered by miRNA transfection and/or overexpression turned out to positively correlate with miRNA expression. For example, Khan et al. [3] identified multiple genes that are upregulated by miRNA transfection. They reasoned that this effect means competition with endogenous miRNAs because upregulated genes were often targeted by endogenous miRNAs. The protein machinery that binds to endogenous miRNAs was occupied by the transfected miRNAs, and as a result, the genes targeted by endogenous miRNAs were upregulated [4]. In addition, Carroll et al. [5] identified a positive correlation between mRNA expression and transfected miRNA. They theorized that the positive correlations are mediated by interactions with transcription factor E2F1.

Despite these findings, to my knowledge, sequence-nonspecific off-target regulation by miRNA transfection has not been extensively studied to date [2]. Most of the miRNA transfection and/or overexpression experiments have been aimed at identifying canonical targets of miRNAs. Most of these experiments have not been analyzed in the context of sequence-nonspecific off-target regulation by miRNA transfection. Although it is unclear why no one has tried to systematically investigate sequence-nonspecific off-target regulation mediated by miRNA transfection, one possible reason is the lack of a suitable methodology. By definition, miRNA transfection experiments cannot be composed of many samples. Typically, a pair of data points consists of miRNA-transfected cells and mock-transfected cells. Although a few more biological and/or technical replicates are possible, the number of samples available is usually less than 10. This number is often too small to detect significantly altered expression of mRNAs whose total number is up to 10^4. In the case when the aim of a study is identification of canonical interactions between miRNA and mRNA, additional information that can reduce the number of mRNAs under study, e.g., bioinformatically predicted mRNAs targeted by transfected miRNAs, is available. This information can enable researchers to identify significant correlations between transfected miRNAs and mRNAs. Nonetheless, this kind of information is usually not available for the analysis of sequence-nonspecific off-target regulation by miRNA transfection.

In this study, with the aim to resolve this difficulty, I applied tensor decomposition (TD)-based unsupervised feature extraction (FE) to miRNA transfection experiments. TD-based unsupervised FE [6–10] is an extension of principal component analysis (PCA)-based unsupervised FE [11–33], which can identify critical genes even when there is only a small number of samples. TD-based unsupervised FE can also identify critical genes by means of a small number of samples available. Because of the use of this methodology, genes whose mRNA expression was likely to be altered by miRNA transfection were identified here in various combinations of cell lines and transfected miRNAs. Most of sets of genes significantly overlapped with one another regardless of transfected cell types and types of miRNA. These genes also showed altered mRNA expression under the influence of a Dicer knockout (KO). This finding suggests that the primary factor that mediated sequence-nonspecific side effects of miRNA transfection is competition for protein machinery with endogenous miRNAs, as suggested by Khan et al. [3]. In addition, these sets of genes significantly overlapped with various sets of genes whose biological functions and significance have been validated experimentally. Thus, sequence-nonspecific off-target regulation caused by miRNA transfection is expected to also play critical roles in various biological processes.

2. Materials and Methods

2.1. Mathematical Formulation of the Tensor and Tensor Decomposition

Because a tensor or tensor decomposition (TD) is not a popular mathematical concept, I briefly formulate them here. Suppose a three-mode tensor, $x_{ijk} \in \mathbb{R}^{N \times M \times K}$, is the expression level of the ith mRNA when the jth miRNA is transfected into the kth sample. Samples are typically composed of biological replicates and/or treated and untreated (or mock-treated, i.e., control) samples, but situations vary. x_{ijk} can also be formulated as two-mode tensor $x_{i(jk)}$, where (jk) represents a pair of a miRNA and a sample, especially when samples are not paired. x_{ijk} can be decomposed to

$$x_{ijk} = \sum_{\ell_1, \ell_2, \ell_3} G(\ell_1, \ell_2, \ell_3) x_{\ell_1 i} x_{\ell_2 j} x_{\ell_3 k},$$

where $G(\ell_1, \ell_2, \ell_3) \in \mathbb{R}^{N \times M \times K}$ is a core tensor, $x_{\ell_1 i} \in \mathbb{R}^{N \times N}, x_{\ell_2 j} \in \mathbb{R}^{M \times M}$ and $x_{\ell_3 k} \in \mathbb{R}^{K \times K}$ are singular value matrices that are orthogonal. Because this construct is obviously overcomplete, there is no unique TD. In this paper, I employed higher-order singular value decomposition [34] (HOSVD) to perform TD.

2.2. Using TD-Based Unsupervised FE for Identification of Genes Whose Expression Is Likely to Be Altered by MiRNA Transfection

To this end, first I need to specify which sample singular value vectors, $x_{\ell_3 k}$, have different values between treated (i.e., miRNA-transfected) samples and control (e.g., mock-transfected) samples. Suppose $x_{\ell_3 k}$ turned out to represent a dissimilarity between a treated sample and control sample in some ways (see Figure 1B as an example). Next, I need to find miRNA singular value vectors, $x_{\ell_2 j}$, that have constant values for all j (see Figure 1A as an example) because I would like to find genes affected constantly by miRNA transfection independently of the type of transfected miRNA, since it should represent sequence-nonspecific off-target regulation. Let us assume that $x_{\ell_2 j}$ fulfilled this requirement. Then, I rank core tensors $G(\ell_1, \ell_2', \ell_3')$ in the order of absolute values (largest at the top). This approach enables me to select ℓ_1 such that $x_{\ell_1 i}$ is associated with constant sequence-nonspecific off-target regulation. After identifying ℓ_1' as those associated with larger absolute values of $G(\ell_1', \ell_2', \ell_3')$, is representing larger absolute values of $x_{\ell_1' i}$ were selected. For this purpose, P-values, P_is (see Figure 1C as an example), were assigned to each i assuming that $x_{\ell_1' i}$ obeys the χ^2 distribution

$$P_i = P_{\chi^2}\left[> \sum_{\ell_1'} \left(\frac{x_{\ell_1' i}}{\sigma_{\ell_1'}} \right)^2 \right]$$

where $P_{\chi^2}[> x]$ is cumulative probability that the argument is greater than x, assuming the χ^2 distribution and that $\sigma_{\ell_1'}$ is the standard deviation. The number of degrees of freedom of the χ^2 distribution is equal to the number of ℓ_1's in the summation. The above equation means that I presume that $x_{\ell_1' i}$ obeys a multiple Gaussian distribution (null hypothesis). Then, P_is were adjusted via the Benjamini–Hochberg (BH) criterion [35]. Genes associated with adjusted $P_i < 0.01$ were finally selected (see the red bin in Figure 1C as an example).

(A)	(B)	(C)

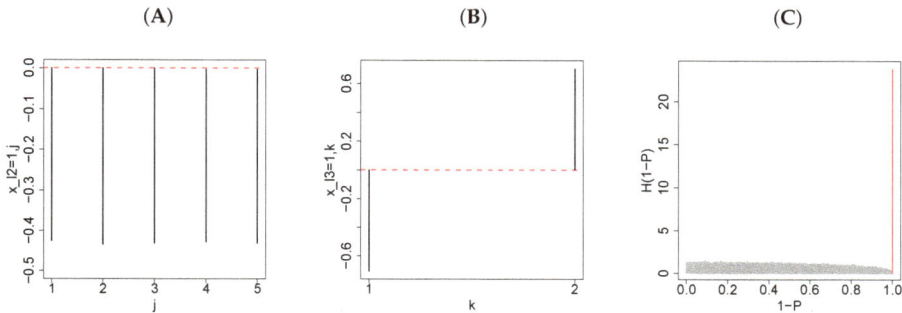

Figure 1. The results on the artificial data: (**A**) $x_{\ell_2=1,j}$ averaged across 100 independent trials. The horizontal red dashed line is $x_{\ell_2=1,j} = 0$ (**B**) $x_{\ell_3=1,k}$ averaged across 100 independent trials. The horizontal red dashed line is $x_{\ell_3=1,k} = 0$ (**C**) A histogram of $1 - P$ computed from $x_{\ell_1=1,i}$. A vertical red segment represents the bin with the smallest P-values.

2.3. Explanatory Discussion of TD-Based Unsupervised FE

Readers may wonder why a simple procedure using TD successfully identifies genes whose expression is likely to be altered by sequence-nonspecific off-target regulation mediated by various transfected miRNAs. This result can be explained as follows. Let us say most x_{ijk}s are a random number while a limited number of x_{ijk}s, e.g., $x_{i'jk}$s are coexpressed, i.e.,

$$x_{i'jk} = x_{jk}, i' = 1, \ldots, N'$$

Then, the contribution of $x_{i'jk}$ has the order of magnitude of N', while that of other $N - N'$ (random) $x_{ijk}, i \neq i'$ has the order of magnitude of $\sqrt{N - N'}$. Thus, even if $N' \ll N$, if $N' \approx \sqrt{N - N'}$, then the contribution of $x_{i'jk}$ outperforms that of $x_{ijk}, i \neq i'$. Thus, the contribution of $x_{i'jk}$ should be detected as a singular value vector, $x_{\ell_1 i}$ within which $x_{\ell_1 i'}$ should have a greater contribution than the others ($x_{\ell_1 i}, i \neq i'$). Given that the contribution of $x_{\ell_1 i}, i \neq i'$ is expected to be a Gaussian distribution, $x_{\ell_1 i'}$ can be detected as outliers that do not follow the Gaussian distribution. This is the possible explanation why simple TD-based unsupervised FE successfully identified genes whose expression was likely to be altered by sequence-nonspecific off-target regulation mediated by various transfected miRNAs.

2.4. Artificial Data

To demonstrate the usefulness of TD-based unsupervised FE, I prepared artificial dataset $x_{ijk} \in \mathbb{R}^{N \times M \times 2}$, where N is the number of genes, and M is the number of samples ($k = 1$ for control and $k = 2$ for treated samples). Each treated sample is supposed to be transfected with distinct miRNA, and each control sample is supposed to be either untransfected or transfected with mock miRNA. In these artificial data, at first $x_{ijk} \in \mathcal{N}(0, 1)$. After that, they are ordered such that $x_{ijk} > x_{i'jk}$ when $i > i'$ with fixed j and k. This situation introduces complete correlations between distinct pairs of js and ks (e.g., rank correlation coefficients between x_{ijk} and $x_{ij'k'}, j \neq j', k \neq k'$ are always equal to 1.0). Then, $x_{ijk}, i > N_0$ were shuffled at fixed j and k to eliminate the correlation. Next, $x_{ijk} \leftarrow (a - 1)x_{ijk} + a\epsilon, \epsilon \in \mathcal{N}(0, 1), a > 0$ for $i \leq N_0$ in order to introduce randomness into correlating rows. After that, $x_{ij2} \leftarrow -x_{ij2}$ to introduce a difference between control and treated samples. Finally, $\frac{N_0}{2} < i \leq N_0$ were shuffled at fixed j to generate a sample-specific difference between control and treated samples. This means that $1 \leq i \leq \frac{N_0}{2}$ correspond to a sample-nonspecific (i.e., independent of j) dissimilarity between control and treated samples that represents sequence-nonspecific off-target regulation, and $\frac{N_0}{2} < i \leq N_0$ correspond to a sample-specific (i.e., dependent on j) dissimilarity between control and treated samples that represents sequence-specific regulation. The task at hand is to identify $1 \leq i \leq \frac{N_0}{2}$ as precisely as possible. Specifically, $N = 2000, M = 5, N_0 = 50, a = 0.5$.

2.5. Gene Expression Profiles

In this subsection, I explain 11 analyzed profiles of gene expression (Table 1) in more detail. All of them were retrieved from Gene Expression Omnibus (GEO) [36]. In some cases (Experiments 1, 2, 3, and 5), I used a two-mode tensor, $x_{i(jk)}$, which is simply denoted as x_{ij}, instead of three-mode tensor x_{ijk}, by expanding the second (j) and the third (k) modes into one column (jk) because matched data were not available. In this case, HOSVD is equivalent to simple singular value decomposition. x_{ijk} and $x_{i(jk)}$ were standardized as $\sum_i x_{ijk} = \sum_i x_{i(jk)} = 0$ and $\sum_i x_{ijk}^2 = \sum_i x_{i(jk)}^2 = N$ before TD was applied.

Table 1. Eleven experiments conducted for this analysis. More detailed information is available in the text.

Exp.	GEO ID	Cell Lines (Cancer)	miRNA	Misc
1	GSE26996	BT549 (breast cancer)	miR-200a/b/c	
2	GSE27431	HEY (ovarian cancer)	miR-7/128	mas5
3	GSE27431	HEY (ovarian cancer)	miR-7/128	plier
4	GSE8501	Hela (cervical cancer)	miR-7/9/122a/128a/132/133a/142/148b/181a	
5	GSE41539	CD1 mice	cel-miR-67,hsa-miR-590-3p,hsa-miR-199a-3p	
6	GSE93290	multiple	miR-10a-5p,150-5p/5p,148a-3p/5p,499a-5p,455-5p	
7	GSE66498	multiple	miR-205/29a/144-3p/5p,210,23b,221/222/223	
8	GSE17759	EOC 13.31 microglia cells	miR-146a/b	(KO/OE)
9	GSE37729	HeLa	miR-107/181b	(KO/OE)
10	GSE37729	HEK-293	miR-107/181b	(KO/OE)
11	GSE37729	SH-SY5Y	181b	(KO/OE)

OE: overexpression.

2.5.1. No. 1: GSE26996

File GSE26996_RAW.tar was downloaded and unpacked. Six files, GSM665046_miR200a_1.txt.gz, GSM665047_miR200b_1.txt.gz, GSM665048_miR200c_1.txt.gz, GSM665049_miR200a_2.txt.gz, GSM665050_miR200b_2.txt.gz, and GSM665051_miR200c_2.txt.gz, were loaded into R using the read.csv function. Then, after the exclusion of probes with ControlType=0, gProcessedSignal and rProcessedSignal were extracted as treated and control samples, respectively. Thus, I have $x_{ij}, 1 \leq j \leq 12, 1 \leq i \leq 43376$. Here, $1 \leq j \leq 6$ and $7 \leq j \leq 12$ are treated and control samples, respectively. For this dataset, gene expression profiles are regarded as a two-mode tensor (matrix). Applying PCA to x_{ij}, I found that $x_{\ell_2=2,j}$ are different between treated and control samples (Figure S1) independently of the type of miRNA transfected. Next, genes were selected by means of $x_{\ell_1=2,i}$.

2.5.2. No. 2: GSE27431

File GSE27431_series_matrix.txt.gz was downloaded. It was loaded into R using the read.csv function. Each column corresponds to individual gene expression profiles named as GSEMXXXXXX, which is the GEO ID. Among those, mas5-processed samples are regarded as No. 2. GSM678153 and GSM678154 are miR-7-treated, GSM678156 and GSM678157 are miR-128-treated, whereas GSM678158, GSM678159, and GSM678160 are control samples. Then, $x_{ij}, 1 \leq i \leq 54675, 1 \leq j \leq 7$ are obtained (two-mode tensor). Applying TD to x_{ij}, I found that $x_{\ell_2=2,j}$ reflects the inverse regulation between miR-128 and miR-7 while the control samples are in between (Figure S2). This finding can be interpreted as follows. Target genes of miR-128 (miR-7) are downregulated, but they are upregulated when miR-7 is transfected because of sequence-nonspecific off-target regulation caused by miRNA transfection. Accordingly, I decided to select genes using $x_{\ell_1=2,i}$ again.

2.5.3. No. 3: GSE27431

GSE27431_series_matrix.txt.gz was again downloaded. It was loaded into R using the read.csv function. Each column corresponds to individual gene expression profiles named as GSMXXXXXX, which is a GEO ID as well. Among those, plier-processed samples are regarded as No. 3. GSM678164 and GSM678165 are miR-7-treated, GSM678167 and GSM678168 are miR-128-treated, and GSM678169, GSM678170, GSM678171, GSM678172, GSM678173, and GSM678174 are control samples. Next, x_{ij}, $1 \leq i \leq 54675, 1 \leq j \leq 10$ are obtained (two-mode tensor). Applying PCA to x_{ij}, I found that $x_{\ell_2=2,j}$ reflects the inverse regulation between miR-128 and miR-7, while the control samples are in between (Figure S3). This result can be interpreted as in No. 2. Thus, I decided to select genes by means of $x_{\ell_1=2,i}$ again.

2.5.4. No. 4: GSE8501

Eighteen raw data files were downloaded from GSMXXXXXX, $210896 \leq XXXXXX \leq 210913$, which are GEO IDs. Columns named as INTENSITY1 and INTENSITY2 are 18 control samples and 18 samples transfected with miR-7/9/122a/128a/132/133a/142/148b/181a (two replicates each), respectively. Then, I generated tensor $x_{ijk}, 1 \leq i \leq 23651, 1 \leq j \leq 18, k = 1, 2$, where js are two replicates of each of nine miRNA-transfected samples and $k = 1, 2$ are control (mock-transfected) and transfected samples, respectively. After applying HOSVD to x_{ijk}, I found that $x_{\ell_3=2,k}$ reflects an inverse relation of expression levels between controls and treated samples, while $x_{\ell_2=1,j}$ reflects constant expression regardless of the type of transfected miRNA (Figure S4). Next, I decided to use $x_{\ell_1 i}$ associated with large absolute values of $G(\ell_1, \ell_2 = 1, \ell_3 = 2)$. Given that $G(\ell_1 = 6, \ell_2 = 1, \ell_3 = 2)$ has the largest absolute value, I decided to use $x_{\ell_1=6,i}$ to select mRNAs.

2.5.5. No. 5: GSE41539

Four files,

GSM1018808_topo_1_empty_trimmed_RNA-Seq.txt.gz,
GSM1018809_topo_2_cel_mir_67_trimmed_RNA-Seq.txt.gz,
GSM1018810_topo_4_mir_590_3p_trimmed_RNA-Seq.txt.gz, and
GSM1018811_topo_3_mir_199a_3p_trimmed_RNA-Seq.txt.gz,

were downloaded from GSE41539. The fourth column ("Unique gene reads") reflected gene expression. Then, I got $x_{ij}, 1 \leq i \leq 36065, 1 \leq j \leq 4$ (two-mode tensor). Applying PCA to x_{ij}, I found that $x_{\ell_2=2,j}$ represents the difference between controls (mock-transfected and cel-miR-67-transfected) and miR-509/199a-3p-transfected samples (Figure S5). After that, I decided to use $x_{\ell_1=2,i}$ for mRNA selection.

2.5.6. No. 6: GSE93290

File GSE93290_RAW.tar was downloaded and unpacked. Sixteen files, from GSM2450420 to GSM2450435, were loaded into R using the read.csv function. In each file, columns named as gProcessedSignal and rProcessedSignal served as controls and treated samples, respectively. Then, I generated three-mode tensor $x_{ijk}, 1 \leq i \leq 62976, 1 \leq j \leq 16, k = 1, 2$, where js are 16 samples and $k = 1, 2$ are control (mock-transfected) and transfected samples. After applying HOSVD to x_{ijk}, I found that $x_{\ell_3=2,k}$ reflects inversely related expression levels between controls and treated samples, whereas $x_{\ell_2=1,j}$ reflects constant expression regardless of the type of transfected miRNAs (Figure S6). Next, I decided to use $x_{\ell_1 i}$ associated with large absolute values of $G(\ell_1, \ell_2 = 1, \ell_3 = 2)$. Because $G(\ell_1 = 7, \ell_2 = 1, \ell_3 = 2)$ has the largest absolute value, I decided to apply $x_{\ell_1=7,i}$ to select mRNAs.

2.5.7. No. 7: GSE66498

File GSE66498_RAW.tar was downloaded and unpacked. Among these data, 19 files were used, i.e., GSM1623420 to GSM1623422 (miR-205 transfected into cell lines PC3, DU145, and C4-2), GSM1623423 and GSM1623424 (miR-29a transfected into cell lines 786O and A498), GSM1623425 to GSM1623427 (miR-451/144-3p/5p transfected into the T24 cell line), GSM1623434 and GSM1623435 (24 and 48 h after miR-210 transfection into the 786O cell line), GSM1623436 to GSM1623439 (miR-145-5p/3p transfected into BOY and T24 cell lines), GSM1623440 (miR-23b transfected into 786O cells), GSM1623444 to GSM1623446 (miR-221/222/223 transfected into the PC3 cell line), and GSM1623447 (miR-223 transfected into the PC3M cell line). In each file, columns named as gProcessedSignal and rProcessedSignal served as controls and treated samples, respectively. I generated three-mode tensor $x_{ijk}, 1 \leq i \leq 62976, 1 \leq j \leq 19, k = 1, 2$, where js are the 19 samples and $k = 1, 2$ are control (mock-transfected) and transfected samples. After applying HOSVD to x_{ijk}, I found that $x_{\ell_3=2,k}$ reflects an inverse relation of expression levels between controls and treated samples, whereas $x_{\ell_2=1,j}$ reflects constant expression regardless of the type of transfected miRNAs (Figure S7). After that, I decided to use $x_{\ell_1 i}$ associated with large absolute values of $G(\ell_1, \ell_2 = 1, \ell_3 = 2)$. Because $G(\ell_1 \in (2,3), \ell_2 = 1, \ell_3 = 2)$ have the largest and almost the same absolute values, I decided to employ $x_{\ell_1=2,i}$ and $x_{\ell_1=3,i}$ to select mRNAs.

2.5.8. No. 8: GSE17759

File GSE17759_RAW.tar was downloaded and unpacked. Among these data, six replicates of miR-146a–overexpressing samples (GSM443535 to GSM443540), four replicates of miR-146b–overexpressing samples (GSM443541 to GSM443544), eight replicates of miR-146a knockout samples (GSM443557 to GSM443564), i.e., in total, 18 files were processed. In each file, columns named as gProcessedSignal and rProcessedSignal served as controls and treated samples, respectively. I generated three-mode tensor $x_{ijk}, 1 \leq i \leq 43379, 1 \leq j \leq 18, k = 1, 2$, where js are the 18 samples and $k = 1, 2$ are control (mock-transfected) and transfected samples. After applying HOSVD to x_{ijk}, I found that $x_{\ell_3=2,k}$ reflects an inverse relation of expression levels between controls and treated samples, while $x_{\ell_2=1,j}$ means constant expression regardless of the type of transfected miRNAs

(Figure S8). Then, I decided to use $x_{\ell_1 i}$ associated with large absolute values of $G(\ell_1, \ell_2 = 1, \ell_3 = 2)$. Given that $G(\ell_1 = 5, \ell_2 = 1, \ell_3 = 2)$ has the largest absolute value, I decided to apply $x_{\ell_1=5,i}$ to select mRNAs.

2.5.9. No. 9: GSE37729

Two files SE37729-GPL6098_series_matrix.txt.gz and GSE37729-GPL6104_series_matrix.txt.gz in the section "Series Matrix File(s)" were downloaded. The two files were merged such that only shared probes were included. Gene expression of HeLa cell lines is considered in Experiment No. 9. GSM926188, GSM926189, GSM926193, GSM926194, GSM926198, and GSM926201 are control samples. GSM926164 and GSM926165 are anti-miR-107-transfected samples. GSM926180, GSM926181, GSM926190, and GSM926191 are miR-107-transfected samples. GSM926162 and GSM926163 are anti-miR-181b-transfected samples. GSM926182, GSM926183, GSM926195, and GSM926196 are miR-181b-transfected samples. Then, I generated three-mode tensor $x_{ijk}, 1 \leq i \leq 9987, 1 \leq j \leq 6$, $1 \leq k \leq 3$, where $k = 1, 2, 3$ correspond to control, miR-7, and miR-181b, respectively. For $k = 2, 3$, $1 \leq j \leq 2$ are anti-miR-transfected samples and $3 \leq j \leq 6$ are miR-transfected samples. After applying HOSVD to x_{ijk}, I found that $x_{\ell_3=2,k}$ reflects an inverse relation of expression levels between controls and treated samples, while $x_{\ell_2=1,j}$ indicates constant expression regardless of the type of transfected miRNA (Figure S9). Next, I decided to employ $x_{\ell_1 i}$ associated with large absolute values of $G(\ell_1, \ell_2 = 1, \ell_3 = 2)$. Because $G(\ell_1 = 2, \ell_2 = 1, \ell_3 = 2)$ has the largest absolute value, I decided to use $x_{\ell_1=2,i}$ to select mRNAs.

2.5.10. No. 10: GSE37729

Two files, SE37729-GPL6098_series_matrix.txt.gz and GSE37729-GPL6104_series_matrix.txt.gz, in the section "Series Matrix File(s)" were downloaded. The two files were merged so that only shared probes are included. Gene expression of HEK 293 cell lines was considered in Experiment No. 10. GSM926206, GSM926207, GSM926211, GSM926212, GSM926216, and GSM926217 are control samples. GSM926168 and GSM926169 are anti-miR-107-transfected samples. GSM926184, GSM926185, GSM926208, and GSM926209 are miR-107-transfected samples. GSM926166 and GSM926167 are anti-miR-181b-transfected samples. GSM926186, GSM926187, GSM926213, and GSM926214 are miR-181b-transfected samples. Next, I generated three-mode tensor $x_{ijk}, 1 \leq i \leq 9987, 1 \leq j \leq 6$, $1 \leq k \leq 3$. $k = 1, 2, 3$ corresponding to control, miR-7, and miR-181b, respectively. For $k = 2, 3$, $1 \leq j \leq 2$ are anti-miR-transfected samples, and $3 \leq j \leq 6$ are miR-transfected samples. After applying HOSVD to x_{ijk}, I determined that $x_{\ell_3=2,k}$ reflects an inverse relation of expression levels between controls and treated samples, while $x_{\ell_2=1,j}$ reflected constant expression regardless of the type of transfected miRNAs (Figure S10). Thus, I decided to use $x_{\ell_1 i}$ associated with large absolute values of $G(\ell_1, \ell_2 = 1, \ell_3 = 2)$. Because $G(\ell_1 = 2, \ell_2 = 1, \ell_3 = 2)$ has the largest absolute value, I decided to employ $x_{\ell_1=2,i}$ to select mRNAs.

2.5.11. No. 11: GSE37729

Two files, SE37729-GPL6098_series_matrix.txt.gz and GSE37729-GPL6104_series_matrix.txt.gz, in the section "Series Matrix File(s)" were downloaded. The two files were merged such that only shared probes were included. Gene expression of SH-SY5Y cell lines was considered in Experiment No. 11. GSM926170, GSM926171, GSM926178, and GSM926179 are control samples. GSM926176 and GSM926177 are anti-miR-181b-transfected samples. GSM926174 and GSM926175 are miR-181b-transfected samples. After that, I generated three-mode tensor $x_{ijk}, 1 \leq i \leq 9987$, $1 \leq j \leq 4, 1 \leq k \leq 2$, where $k = 1, 2$ correspond to control and miR-181b, respectively. For $k = 2, 3$, $1 \leq j \leq 2$ are anti-miR-transfected samples, whereas $3 \leq j \leq 6$ are miR-transfected samples. After applying HOSVD to x_{ijk}, I found that $x_{\ell_3=2,k}$ reflects an inverse relation of expression levels between controls and treated samples, whereas $x_{\ell_2=1,j}$ reflects constant expression independently of transfected miRNAs (Figure S11). Then, I decided to use $x_{\ell_1 i}$ associated with large absolute values of

$G(\ell_1, \ell_2 = 1, \ell_3 = 2)$. Because $G(\ell_1 = 2, \ell_2 = 1, \ell_3 = 2)$ has the largest absolute value, I decided to use $x_{\ell_1=2,i}$ to select mRNAs.

3. Results

To demonstrate the usefulness of TD-based unsupervised FE, I applied it to artificial data composed of a three-mode tensor, $x_{ijk} \in \mathbb{R}^{N \times M \times 2}$, which is the expression level of the ith gene of the jth sample, where $k = 1$ is control and $k = 2$ is a treated (transfected with distinct miRNAs) sample. Among N genes, N_0 genes are affected by miRNA transfection while the other $N - N_0$ genes are not affected. Among the N_0 genes affected by miRNA transfection, $\frac{N_0}{2}$ genes are supposed to be regulated independently of samples (hence, a sequence-nonspecific off-target effect) while the other $\frac{N_0}{2}$ genes vary from sample to sample (i.e., miRNA-specific regulation). Applying TD-based unsupervised FE to the artificial dataset (averaged across 100 independent trials), I got a result (Figure 1). $x_{\ell_2=1,j}$ (Figure 1A) and $x_{\ell_3=1,k}$ (Figure 1B), which are always associated with core tensor $G(1,1,1)$ with the largest absolute values, represent constant gene expression across M samples and inverted expression levels between control ($k = 1$) and treated ($k = 2$) samples. Accordingly, genes associated with these two are expected to represent sample- or transfected miRNA-independent (thus, sequence-nonspecific off-target) regulation. Given that $x_{\ell_1=1,i}$ is always associated with core tensor $G(1,1,1)$ with the largest absolute values, P-values (Figure 1C) are computed using $x_{\ell_1=1,i}$. It is obvious that there is a sharp peak at the smallest P-values in the histogram of $1 - P$, which presumably does not correspond to the null hypothesis (that $x_{\ell_1=1,i}$ follows the normal distribution). To test whether genes associated with these much smaller P-values include genes $i \leq \frac{N_0}{2}$, the probabilities to be selected by TD-based unsupervised FE are averaged across $i \leq \frac{N_0}{2}$ and $i > \frac{N_0}{2}$, respectively. Then, the former is as large as 0.86, while the latter is as small as 0. (This means that genes $i > \frac{N_0}{2}$ have never been selected by TD-based unsupervised FE.) This observation suggests that TD-based unsupervised FE is effective at sorting out genes—that are expressed independently of samples—from genes expressed only in a limited number of samples and genes not expressed at all. To determine whether TD-based unsupervised FE can outperform the conventional supervised method, the t test and significance analysis of microarrays (SAM) [37] were carried out. For these two methods, P-values obtained were also corrected by means of the BH criterion, and genes associated with adjusted P-values less than 0.01 were selected. Then, the average probability for $i \leq \frac{N_0}{2}$ is 0.43 according to the t test and 0.62 according to SAM. The average probability for $i > \frac{N_0}{2}$ is 2×10^{-4} according to the t test and 3×10^{-5} according to SAM. Therefore, TD-based unsupervised FE is more effective than either the t test or SAM.

To identify genes whose expression alteration is likely to be mediated by sequence-nonspecific off-target regulation caused by miRNA transfection, integrated analysis of gene expression profiles after transfection of various miRNAs was performed. Simultaneous analysis of multiple experiments each of which employs single miRNA transfection will blur sequence-specific regulation of mRNAs while a sequence-nonspecific off-target effect will remain. Furthermore, to avoid biases due to research groups or individual studies, 11 experiments collected from studies involving distinct combinations of transfected miRNAs and cell lines were analyzed simultaneously (Table 1).

Genes—whose expression alteration is likely to be caused by sequence-nonspecific off-target regulation that was induced similarly by various transfected miRNAs—were selected using TD-based unsupervised FE in each of the 11 experiments. The reason why TD-based unsupervised FE was employed is as follows. First, PCA-based unsupervised FE, from which TD-based unsupervised FE was developed, is known to function even when only a small number of samples is available [16,26]. Tensor representation is also more suitable for the present experiments, where multiple genes, multiple miRNAs and controls, or transfected samples are simultaneously considered (they can be represented as a three-mode tensor; see Methods). Second, we can check whether there are miRNAs associated with a common expression pattern among all the miRNA transfection experiments by studying outcomes; we have opportunities to exclude experiments not associated with sequence-nonspecific off-target regulation caused by miRNA transfection.

These 11 analyzed experiments—in which mRNAs associated with sequence-nonspecific off-target regulation caused by miRNA transfection were successfully identified—deal with distinct cell lines into each of which distinct miRNAs were transfected. Nevertheless, gene sets identified in individual experiments not only significantly overlapped with one another but also were associated with a large enough odds ratio (from 300 to 500, Table 2), although the number of genes detected in each experiment varied from ~100 to ~800 ("#" in Table 2). This finding suggests that there are some sets of genes whose expression was robustly altered via sequence-nonspecific off-target regulation that was induced similarly by various transfected miRNAs.

Table 2. Fisher's exact tests for coincidence among 11 miRNA transfection experiments. Upper triangle: P-value; lower triangle: odds ratio.

Exp.	1	2	3	4	5	6	7	8	9	10	11
#	232	711	747	441	123	292	246	873	113	104	120
1 232		4.14×10^{19}	6.59×10^{22}	3.96×10^{41}	4.12×10^{71}	9.41×10^{70}	2.90×10^{60}	1.34×10^{17}	1.15×10^{27}	6.84×10^{26}	2.66×10^{7}
2 711	7.68		0.00	1.89×10^{18}	4.93×10^{27}	5.59×10^{20}	2.69×10^{32}	4.62×10^{13}	9.23×10^{16}	8.66×10^{12}	1.37×10^{3}
3 747	8.30	345.52		3.63×10^{20}	7.96×10^{21}	5.70×10^{12}	1.82×10^{27}	9.52×10^{12}	1.18×10^{14}	1.01×10^{12}	3.90×10^{6}
4 441	18.23	5.19	5.34		6.14×10^{41}	1.01×10^{34}	1.44×10^{69}	4.61×10^{11}	2.16×10^{30}	4.09×10^{28}	1.35×10^{10}
5 123	53.86	9.04	7.27	17.48		2.9×10^{179}	1.27×10^{63}	6.24×10^{15}	3.16×10^{25}	2.37×10^{17}	4.69×10^{9}
6 292	61.50	8.15	5.52	17.71	204.39		3.53×10^{53}	2.57×10^{15}	6.65×10^{22}	1.65×10^{12}	5.60×10^{5}
7 246	20.27	5.35	4.67	12.39	20.11	22.03		6.91×10^{42}	1.77×10^{36}	4.50×10^{31}	2.78×10^{14}
8 873	18.61	7.22	6.51	8.29	15.61	18.53	20.73		1.81×10^{7}	1.37×10^{6}	2.76×10^{2}
9 113	39.34	9.87	8.77	25.98	32.44	34.90	21.94	16.02		3.7×10^{125}	9.27×10^{18}
10 104	40.29	8.22	8.27	26.64	23.34	20.86	21.56	15.18	517.87		6.82×10^{16}
11 120	10.15	3.19	4.43	9.19	11.55	8.11	8.28	4.92	19.57	18.70	

#: the number of genes selected for each of 11 experiments via TD- or PCA-based unsupervised FE.

Although this finding itself is remarkable enough to be reported, the observed coincidences may be accidental for some unknown reason and may not be associated with anything biologically valid. To validate biological significance of the identified genes, they were uploaded to Enrichr [38], which is an enrichment analysis server validating various biological terms and concepts. As a result, these genes were found to be enriched with various biological terms and concepts (see below).

First, the identified gene sets mostly included various target genes of transcription factors (TFs) (Table 3). Although the number of TFs detected varied from ~10 to ~100 (#2 in Table 3), the detection of multiple instances of enrichment with genes that TFs target may be evidence that these genes cooperatively function in the cell because common TFs' target genes often have shared biological functions [39,40]. In particular, the most frequent cases of enrichment of TFs are common among the 11 experiments analyzed. These include EKLF, MYC, NELFA, and E2F1. These data can be biologically interpreted as follows.

Table 3. In each of 11 experiments, 20 top-ranked significant TFs whose sets of target genes significantly overlap with the set of genes selected for each experiment were identified. Then, EKLF, MYC, NELFA, and E2F1 turned out to be among the 20 top-ranked significant TFs for all 11 experiments.

			EKLF		MYC		NELFA		E2F1	
Exp.	#1	#2	OL	adj. *P*-Value	OL	adj. *P*-Value	OL	adj. *P*-Value	OL	adj. *P*-Value
1	232	30	40/1239	2.16×10^7	53/1458	6.22×10^{12}	59/2000	8.94×10^{10}	61/1529	1.30×10^{15}
2	711	77	94/1239	1.51×10^{10}	106/1458	1.01×10^{10}	134/2000	3.26×10^{11}	100/1529	6.14×10^8
3	747	97	100/1239	2.28×10^{11}	98/1458	2.96×10^7	152/2000	1.93×10^{15}	108/1529	3.89×10^9
4	441	43	83/1239	4.77×10^{18}	99/1458	2.08×10^{22}	105/2000	1.06×10^{15}	85/1529	9.29×10^{14}
5	123	45	26/1239	2.16×10^6	25/1458	9.12×10^5	31/2000	4.26×10^5	28/1529	7.41×10^6
6	292	19	51/1239	2.38×10^9	65/1458	5.54×10^{14}	63/2000	2.72×10^7	69/1529	8.31×10^{15}
7	246	11	37/1239	5.11×10^5	48/1458	5.97×10^8	46/2000	1.45×10^3	64/1529	7.02×10^{16}
8	873	55	188/1239	8.33×10^{52}	189/1458	8.58×10^{42}	222/2000	3.52×10^{39}	157/1529	8.24×10^{23}
9	113	36	24/1239	4.47×10^6	30/1458	2.86×10^8	32/2000	2.33×10^6	40/1529	6.84×10^{15}
10	104	22	27/1239	1.16×10^8	25/1458	4.83×10^6	36/2000	1.07×10^9	35/1529	3.63×10^{12}
11	120	22	21/1239	8.02×10^4	27/1458	2.25×10^5	29/2000	4.68×10^4	25/1529	3.57×10^4

#1: the number of genes selected for each of 11 experiments via TD- or PCA-based unsupervised FE; #2: the number of TFs whose sets of target genes significantly (adjusted *P*-values < 0.01) overlap with the set of genes selected for each experiment; OL: overlaps, (the number of genes coinciding with the genes selected for each experiment)/(genes listed in Enrichr as TF target genes).

The apparent significant alteration of expression of MYC target genes may be due to miRNAs regulating *MYC* [41–43]. The apparent alteration of expression of E2F1 target genes may be explained similarly because these miRNAs also target *E2F1* [41,43]. In actuality, 1551 genes targeted by only one miRNA but associated with altered gene expression caused by miRNA transfection are significantly targeted by MYC and E2F1. Enrichment with EKLF target genes among these 1551 genes not targeted by more than one miRNA—but showing altered expression caused by transfection with one of miRNAs—is a puzzle, although Yien and Beiker suggested that some miRNAs are likely also regulated by EKLF [44]. Identification of NELFA target genes is explained by the tight interaction between NELFA and c-MYC [45,46]

Additionally, I checked "TargetScan microRNA" in Enrichr to test whether enrichment with genes targeted by the transfected miRNAs would be detected. As a result, only for two of the 11 experiments (Experiments No. 2 and 3), enrichment with genes targeted by nonzero miRNAs was detected. This is possibly because $x_{\ell_2=2,j}$ for these two experiments also detected genes targeted by transfected miRNAs (Figures S2 and S3). In any case, the fact that most of experiments (9 out of 11) did not show enrichment with genes targeted by transfected miRNAs is consistent with the hypothesis that gene expression alteration caused by miRNA transfection is primarily due to the competition for protein machinery between endogenous miRNAs and transfected miRNAs; this pattern can remain unchanged among transfection experiments with different miRNAs.

Next, I checked KEGG pathway enrichment data. Primary enriched KEGG pathways among the identified genes are related to diseases (Table 4). Seven ((ii), (iii), (v), (vi), (vii), (viii), and (x)) out of 10 most frequently enriched KEGG pathways in the 11 experiments are directly related to various diseases. Among the remaining three ((i), (iv), and (ix)), oxidative phosphorylation is a disease-related KEGG pathway because its malfunction causes combined oxidative phosphorylation deficiency (https://www.omim.org/entry/609060). Another one, "endoplasmic reticulum", is also a disease-related pathway because its malfunction is observed in neurological diseases [47]. Therefore, these genes may also contribute to the onset or progression of various diseases and could be therapeutic targets. As a result, sequence-nonspecific off-target regulation caused by miRNA transfection may be a therapeutic method.

Table 4. In each of 11 experiments, 20 top-ranked significant KEGG pathways whose associated genes significantly match some genes selected for each experiment were identified. Thus, the following KEGG pathways are most frequently ranked within the top 20.

Exp.	#	(i)	(ii)	(iii)	(iv)	(v)	(vi)	(vii)	(viii)	(ix)	(x)
1	(232)	31/137	7/168	10/142	6/133	9/55	9/193			7/169	8/203
	[10]	3.69×10^{29}	3.18×10^{2}	1.66×10^{4}	3.45×10^{2}	1.02×10^{6}	6.85×10^{3}			3.18×10^{2}	3.01×10^{2}
2	(711)	36/137	18/168	14/142	12/133	13/55				16/169	18/203
	[12]	3.43×10^{19}	1.48×10^{3}	1.05×10^{2}	3.20×10^{2}	5.92×10^{6}				8.12×10^{3}	8.12×10^{3}
3	(747)	23/137	15/168			14/55				18/169	19/203
	[15]	3.58×10^{7}	1.94×10^{2}			1.20×10^{6}				2.02×10^{3}	4.78×10^{3}
4	(441)	50/137	15/168	19/142	18/133	6/55	19/193	7/78	12/151	9/169	
	[10]	2.92×10^{45}	1.91×10^{4}	3.97×10^{8}	6.42×10^{8}	2.49×10^{2}	3.40×10^{6}	2.74×10^{2}	4.44×10^{3}	1.29×10^{1}	
5	(123)	9/137						8/78		6/169	8/203
	[23]	2.97×10^{6}						6.08×10^{7}		4.29×10^{3}	3.03×10^{4}
6	(292)	45/137	20/168	19/142	18/133	4/55	19/193	11/78	12/151		
	[14]	1.35×10^{46}	3.32×10^{11}	2.27×10^{11}	4.00×10^{11}	7.95×10^{2}	2.24×10^{9}	4.90×10^{7}	4.87×10^{5}		
7	(246)	40/137	9/168	10/142	9/133		11/193	4/78	7/151		6/203
	[6]	5.61×10^{42}	6.60×10^{3}	5.80×10^{4}	1.32×10^{3}		1.16×10^{3}	2.57×10^{1}	6.31×10^{2}		4.52×10^{1}
8	(873)	75/137	30/168	32/142	32/133		36/193	14/78	24/151	25/169	
	[24]	5.59×10^{63}	2.09×10^{9}	9.32×10^{13}	1.89×10^{13}		7.51×10^{12}	1.39×10^{4}	1.62×10^{6}	3.11×10^{6}	
9	(113)	18/137	11/168	12/142	10/133	6/55	12/193	4/78	11/151		
	[20]	8.24×10^{18}	7.10×10^{8}	1.66×10^{9}	8.42×10^{8}	8.85×10^{6}	2.96×10^{8}	6.64×10^{3}	2.96×10^{8}		
10	(104)	11/137	8/168	9/142	8/133	5/55	10/193		8/151		
	[20]	1.98×10^{8}	6.68×10^{5}	3.23×10^{6}	1.71×10^{5}	1.56×10^{4}	3.23×10^{6}		3.60×10^{5}		
11	(120)	6/137		4/142		5/55					5/203
	[3]	9.04×10^{3}		8.49×10^{2}		2.98×10^{3}					6.83×10^{2}

(i) Ribosome: hsa03010; (ii) Alzheimer's disease: hsa05010; (iii) Parkinson's disease: hsa05012; (iv) Oxidative phosphorylation: hsa00190; (v) Pathogenic *Escherichia coli* infection:hsa05130; (vi) Huntington's disease: hsa05016; (vii) Cardiac muscle contraction: hsa04260; (viii) Nonalcoholic fatty liver disease (NAFLD): hsa04932; (ix) Protein processing in endoplasmic reticulum: hsa04141; and (x) Proteoglycans in cancer: hsa05205. (numbers): gene; [numbers]: KEGG pathways. Upper rows in each exp: (the number of genes coinciding with the genes selected for each experiment)/(genes listed in Enrichr in each category). Lower rows in each exp: adjusted *P*-values provided by Enrichr.

Actually, gene expression alteration mediated by sequence-nonspecific off-target regulation is analogous to treatments with various candidate drugs (Tables 5 and 6). Thus, combinatorial transfection with miRNAs may replace drug treatment in some conditions and can be used for therapeutic purposes, too. For example, LDN-192189 ((ii) in Table 5) is reported to improve neuronal conversion of human fibroblasts [48] and was proposed as a therapy for Alzheimer's disease [49]. GSK-1059615b ((i) and (iii) in Table 5) was once considered a PI3K-kt pathway inhibitor in clinical development for the treatment of cancers [50]. WYE-125132 ((iv) in Table 5) is also known to suppress tumor growth [51] (although the name WYE-125132 does not appear in that article, compound 8a was named WYE-125132 later). Afatinib ((v) in Table 5) is a famous drug for non-small cell lung cancer [52]. PI-103 ((vi) in Table 5) is a drug for acute myeloid leukemia [53]. PD-0325901 was reported to affect heart development [54]. Chelerythrine chloride ((viii) in Table 5) has been reported to affect embryonic chick heart cells [55]. GDC-0980 ((i) in Table 6) is a known anticancer drug [56]. PLX-4720 ((ii) in Table 6) has been considered for both cancer and heart disease treatment [57]. Dinaciclib is another anticancer drug [58]. Because these are related to diseases reported in Table 4, sequence-nonspecific off-target regulation caused by miRNA transfection may be a therapeutic strategy.

Table 5. In each of 11 experiments, 20 top-ranked significant treatments with compounds whose downregulated genes significantly coincide with some genes selected for each experiment were identified. Thus, treatments with the following compounds are most frequently ranked within top 20.

Exp.	#		(i)	(ii)	(iii)	(iv)	(v)	(vi)	(vii)	(viii)
1	(232)	OL	10/85				16/154			
	[129]	adj *P*-value	6.14×10^5				6.59×10^7			
2	(711)	OL								
	[329]	adj *P*-value								
3	(747)	OL								
	[417]	adj *P*-value								
4	(441)	OL	16/85		14/105	15/109		16/144		
	[67]	adj *P*-value	8.56×10^7		1.15×10^4	5.10×10^5		1.56×10^4		
5	(123)	OL		17/141			12/154			
	[219]	adj *P*-value		2.87×10^{13}			2.26×10^7			
6	(292)	OL			13/105					19/137
	[132]	adj *P*-value			9.69×10^6					3.04×10^9
7	(246)	OL	9/85				12/154			
	[61]	adj *P*-value	1.41×10^3				8.88×10^4			
8	(873)	OL	30/85	46/141					35/162	
	[255]	adj *P*-value	1.19×10^{15}	1.62×10^{23}					5.21×10^{12}	
9	(119)	OL	9/85	11/141		8/109		8/144	8/162	12/137
	[74]		6.25×10^6	3.48×10^6	3.50×10^4		1.45×10^3	2.33×10^3	2.76×10^7	
10	(104)	OL	9/85		9/105	9/109		10/144		12/137
	[155]	adj *P*-value	1.87×10^6		4.96×10^6	6.22×10^6		4.96×10^6		7.01×10^8
11	(120)	OL		10/141					9/162	
	[127]	adj *P*-value		4.74×10^5					5.26×10^4	

(i) LJP005_BT20_24H-GSK-1059615-3.33; (ii) LJP005_HS578T_24H-LDN-193189-10; (iii) LJP005_MCF10A_24H-GSK-1059615-10; (iv) LJP006_BT20_24H-WYE-125132-10; (v) LJP006_HS578T_24H-afatinib-10; (vi) LJP006_MCF10A_24H-PI-103-10; (vii) LJP007_HT29_24H-PD-0325901-0.12; and (viii) LJP009_HEPG2_24H-chelerythrine_chloride-10. #: (numbers): genes; [numbers]: compounds. Upper rows in each exp: OL, overlaps, (the number of genes coinciding with the genes selected for each experiment)/(genes listed in Enrichr in each category). Lower rows in each exp: adjusted *P*-values provided by Enrichr.

Table 6. In each of the 11 experiments, 20 top-ranked significant treatments with compounds whose sets of upregulated genes significantly overlap with the set of genes selected for each experiment were identified. Therefore, treatment with the following compounds is most frequently ranked within the top 20.

Exp.	#	(i)	(ii)	(iii)
1	(232)			16/166
	[191]			2.16×10^7
2	(711)	23/125	30/163	
	[450]	1.58×10^7	1.11×10^9	
3	(747)		34/163	
	[559]		6.14×10^{12}	
4	(441)	15/125		
	[85]	2.12×10^4		
5	(123)			
	[116]			
6	(292)		18/163	
	[190]		2.46×10^7	
7	(246)			
	[145]			
8	(873)	22/125		31/166
	[69]	6.56×10^5		1.80×10^7
9	(119)			8/166
	[0]			1.95×10^2
10	(104)			
	[0]			
11	(120)			
	[33]			

(i) LJP005_A375_24H-GDC-0980-0.37; (ii) LJP005_HEPG2_24H-PLX-4720-10; and (iii) LJP007_MCF7_24H-dinaciclib-0.12 #: (numbers): genes; [numbers]: compounds. Upper rows in each exp: OL, overlaps (the number of genes coinciding with the genes selected for each experiment)/(genes listed in Enrichr in each category). Lower rows in each exp: adjusted *P*-values provided by Enrichr.

Next, I studied tissue specificity of the gene expression alteration caused by sequence-nonspecific off-target regulation that was induced by miRNA transfection. Associations with GTEx tissue samples were also observed. For example, in GTEx up (Table 7), enrichment cases were observed in the brain and testes, both of which were reported to be outliers in clustering analysis [59]. Thus, it is not surprising that cases of enrichment in these two tissues were identified primarily. On the other hand, in GTEx down (Table 8), enrichment instances in the skin were mostly observed. Readers may wonder how sequence-nonspecific off-target regulation caused by miRNA transfection can contribute to the tissue specificity of gene expression profiles. Võsa et al. [60] observed that polymorphisms in miRNA response elements (MRE-SNPs) that either disrupt a miRNA-binding site or create a new miRNA-binding site can affect the allele-specific expression of target genes. Therefore, sequence-nonspecific off-target regulation caused by miRNA transfection may have the ability to contribute to tissue specificity of the gene expression profiles through distinct functionality of genetic variations in distinct tissues. Despite these coincidences, how the sequence-nonspecific off-target effect contributes to differentiation is unclear. These coincidences may simply be consequences, not causes. More studies are needed to directly implicate the sequence-nonspecific off-target effect in differentiation itself.

Table 7. In each of the 11 experiments, 20 top-ranked significant tissues whose set of upregulated genes significantly overlapped with the set of genes selected for each experiment were identified.

Exp.	#	(i)	(ii)	(iii)	(iv)	(v)	(vi)	(vii)	(viii)	(ix)	(x)
1	(232)	68/1509	51/1066	65/1384	74/1773	77/1764		69/1770		53/1174	68/1710
	[216]	1.11×10^{20}	3.65×10^{16}	1.28×10^{20}	8.31×10^{21}	8.66×10^{23}		8.65×10^{18}		8.12×10^{16}	7.14×10^{18}
2	(711)						78/625				
	[470]						3.13×10^{20}				
3	(747)						72/625				
	[441]						1.93×10^{15}				
4	(441)	80/1509	61/1066	82/1384		87/1764	54/625		54/525		
	[151]	3.20×10^{11}	1.01×10^{9}	7.82×10^{14}		1.08×10^{5}	6.23×10^{15}		6.83×10^{18}		
5	(123)	31/1509	29/1066	32/1384	33/1773			34/1770		26/1174	33/1710
	[150]	2.48×10^{7}	9.29×10^{9}	1.75×10^{8}	5.68×10^{7}			1.99×10^{7}		9.17×10^{7}	2.90×10^{7}
6	(292)	80/1509	66/1066	70/1384	80/1773	85/1764	46/625	78/1770	50/525	63/1174	76/1710
	[196]	7.02×10^{22}	3.72×10^{21}	4.91×10^{18}	4.91×10^{18}	4.97×10^{21}	2.53×10^{17}	4.73×10^{17}	4.99×10^{23}	2.53×10^{17}	7.45×10^{17}
7	(246)	64/1509	53/1066	57/1384	68/1773	68/1764	34/625	63/1770	35/525	52/1174	63/1710
	[135]	7.77×10^{16}	1.13×10^{15}	1.02×10^{13}	3.42×10^{15}	3.28×10^{15}	5.23×10^{11}	1.16×10^{12}	1.03×10^{13}	1.10×10^{13}	2.75×10^{13}
8	(873)	164/1509	131/1066	156/1384	155/1773	168/1764		161/1770		128/1174	157/1710
	[178]	1.61×10^{25}	8.18×10^{25}	1.61×10^{25}	2.25×10^{15}	4.48×10^{20}		2.02×10^{17}		1.31×10^{19}	2.38×10^{17}
9	(119)	39/1509	28/1066	40/1384	38/1773	37/1764	24/625		25/525		
	[225]	1.13×10^{13}	8.60×10^{10}	2.22×10^{15}	5.07×10^{11}	1.92×10^{10}	3.31×10^{11}		1.13×10^{13}		
10	(104)	29/1509	22/1066	30/1384	31/1773	29/1764		28/1770	15/525	23/1174	27/1710
	[156]	2.83×10^{7}	5.82×10^{6}	1.93×10^{8}	4.52×10^{7}	4.31×10^{6}		1.19×10^{5}	1.35×10^{5}	5.82×10^{6}	1.51×10^{5}
11	(120)						13/625		12/525		
	[5]						1.54×10^{2}		1.38×10^{2}		

(i) GTEX-QDT8-0011-R10A-SM-32PKG_brain_female_30-39_years; (ii) GTEX-QMR6-1426-SM-32PLA_brain_male_50-59_years; (iii) GTEX-TSE9-3026-SM-3DB76_brain_female_60-69_years; (iv) GTEX-PVOW-0011-R3A-SM-32PKX_brain_male_40-49_years; (v) GTEX-PVOW-2526-SM-2XCF7_brain_male_40-49_years; (vi) GTEX-XAJ8-1326-SM-47JYT_testis_male_40-49_years; (vii) GTEX-N7MS-0011-R3a-SM-33HC6_brain_male_60-69_years; (viii) GTEX-OHPM-2126-SM-3LK75_testis_male_50-59_years; (ix) GTEX-PVOW-0011-R5A-SM-32PL7_brain_male_40-49_years; and (x) GTEX-PVOW-2626-SM-32PL8_brain_male_40-49_years. (numbers): gene; [numbers]: tissues. *P*-values are the ones adjusted by Enrichr.

Table 8. In each of the 11 experiments, 20 top-ranked significant tissues whose set of downregulated genes significantly overlapped with the set of genes selected for each experiment were identified.

Exp.	#	(i)	(ii)	(iii)	(iv)	(v)	(vi)
1	(232)	57/1709	54/1488	53/1027	53/1121	55/1599	52/1103
	[201]	4.88×10^{11}	1.40×10^{11}	4.65×10^{17}	1.13×10^{15}	4.65×10^{11}	1.96×10^{15}
2	(711)	166/1709			133/1121	175/1599	
	[414]	4.72×10^{32}			1.27×10^{33}	2.48×10^{40}	
3	(747)	174/1709				165/1599	
	[365]	1.44×10^{33}				2.20×10^{32}	
4	(441)	98/1709	89/1488	73/1027		98/1599	70/1103
	[219]	2.41×10^{16}	8.07×10^{16}	2.17×10^{16}		5.24×10^{18}	1.38×10^{13}
5	(123)			35/1027	35/1121		37/1103
	[486]			9.87×10^{15}	1.04×10^{13}		2.03×10^{15}
6	(292)		69/1488	60/1027	53/1121		61/1103
	[171]		2.30×10^{15}	3.96×10^{17}	5.96×10^{12}		9.15×10^{17}
7	(246)		55/1488	57/1027	49/1121	53/1599	51/1103
	[113]		2.46×10^{11}	7.80×10^{19}	1.81×10^{12}	3.39×10^{9}	1.38×10^{13}
8	(873)	188/1709		137/1027	136/1121	182/1599	138/1103
	[185]	1.15×10^{30}		8.37×10^{30}	1.34×10^{25}	3.79×10^{31}	3.16×10^{27}
9	(119)	31/1709	36/1488				
	[224]	4.11×10^{7}	6.46×10^{11}				
10	(104)	31/1709	33/1488				
	[156]	3.24×10^{8}	1.64×10^{10}				
11	(120)						
	[20]						

(i) GTEX-Q2AH-0008-SM-48U2J_skin_male_40-49_years; (ii) GTEX-O5YT-0126-SM-48TBW_skin_male_20-29_years; (iii) GTEX-P4PQ-0008-SM-48TDX_skin_male_60-69_years; (iv) GTEX-R55D-0008-SM-48FEV_skin_male_50-59_years; (v) GTEX-R55E-0008-SM-48FCG_skin_male_20-29_years; and (vi) GTEX-RU72-0008-SM-46MV8_skin_female_50-59_years. (numbers): gene; [numbers]: tissues. Upper rows in each exp: OL, overlaps (the number of genes coinciding with the genes selected for each experiment)/(genes listed in Enrichr in each category). Lower rows in each exp: adjusted P-values provided by Enrichr.

Genes whose expression alteration is likely to be caused by sequence-nonspecific off-target regulation that miRNA transfection induces are also enriched in pluripotency (Table 9). Because miRNAs are known to mediate reprogramming [61], sequence-nonspecific off-target regulation caused by transfection of multiple miRNAs may contribute to reprogramming too. In actuality, primary binding TFs include MYC ((i) and (iv) in Table 9) and KLF4 ((vi) and (vii) in Table 9), which are two of four Yamanaka factors that mediate pluripotency. Other TFs in Table 9 are DMAP1 (iii) and TIP60 (v). DMAP1 is a member of the TIP60-p400 complex that maintains embryonic stem cell pluripotency [62]. The remaining one, ZFX (x), has been reported to control the self-renewal of embryonic and hematopoietic stem cells [63]. In addition, two gene KO experiments have been conducted ((viii) and (ix)). One of the genes in question, *SUZ12*, encodes a polycomb group protein that mediates differentiation [64], whereas the other, *ZFP281*, is a known pluripotency suppressor [65]. There are no known widely accepted mechanisms by which miRNA transfection can induce pluripotency. Because sequence-nonspecific off-target regulation seems to alter expression of genes critical for pluripotency, it may contribute to the mechanism.

Table 9. In each of the 11 experiments, 20 top-ranked significant terms in the Embryonic Stem Cell Atlas from Pluripotency Evidence (ESCAPE) whose set of associated genes significantly overlapped with the set of genes selected for each experiment were identified.

Exp.	#	(i)	(ii)	(iii)	(iv)	(v)	(vi)	(vii)	(viii)	(ix)	(x)
1	(232)	53/1458	90/2469	55/1789	56/1200	24/705	38/1700	40/1502	15/315	17/186	66/3249
	[15]	3.85×10^{12}	1.63×10^{22}	7.36×10^{10}	1.35×10^{17}	7.38×10^{5}	1.40×10^{3}	2.40×10^{5}	1.13×10^{4}	2.84×10^{9}	6.10×10^{5}
2	(711)	106/1458	184/2469	96/1789	90/1200				33/315		182/3249
	[22]	1.29×10^{10}	2.83×10^{21}	5.46×10^{4}	1.09×10^{9}				1.43×10^{6}		4.08×10^{9}
3	(747)	98/1458	199/2469	106/1789	93/1200	55/705			32/315	19/186	184/3249
	[28]	4.59×10^{7}	7.29×10^{25}	2.60×10^{5}	1.88×10^{9}	8.48×10^{6}			1.09×10^{5}	1.17×10^{3}	9.00×10^{8}
4	(441)	99/1458	153/2469	109/1789	95/1200	51/705	64/1700	60/1502		23/186	145/3249
	[18]	1.37×10^{22}	1.91×10^{32}	1.74×10^{21}	2.31×10^{26}	3.51×10^{12}	3.44×10^{4}	1.21×10^{4}		8.71×10^{10}	1.39×10^{16}
5	(123)	25/1458	46/2469	26/1789					19/315		
	[32]	3.65×10^{5}	9.91×10^{11}	3.38×10^{4}					1.66×10^{11}		
6	(292)	65/1458	108/2469	54/1789	59/1200	24/705	47/1700	37/1502	13/315	19/186	68/3249
	[8]	2.14×10^{14}	3.68×10^{25}	1.03×10^{5}	1.26×10^{14}	3.96×10^{3}	6.28×10^{4}	2.48×10^{2}	1.89×10^{2}	2.10×10^{9}	2.34×10^{2}
7	(246)	48/1458	78/2469	43/1789	49/1200	17/705	31/1700	33/1502	13/315	15/186	
	[6]	2.16×10^{8}	1.87×10^{13}	6.95×10^{4}	8.53×10^{12}	1.48×10^{1}	2.48×10^{1}	2.70×10^{2}	5.72×10^{3}	6.01×10^{7}	
8	(873)	189/1458	303/2469	184/1789	159/1200	89/705	130/1700	116/1502	38/315	38/186	247/3249
	[28]	5.44×10^{42}	6.58×10^{67}	1.81×10^{27}	6.07×10^{36}	4.59×10^{18}	4.78×10^{9}	2.74×10^{8}	3.67×10^{7}	4.46×10^{14}	1.89×10^{18}
9	(119)	30/1458	54/2469	37/1789	28/1200	16/705	23/1700	28/1502		9/186	32/3249
	[14]	1.76×10^{8}	4.54×10^{18}	1.22×10^{10}	6.09×10^{9}	6.01×10^{5}	1.46×10^{3}	5.60×10^{7}		3.47×10^{5}	1.15×10^{2}
10	(104)	25/1458	56/2469	33/1789	21/1200	13/705	23/1700	19/1502		10/186	
	[11]	3.77×10^{6}	3.80×10^{22}	4.45×10^{9}	2.91×10^{5}	1.53×10^{3}	4.39×10^{4}	4.56×10^{3}		2.87×10^{6}	
11	(120)	27/1458	44/2469		17/1200	13/705	23/1700	25/1502	7/315		
	[14]	1.28×10^{5}	1.60×10^{9}		1.20×10^{2}	5.75×10^{3}	4.08×10^{3}	1.97×10^{4}	3.00×10^{2}		

(i) CHiP_MYC-19079543; (ii) mESC_H3K36me3_18692474; (iii) CHiP_DMAP1-20946988; (iv) CHiP_MYC-18555785; (v) CHiP_TIP60-20946988; (vi) CHiP_KLF4-18358816; (vii) CHiP_KLF4-19030024; (viii) SUZ12-17339329_UP; (ix) ZFP281-21915945_DOWN; and (x) CHiP_ZFX-18555785. (numbers): gene; [numbers]: TF binding, histone modification and a gene KO or overexpression. Upper rows in each exp: OL, overlaps (the number of genes coinciding with the genes selected for each experiment)/(genes listed in Enrichr in each category). Lower rows in each exp: adjusted *P*-values provided by Enrichr.

Besides, I checked whether protein–protein interactions (PPIs) are enriched in each of the 11 gene sets selected for each of the 11 experiments (Table 10). Gene sets were uploaded to the STRING server [66]. For all the gene sets, instances of PPI enrichment were highly significant. Given that proteins rarely function alone and often function in groups, this is evidence that our analysis is biologically reliable.

Table 10. PPI enrichment by the STRING server. Column "genes" shows numbers of genes recognized by the STRING server.

Exp.	Genes	Edges		P-Values
		Observed	Expected	
1	195	1638	591	0
2	623	4271	2577	0
3	658	4506	2920	0
4	392	3418	1273	0
5	118	539	197	0
6	276	2048	569	0
7	182	1167	303	0
8	640	10176	4342	0
9	112	464	165	0
10	103	323	127	0
11	118	143	104	1.58×10^{4}

I demonstrated enrichment of various biological processes in gene sets. Although there were more cases of enrichment of biological terms in Enrichr, it is impossible to consider and discuss all of them. Instead, I discuss the enrichment cases identified in GeneSigDBs in Enrichr that include various biological properties (Table 11). Numerous GeneSigDBs that reflect a wide range of functional genes were enriched too. This result also suggested that the identified genes may contribute to a wide range of biological activities.

Table 11. In each of the 11 experiments, 20 top-ranked significant terms in GeneSigDB whose set of associated genes significantly overlapped with the set of genes selected for each experiment were identified.

			(i)		(ii)		(iii)		(iv)	
Exp.	#1	#2	OL	adj. *P*-Value	OL	adj. *P*-Value	OL	adj. *P*-Value	OL	adj. *P*-Value
1	232	154	97/1548	7.60×10^{44}	36/663	1.68×10^{12}	77/2585	1.72×10^{13}	36/238	6.63×10^{27}
2	711	194	152/1548	4.03×10^{29}	72/663	3.28×10^{15}	222/2585	8.96×10^{36}	44/238	3.05×10^{17}
3	747	285	149/1548	4.32×10^{25}	87/663	1.58×10^{22}	222/2585	4.01×10^{32}	30/238	2.05×10^{7}
4	441	106	146/1548	7.43×10^{52}	62/663	6.57×10^{20}	145/2585	1.51×10^{25}	46/238	2.15×10^{27}
5	123	183	44/1548	4.46×10^{16}	13/663	1.84×10^{3}	37/2585	1.34×10^{5}	24/238	1.54×10^{19}
6	292	106	103/1548	1.21×10^{38}	36/663	1.79×10^{9}	94/2585	1.51×10^{15}	39/238	5.50×10^{27}
7	246	95	81/1548	5.23×10^{28}	30/663	1.34×10^{7}	66/2585	3.88×10^{7}	30/238	2.86×10^{19}
8	873	269	256/1548	1.82×10^{81}	100/663	8.26×10^{26}	229/2585	4.03×10^{25}	86/238	7.47×10^{53}
9	119	93	60/1548	6.98×10^{34}	30/663	9.55×10^{17}	48/2585	9.73×10^{13}	22/238	4.26×10^{18}
10	104	77	52/1548	1.67×10^{27}	25/663	1.59×10^{12}	45/2585	4.60×10^{12}	15/238	2.49×10^{10}
11	120	68	36/1548	6.11×10^{10}	17/663	4.25×10^{5}	40/2585	1.07×10^{6}	13/238	6.73×10^{7}

(i) A multiclass predictor based on a probabilistic model, i.e., application to gene expression profiling-based diagnosis of thyroid tumors; (ii) A redox signature score identifies diffuse large B-cell lymphoma patients with a poor prognosis; (iii) A comparison of the gene expression profile of undifferentiated human embryonic stem cell lines and differentiating embryoid bodies; and (iv) A gene expression profile of rat left ventricles reveals persisting changes after a chronic mild-exercise protocol: implications for cardioprotection. #1: genes; #2: terms. OL: overlaps (the number of genes coinciding with the genes selected for each experiment)/(genes listed in Enrichr in each category).

As readers can see, top four most frequently significant terms are related to either diseases or differentiation, which are often said to be biological concepts to which miRNAs contribute to. Although canonical target genes of miRNA have mainly been sought to understand biological functions of miRNAs, sequence-nonspecific off-target regulation may be important as well.

Figure 2 shows the summary of results obtained in this section.

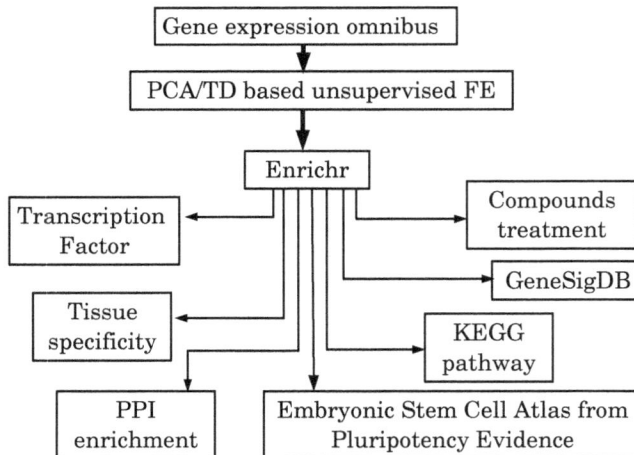

Figure 2. A schematic diagram that summarizes the obtained results.

4. Discussion

Many biological conceptual cases of enrichment were observed in the identified sets of genes across the 11 experiments. Nonetheless, detailed mechanisms by which sequence-nonspecific off-target effects (that transfected miRNAs produce) can regulate expression of these genes are unclear. To identify such a mechanism, enrichment of genes associated with the altered expression pattern caused by a Dicer KO within these 11 gene sets was studied by means of Enrichr. Among 16 experiments included in Enrichr ("Single Gene Perturbations from GEO up" and "Single Gene Perturbations from GEO down"), most of them are associated with enrichment of the identified genes in 11 miRNA transfection experiments (Table 12). In addition, these sets of genes significantly overlap with the set of genes associated with binding to Dicer according to immunoprecipitation (IP) experiments (Table 12). There are also data supporting the hypothesis that sequence-nonspecific off-target regulation is due to the competition for protein machinery between transfected miRNAs and endogenous miRNAs in the cell.

Table 12. GEO DICER KO: The number of experiments among the 16 experiments included in Enrichr whose set of listed genes significantly overlapped with the set of genes identified in each of the 11 experiments. IP: Fisher's exact test for the overlap between the set of genes that bind to Dicer in immunoprecipitation (IP) experiments and the set of genes selected in each of the 11 experiments.

Experiments		1	2	3	4	5	6
GEO DICER KO	up	12/16	12/16	12/16	12/16	14/16	11/16
	down	13/16	12/16	12/16	13/16	14/16	10/16
IP	*P*-value	2.49×10^{23}	7.22×10^{22}	1.31×10^{17}	5.55×10^{29}	5.21×10^{35}	1.78×10^{20}
	odds	47.4	20.6	15.9	38.7	64.2	41.2
Experiments			7	8	9	10	11
GEO DICER KO	up		12/16	14/16	12/16	13/16	12/16
	down		12/16	12/16	14/16	14/16	10/16
IP	*P*-value		4.72×10^{32}	4.29×10^{16}	2.19×10^{11}	3.96×10^{10}	4.64×10^{8}
	odds		37.0	41.4	42.6	39.6	27.3

On the other hand, the number of miRNAs that target genes whose expression was likely to be altered by miRNA transfection significantly correlated with the number of experiments where individual genes were selected among the 11 conducted experiments (see Figure 3. Pearson's and Spearman's correlation coefficients are 0.13, $P = 3.9 \times 10^{-11}$ and 0.29, $P < 2.2 \times 10^{-16}$, respectively). Genes targeted by a greater number of individual miRNAs are likely to be affected by miRNA transfection, which occupies the protein machinery binding to transcripts of these genes. Therefore, the significant correlation is consistent with the hypothesis that sequence-nonspecific off-target regulation is due to competition for protein machinery between a transfected miRNA and endogenous miRNAs in cells.

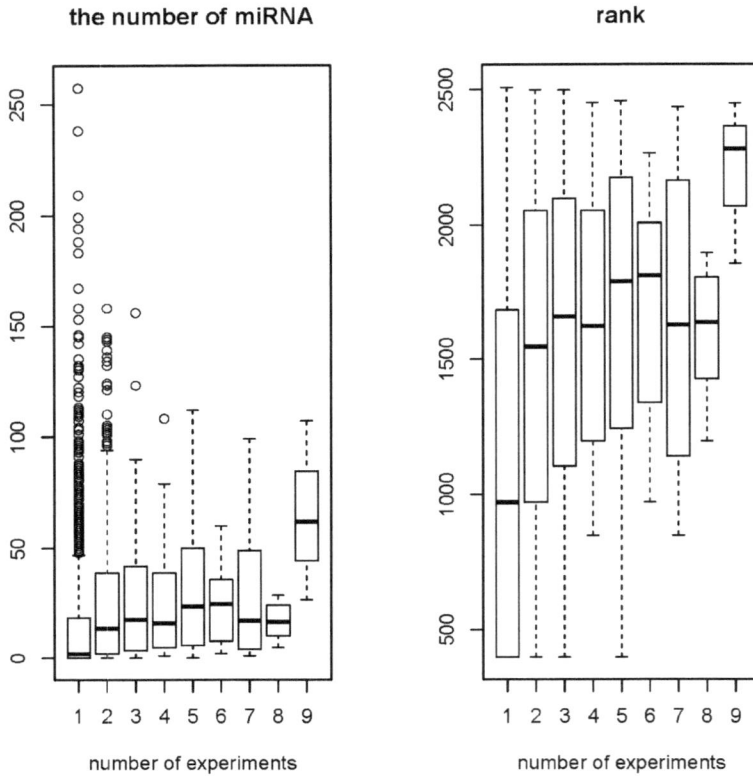

Figure 3. A boxplot of the number of miRNAs that target individual genes as a function of the number of experiments that select individual genes within 11 experiments (most frequently selected genes were selected in nine experiments): (**Left**) raw numbers (Pearson's correlation coefficient = 0.13, $P = 3.9 \times 10^{-11}$); and (**Right**) ranks of numbers (Spearman's correlation coefficient = 0.29, $P < 2.2 \times 10^{-16}$)

Thus, primarily, gene expression alteration by sequence-nonspecific off-target regulation caused by miRNA transfection is likely due to suppression of miRNA functionality because of a reduction in the amount of available protein machinery owing to occupation by transfected miRNAs.

Despite the above arguments, it cannot be proven that competition for protein machinery is the primary cause of sequence-nonspecific off-target regulation. Upregulation of genes can also be caused by indirect effects, e.g., genes that suppress expression of other genes are repressed by transfected miRNA, although this mechanism is unlikely to cause upregulation of common genes regardless of the transfected miRNAs. Additional experimental validation will clarify which one is the correct scenario.

Furthermore, I compared the performance of TD-based unsupervised FE with the performance of major supervised methods. Previously, when PCA-based unsupervised FE, on which TD-based unsupervised FE is based, has been applied to various problems, PCA-based unsupervised FE often outperformed other (supervised) methods. For example, when PCA-based unsupervised FE was successfully used to identify genes commonly associated with aberrant promoter methylation among three autoimmune diseases [19], no supervised methods—except for PCA-based unsupervised

FE—could identify common genes. On the other hand, when PCA-based unsupervised FE was applied to identify common HDACi target genes between two independent HDACis [16], two supervised methods (Limma [67] and categorical regression, i.e. analysis of variance (ANOVA)) identified the same number of common genes as did PCA-based unsupervised FE. Nevertheless, biological validation of the selected genes supported the superiority of PCA-based unsupervised FE. In contrast to the many instances of biological-term enrichment that were observed among genes selected by PCA-based unsupervised FE, such cases of enrichment among the genes selected by a supervised method were not detected.

Although these are only two examples, this kind of advantages of PCA-based unsupervised FE have often been observed. To confirm the superiority of TD-based unsupervised FE too, I consider Experiments No. 9 and No. 10 in Table 1 for the comparisons with other (supervised) methods. The reason for this choice is as follows. At first, these two were taken from the same dataset (GSE37729) and the same miRNAs (miR-107/181b) were transfected; thus, these two experiments are expected to have a greater number of common genes selected than do other pairs of experiments in Table 1. Second, the pair No. 9 and No. 10 has the highest odds ratio in Table 2, as expected. Thus, these data are suitable for the comparison with another method.

When using ANOVA and SAM [37], I found that P-values and the odds ratio computed by Fisher's exact test are 0.09 and 2.9, respectively, for both methods (the number of genes selected is taken to be 103 and 104 for Experiments No. 9 and No. 10, respectively, using ranking based upon P-values assigned to each gene because these numbers are the same as those selected by TD-based unsupervised FE, see "#" in Table 2).

A more sophisticated and advanced supervised method may show somewhat better performance than do SAM and categorical regression. Because TD-based unsupervised FE highly outperformed these two conventional and frequently used supervised methods, it is unlikely that another supervised method can compete with TD-based unsupervised FE (advantages of PCA-based unsupervised FE over various conventional supervised methods have been reported repeatedly too [14–33]). More comprehensive performance comparisons with the t test are provided in the Supplementary Materials.

Readers may also wonder whether the null assumption that $x_{\ell_1,i}$ obeys a normal distribution is appropriate because $x_{\ell_1,i}$ is not proven to follow a normal distribution. This approach is not problematic for the following reasons. First, the null hypothesis that $x_{\ell_1,i}$ obeys a normal distribution is supposed to be rejected later. Thus, even if $x_{\ell_1,i}$ does not follow the normal distribution, this is not a problem. Second, it is reasonable to assume that $x_{\ell_1,i}$ follows the normal distribution under the assumption that x_{ijk} is drawn from a random number (Figure 1C); this statement is suitable as the null hypothesis. Be that as it may, a question may arise whether the null hypothesis that $x_{\ell_1,i}$ obeys the normal distribution is not suitable if this assumption is mostly violated. To evaluate how well the null hypothesis is fulfilled, I demonstrate the result for GSE26996 as a typical example. Figure 4A presents the scatter plot of $x_{\ell_1,i}$, $\ell_1 = 1, 2$ where $x_{\ell_1=2,i}$ served for selection of genes as mentioned in the Section 2.5.1. Considering the fact that the total number of probes in the microarray is more than 20,000 and the number of probes selected is 379, these 379 probes are obviously outliers along the direction of $x_{\ell_1=2,i}$. Figure 4B depicts the histogram of $1 - P$ under the null hypothesis that $x_{\ell_1=2,i}$ follows the normal distribution. Although smaller $1 - P$ (<0.3) s deviate from constant values that are expected under the null hypothesis, a sharp peak that includes the selected 379 probes is evidently located at the largest $1 - P$. Consequently, it is satisfactory for identifying genes i associated with much larger (that is, invalidating the null hypothesis) absolute $x_{\ell_1=2,i}$ values.

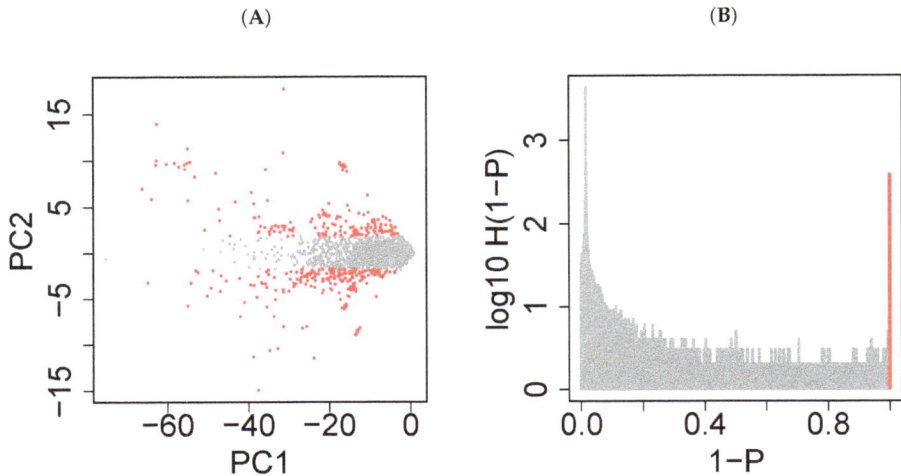

Figure 4. (**A**) The scatter plot of $x_{\ell_1,i}$, $\ell_1 = 1$ (horizontal axis), 2 (vertical axis) for GSE26996; 379 red dots are selected probes; and (**B**) a semilogarithmic plot of the histogram of $1 - P$ under the null hypothesis that $x_{\ell_1=2,i}$ obeys a normal distribution. A sharp peak is observed in the red bin with the largest $1 - P$, which includes all the probes selected in (**A**).

Finally, I would like to consider possible objections to the hypothesis proposed in this study: sequence-nonspecific off-target regulation of mRNA mediated by miRNA transfection is primarily mediated by competition for the protein machinery. The first possible objection is that some miRNA can bind to a promoter region directly. This process in not mentioned in the above discussion. For example, Kim et al. [68] found that miR-320 can bind to the promoter region of *POLR3D*. Nevertheless, this kind of direct binding to DNA by transfected miRNA cannot be the interpretation of the present findings, because it is still sequence specific. Thus, direct binding to DNA cannot be an alternative interpretation of the sequence-nonspecific regulation presented in Table 2. The second objection is that miRNA can sometimes bind to mRNA with insufficient support by proteins. For example, Lima et al. [69] found that single-stranded siRNAs can bind to mRNA. Given that single-stranded siRNAs do not have to be processed by the DICER that is mentioned in Table 12, this topic apparently seems to be outside the scope of this study. Nonetheless, single-stranded siRNAs that Lima et al. identified still need the AGO2 protein. Thus, the regulation of mRNA expression by single-stranded siRNAs can still be under the control of competition for protein machinery. Therefore, it is hard to say whether the process identified by Lima et al. is independent of protein machinery competition. The third objection to the scenario proposed in this study is that miRNA can often upregulate target mRNAs [70,71]. This means that observation of upregulation caused by miRNA transfection—which was brought up as one of side proofs for protein machinery competition in the text above—does not always have to be mediated by competition for protein machinery. On the other hand, this process is also sequence specific. Thus, direct upregulation by transfected miRNA still cannot explain the sequence-nonspecific regulation of mRNAs presented in Table 2. Therefore, at the moment, protein machinery competition is the only possible explanation of the sequence-nonspecific regulation of mRNAs shown in Table 2.

5. Conclusions

In this study, I applied recently proposed PCA- and TD-based unsupervised FE to mRNA profiles of miRNA-transfected cell lines. mRNAs associated with significant dysregulation turned out to be independent of transfected miRNAs to some extent. This sequence-nonspecific off-target regulation is associated with various biological functions according to enrichment analysis. It is also likely to be caused by protein machinery competition between endogenous miRNAs and transfected miRNAs.

Supplementary Materials: The following files are available online at www.mdpi.com/2073-4409/7/6/54/s1, Supplementary Document: the *t* test applied to a real dataset; Supplementary Data: a full list of the identified genes; Supplementary Figures: Figures S1–S11; and Video Abstract.

Acknowledgments: The author thanks reviewers for pointing out the references that are useful for the discussion about possible objections.

Funding: This research was funded by Japan Society for the Promotion of Science under the grant number KAKENHI 17K00417. APC was sponsored by MDPI.

Conflicts of Interest: The author declares no conflict of interest.

References

1. Bartel, D.P. MicroRNAs: Target Recognition and Regulatory Functions. *Cell* **2009**, *136*, 215–233. [CrossRef] [PubMed]
2. Olejniczak, M.; Galka, P.; Krzyzosiak, W.J. Sequence-non-specific effects of RNA interference triggers and microRNA regulators. *Nucleic Acids Res.* **2010**, *38*, 1–16. [CrossRef] [PubMed]
3. Khan, A.A.; Betel, D.; Miller, M.L.; Sander, C.; Leslie, C.S.; Marks, D.S. Transfection of small RNAs globally perturbs gene regulation by endogenous microRNAs. *Nat. Biotechnol.* **2009**, *27*, 549–555. [CrossRef] [PubMed]
4. Nagata, Y.; Shimizu, E.; Hibio, N.; Ui-Tei, K. Fluctuation of global gene expression by endogenous miRNA response to the introduction of an exogenous miRNA. *Int. J. Mol. Sci.* **2013**, *14*, 11171–11189. [CrossRef] [PubMed]
5. Carroll, A.P.; Tran, N.; Tooney, P.A.; Cairns, M.J. Alternative mRNA fates identified in microRNA-associated transcriptome analysis. *BMC Genom.* **2012**, *13*, 561. [CrossRef] [PubMed]
6. Taguchi, Y.H. Tensor decomposition-based unsupervised feature extraction identifies candidate genes that induce post-traumatic stress disorder-mediated heart diseases. *BMC Med. Genom.* **2017**, *10*, 67. [CrossRef] [PubMed]
7. Taguchi, Y.H. One-class Differential Expression Analysis using Tensor Decomposition-based Unsupervised Feature Extraction Applied to Integrated Analysis of Multiple Omics Data from 26 Lung Adenocarcinoma Cell Lines. In Proceedings of the 2017 IEEE 17th International Conference on Bioinformatics and Bioengineering (BIBE), Washington, DC, USA, 23–25 October 2017; pp. 131–138.
8. Taguchi, Y.H. Identification of candidate drugs using tensor-decomposition-based unsupervised feature extraction in integrated analysis of gene expression between diseases and DrugMatrix datasets. *Sci. Rep.* **2017**, *7*, 13733. [CrossRef] [PubMed]
9. Taguchi, Y.H. Tensor decomposition-based unsupervised feature extraction applied to matrix products for multi-view data processing. *PLoS ONE* **2017**, *12*, e0183933. [CrossRef] [PubMed]
10. Taguchi, Y.H. Identification of candidate drugs for heart failure using tensor decomposition-based unsupervised feature extraction applied to integrated analysis of gene expression between heart failure and drugmatrix datasets. In *Intelligent Computing Theories and Application*; Huang, D.S., Jo, K.H., Figueroa-García, J.C., Eds.; Springer International Publishing: Cham, Switzerland, 2017; pp. 517–528.
11. Taguchi, Y.H.; Wang, H. Exploring microRNA Biomarker for Amyotrophic Lateral Sclerosis. *Int. J. Mol. Sci.* **2018**, *19*, 1318. [CrossRef] [PubMed]
12. Taguchi, Y.H.; Wang, H. Genetic association between amyotrophic lateral sclerosis and cancer. *Genes* **2017**, *8*, 243. [CrossRef] [PubMed]
13. Taguchi, Y.H.; Iwadate, M.; Umeyama, H.; Murakami, Y. Principal component analysis based unsupervised feature extraction applied to bioinformatics analysis. In *Computational Methods with Applications in Bioinformatics Analysis*; World Scientific: Singapore, 2017; Chapter 8, pp. 153–182.
14. Taguchi, Y.H. microRNA-mRNA Interaction identification in wilms tumor using principal component analysis based unsupervised feature extraction. In Proceedings of the 2016 IEEE 16th International Conference on Bioinformatics and Bioengineering (BIBE), Taichung, Taiwan, 31 October–2 November 2016; pp. 71–78.

15. Taguchi, Y.H. Principal Components Analysis Based Unsupervised Feature Extraction Applied to Gene Expression Analysis of Blood from Dengue Haemorrhagic Fever Patients. *Sci. Rep.* **2017**, *7*, 44016. [CrossRef] [PubMed]
16. Taguchi, Y.H. Principal component analysis based unsupervised feature extraction applied to publicly available gene expression profiles provides new insights into the mechanisms of action of histone deacetylase inhibitors. *Neuroepigenetics* **2016**, *8*, 1–18. [CrossRef]
17. Taguchi, Y.H.; Iwadate, M.; Umeyama, H. Principal component analysis-based unsupervised feature extraction applied to in silico drug discovery for posttraumatic stress disorder-mediated heart disease. *BMC Bioinform.* **2015**, *16*, 139. [CrossRef] [PubMed]
18. Taguchi, Y.H.; Okamoto, A. Principal Component Analysis for Bacterial Proteomic Analysis. In *Pattern Recognition in Bioinformatics*; Shibuya, T., Kashima, H., Sese, J., Ahmad, S., Eds.; Springer: Berlin/Heidelberg, Germany, 2012; Volume 7632, *LNCS*, pp. 141–152.
19. Ishida, S.; Umeyama, H.; Iwadate, M.; Taguchi, Y.H. Bioinformatic Screening of Autoimmune Disease Genes and Protein Structure Prediction with FAMS for Drug Discovery. *Protein Pept. Lett.* **2014**, *21*, 828–839. [CrossRef] [PubMed]
20. Kinoshita, R.; Iwadate, M.; Umeyama, H.; Taguchi, Y.H. Genes associated with genotype-specific DNA methylation in squamous cell carcinoma as candidate drug targets. *BMC Syst. Biol.* **2014**, *8*, S4. [CrossRef] [PubMed]
21. Taguchi, Y.H.; Murakami, Y. Principal component analysis based feature extraction approach to identify circulating microRNA biomarkers. *PLoS ONE* **2013**, *8*, e66 714. [CrossRef]
22. Taguchi, Y.H.; Murakami, Y. Universal disease biomarker: Can a fixed set of blood microRNAs diagnose multiple diseases? *BMC Res. Notes* **2014**, *7*, 581. [CrossRef] [PubMed]
23. Murakami, Y.; Toyoda, H.; Tanahashi, T.; Tanaka, J.; Kumada, T.; Yoshioka, Y.; Kosaka, N.; Ochiya, T.; Taguchi, Y.H. Comprehensive miRNA expression analysis in peripheral blood can diagnose liver disease. *PLoS ONE* **2012**, *7*, e48366. [CrossRef] [PubMed]
24. Murakami, Y.; Tanahashi, T.; Okada, R.; Toyoda, H.; Kumada, T.; Enomoto, M.; Tamori, A.; Kawada, N.; Taguchi, Y.H.; Azuma, T. Comparison of Hepatocellular Carcinoma miRNA Expression Profiling as Evaluated by Next Generation Sequencing and Microarray. *PLoS ONE* **2014**, *9*, e106314. [CrossRef] [PubMed]
25. Murakami, Y.; Kubo, S.; Tamori, A.; Itami, S.; Kawamura, E.; Iwaisako, K.; Ikeda, K.; Kawada, N.; Ochiya, T.; Taguchi, Y.H. Comprehensive analysis of transcriptome and metabolome analysis in Intrahepatic Cholangiocarcinoma and Hepatocellular Carcinoma. *Sci. Rep.* **2015**, *5*, 16294. [CrossRef] [PubMed]
26. Umeyama, H.; Iwadate, M.; Taguchi, Y.H. TINAGL1 and B3GALNT1 are potential therapy target genes to suppress metastasis in non-small cell lung cancer. *BMC Genom.* **2014**, *15*, S2. [CrossRef] [PubMed]
27. Taguchi, Y.H.; Iwadate, M.; Umeyama, H. Heuristic principal component analysis-based unsupervised feature extraction and its application to gene expression analysis of amyotrophic lateral sclerosis data sets. In Proceedings of the 2015 IEEE Conference on Computational Intelligence in Bioinformatics and Computational Biology (CIBCB), Niagara Falls, ON, Canada, 12–15 August 2015; pp. 1–10.
28. Taguchi, Y.H.; Iwadate, M.; Umeyama, H.; Murakami, Y.; Okamoto, A. Heuristic principal component analysis-aased unsupervised feature extraction and its application to bioinformatics. In *Big Data Analytics in Bioinformatics and Healthcare*; Wang, B., Li, R., Perrizo, W., Eds.; IGI Global: Hershey, PA, USA, 2015; pp. 138–162.
29. Taguchi, Y.H. Integrative analysis of gene expression and promoter methylation during reprogramming of a non-small-cell lung cancer cell line using principal component analysis-based unsupervised feature extraction. In *Intelligent Computing in Bioinformatics*; Huang, D.S., Han, K., Gromiha, M., Eds.; Springer: Berlin/Heidelberg, Germany, 2014; Volume 8590, *LNCS*, pp. 445–455.
30. Taguchi, Y.H. Identification of aberrant gene expression associated with aberrant promoter methylation in primordial germ cells between E13 and E16 rat F3 generation vinclozolin lineage. *BMC Bioinform.* **2015**, *16*, S16. [CrossRef] [PubMed]
31. Taguchi, Y.H. Identification of More Feasible MicroRNA-mRNA Interactions within Multiple Cancers Using Principal Component Analysis Based Unsupervised Feature Extraction. *Int. J. Mol. Sci.* **2016**, *17*, 696. [CrossRef] [PubMed]

32. Taguchi, Y.H. Principal component analysis based unsupervised feature extraction applied to budding yeast temporally periodic gene expression. *BioData Min.* **2016**, *9*, 22. [CrossRef] [PubMed]

33. Taguchi, Y.H.; Iwadate, M.; Umeyama, H. SFRP1 is a possible candidate for epigenetic therapy in non-small cell lung cancer. *BMC Med. Genom.* **2016**, *9*, 28. [CrossRef] [PubMed]

34. Lathauwer, L.D.; Moor, B.D.; Vandewalle, J. A multilinear singular value decomposition. *SIAM J. Matrix Anal. Appl.* **2000**, *21*, 1253–1278. [CrossRef]

35. Benjamini, Y.; Hochberg, Y. Controlling the False Discovery Rate: A Practical and Powerful Approach to Multiple Testing. *J. R. Stat. Soc. Ser. B Methodol.* **1995**, *57*, 289–300.

36. Barrett, T.; Wilhite, S.E.; Ledoux, P.; Evangelista, C.; Kim, I.F.; Tomashevsky, M.; Marshall, K.A.; Phillippy, K.H.; Sherman, P.M.; Holko, M.; et al. NCBI GEO: Archive for functional genomics data sets–update. *Nucleic Acids Res.* **2013**, *41*, D991–D995. [CrossRef] [PubMed]

37. Tusher, V.G.; Tibshirani, R.; Chu, G. Significance analysis of microarrays applied to the ionizing radiation response. *Proc. Natl. Acad. Sci. USA* **2001**, *98*, 5116–5121. [CrossRef] [PubMed]

38. Kuleshov, M.V.; Jones, M.R.; Rouillard, A.D.; Fernandez, N.F.; Duan, Q.; Wang, Z.; Koplev, S.; Jenkins, S.L.; Jagodnik, K.M.; Lachmann, A.; et al. Enrichr: A comprehensive gene set enrichment analysis web server 2016 update. *Nucleic Acids Res.* **2016**, *44*, W90–W97. [CrossRef] [PubMed]

39. Allocco, D.J.; Kohane, I.S.; Butte, A.J. Quantifying the relationship between co-expression, co-regulation and gene function. *BMC Bioinform.* **2004**, *5*, 18. [CrossRef] [PubMed]

40. Karczewski, K.J.; Snyder, M.; Altman, R.B.; Tatonetti, N.P. Coherent functional modules improve transcription factor target identification, cooperativity prediction, and disease association. *PLoS Genet.* **2014**, *10*, e1004122. [CrossRef] [PubMed]

41. Suarez, Y.; Fernandez-Hernando, C.; Yu, J.; Gerber, S.A.; Harrison, K.D.; Pober, J.S.; Iruela-Arispe, M.L.; Merkenschlager, M.; Sessa, W.C. Dicer-dependent endothelial microRNAs are necessary for postnatal angiogenesis. *Proc. Natl. Acad. Sci. USA* **2008**, *105*, 14082–14087. [CrossRef] [PubMed]

42. Dews, M.; Homayouni, A.; Yu, D.; Murphy, D.; Sevignani, C.; Wentzel, E.; Furth, E.E.; Lee, W.M.; Enders, G.H.; Mendell, J.T.; et al. Augmentation of tumor angiogenesis by a Myc-activated microRNA cluster. *Nat. Genet.* **2006**, *38*, 1060–1065. [CrossRef] [PubMed]

43. O'Donnell, K.A.; Wentzel, E.A.; Zeller, K.I.; Dang, C.V.; Mendell, J.T. c-Myc-regulated microRNAs modulate E2F1 expression. *Nature* **2005**, *435*, 839–843. [CrossRef] [PubMed]

44. Yien, Y.Y.; Bieker, J.J. EKLF/KLF1, a tissue-restricted integrator of transcriptional control, chromatin remodeling, and lineage determination. *Mol. Cell. Biol.* **2013**, *33*, 4–13. [CrossRef] [PubMed]

45. Rahl, P.B.; Lin, C.Y.; Seila, A.C.; Flynn, R.A.; McCuine, S.; Burge, C.B.; Sharp, P.A.; Young, R.A. c-Myc regulates transcriptional pause release. *Cell* **2010**, *141*, 432–445. [CrossRef] [PubMed]

46. Das, P.P.; Shao, Z.; Beyaz, S.; Apostolou, E.; Pinello, L.; De Los Angeles, A.; O'Brien, K.; Atsma, J.M.; Fujiwara, Y.; Nguyen, M.; et al. Distinct and combinatorial functions of Jmjd2b/Kdm4b and Jmjd2c/Kdm4c in mouse embryonic stem cell identity. *Mol. Cell* **2014**, *53*, 32–48. [CrossRef] [PubMed]

47. Roussel, B.D.; Kruppa, A.J.; Miranda, E.; Crowther, D.C.; Lomas, D.A.; Marciniak, S.J. Endoplasmic reticulum dysfunction in neurological disease. *Lancet Neurol.* **2013**, *12*, 105–118. [CrossRef]

48. Ladewig, J.; Mertens, J.; Kesavan, J.; Doerr, J.; Poppe, D.; Glaue, F.; Herms, S.; Wernet, P.; Kogler, G.; Muller, F.J.; et al. Small molecules enable highly efficient neuronal conversion of human fibroblasts. *Nat. Methods* **2012**, *9*, 575–578. [CrossRef] [PubMed]

49. Hu, W.; Qiu, B.; Guan, W.; Wang, Q.; Wang, M.; Li, W.; Gao, L.; Shen, L.; Huang, Y.; Xie, G.; et al. Direct Conversion of Normal and Alzheimer's Disease Human Fibroblasts into Neuronal Cells by Small Molecules. *Cell Stem Cell* **2015**, *17*, 204–212. [CrossRef] [PubMed]

50. Engelman, J.A. Targeting PI3K signalling in cancer: Opportunities, challenges and limitations. *Nat. Rev. Cancer* **2009**, *9*, 550–562. [CrossRef] [PubMed]

51. Curran, K.J.; Verheijen, J.C.; Kaplan, J.; Richard, D.J.; Toral-Barza, L.; Hollander, I.; Lucas, J.; Ayral-Kaloustian, S.; Yu, K.; Zask, A. Pyrazolopyrimidines as highly potent and selective, ATP-competitive inhibitors of the mammalian target of rapamycin (mTOR): Optimization of the 1-substituent. *Bioorg. Med. Chem. Lett.* **2010**, *20*, 1440–1444. [CrossRef] [PubMed]

52. Morin-Ben Abdallah, S.; Hirsh, V. Epidermal Growth Factor Receptor Tyrosine Kinase Inhibitors in Treatment of Metastatic Non-Small Cell Lung Cancer, with a Focus on Afatinib. *Front. Oncol.* **2017**, *7*, 97. [CrossRef] [PubMed]

53. Park, S.; Chapuis, N.; Bardet, V.; Tamburini, J.; Gallay, N.; Willems, L.; Knight, Z.A.; Shokat, K.M.; Azar, N.; Viguie, F.; et al. PI-103, a dual inhibitor of Class IA phosphatidylinositide 3-kinase and mTOR, has antileukemic activity in AML. *Leukemia* **2008**, *22*, 1698–1706. [CrossRef] [PubMed]

54. Anastasaki, C.; Rauen, K.A.; Patton, E.E. Continual low-level MEK inhibition ameliorates cardio-facio-cutaneous phenotypes in zebrafish. *Dis. Models Mech.* **2012**, *5*, 546–552. [CrossRef] [PubMed]

55. Wei, H.; Mei, Y.A.; Sun, J.T.; Zhou, H.Q.; Zhang, Z.H. Regulation of swelling-activated chloride channels in embryonic chick heart cells. *Cell Res.* **2003**, *13*, 21–28. [CrossRef] [PubMed]

56. Wallin, J.J.; Edgar, K.A.; Guan, J.; Berry, M.; Prior, W.W.; Lee, L.; Lesnick, J.D.; Lewis, C.; Nonomiya, J.; Pang, J.; et al. GDC-0980 is a novel class I PI3K/mTOR kinase inhibitor with robust activity in cancer models driven by the PI3K pathway. *Mol. Cancer Ther.* **2011**, *10*, 2426–2436. [CrossRef] [PubMed]

57. Bronte, E.; Bronte, G.; Novo, G.; Bronte, F.; Bavetta, M.G.; Lo Re, G.; Brancatelli, G.; Bazan, V.; Natoli, C.; Novo, S.; et al. What links BRAF to the heart function? New insights from the cardiotoxicity of BRAF inhibitors in cancer treatment. *Oncotarget* **2015**, *6*, 35589–35601. [CrossRef] [PubMed]

58. Kumar, S.K.; LaPlant, B.; Chng, W.J.; Zonder, J.; Callander, N.; Fonseca, R.; Fruth, B.; Roy, V.; Erlichman, C.; Stewart, A.K. Dinaciclib, a novel CDK inhibitor, demonstrates encouraging single-agent activity in patients with relapsed multiple myeloma. *Blood* **2015**, *125*, 443–448. [CrossRef] [PubMed]

59. Ardlie, K.G.; Deluca, D.S.; Segrè, A.V.; Sullivan, T.J.; Young, T.R.; Gelfand, E.T.; Trowbridge, C.A.; Maller, J.B.; Tukiainen, T.; Lek, M.; et al. The Genotype-Tissue Expression (GTEx) pilot analysis: Multitissue gene regulation in humans. *Science* **2015**, *348*, 648–660.

60. Vosa, U.; Esko, T.; Kasela, S.; Annilo, T. Altered Gene Expression Associated with microRNA Binding Site Polymorphisms. *PLoS ONE* **2015**, *10*, e0141351. [CrossRef] [PubMed]

61. Anokye-Danso, F.; Trivedi, C.M.; Juhr, D.; Gupta, M.; Cui, Z.; Tian, Y.; Zhang, Y.; Yang, W.; Gruber, P.J.; Epstein, J.A.; et al. Highly efficient miRNA-mediated reprogramming of mouse and human somatic cells to pluripotency. *Cell Stem Cell* **2011**, *8*, 376–388. [CrossRef] [PubMed]

62. Mohan, K.N.; Ding, F.; Chaillet, J.R. Distinct roles of DMAP1 in mouse development. *Mol. Cell. Biol.* **2011**, *31*, 1861–1869. [CrossRef] [PubMed]

63. Galan-Caridad, J.M.; Harel, S.; Arenzana, T.L.; Hou, Z.E.; Doetsch, F.K.; Mirny, L.A.; Reizis, B. Zfx controls the self-renewal of embryonic and hematopoietic stem cells. *Cell* **2007**, *129*, 345–357. [CrossRef] [PubMed]

64. Pasini, D.; Bracken, A.P.; Hansen, J.B.; Capillo, M.; Helin, K. The polycomb group protein Suz12 is required for embryonic stem cell differentiation. *Mol. Cell. Biol.* **2007**, *27*, 3769–3779. [CrossRef] [PubMed]

65. Fidalgo, M.; Shekar, P.C.; Ang, Y.S.; Fujiwara, Y.; Orkin, S.H.; Wang, J. Zfp281 functions as a transcriptional repressor for pluripotency of mouse embryonic stem cells. *Stem Cells* **2011**, *29*, 1705–1716. [CrossRef] [PubMed]

66. Szklarczyk, D.; Franceschini, A.; Wyder, S.; Forslund, K.; Heller, D.; Huerta-Cepas, J.; Simonovic, M.; Roth, A.; Santos, A.; Tsafou, K.P.; et al. STRING v10: Protein-protein interaction networks, integrated over the tree of life. *Nucleic Acids Res.* **2015**, *43*, D447–D452. [CrossRef] [PubMed]

67. Ritchie, M.E.; Phipson, B.; Wu, D.; Hu, Y.; Law, C.W.; Shi, W.; Smyth, G.K. limma powers differential expression analyses for RNA-sequencing and microarray studies. *Nucleic Acids Res.* **2015**, *43*, e47. [CrossRef] [PubMed]

68. Kim, D.H.; Sætrom, P.; Snøve, O.; Rossi, J.J. MicroRNA-Directed Transcriptional Gene Silencing in Mammalian Cells. *Proc. Natl. Acad. Sci. USA* **2008**, *105*, 16230–16235. Available online: http://www.pnas.org/content/105/42/16230.full.pdf (accessed on 1 June 2018). [CrossRef] [PubMed]

69. Lima, W.F.; Prakash, T.P.; Murray, H.M.; Kinberger, G.A.; Li, W.; Chappell, A.E.; Li, C.S.; Murray, S.F.; Gaus, H.; Seth, P.P.; et al. Single-Stranded siRNAs Activate RNAi in Animals. *Cell* **2012**, *150*, 883–894. [CrossRef] [PubMed]

70. Vasudevan, S.; Tong, Y.; Steitz, J.A. Switching from Repression to Activation: MicroRNAs can up-regulate translation. *Science* **2007**, *318*, 1931–1934. [CrossRef] [PubMed]
71. Ghosh, T.; Soni, K.; Scaria, V.; Halimani, M.; Bhattacharjee, C.; Pillai, B. MicroRNA-mediated up-regulation of an alternatively polyadenylated variant of the mouse cytoplasmic β-actin gene. *Nucleic Acids Res.* **2008**, *36*, 6318–6332. [CrossRef] [PubMed]

Review

MicroRNAs at the Interface between Osteogenesis and Angiogenesis as Targets for Bone Regeneration

Leopold F. Fröhlich

Department of Cranio-Maxillofacial Surgery, University of Münster, Albert-Schweitzer-Campus 1, 48149 Münster, Germany; leopold.froehlich@ukmuenster.de; Tel.: +49-251-834-7007

Received: 20 December 2018; Accepted: 30 January 2019; Published: 3 February 2019

Abstract: Bone formation and regeneration is a multistep complex process crucially determined by the formation of blood vessels in the growth plate region. This is preceded by the expression of growth factors, notably the vascular endothelial growth factor (VEGF), secreted by osteogenic cells, as well as the corresponding response of endothelial cells, although the exact mechanisms remain to be clarified. Thereby, coordinated coupling between osteogenesis and angiogenesis is initiated and sustained. The precise interplay of these two fundamental processes is crucial during times of rapid bone growth or fracture repair in adults. Deviations in this balance might lead to pathologic conditions such as osteoarthritis and ectopic bone formation. Besides VEGF, the recently discovered important regulatory and modifying functions of microRNAs also support this key mechanism. These comprise two principal categories of microRNAs that were identified with specific functions in bone formation (osteomiRs) and/or angiogenesis (angiomiRs). However, as hypoxia is a major driving force behind bone angiogenesis, a third group involved in this process is represented by hypoxia-inducible microRNAs (hypoxamiRs). This review was focused on the identification of microRNAs that were found to have an active role in osteogenesis as well as angiogenesis to date that were termed "CouplingmiRs (CPLGmiRs)". Outlined representatives therefore represent microRNAs that already have been associated with an active role in osteogenic-angiogenic coupling or are presumed to have its potential. Elucidation of the molecular mechanisms governing bone angiogenesis are of great relevance for improving therapeutic options in bone regeneration, tissue-engineering, and the treatment of bone-related diseases.

Keywords: bone angiogenesis; osteogenesis; angiogenic-osteogenic coupling; microRNAs; bone regeneration; bone formation; bone tissue-engineering; angiomiRs; osteomiRs; hypoxamiRs

---

1. Introduction

The replacement of large bone defects and the availability of adequate tissue-engineered bone remains a major clinical challenge. This tremendous demand results from the high incidence of large segmental bone defects due to trauma, congenital malformations, ageing, or bone-related diseases such as osteoporosis, inflammation or tumors [1]. However, the development of gene therapy approaches in recent years demonstrated that tissue-engineered bone offers new therapeutic strategies to repair tissue defects. Thereby, one of the major disadvantages in the clinical use of engineered bone constructs so far, i.e., the inability to provide sufficient blood supply in the initial phase after implantation which leads to insufficient cell integration and cell death, could be overcome [2]. As an explanation for this shortcoming, the function of angiogenesis—the process of forming new blood vessels from pre-existing vasculature—in bone regeneration is still poorly defined, and the molecular mechanisms that regulate angiogenesis in bone are only just starting to be unraveled. Angiogenesis, a term that was coined in 1935 to describe the formation of blood vessels in the placenta, occurs during normal vertebrate embryogenesis but also as a response to pathophysiological circumstances during the processes of tumor formation, wound healing and tissue regeneration [3,4]. Elucidating the wide orchestrating variety of signal pathways and stimuli linking angiogenesis and bone formation on the molecular level is therefore of great interest. In this context, the gained information will yield improved bone replacement in fracture healing, or prevention of bone loss in osteoporosis and therapeutically-induced reparative angiogenesis.

Bone, with its main function in supporting and withstanding mechanical forces, is a mineralized mesenchymal tissue that also possesses an important role in maintaining mineral homeostasis and the energy metabolism of the organism [5]. The formation of bone, which is either developed by the endochondral or by the intramembranous ossification program, is a complex process which depends on physiological interaction with blood vessels that are simultaneously formed [6–9]. Endochondral ossification designates the process of long bone formation which results from the intermediate generation of a primordial cartilage skeleton composed of mesenchymal stem cells (MSCs) that differentiate into chondrocytes and only at a later stage gradually transform into mature bone recruited by different types of bone-forming cells [6]. Intramembranous ossification, in contrast, denominates the process of establishing flat bones where condensed MSCs directly differentiate into osteoblasts, forming an ossification center. This pathway leads to the formation of craniofacial, calvarial, and clavicle bones [10]. In either type of bone development, angiogenesis—the invasion of small blood vessels derived from preexisting blood vessels—is required [11–13]. Bone angiogenesis is induced by growth factors expressed by osteogenic cells such as hypertrophic chondrocytes and osteoblasts at an early stage during osteogenesis [14]. Thereby, the transport of oxygen and nutrients, as well as the further recruitment of osteoprogenitor cells (MSCs) and osteoblasts, are facilitated. In the later phase of bone formation, angiogenesis is essential for trabecular (spongy, cancellous) bone formation and for maturation of the newly-formed bone by close coordination of mineralization and vascularization in either type of bone. In addition, endochondral angiogenesis is particularly important for the replacement of cartilaginous structures at the primary ossification center which generates the bone marrow cavity, and, at a later stage, in establishing the secondary ossification center at the epiphyses (the distal end of long bones) [15,16]. Recent in vitro experiments in transgenic mice demonstrated that this task is accomplished by a specialized type of blood vessel, i.e., the so-called H-type that is present in long bones [17]. This underlines the fact that, comparable to other organs, the acquisition and maintenance of specialized properties by endothelial cells (ECs) is very important for the functional homeostasis in bone.

Bone is a tissue that has to undergo permanent remodeling and requires a counterbalanced process between the anabolic activities of bone formation (osteogenic) cells and the catabolic activities of bone resorption (osteoclast) cells. This activity enforces continuous self-renewal of bone, thereby maintaining an appropriate bone mass and calcium equilibrium. Interference with bone homeostasis could prevent tissue formation, leading to immature or abnormal bone formation [18]. Impaired blood vessel formation, as found in age-related or disease-induced bone loss, therefore, could also result in imbalanced or defective bone formation [19]. In support of this, a reduced density of blood vessels altering the microcirculation was found to be present in osteoporotic bone in mice and humans, that may lead to local abnormal bone metabolism and provoke an increased risk of fracturing [20,21]. Furthermore, the genetic program of bone angiogenesis needs reactivation during callus formation in fracture healing, which represents a complex process that has not yet been fully elucidated [22].

2. Molecular Regulation of Bone Angiogenesis

Bone angiogenesis is mainly governed by a spectrum of transcription factors and growth factors which have been mostly elaborated for endochondral ossification so far [23,24]. It involves interactive signaling between cells of the skeletal system, namely chondrocytes and osteoblasts, and cells derived from the bone vascular system, primarily ECs. Initially, the formation of blood vessels is promoted by osteogenic cells producing pro-angiogenic factors which, in turn, later support the settlement of osteoprogenitor cells [25]. One of the major driving forces behind angiogenic-osteogenic coupling in bone, where oxygen concentrations below 1% are encountered, is the necessity of supplying the tissue with oxygen [15,26,27]. Via tightly hypoxia-regulated induction of the transcriptional activator hypoxia inducible factor (HIF), a cascade of target genes which are involved in a wide variety of biological processes including energy metabolism, erythropoiesis, cell survival, apoptosis and angiogenesis including the major angiogenic regulator vascular endothelial growth factor (VEGF) are expressed in osteogenic cells [28]. The direct mediator of oxygen-dependent modifications of the HIF factors is the von Hippel-Lindau tumor suppressor protein (pVHL), an E3 ubiquitin ligase that targets HIF-1a for proteasomal degradation. During endochondral ossification, VEGF-A is produced by both chondrocytes, particularly in a later stage in which they undergo hypertrophy, and by osteoblasts [29]. As proof of principle, it has been demonstrated that in mice which produce only altered expression levels of a soluble form of VEGF (VEGF120), delayed blood vessel invasion into the primary ossification center and altered osteoblast differentiation in vitro occurs [11]. In addition to HIFs also the fibroblast growth factor (FGF) family of signaling molecules with the members FGF2 and FGF9, and their receptors FGFR1 and 2, were found to be involved in the transcriptional regulation of VEGFA and VEGFR2 expression during formation of blood vessels in bone [30]. Recently, Notch signaling has been implicated as a response to VEGF signaling in ECs in osteogenic-angiogenic coupling [31]. While in other tissues NOTCH expression generally negatively regulates angiogenesis, it seems to have the opposite role in enchondral bone formation by promoting EC proliferation and vessel growth. In a reciprocal fashion, ECs respond by an angiocrine release of NOGGIN, an antagonist of the bone morphogenetic protein (BMP) pathway that stimulates the maturation of hypertrophic chondrocytes expressing VEGF in the growth plate. This could be nicely demonstrated in mice lacking Notch specifically in ECs, which demonstrated reduced bone angiogenesis due to a loss of type-H blood vessels, as well as mutant bone formation and VEGF expression. Other major players are represented by members of matrix metalloproteases (MMPs) due to their proteolytic activity originating from osteoclasts and vascular cells [32]. MMPs also seem to mediate intracellular signaling involving extracellular matrix-integrin interactions necessary during bone angiogenesis and bone remodeling. Besides VEGFA, the placental growth factor (PlGF/PGF), a member of the VEGF family which binds to VEGFR1, has been found to play an exclusively important role in callus remodeling during fracture healing [33,34]. Other known modulating factors of angiogenesis during bone repair include transforming growth factor beta (TGF-β), BMPs, and growth differentiation factor (GDF) [35].

3. The Role of MicroRNAs

MicroRNAs (miRs/miRNAs) are a newly discovered expanding class of endogenous small, non-coding RNAs that positively or negatively regulate gene expression and cellular processes via the RNA interference pathway [5,36–44]. By targeting messenger RNA transcripts post-translationally, they provoke either translational repression or degradation, depending on the degree of sequence homology. Upon precursor transcription from intronic or polycistronic genomic loci by RNA polymerase II, biogenesis of the primary miRNA (pri-miRNA) transcript takes place by a two-step processing mechanism involving the RNAses Drosha and DICER (DGCR8 RNase III complex) [45–47]. Thereby, single or multiple miRNAs that form hairpin-like structures are exported to the cytoplasm by an exportin 5- and RAN-GTP-dependent process, and cleaved. However, in an alternative non-canonical pathway, miRNAs can be configured by direct transcription or refolded spliced introns as endo-shRNAs (endogenous short hairpin RNAs) or as mirtrons, respectively [48]. Subsequently, the targeting strand of the double-stranded mature miRNAs that are 18 to 22 nucleotides in length is integrated into the RNA-induced silencing complex (RISC) with the help of Argonaute proteins. RISC can finally bind specific target (or so-called "seed") sequences of mRNAs represented by 2 to 8 nucleotides located mostly in their 3′-untranslated regions (UTRs) [49,50]. Up- or down- regulation of the miRNA itself by stage- and tissue-specific expression patterns during development can lead to modified expression of its target genes. Thus, miRNAs function as decisive regulatory molecules in many different cellular activities such as development, proliferation, and differentiation, metabolism, or apoptotic cell death, or even cell fate determination and maintenance (e.g., pluripotency control of embryonic stem cells) [51–60]. Moreover, relevant to bone biology, miRNAs have also been identified to acquire endocrine or paracrine functions by their secretion into the blood stream where they subsequently circulate [61]. Consistent with this finding, it was determined that a single miRNA can be involved in coordinating genetic networks by simultaneously regulating the endogenous expressions of multiple target genes. While miRNAs seem to function rather as auxiliary factors during normal physiological processes, their task seems to become more important under stress or disease-related conditions. Accordingly, disturbed miRNA expression is increasingly identified in a number of pathological conditions such as tumorigenesis or viral infection [53,62].

4. MicroRNAs in Bone Angiogenesis: OsteomiRs, AngiomiRs, and HypoxamiRs

It is now well established that miRNAs are physiologically relevant to all steps of bone as well as blood vessel formation during embryonic development and in maintenance during adulthood [63]. OsteomiRs have been identified to regulate chondrocyte, osteoblast, and osteoclast differentiation by positively targeting the principal osteogenic transcription factors and signaling molecules of osteogenesis [5,38,64–69]. In addition to regulating MSC commitment—i.e., the differentiation of precursor cells into chondrocytic and osteoblastic lineages—several studies showed that miRNAs also contribute to the maturation and function of these cells, suggesting also important roles in bone regeneration. Nevertheless, the exact mechanisms of skeletal miRNAs governing the complex interactions and signaling pathways of different bone-forming cells are only beginning to be elucidated. Deregulated miRNAs expression or even genetic variation by mutations or single nucleotide polymorphisms in miRNAs or their binding sites have been identified furthermore in bone disorders such as osteoporosis, osteosarcoma, osteopetrosis, osteogenesis imperfecta, osteoarthritis, and furthermore in bone fracture [63,70–74].

Increasing evidence further indicates that miRNAs act as pro- and anti-angiogenic regulators of adaptive blood vessel growth in normal cardiovascular development and in tumor angiogenesis [75–80]. The role of these so-called angiomiRs or vascular microRNAs in angiogenic development was initially discovered by the detection of severely disrupted blood vessel formation and delayed angiogenic capabilities of ECs in mid-gestational lethal mouse mutants for the miRNA precursor-processing enzyme Dicer [81]. These mutants die around embryonic day 12 to 14 of development due to vascular defects in the embryo and the yolk sac. Furthermore, a smooth muscle-specific *Dicer* deletion in the mouse exerted late embryonic lethality associated with extensive internal hemorrhage which could be explained by a significant loss of vascular contractile function, smooth muscle cell (SMC) differentiation, and vascular remodeling [82]. Knockdown experiments of *Dicer* in zebrafish moreover provoked a phenotype of pericardial edema and inadequate circulation. But also, loss-of-function of the EC-specific miR-126 in homozygous deficient mice caused defects in vascular integrity and angiogenesis [83]. These findings suggested that angiomiRs modulate crucial target genes in cells derived from angioblastic precursor cells and SMC, which are indispensable during embryonic angiogenesis. By investigating the function of Dicer in adult mice and human cells, considerable dysregulated angiogenesis related to growth factor release, ischemia, and wound healing could be revealed, reflecting important postnatal angiogenic functions [80,84,85]. To date, miRNA have been implicated in a long list of cardiovascular diseases comprising myocardial infarction, heart failure, stroke, peripheral and coronary artery disease and several more [86,87]. Nevertheless, the pathological implications of angiomiRs surfaced also with the help of endothelium-specific Dicer-deficient mice, as the ablation led to reduced tumor progression due to diminished angiogenesis, which is a prerequisite for tumor development [88]. For example, two miRNAs induced by VEGF expression (miRs-296, miRs-132) have been identified as candidates supporting the angiogenic switch during tumor formation i.e., the transition from a pre-vascular to a vascularized tumor phenotype [89,90]. In conclusion, the combination of Dicer-deficient angiogenic phenotypes suggests crucial roles for miRNAs in regulating structure and function of embryonic and postnatal blood vessel development.

In the context of angiogenesis, an additional, very important category is a specialized subset of hypoxia-inducible miRNAs, whose increasing number of representatives was also termed hypoxamiRs [91–96]. Thus, reduced oxygen supply in ossification centers of bone stimulate the expression of VEGF and other angiogenic factors that lead to the development of blood vessel structures [97]. Additionally, hypoxia-regulated pathways have been attributed to regulatory functions such as smooth muscle cell proliferation and contractility, cardiac remodeling, cardiac metabolism and ischemic cardiovascular diseases [94]. Together with a variety of other target genes which are important for physiological low oxygen adaption, their expression is initiated by upregulation of the transcription factor hypoxia-inducible factor alpha (HIF) [98]. One group of hypoxamiRs are therefore upregulated following HIF expression (HIF-dependent hypoxamiRs), with the master hypoxamiR-210 being the most prominent example [99,100]. Hypoxia-dependently expressed miRNAs that affect HIF expression itself also belong to hypoxamiRs. Thus, for the adaptation to low oxygen conditions and induction of angiogenesis, HIF displays a unique role by controlling further upregulation of hypoxamiR-424 in ECs, which promotes its own protein stabilization [101]. A last group of hypoxamiRs, moreover, influences HIF expression in the absence of hypoxia. As an example, miR-31 decreases HIF-1α expression via the "factor-inhibiting HIF (FIH)" while the miR17-92 cluster suppresses HIF-1α upon c-MYC induction [102,103].

5. Specific MicroRNAs Implicated in Angiogenic-Osteogenic Coupling

Taken together, the functions of osteomiRs, angiomiRs, and hypoxamiRs suggest the possibility that miRNAs will also have crucial roles in bone angiogenesis. Subsequently, miRNAs will be outlined that were found to have a significant function in osteogenesis as well as angiogenesis, and therefore represent miRNAs that have already been identified to have an active role in angiogenic-osteogenic coupling or are presumed to have its potential (Figure 1, Table 1). Collectively, these may also be referred to as "CouplingmiRs/CPLGmiRs". MiRNAs with a confirmed function in this process could be employed as therapeutic targets in bone regeneration. Consequently, they could improve the coordination and enhancement of the endogenous osteogenesis and angiogenesis process. Elucidation of the molecular mechanisms governing osteoblast differentiation and angiogenesis are furthermore of great importance for improving the treatment of bone-related diseases.

5.1. MiR-9

miR-9 is a highly conserved microRNA, but exhibits a divergent expression pattern, and seems to modulate different targets in a cellular context- and developmental stage-specific manner [104]. The studies of Han et al. have demonstrated a regulatory role for miR-9 in the development and differentiation of human bone marrow derived MSCs (hBM-MSCs) and neural progenitor cells on proliferation, migration and differentiation [105]. In this regard, cell-autonomous effects have been described in vertebrates by effecting Notch, Wnt, and BAF53a expression. Furthermore, miR-9 has also been related to tissue repair processes involving human MSCs (hMSCs). Qu et al. provided evidence that increased miR-9 expression levels closely correlate with enhanced differentiation of MC3T3-E1 osteoblasts [106]. The effects of miR-9 on angiogenesis were studied by the same authors by using human umbilical vein endothelial cells (HUVECs). Here, miR-9 mimics transfection effectively increased VEGF, VE-cadherin, and FGF concentrations in the culture medium, leading to increased EC migration and capillary tube formation in vitro. The underlying molecular mechanism for both the regulation of osteoblast differentiation and angiogenesis was found in the activation of AMP-activated (AMPK) signaling. Conclusively, the results of Qu et al. imply a potential important role of miR-9 in regulating the process of bone injury repair, and therefore, a potential therapeutic target for the treatment of bone injury-related diseases.

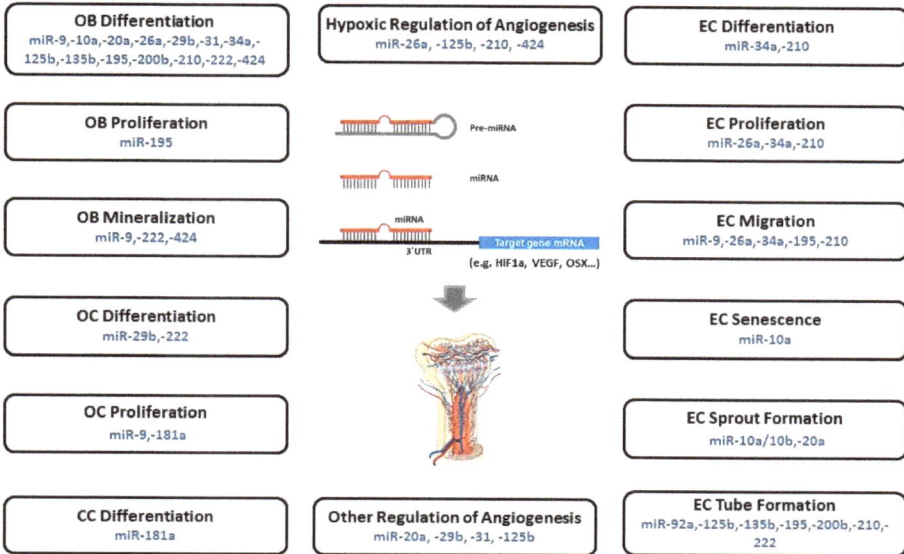

Figure 1. MicroRNAs (miRs/miRNAs) involved in the regulation and coupling of bone angiogenesis ("CouplingmiRs/CPLGmiRs"). Reported miRNAs contributing to the formation of blood vessels during the processes of formation, repair and regeneration of bone were allocated with the individual functions of their target genes during osteogenesis, angiogenesis, or hypoxic regulation of bone angiogenesis. OB, osteoblast; OC, osteoclast; CC, chondrocyte; EC, endothelial cell.

A different study that analyzed the function of miR-9 in C2C12 mesenchymal cells further supported the role of miR-9 in osteoblast differentiation. By significantly decreasing the expression of DKK1 protein, but not of its mRNA, miR-9 stimulated alkaline phosphatase (ALP) activity and osteoblast mineralization, as well as the expression of several osteoblast marker genes, such as COL I (collagen I), OCN (osteocalcin), and BSP (bone sialoprotein) [107]. MiR-9 furthermore was detected as a tumor-secreted pro-angiogenic miRNA that promoted EC migration and tumor angiogenesis in vitro. Mechanistically, this was explained by interference with the expression levels of SOCS5, and thereby, the activation of the JAK-STAT pathway [108].

Moreover, altered expression of miR-9 was found during screening osteoarthritis cartilage involved in the control of tumor necrosis factor α (TNFα) expression. Here, miR-9 has been implicated as a key regulator in the process of endochondral ossification, since its expression varied significantly between the early and late stage of chondrocyte development [73,109]. Functional experiments in a mouse tibial plateau fracture model implicated that miR-9 and miR-181a significantly downregulated Bim concentration. Thereby, osteoclast survival was stimulated and the migration ability of osteoclasts was effected [110].

5.2. MiR-10a

The function of miR-10a during osteoblast differentiation and angiogenesis in vitro was analyzed in MC3T3-E1 and MUVEC (mouse umbilical vein endothelial cells) by the research group of Li et al., respectively [111]. Upon BMP2-induced osteoblast differentiation of MC3T3-E1 cells, miR-10a was downregulated. In contrast, when miR-10a was overexpressed, a suppressive effect on β-catenin and LEF1 expression could be demonstrated. Overexpression inhibited osteogenic differentiation, as demonstrated by reduced expression of the osteoblast-differentiation markers ALP, runt-related transcription factor 2 (RUNX2), Osterix (OSX), and distal-less homeobox 5 gene (DLX5). Moreover, it led to a decrease in MUVEC proliferation, migration and tube formation in combination with reduced concentrations of the angiogenesis-related genes VEGF, VE-cadherin, cyclin D1, and MMP2. As the canonical WNT/β-catenin signaling pathway was found to play an important role in osteogenic cell proliferation, differentiation, and bone regeneration, miR-10a offers a potential therapeutic target for the treatment of bone regeneration and bone-related diseases [112].

A further report referred to the role of miR-10a in regulating endothelial progenitor cell (EPC) senescence in the mouse [113]. Zhu et al. provided evidence that upon upregulation of miR-10A and miR-21, Hmga2 (High-mobility group AT-hook 2) expression is gradually decreased during aging in bone-marrow cells that were enriched for EPCs. Suppression of miR-10A* and miR-21 in aged EPCs, on the other hand, increased Hmga2 expression and improved EPC angiogenesis in vitro and in vivo. This could be demonstrated by rejuvenated EPCs, which resulted in decreased senescence-associated β–galactosidase expression, increased self-renewal potential and decreased p16Ink4a/p19Arf expression. In conclusion, the study demonstrates that the miR-10A*/miR-21–Hmga2–P16Ink4A/P19Arf axis controls EPC senescence and angiogenesis and may represent a potential therapeutic intervention target for improving EPC-mediated angiogenesis and vascular repair.

5.3. MiR-10a/10b

Hassel et al. investigated the function of miR-10 in zebrafish blood vessel formation. During embryogenesis, the knockdown of miR-10a/10b impaired blood vessel outgrowth due to an altered tip cell differentiation behavior, and led to defects of intersegmental vessel growth by modulating fms-related tyrosine kinase 1 (flt1) levels post-transcriptionally [114]. However, as the knockdown of flt1 did not fully rescue the angiogenic phenotypes in miR-10 mutant zebrafish, as well as in miR-10-deficient HUVECs, the authors concluded that in ECs, flt1 could not represent a direct exclusive target of miR-10. They provided evidence that miR-10a/10b regulated angiogenesis in a Notch-dependent manner by directly targeting mib1 (mindbomb E3 ubiquitin protein ligase 1) in zebrafish ECs. Inhibition of mib1 and Notch signaling partially rescued the angiogenic defects in miR-10 morphants, suggesting that the observed angiogenic defects in miR-10a/10b morphants are caused by up-regulation of Notch signaling [115].

5.4. MiR-20a

As a member of the extensively studied miR-17-92 cluster with prominent roles in tissue and organ development, the role of miR-20a was elucidated during osteogenesis by Zhang et al. [116]. They disclosed that, together with several osteoblast markers (BMP2, BMP4, RUNX2, OSX, OCN and OPN), miR-20a was upregulated during osteogenic differentiation of hBM-MSCs derived from bone marrow from differently aged persons. Adipocyte markers, however, such as PPARγ and the osteoblast antagonists, BAMBI and CRIM1, were down-modulated. By introducing miR-20a mimics and lentiviral-miR20a-expression vectors into hBM-MSCs, they verified that miR-20a enhances osteogenic differentiation. Simultaneous direct interaction with all the aforementioned positive and negative effectors of BMP/RUNX2 signaling could be confirmed.

In a report by Doebele et al., miR-20a was investigated with respect to its cell-intrinsic angiogenic activity in EC, as different members of the miR-17-92 cluster that are highly expressed in tumor cells were found to be expressed at increased levels during ischemic conditions [117]. They determined that in vitro overexpression of miR-20a (together with miR-17, -18a, -19a) rigorously inhibited EC sprout formation, whereas their inhibition using antagomiR treatment led to an increase of spheroid sprouting (irrespective of miR-19a). Interestingly, in vivo matrigel plug assays employing antagomiRs for miRs-17-20a as inhibitors demonstrated enhanced angiogenic sprouting, but in contrast to in vitro results, they were unaltered in tumor angiogenesis, indicating context-sensitive regulation. In particular, the pro-angiogenic target gene Janus kinase 1, but also the cell cycle inhibitor p21 and the S1P receptor EDG, were shown to be downregulated by miR-17/20a.

Deng et al. provide evidence that miR-20a and miR-31 serve as stimulators of angiogenesis [118]. As an underlying mechanism, they found that the expression of both miRNA molecules is upregulated via AKT and ERK signals that are themselves activated by the angiogenic factor VEGF. As target genes, miR-20a and miR-31 were found to directly associate with the 3′-UTR of the tumor necrosis factor superfamily-15 (TNFSF15) gene, thereby clarifying its by then unknown mechanistic role in vascular homeostasis. TNFSF15 is expressed in ECs of mature vasculature, and is a known inhibitor of angiogenesis. Interestingly, VEGF-stimulated downregulation of TNFSF15 could be attenuated by treatment of HUVECs with AKT inhibitor LY294002, leading to reduced miR-20a and miR-31 levels, while ERK inhibitor U0126 prevented VEGF-induced expression of miR-20a only. In contrast, inactivation of either ERK or AKT signals restored TNFSF15 gene expression and elevated miR-20a or miR-31 levels which led to enhanced capillary-like tube formation in an in vitro angiogenesis assay.

5.5. MiR-26a/b

Luzi et al. investigated and confirmed an important function of miR-26a during human adipose-derived MSCs (hADSCs) differentiation towards the osteogenic lineage induced by treatment with dexamethasone, ascorbic acid, and beta-glycerol phosphate [119]. Upon inhibition of miR-26a by antisense RNA, upregulation of the transcription factor SMAD1—which was predicted in silico—and its regulated osteogenic differentiation marker genes could be observed in treated osteoblasts. In a follow-up study of Luzi et al., these results were extended to include the interaction between menin and miR-26a as regulators of osteogenic differentiation in hADSCs [120]. Menin is a presumable transcriptional regulator that modulates mesenchymal cell commitment to the myogenic or osteogenic lineages. It is encoded by the *MEN1* oncosuppressor gene which causes the multiple endocrine neoplasia type-1 syndrome. The results demonstrated orchestrated down-regulation of *MEN1* mRNA and miR-26a, with a consequent up-regulation of SMAD1 protein in hADSCs.

Su et al., however, reported an opposing role of miR-26 in hBM-MSC during osteogenic differentiation, suggesting distinct post-transcriptional regulation of tissue-specific hMSC differentiation [121]. Using bioinformatics and functional assays, they confirmed that miR-26a directly regulates *SMAD1*, but added GSK3β as a target to regulate BMP and WNT signaling pathways. The distinct activation pattern and comparative analysis revealed that miR-26a significantly inhibited *SMAD1* to suppress BMP signaling for interfering with the osteogenic differentiation of hADSCs, whereas it targeted on GSK3β to activate WNT signaling for promoting osteogenic differentiation of hBM-SCs. Overall, they concluded that the BMP pathway was more essential for promoting osteogenic differentiation of hADSCs, whereas WNT signaling was enhanced more potently and played a more important role than BMP signaling in osteogenic differentiation of hBM-SCs. In conclusion, although miR-26a enhances osteogenic differentiation in both cell types, different signaling pathways were employed in hBM-MSCs and hADSCs.

In to addition unrestricted somatic stem cells (USSC), the studies of Trompeter et al. investigated a rare population in human cord blood with respect to osteogenic differentiation. Gene expression profiling of two different USSC cell lines (SA5/73 and SA8/25) identified, among other candidates, miRs-26a/b and miR-29b to be consistently upregulated during osteogenic differentiation [122]. As osteo-inhibitory targets of these miRNAs, CDK6 and HDAC4 were evaluated that were downregulated during osteogenic differentiation of USSC, whereas SMAD1 was found as an osteo-promoting target. During osteogenic differentiation of USSC or following ectopic expression of miR-26a/b, SMAD1 exhibited an unchanged expression level, however.

Entangling of miR-26a in pathological and physiological angiogenesis was investigated by Icli et al. in ECs [123]. They studied the effects of modifying the expression of BMP/SMAD1 signaling. Upregulation of miR-26a led to EC cycle arrest, inhibited EC migration, sprouting angiogenesis, and network tube formation in matrigel. Upon inhibiting miR-26a expression, a contrasting phenotype could be detected. At the molecular level, Icli et al. demonstrated direct binding of miR-26 to the 3′-UTR of *SMAD1* thereby reducing its mRNA levels, which subsequently suppressed *ID1* expression and increased *p21WAF/CIP* and *p27* protein expression.

5.6. MiR-29b

MiRNA profiling of MC3T3 preosteoblastic cells derived from fetal mouse calvaria and differentiated to osteoblasts led to the identification of miR-29b, among other members of the miR-29, miR-let-7, and miR-26 families by Li et al. [124]. Versatile effects of miR-29b were found to promote osteoblastogenesis at multiple stages as a key regulator. One mechanism pursues the silencing of negative regulators of osteogenic differentiation, such as TGF-β3, HDAC4, ACTVR2A, CTNNBIP1, and DUSP2 that involve particularly the osteogenic function of RUNX2, as well as the SMAD, ERK, p38 MAPK, and WNT signaling pathways. A second path seeks the suppression of extracellular matrix protein synthesis relevant to bone development (such as COL1A1, COL5A3, and COL4A2) to preserve the differentiated phenotype during mineralization of mature osteoblasts. This alternative mechanism seems to enhance mineral deposition and to prevent fibrosis.

Rossi et al. also disclosed a link of miR-29b to osteoclastogenesis and proposed it for the treatment of multiple myeloma-related bone disease as its expression declined increasingly during human osteoclast differentiation and affected proper bone resorption [125]. Several findings indicated that miR-29b is a negative regulator of human OCL differentiation and activity. Thus, lentiviral transduction of miR-29b into OCLs was associated with diminished tartrate acid phosphatase expression, lacunae generation, and collagen degradation. Attenuated resorptive osteoclast capabilities, due to miR-29b inhibition of proteolytic enzymes, were documented by reduced cathepsin K, MMP-9, and MMP-2 expression. Overall, downstream phenotypic effects along the M-CSF and RANK-L axes that led to impaired action of the master transcription factor NAFTc-1 were explained by miR-29b targeting of c-FOS.

Zhang et al. found inhibitory activity of miR-29b on VEGF secretion via the anti-angiogenic cytokine TNFSF15 (VEGI; TL1A) in the mouse EC line bEnd.3, which defines a new angiogenesis-related signaling pathway [126]. In contrast, down-modulation of TNFSF15 activity by a specific siRNA against its receptor DR3/TNFRSF25, or a neutralizing antibody against TNFSF15, reinstated VEGF generation but suppressed miR-29b expression. TNFSF15-enhanced activation of the JNK-GATA3 signaling pathway was furthermore able to stimulate miR-29b expression but silenced VEGF production, as demonstrated by a specific JNK inhibitor or siRNA.

Li et al. described an anti-angiogenic and anti-tumorigenic role for miR-29b by the regulation of AKT3 expression [127]. AKT is known to induce tumor vascularization via VEGF, and cancer cell activity via c-MYC arrest, in breast tumor. In vitro and in vivo ectopic expression of miR-29b therefore blocked angiogenesis, as well as tumor cell formation, evidencing it as a potential useful anti-cancer therapeutic agent.

5.7. MiR-31

Granchi et al. detected miR-31 by profiling miRNA expression during osteogenic differentiation and mineralization of hBM-MSCs that were derived from three individual donors [128]. As an identified direct target gene, miR-31 differentially modulated the expression of the bone-specific transcription factor OSX during osteogenic differentiation [129]. This could be demonstrated by an inverse miRNA-target expression ratio in osteosarcoma cell lines and an increase in OSX expression upon specific miR-31 inhibition.

Deng et al. provided evidence that the expression of miR-31 increasingly declined during the osteogenic differentiation of hBM-MSC cells [130]. This regulation coincided with increased ALP activity, mineralization of hBM-MSC cultures and expression of the osteogenic transcription factors OPN, BSP, OSX, and OCN, with the exception of RUNX2. Mechanistically, they uncovered a RUNX2, SATB2, and miR-31 regulatory feedback loop that determined hBM-MSC differentiation using inhibitors and mimics of miR-31. RUNX2 directly regulates miR-31 expression levels which itself controls the translation of SATB2 protein.

In a study of Suarez et al. that investigated TNF-mediated induction of endothelial adhesion molecules, the expression of miR-155, -31, -17, and -191 were found to be increased without a change of miR-20a, -222, and -126 levels in HUVECs [84]. By miRNA target prediction algorithms, the 3'-UTR of E-selectin was identified as target molecule, which was verified experimentally by reporter assays using miR-31 mimics. miR-31-mediated regulation of E-selectin expression not only regulates binding of neutrophil granulocytes to HUVECs, but is also involved in inhibition of angiostatin-induced angiogenesis by affecting cell migration [131].

An angiogenesis-related function of miR-31 reported by Deng et al. has already been described above in the subsection of miR-20a [118].

5.8. MiR-34a

Opposing roles for miR-34a in differentiation of hMSC towards osteoblast have been reported. Chen et al. identified miR-34a in a microarray screening, and stated that it inhibits osteoblast differentiation in hMSC and in vivo bone formation in a preclinical model of heterotopic bone formation in mice [132]. Thus, when miR-34a was overexpressed in hMSC, it inhibited early and late OB commitment, as well as differentiation and hMSC proliferation, while anti-miR oligonucleotide treatment reversed these effects. In addition to several cell cycle regulator and cell proliferation proteins (including CDK4, CDK6, and Cyclin D1), JAGGED1, a NOTCH1 receptor ligand, was elicited as a target gene of miR-34a that was regulated at both the transcriptional and translational levels, as determined by RNA interference. JAGGED1 has previously been implicated in human bone biology, as its deficiency causes skeletal abnormalities in the Alagille syndrome.

Kang et al., however, claimed that miR-34a-5p-induced activation of the Notch signaling pathway is a positive regulator of glucocorticoid-mediated osteogenic differentiation of hMSCs [133]. They demonstrated dexamethasone-inhibited osteoblastic differentiation of murine BM-MSC via miR-34a-5p-mediated gene silencing of coincidently identical target genes, as published by Chen et al. Differences in both reports were discussed to be due to different roles of the Notch signaling pathway in osteogenic differentiation of hMSCs that were derived from different species.

A more recent publication by Fan et al. found miR-34a upregulation during osteogenic differentiation of hADSCs [134]. Elevated levels of miR-34a in hADSCs promoted mineralization, ALP activity, and the expression of the key regulatory osteogenic transcription factor *RUNX2* by targeting the retinoblastoma binding protein 2 (RBP2); furthermore, heterotopic bone formation was enhanced in vivo. Expression of *NOTCH1* and *Cyclin D1* genes, that were also involved in this coregulatory network, were found to be downregulated on the other hand, which facilitates cell cycle exit [135,136]; this is the consequence of suppressed proliferation but enforced terminal maturation of osteoblasts by RUNX2.

When investigating the role of miR-34a in glucocorticoid-induced osteonecrosis of the femoral head (GIOFH), Zha et al. verified the findings of Kang et al., but extended this knowledge [137]. By investigating dexamethasone-treated rat subjected to miR-34a-overexpressing lentiviruses, decreased blood vessel development was observed, indicating that VEGF presents a regulatory target of miR-34a. In vitro, miR-34a overexpression enhanced the inhibitory effects of dexamethasone on the viability and activity of ECs and downregulated VEGF protein expression levels.

A study by Zhao et al. investigated the angiogenic role of miR-34a in EPCs derived from adult male Spraque-Dawley rats [138]. The rationale behind this project was the previous implication of miR-34a in targeting silent information regulator 1 (*SIRT1*), provoking cell cycle arrest or apoptosis. The results confirmed the inhibitory effects of raised miR-34a expression levels on *SIRT1* and, thus, on EPC-mediated angiogenesis by inducing senescence. Mechanistic causes could be found in increased acetylated levels of the FOXO1 transcription factor, regulated by *SIRT1*, which could be demonstrated by knockdown of *SIRT1*. The angiogenesis-promoting role of miR-34a and Sirt1 in this context seems to lie in its previously discovered function in vascular endothelial homeostasis by preventing stress-induced senescence in health and disease [139].

Table 1. Summary of microRNAs with a presumed role in osteogenic-angiogenic coupling paving the way for bone angiogenesis ("CouplingmiRs/CPLGmiRs").

MicroRNAs	Targets [1]	Regulatory Role	Effects	Study Models	Ref.
MiR-9	VEGF, VE-CAD (CD144)	AMPK signaling pathway	Enhanced osteogenic diff. & mineral.; increased angiogenesis	MC3T3-E1	[106]
	DKK1	COL1, OCN, BSP; ALP activity	OB diff. & mineralization	C2C12 cells	[107]
	SOCS5	JAK-STAT signaling pathway	Promotion of EC migration & angiogenesis	Primary microvascular ECs, HUVECs	[108]
	Cbl	Bim ubiquitination, apoptosis	Promotion of OC survival	OC, OC precursor cells (RAW264.7)	[110]
MiR-10a	β-catenin, LEF1; VEGF, VE-CAD (CD144), cyclin D1, MMP2	Wnt signaling; angiogenesis-related gene expression	Inhibition of osteogenic diff. & blood vessel formation	MC3T3-E1 MUVECS	[111]
	HMGA2	β-galactosidase expr.; p16Ink4a/p19Arf expression	EPC senescence & angiogenesis; self-renewal potential	lin−BM-MSCs	[113]
MiR-10a/10b	MIB1	Notch signaling	Regulating blood vessel outgrowth/tip cell behavior	HUVECs	[115]
MiR-20a	BMP2, BMP4, RUNX2	Effects BMP/RUNX2 signaling positively; blocks OB inhibitors & PPARγ	Enhances osteogenic differentiation; suppresses adipogenesis	hBM-MSC	[116]
	JAK1; p21, S1P receptor EDG	Downregulation of proangiogenic JAK 1 & cell cycle inhibitors	Inhibits EC sprout formation	HUVECs	[117]
	TNFSF15	VEGF-AKT/ERK −miR20a/31 signaling	Stimulation of angiogenesis	HUVECs	[118]
	VEGF, ANG1, RUNX2, BMP2 OCN, ALP; GSK3β	WNT signaling activation	Enhanced angiogenesis & bone regeneration	Primary hBM-MSC, MC3T3-E1	[76,98]
MiR-26a	VEGF	PIK3C2α/AKT/HIF-α/VEGFA pathway	Inhibition of angiogenesis;	HUVECs	[141]
	SMAD1	BMP signaling inhibition	OB differentiation	hADSCs	[119]
	SMAD1	BMP signaling	Inhibits EC growth, proliferation, migration; regulates early angiogenesis	HUVECs	[123]
MiR-29b	TGF-β3, HDAC4, ACTVR2A, CTNNBIP1, DUSP2; COL1A1, 5A3, 4A2	Silences neg. osteogenic regulators suppresses ECM protein synthesis	Promotes osteoblastogenesis at multiple stages	MC3T3 pre-OB	[124]
	c-FOS	Reduced TRAP expr., lacunae generation, collagen degradation	Neg. regulator of human OC differentiation and activity	OC (CD14+)	[125]

Table 1. *Cont.*

MicroRNAs	Targets [1]	Regulatory Role	Effects	Study Models	Ref.
	TNFSF15	TNFSF15-enhanced JNK-GATA3 signal. & VEGF inhibition	Suppression of VEGF secretion	Mouse EC line bEnd.3	[126]
	AKT3	Inhibition of tumor vascularization via VEGF & cancer cell activity via c-MYC	Anti-angiogenic and anti-tumorigenic role	HUVECs, Breast cancer cells	[127]
	OSX	Downregulation of OSX	Influences osteogenic differentiation	hMSC; Osteosarcoma cell	[129]
MiR-31	Satb2 protein	Inhibition by RUNX2; Upregulation of Satb2 protein & osteogenic TF	Induces BM-MSC osteogenic differentiation	hBM-MSC	[130]
	E-selectin	Regulation of E-selectin expression	Inhibition of angiostatin-induced angiogenesis; TNF-mediated induction of endothelial adhesion	HUVECs	[84]
	TNFSF15	VEGF-AKT/ERK –miR20a/31 signaling	Stimulation of angiogenesis	HUVECs	[118]
	Jagged1	Regulation of cell cycle regulator & proliferation proteins & Jagged1	Inhibition of osteoblast differentiation	hMSC; mouse heterotopic bone formation model	[132]
	JAGGED1	Activation of Notch signaling	Induction of glucocorticoid-mediated osteogenic differentiation	hMSC	[133]
MiR-34a	RBP2	Promotes mineral, ALP activity & RUNX2 expression; downreg. NOTCH1 & Cyclin D1 expr.	Promotion of osteogenic differentiation; enhanced heterotopic bone formation	hADSCs; mouse heterotopic bone formation model	[134]
	VEGF	Inhibitory effects of dexamethasone on EC viability & VEGF	Decreased blood vessel development	Rat Glucocorticoid-induced osteonecrosis	[137]
	SIRT1	Increased SIRT1 expr. & FOXO1 acetylation regulating vascular EC homeostasis	Inhibition of EPC-mediated angiogenesis	Rat EPC	[138]
	E2F3a, survivin	Interference with VEGF secretion, EC proliferation & migration	Dysregulated tumor angiogenesis	HNSCC tumors & cells	[140]
	?	?	Enhanced fracture healing & inhib. of neovascularization	Mice with femoral fracture	[142]
MiR 92a	HGF, ANGPT1	ITGA5, MEK4	Inhibition of tube formation by HUVECs	hADSCs	[143]
	?	integrin a5, sirtuin1, eNOS	Attenuates neointimal lesion by accelerating re-endothelialization	MiR-92a knockout mice	[144]

Table 1. *Cont.*

MicroRNAs	Targets [1]	Regulatory Role	Effects	Study Models	Ref.
MiR-125b	OSX	RUNX2, a-SMC, ALP, matrix mineralization	Calcification of vascular smooth muscle cells	HCASMCs	[145]
	ErbB2	?	Inhibits OB diff by downreg. of cell proliferation	ST2 cells (mMSCs)	[146]
	VEGF, ERBB2		Regulation of angiogenesis during wound healing	HUVECS	[147]
	Cbf-beta	ALP, OCN, OPN	Inhibition of osteogenic differentiation	C3H10T1/2	[148]
	SMAD4	ALP, RUNX2	Downregulation of osteogenic differentiation	hMSCs	[149]
	VE-Cadherin		Inhibition of blood vessel (tube) formation	HUVECs	[150]
MiR-135b	?	?	OB differentiation	hBM-SCs	[151]
	HIF-1	?	Enhanced endothelial tube formation	Human MM cells; HUVECs	[152]
	SMAD5	?	Impaired osteogenic differentiation	hMSCs	[153]
	?	CCN1, aggrecan	Maintaining homeostasis of chondrocytes	Human HCS-2/8 cells	[154]
	COL10A1		Chondrocyte differentiation	hMSC	[155]
MiR-181a	RGS16	CXCR4 signaling; VEGF, MMP1	Angiogenesis & metastasis in chondrosarcoma	Xenograft mice; JJ chondrosarc. cells	[156]
	?	VEGF expression	Chondrosarcoma-associated angiogenesis	JJ chondrosarc. cell line	[157]
	Cbl	Bim ubiquitination, apoptosis	promote OC survival	OC, OC precursor cells (RAW264.7)	[110]
MiR-195	?	VEGF	Osteogenic diff. & proliferation; control of angiogenesis	hMSC(MC3T3) chick chorio-allantoic membrane	[158]
	?	VEGF, VAV2 CDC42	HCC-associated angiogenesis & metastasis; migration & capillary tube form. of ECs	QGY-7703, MHCC-97H HCC cells; HUVECs	[159]

Table 1. *Cont.*

MicroRNAs	Targets [1]	Regulatory Role	Effects	Study Models	Ref.
MiR-200b	ZEB1	ZEB1-TF target genes	Inhibits proliferation, migration & invasion of osteosarcoma cells	Osteosarcoma U2OS, Saos2, HOS, MG63	[160]
	VEGF-A; ZEB2, ETS1, KDR,GATA2	Decreases VEGF-A expression & TF-target genes	Inhibition of VEGF-A induced osteogenesis; Inhibition of TF-activated angiogenesis	Rat BM-MSC & HUVEC coculture	[161]
	VEGF, FLT-1, and KDR	VEGF-induced phosph. of ERK1/2	Inhibition of angiogenesis; red. capillary formation	A549 cells, HUVECs	[162]
	AcvR1b	Inhib. of TGFb/activin signaling	Promotes OB differentiation	ST2 stromal cells	[163]
MiR-210	VEGF	PPARgamma, ALP, OSX	Promoteion of OB diff., inhibition of adipocyte diff.	hBM-SCs, 17β-estradiol (E2)treated OB	[164]
	EFNA3	VEGF-expression mediated angiogenesis	EC survival, diff., migration; stim. of tubulogen. & chemotaxis	HUVECs	[165]
	SMAD 1, 5, 8 protein & phosphoryl.	Decreased SMAD5-RUNX2 signaling & OSX, ALP, and OC levels & mineral.	Neg. regulator of osteogenic differentiation	hBM-SC	[166]
MiR-222	c-Src, Dcstamp	RANKL-induced expression of TRAP & cathepsin K	Inhibitory regulator of c-Src-mediated osteoclastogenesis	RAW264.7 pre-OC cells	[167]
	c-KIT	Suppression of tube formation, wound healing, cell migration via SCF	Inhibitory regulation of in vitro angiogenesis	HUVEC	[168]
MiR-424	RUNX, CBFβ, BMP	Osteogenic diff. of hMSCs	Bone formation	hMSCs	[169]
	MAPK, WNT & insulin signal.	OB differentiation of hMSCs	Bone formation	hMSCs	[170]
	FGF-2; via FOXO1	Decrease of ALP, mineralization & osteog. markers	Enhances proliferation & osteogenic differentiation of hMSCs	Pigs, cellular oxidative stress model	[171]
	CUL2; via RUNX-1 → C/EBPα→ PU.1	Stabilization of HIF-1α	Regulation of Angiogenesis	ECs, ischemic tissues	[101]

[1] Identified target genes or downstream effectors; HUVECs, human umbilical vein endothelial cells; ?, unknown molecular target(s) and/or regulatory role; MUVECs, mouse umbilical vein endothelial cells; EC, endothelial cells; EPC, endothelial progenitor cells; hMSCs, human mesenchymal stem cells; hBM-SC, bone-marrow derived stem cells; hADSCs, human adipose-derived mesenchymal stem cells; TNFSF15; cytokine tumor necrosis factor superfamily 15; OB, osteoblasts; OC, osteoclasts; TF, transcription factor.

A negative regulatory role for miR-34a could also be documented by Kumar et al. in tumor angiogenesis in head and neck squamous cell carcinoma (HNSCC) [140]. MiR-34a expression was markedly decreased in HNSCC tumors and cell lines, and ectopic expression of the miR-34a therefore affected cell proliferation and migration in HNSCC cell lines and in a SCID mouse xenograft model. These effects seemed to be mediated by the regulation of survivin expression via the transcription factor E2F3a that is critical for cell cycle progression. Furthermore, tumor angiogenesis was found to be dysregulated by interference of miR-34a with VEGF secretion in tumor cells, as well as EC proliferation, migration, and tube formation, by downregulating a number of key proteins including E2F3, SIRT1, survivin, and CDK4.

5.9. MiR-92a

The autosomal dominant Feingold syndrome that is characterized by microcephaly, short stature, and digital anomalies was identified in individuals carrying hemizygous deletions of the miR-17-92 cluster [172]. To dissect this complex phenotype that could be partially mirrored in mice deficient of the miR-17-92 cluster, and due to postnatal lethality, a single miR-92a targeted mouse line was established by Penzkofer et al. [173]. Surviving mice exhibited reduced body weight and skeletal defects represented by reductions in body length, skull, and tibia length, as well as metacarpal bone size; these defects were presumably also the result of osteoblast proliferation and differentiation phenotypes, as found in hemizygous miR-17-92 mouse mutants [174]. In contrast, however, single miR-92a deletion does not cause low bone mineral density attributed to reduced type-I collagen mRNA, ALP activity, and mineralization ability. Phenotypic differences could be explained by the design of the genomic deletion of miR-92a that partially attenuated expression levels of miR-18a in skeletal tissue. The direct molecular targets of miR-92a responsible for regulation of osteogenesis have not yet been elicited.

A recent study of Mao et al., however, identified aggrecanase-1 and aggrecanase-2 (ADAMTS4/5) as direct miR-92a targets in chondrogenic hMSCs and human chondrocytes by reporter assays [175]. These two members of the ADAMTS family represent MMPs that are important for normal chondrocyte differentiation, but which also promote the progression of osteoarthritis by cartilage degeneration. In comparison to normal cartilage, real-time-PCR analyses also revealed higher miR-92a-3p expression levels in chondrogenic hMSCs, whereas markedly reduced miRNA expression could be detected in ostearthritis cartilage. Moreover, miR-92a-3p-modified expression of ADAMTS-4/5 could be downregulated by IL-1β transfected primary human chondrocytes. The study was preceded by previous findings about miR-92a-3p upregulation in hADSC-derived chondrocytes and chondrogenic hMSCs, as well as osteoarthritis cartilage, where histone deacetylase 2 (HDAC2) was identified as a target gene of miR-92a-3p [176,177].

MiR-92a was first reported as a part of the miR-17-92 cluster by Bonauer et al., and was found to be abundant in human ECs [178]. MiR-92 was established as a suppressor of angiogenesis which targets the expression of several proangiogenic proteins, including the integrin subunit alpha 5 (ITGA5) that is well known for severe vascular defects in gene targeted mice. Systemic administration of miR-92a antagomiR therefore improved the growth of blood vessels and functional recovery of damaged tissue in mouse models of limb ischemia and myocardial infarction, possibly by indirectly inhibiting apoptosis [58]. Thus, miR-92a may serve as a valuable therapeutic target in the setting of ischemic disease.

Daniel et al. further elaborated the function of miR-92a upon re-endothelialization and neointimal formation after wire-induced injury of the femoral artery in mice [144]. By using specific LNA (locked nucleic acids)-based antimiRs as well as miR-92a-deficient ECs in the mouse, they found enhanced re-endothelialization and inhibited neointimal formation induced by de-repression of the miR-92a targets integrin a5 and sirt1 [179]. Thus, an important role can be attributed to miR-92a in blood vessel regeneration in ischemic tissues and after vascular injury.

In a subsequent study, Zhang et al. investigated the ability of miR-92a to influence apoptosis and angiogenesis in ECs in the presence of oxidative stress [180]. It was previously determined that senescent ECs that undergo apoptosis, and which are frequent in ageing and atherosclerosis, have diminished miR-92a levels [181]. The results provided evidence that pre-miR-92a treatment of HUVECs prevented oxidative stress-induced apoptosis in EC, whereas capillary tube formation i.e., angiogenesis, was maintained. Mechanistically, it was determined that pre-miR-92 directly repressed PTEN, and thereby, activated the AKT signaling pathway that regulates EC apoptosis and angiogenesis.

In a study of Kalinina et al., paracrine effects of miR-92a on hMSCs were studied [143]. Conditioned medium of hMSCs transfected with pre-miR-92a prevented tube formation by HUVECs, which could be attributed to significantly lower secretion of hepatocyte growth factor (HGF) and angiopoetin-1 independent of VEGF secretion. HGF is a factor required for stimulation of proliferation of ECs, whereas angiopoietin-1 regulates vessel stabilization and maturation. As neither gene was predicted as direct miR-92a targets, these still have to be identified. Restoration of tube formation was achieved by replenishment of HGF, but not with anti-miR-92a treatment of hMSCs. This led to the conclusion that miR-92a suppresses hMSC-induced angiogenesis by downregulating the secretion of HGF and, therefore, is involved in the control of anti-angiogenic activities in hMSCs.

5.10. MiR-125b

MiR-125b is an early discovered miRNA that plays a key role in cellular functions. Experiments by Mizuno et al. using BMP-4-induced or exogenous miRNA-transfected murine MSC ST2 cells found miR-125b to inhibit osteoblastic (and adipogenic) differentiation through modulating cell proliferation [146]. Further experiments with transfection of siRNA identified ErbB2 as a target gene. miR125b-gene targeted mice, generated by Lee et al., identified further supported the important function of miR-125b in regulating stem cell directional differentiation, but did not reveal its molecular role [182].

Decreasing levels of miR-125b were also identified by Huang et al. during the differentiation of C3H10T1/2 cells, and reporter gene assays led to the identification of a putative target binding site in the 3′-UTR of the Cbfβ gene, a master regulatory gene of osteogenesis [148]. Therefore, silencing of miR-125b increased the mRNA levels of Cbfβ and of osteoblastic marker genes ALP, OCN, and OPN. Conclusively, RUNX2 is considered as an indirect target of miR-125b as well.

Transient expression of miR-125b expression in hypoxic VEGF-or or bFGF stimulated ECs was moreover shown by Muramatsu et al. to directly bind and block the translation of vascular endothelial (VE)-cadherin mRNA and therefore inhibit in vitro tube formation by ECs [150]. Thus, miR-125b may be tested in tumor therapy for inducing disruption of blood vessel formation.

5.11. MiR-135b

During the osteogenic differentiation of several USSC, miR-135b was identified as the most consistently down-regulated candidate by Schaap-Oziemlak et al. [151]. Retroviral overexpression resulting in decreased mineralization confirmed a function of miR-135b in osteogenesis of USSC. Furthermore, quantitative RT-PCR analysis of USSC that overexpressed miR-135b showed decreased expression of the bone mineralization markers IBSP and OSX.

In a profiling study by Xu et al. that investigated the exosomal content of hBM-MSCs during osteogenic differentiation, miR-135b was found to be significantly increased [183]. Bioinformatic analysis led to the conclusion that several important pathways related to osteoblastic differentiation were engaged, and that exosomal miRNA is a regulator thereof.

By establishing hypoxia-resistant multiple myeloma (MM) cells through exposure to chronic hypoxia, Umezu et al. presumably identified a new mechanism of hypoxia-induced angiogenesis targeting the FIH-1/HIF-1 signaling pathway via exosome-contained miRNAs [152,184]. In contrast to the transiently hypoxia-upregulated miR-210 transcriptional levels which decline gradually under normoxic conditions, miR-135b levels were maintained high, even under normoxic conditions. By delivering exosomes to HUVECs, miR-135b is enabled to target the factor-inhibiting hypoxia-inducible factor-1 gene (FIH-1) that encodes an asparaginyl hydroxylase enzyme which inhibits HIF-1a. The positive correlation between miR-135b, HIF-1a, and microvessel density was initially identified by Zhang et al. in a HNSCC model [185]. As a result, endothelial tube formation could be promoted under hypoxic as well as normoxic environments.

5.12. MiR-181a

Among other functions, the miR-181 family has been implicated in genetic regulation of early hematopoiesis and lymphangiogenesis [60,186,187]. Microarray analysis of human HCS-2/8 cells led to the characterization of miR-181a. Sumiyoshi et al. accredited miR-181a with an important function in the maintenance of cartilaginous metabolism by a negative feedback system employing repression of the CCN family member 1 (CCN1) and aggrecan (ACAN) genes, which are both known to be involved in chondrocyte differentiation [154].

In another study, the examination of synovial fluid cells in a mouse model of tibial plateau fractures led to the detection of two downregulated miRNAs, miR-9 and miR-181a [110]. Cbl, an important E3 ubiquitin ligase for bone resorption that was tested as a putative target gene of miR-9 and miR-181a, elicited increased amounts of ubiquitinated Bim, a pro-apoptotic gene in mouse primary osteoclast cells. Therefore, Wang et al. concluded that upregulated Cbl might regulate the survival rate of primary mouse osteoclast cells, as previously Cbl-dependent apoptosis via ubiquitinated Bim was reported [188].

Sun et al. implicated miR-181a as a potential oncomiR (cancer-associated miRNA) in the angiogenesis and metastasis of chondrosarcoma in a xenograft mouse model [156]. They found that miR-181 increases VEGF and MMP1 expression, as well as CXCR4 signaling, via negatively regulating RGS16 (regulator of G-protein signaling 16) under hypoxic cell culture conditions. RGS16 is an inhibitor of CXCR4 which regulates, upon amplified signaling, angiogenesis, invasion, and metastasis in chondrosarcoma. The therapeutic usefulness of this mechanism was proven by miR-181a antagomiR treatment, which decreased proangiogenic gene expression as well as tumor growth and metastasis in a xenograft mouse model.

5.13. MiR-195

Together with miR-497, miR-195 was downregulated in a microarray screen of primary hMSCs performed by Almeida et al. [158]. hMSCs underwent induced osteogenic differentiation with the aim of identifying candidates that are capable of contributing to bone fracture repair. Osteogenic markers were therefore found to be diminished upon overexpression or increased upon inhibition of miR-195 in hMSCs. Using the chicken CAM assay, studying the paracrine effects of hMSCs, the authors furthermore demonstrated decreased blood vessel formation in vivo. VEGF was identified as the target gene that mediates this phenotype, at least in part. MiR-195 interacted with the VEGF 3′-UTR in bone cancer cells and also regulated mRNA and protein expression levels.

Reduced expression of miR-195 furthermore resulted in increased angiogenesis and metastasis in HCC tissues, whereas either loss-of-function or gain-of-function abrogated the ability of HCC cells to migrate and induce capillary tube formation of ECs, as reported by Wang et al. [159]. In xenograft tumors, upregulated miR-195 expression provoked reduced microvessel densities and the formation of metastases. Detailed molecular investigations disclosed VEGF and the prometastatic factors VAV2 and CDC42 as direct targets of miR-195, which could be proven by mirroring the phenotype either by knockdown or overexpression of these target genes. The group also demonstrated that higher VEGF levels due to miR-195 down-regulation promoted EC-mediated tumor angiogenesis by the involvement of VEGF receptor 2 signaling.

5.14. MiR-200b

miR-200b was identified by microarray screening as a miRNA that exhibits downregulated expression upon exposure of osteoblasts to collagen and to silicate-based periodontal grafting material (PerioGlas, P-15) that is used to promote bone formation [189]. Rønbjerg et al. demonstrated miR-200b as a potent regulator of the target gene ZFHX1B via direct interaction [190]. ZFHX1B is a transcriptional repressor involved in the regulation of the TGFβ signaling pathway and epithelial-mesenchymal transitions by E-cadherin, a mediator of cell-cell adhesion, in mesenchymal cells.

In addition to the known communication of interacting cells by the exosomal release of miRNAs (e.g., mir-135b [153,191]), Fan et al. discovered angiogenic-osteogenic cell coupling and reciprocal interactions via gap junctions [161,192]. They provided evidence that in a direct co-culture, miR-200b was transferred in a TGF-β-stimulated process from rat BM-MSCs to HUVECs through gap junctions formed of connexin 43. By this transfer and decrease of miR-200b, VEGF-A-induced expression enhanced osteogenic differentiation in BM-MSCs. In HUVECs, increasing miR-200b levels down-modulated ZEB2, ETS1, KDR and GATA2 transcription factors, which led to a decline of the angiogenic potential of HUVECs, in contrast. In vitro angiogenesis in this co-culture could therefore be partially rescued by employing the TGF-β inhibitor SB431542 or TGF-β-neutralizing antibody. These findings could provide a new strategy for cell-based bone regeneration.

In lung epithelial carcinoma cells (A549 cells), direct negative regulation of VEGF, VEGFR1 (Flt-1) and VEGFR2 (KDR) by binding of miR-200b to the corresponding 3′-UTR could be demonstrated by Liu et al. [193]. This interaction could be furthermore confirmed in an in vitro angiogenesis assay by transfection of HUVECs, which resulted in reduced capillary tube formation and significantly reduced VEGF-induced phosphorylation of ERK1/2. In addition, miR-200b targets the Ets-1 transcription factor, that might concomitantly downregulate VEGFR2, as found in human mammalian epithelial cells by Chan et al. [194]. Thus, miR-200b may be used as a therapeutic angiogenesis inhibitor.

5.15. MiR-210

Upregulated expression of miR-210 was detected in BMP-4-induced osteoblastic differentiation of murine stromal BM-MSC ST2 cells by Mizuno et al. [163]. Transfection experiments of sense and antisense miR-210 therefore promoted or repressed osteogenesis, respectively. As a target gene mediating the positive regulation, the activin A receptor type-1B (AcvR1b) gene was elicited that effects the TGF-β/activin signaling pathway negatively.

Studies of Fasanaro et al. proved that the expression of miR-210 progressively increases upon exposure to hypoxia. The overexpression of miR-210 in HUVECs using miRNA mimics stimulated the formation of capillary-like structures and enhanced *VEGF*-induced cell chemotaxis [165,195,196]. Thus, miR-210 up-regulation is a crucial hypoxia response element of ECs, affecting cell survival, migration, and differentiation and might participate in the modulation of the angiogenic response to ischemia [99,197,198]. As a consistently reported target gene mediating these effects, Ephrin-A3 receptor tyrosine kinase ligand (*EFNA3*) was validated. *EFNA3* was previously shown to play a role in the regulation of angiogenesis and VEGF signaling during vascular development and remodeling via EFNA1/EphA2 interaction [199,200]. Overexpression of an Ephrin-A3 allele that is not targeted by miR-210 therefore prevented miR-210-mediated stimulation of tube formation and EC migration.

In its function of promoting osteoblast differentiation by increased VEGF, ALP and OSX expression in rat MSCs and suppression of adipocyte differentiation, due to decreased PPAR-γ in vitro, miR-210 was also implicated in the regulation of postmenopausal osteoporosis [164]. Although the exact mechanism still needs to be elaborated, Liu et al. also found that HIF-1α and VEGF expression was increased in 17β-estradiol (E2)-treated osteoblasts.

5.16. MiR-222

Yan et al. identified miR-222-3p as a negative regulator involved in osteogenic differentiation of hBM-MSCs [166]. Enhancement of miR-222-3p function in hBM-MSCs was blocking protein levels of SMAD5 and RUNX2. miR-222-3p-specific inhibition via lentivirus infection, in contrast, led to their enhanced expression and increased phosphorylation of Smad proteins (1, 5, 8) that are responsible for expression of osteogenic genes. Thus, in addition to the Smad5-RUNX2 signaling pathway elevated levels of the osteoblast markers OSX, ALP, and OC, as well as increased matrix mineralization, could be detected.

The role of miR-222-3p in c-Src-mediated regulation of osteoclastogenesis was proven by Takigawa et al. [167]. Depending on the use of miR-222-3p inhibitor or mimics in RAW264.7 pre-osteoclastic cells, either upregulation, or downregulation of the mRNA expression levels of osteoclast marker genes NFATc1 or TRAP, were observed, respectively. Inhibition of of c-Src activity and activation of osteoclastogenesis via miR-222-3p was implemented by increased amounts of RANKL-induced expression of TRAP and cathepsin K protein levels. Thereby, the number of multi-nucleated osteoclasts and their pit formation was reduced.

Poliseno et al. investigated the role of miR-221 and miR-222 during in vitro angiogenesis [168]. Both miRNAs were identified among 15 upregulated miRNAs that allegedly target receptors of angiogenic factors in a large-scale screen of HUVECs. Mechanistically, they were found to post-translationally modify the angiogenic effects of stem cell factor (SCF) by targeting its receptor, c-KIT. Together with other angiogenic growth factors, such as VEGF, bFGF, or HGF, SCF is able to stimulate proliferation and migration of ECs and induce capillary-like tube formation when performing in vitro angiogenesis assays [201]. Consistently, tube formation, wound healing, or cell migration was suppressed in miR-221/miR-222– transfected and SCF-treated HUVECs.

5.17. MiR-424

miR-424 was identified by differential screening in a miRNA microarray of isolated primary hMSCs from four individuals that were, or were not, osteo-differentiated using osteogenic differentiation medium. In combination with miR-31, miR-106a, and miR-148a, Gao et al. found that miR-424 was suppressed, and predicted it to target RUNX2, CBFB (core-binding factor, beta subunit), and BMPs [169]. In a similar experimental setting of Vimalraj et al., miR-424, together with miR-106a, miR-148a, let-7i and miR-99a, were detected as hMSC-specific miRNAs that were found to be expressed only in undifferentiated hMSCs [170]. Here, bioinformatics analysis mostly predicted the MAPK, WNT, and insulin signaling pathways as targets.

A recent study of Li et al. elucidated further specific functional mechanisms of miR-424 during bone formation and oxidative stress [171]. It was reported that miR-424 was downregulated by the transcription factor FOXO1 using consensus binding sites in the promoter. Subsequently, via miR-424 mediated upregulation of FGF2 under oxidative stress, proliferation and osteogenic differentiation of hMSCs was accomplished. Uncovering the miR-424/FGF2 pathway revealed a new mechanism how FOXO1 promotes bone formation and could potentially enhance bone repair.

miR-424 was furthermore recognized by Gosh et al. as an important hypoxamiR in human ECs by revealing a novel pathway for HIF regulation [101]. Under hypoxic conditions levels miR-424 or its rodent homolog, mu-miR-322, were found to be elevated in ECs and in ischemic tissues undergoing vascular remodeling in an experimental myocardial infarction rat model. It was elicited that the target of miR-424 is represented by cullin 2 (CUL2), which serves the purpose of stabilizing the hypoxic transcription factor HIF-1α. CUL2 is an essential component for assembling the ubiquitin ligase system, normally leading to its continuous degradation. MiR-424 expression was experimentally evaluated to be regulated by the transcription factor PU.1 that itself was found to be regulated by RUNX-1 and C/EBPα transcription factors. Ectopic expression of miR-424 in retrovirally-transduced HUVECs therefore stimulated angiogenesis in vivo in athymic mice that were subcutaneously implanted.

6. Outlook and Future Directions: MiRNAs in Therapeutic Applications

Bone is a highly vascularized tissue, and is thus reliant on the coordinated interaction between osteogenesis and angiogenesis which occurs between osteoblastic and ECs during development, remodeling and regeneration of the skeletal system [22,27,202]. Deciphering the molecular nature of the mechanisms that couple bone formation to blood vessel formation should therefore be of great interest to enhance bone formation capabilities in vitro and in vivo [203]. These specific underlying mechanisms are gaining growing attention under physiologic and pathologic conditions such as fracture healing, prevention of osteoporosis, tissue engineering or bone regeneration. Thus far, mostly a combination of different growth factors controlling osteogenesis and angiogenesis (e.g., such VEGF, angiopoietins, BMPS, RUNX2) have succeeded in achieving bone formation to a certain degree [33,204–206]. Nevertheless, each of these factors has individual roles at certain stages of development, and may disturb the subtle orchestration of required regulation steps.

MicroRNAs provide potent modifiers which coordinate a broad spectrum of biological processes. In contrast to transcription factors, single miRNAs may only modestly affect individual mRNA target expression. However, on the one hand a specific target mRNA can experience increased repression if it has multiple binding sites in the 3'-UTR. On the other hand, the same 3'-UTR of a gene can be targeted by many different miRNAs simultaneously that intervene with its regulation of expression [207]. Bioinformatic analysis of target prediction databases, such as TargetScan, miRanda and Pictar, suggest that a single average miRNA is capable of modulating up to several hundred target genes by perfect or imperfect base-pairing in combination with tissue-specific expression [208–211]. Additionally, by feedforward or feedback loops that form a regulatory network, miRNA effects can be amplified [212]. And it is estimated that approximately one-third of genome-encoded proteins are effected by miRNA regulation [213]. Therefore, these regulatory RNAs which control multiple endogenous signaling processes simultaneously possess a unique capacity to interfere with all cellular processes and have become an important tool in biological and medical research. As the application of miRNA-based methods for the treatment and monitoring of different pathological conditions is constantly becoming more prevalent, their therapeutic engagement could establish a refined method to stimulate inartificial bone development.

There is increasing evidence that miRNAs play important roles in controlling osteogenesis and angiogenesis. One study identified mutations in miR-2861 in two related adolescents that likely contributed to primary osteoporosis [214]. miR-2861 regulates osteoblast differentiation by targeting HDAC5, which enhances RUNX2 degradation. Moreover, a recent study showed accelerated osteoblast differentiation of hMSCs in a three-dimensional scaffold in vitro through manipulation of miR-148b and miR-489 expression [215]. Several studies also explored the use of miRNA mimicking or inhibitory agents for bone regeneration in animal studies. However, only a few studies have found that miRNAs are positive regulators of these processes. As outlined, miR-29b has been reported to promote osteoblast differentiation by targeting several well characterized inhibitors of bone formation in vitro [124]. Single site-specific delivery of miR-29b into a two-week post fracture callus by Lee et al. significantly improved mouse femoral fracture healing [216]. This was documented radiographically by a decrease of callus width and area; histomorphometrical and micro-computed tomographical analyses demonstrated increased bone volume fraction and bone mineral density of the callus. A single report further delineated the use of miRNAs as a therapeutic strategy to modulate angiogenesis-osteogenesis coupling during bone regeneration or repair in a subcutaneous assay in the mouse. The studies of Li et al. preferred miR-26a over miR-21 and miR-29b, which were all identified as the most potent candidates to mediate both angiogenesis and osteogenesis by microarray profiling of primary osteoblasts [217]. miR-26a was found to be a factor that is upregulated in newly-formed bone, and itself stimulates the expression of osteoblast-specific makers RUNX2, ALP, and OCN, mineralization of osteoblasts as well as VEGF secretion in murine primary BM-MSCs and in MC3T3-E1 cells in vitro. The enhancement of miR-26a expression in vivo by transfection of BM-MSCs with miR-26a mimicking agents led to complete repair of a critical-size calvarial bone defect, mainly due to simultaneously regulating endogenous angiogenesis-osteogenesis coupling. To date, its ability to integrate multiple signaling pathways thus makes miR-26a an ideal candidate for regenerating bone by miRNA-based therapy. Murata et al. detected significantly decreased levels of miR-92a in patients with trochanteric fractures, and investigated its significance in a mouse femoral fracture model [142]. Systemic as well as local administration of antimir-92a via LNA-stabilized oligonucleotide increased callus volume and enhanced fracture healing in the early phase by promoting neovascularization in the mouse femur. Nevertheless, as outlined previously, the direct osteogenesis-related molecular targets for this mechanism, which offers therapeutic potential for repairing bone, still remain to be clarified. In addition to the basic osteogenic functions of miR-31, Deng et al. investigated the therapeutic potential of hAD-MSC combined with beta-calcium triphosphate scaffolds in repairing a rat critical-sized calvarial defect [130,217]. The group reported that a knockdown of miR-31 significantly enhanced the repair of the defect, as could be noticed by an increased bone volume and mineral density in combination with a decreased scaffold. As the molecular basis of the osteogenic differentiation and bone regeneration program, in vitro results using lentiviral expression constructs revealed a BMP-2-inducible regulatory loop between Runx2, miR-31, as well as the miR-31 target gene, AT-rich sequence-binding protein 2 (Satb2). As a negative modulator of angiogenesis, Yoshizuka et al. locally administered miR-222 inhibitor mixed with atelocollagen to a rat femoral transverse fracture with the purpose of enhancing bone healing by stimulating osteogenesis, chondrogenesis, as well as angiogenesis [218]. Bone union at the fracture site with increased capillary density could be confirmed by radiographic, μCT and histological evaluation at 8 weeks after administration. Inhibition of miR-222 promoted osteogenic (RUNX2, COL1A1, OCN) or chondrogenic (COL2A1, aggrican, SOX9) differentiation in hMSCs, as determined by expression of osteogenic or chondrogenic markers, respectively.

In this review, miRNAs or "CouplingmiRs/CPLGmiRs" were identified that are known so far to regulate, either positively or negatively, angiogenesis as well as osteogenesis function—as potent molecular managers—that may simultaneously regulate multiple endogenous signaling cascades. This has been summarized in Figure 2, where the described miRNAs have also been depicted with regard to their allocation during the developmental fate of osteogenic or angiogenic cells. As for many miRNAs in general, also for coupling miRs/CPLGmiRs, most mechanisms of action are negative/inhibitory in nature for the regulation of osteogenesis as well as angiogenesis. Also noteworthy, but not surprising, is the finding that most miRNAs potentially involved in angiogenic-osteogenic coupling are encountered in the osteoblastic lineage rather than in chondroblasts or osteoclasts. And miRNAs are particularly present during the differentiation of osteoblasts, while they exert their modulatory effects during angiogenesis mostly in mature endothelial cells. Delivery of miRNAs may provide a way to maximally mimic the native bone development environment, and thus possess the therapeutic potential to enhance bone regeneration and repair. In contrast, miR inhibitors, i.e., antisense oligonucleotides directed against miRNAs, can be applied in the form of antagomiRs or LNA-antimiRs [58,179,219]. RNA/DNA hybrids modified by this locked nucleic acid-technology represent single stranded modified antisense oligonucleotides with increased stability and affinity and also facilitate cell penetration. Furthermore, they are biocompatible, as they cause no cellular immune response, are highly soluble, and have already been tested by various in vitro and in vivo studies as well as clinical trials (e.g., miR-122 LNA anti-miRNA oligonucleotides for Hepatitis C virus or miR-34 for solid cancers) [220,221]. Additionally, any new findings in this field of research might additionally yield considerable anti-angiogenic targets for tumor therapies, as tumor angiogenesis—i.e., the formation of tumor-associated angiogenic vessels—is a key requirement in tumor growth, progression, and metastasis. As an example, it would be very favorable to identify angiomiRs that negatively regulate VEGF expression in combination with current anti-VEGF therapies. Inversely, the identification of novel oncomiRs could also pave the way and lead to the description of unknown miRNAs with a function in bone angiogenesis [222–224]. Currently, microRNA-targeting therapy is still in development due to new challenges compared to conventional drugs, and results of experimental studies in animal models need to be transferred to clinical applications. Nevertheless, in addition to their role as potential angiogenic or bone regenerative therapeutic targets, miRNAs are emerging as distinguished disease biomarkers relevant to specific physiologic or pathologic conditions which could serve for immediate use in cardiovascular and bone diseases.

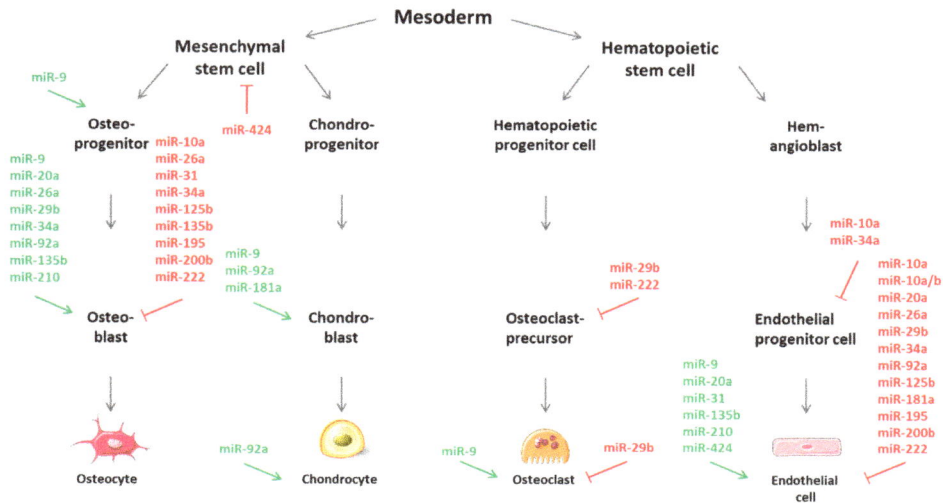

Figure 2. Regulating effects of microRNAs (miRs/miRNAs) involved in the regulation and coupling of bone angiogenesis ("CouplingmiRs") during cell fate determination. MiRNAs were assigned with their individual positive/stimulatory (green colored) or negative/inhibitory (red colored) function and occurrence during the specific differentiation steps of osteogenic and angiogenic cells. Cell images from Servier Medical Art by Servier licensed under a Creative Commons Attribution 3.0.

Author Contributions: L.F.F. was responsible for collecting literature, drafting, writing, reviewing, and illustration of the manuscript.

Funding: We acknowledge support by Open Access Publication Fund of University of Münster.

Conflicts of Interest: The author declares no conflict of interest.

References

1. Crane, G.; Ishaug, S.; Mikos, A. Bone tissue engineering. *Nat. Med.* **1995**, *1*, 1322–1324. [CrossRef] [PubMed]
2. Rouwkema, J.; Rivron, N.C.; Blitterswijk, C.A. Vascularization in tissue engineering. *Trends Biotechnol.* **2008**, *26*, 434–441. [CrossRef] [PubMed]
3. Hertig, A. Angiogenesis in the early human chorion and the primary placenta of the macaque monkey. *Contrib. Embryol.* **1935**, *25*, 37–81.
4. Chung, A.S.; Ferrara, N. Developmental and pathological angiogenesis. *Annu. Rev. Cell Dev. Biol.* **2011**, *27*, 563–584. [CrossRef]
5. Lian, J.B.; Stein, G.S.; van Wijnen, A.J.; Stein, J.L.; Hassan, M.Q.; Gaur, T.; Zhang, Y. MicroRNA control of bone formation and homeostasis. *Nat. Rev. Endocrinol.* **2013**, *8*, 212–227. [CrossRef] [PubMed]
6. Kronenberg HM Developmental regulation of the growth plate. *Nature* **2003**, *423*, 332–336. [CrossRef] [PubMed]
7. Berendesen, A.; Olsen, B. Bone development. *Bone* **2015**, *80*, 14–18. [CrossRef]
8. Olsen, B.; Reginato, A.; Wang, W. Bone development. *Annu. Rev. Cell Dev. Biol.* **2000**, *16*, 191–220. [CrossRef]
9. Karsenty, G. The complexities of skeletal biology. *Nature* **2003**, *423*, 316–318. [CrossRef]
10. Helms, J.; Schneider, R. Cranial skeletal biology. *Nature* **2003**, *423*, 326–331. [CrossRef]

11. Maes, C.; Carmeliet, P.; Moermans, K.; Stockmans, I.; Smets, N. Impaired angiogenesis and endochondral bone formation in mice lacking the vascular endothelial growth factor isoforms VEGF 164 and VEGF 188. *Mech Dev.* **2002**, *111*, 61–73. [CrossRef]
12. Gerber, H.; Ferrara, N. Angiogenesis and Bone Growth. *TCM* **2000**, *10*, 223–228. [CrossRef]
13. Brandi, M.; Collin-Osdoby, P. Vascular biology and the skeleton. *J. Bone Miner. Res.* **2006**, *21*, 183–192. [CrossRef] [PubMed]
14. Sivaraj, K.K.; Adams, R.H. Blood vessel formation and function in bone. *Development* **2016**, *143*, 2706–2715. [CrossRef] [PubMed]
15. Schipani, E.; Maes, C.; Carmeliet, G.; Semenza, G.L. Regulation of osteogenesis-angiogenesis coupling by HIFs and VEGF. *J. Bone Miner. Res.* **2009**, *24*, 1347–1353. [CrossRef] [PubMed]
16. Schipani, E.; Wu, C.; Rankin, E.B.; Giaccia, A.J. Regulation of Bone Marrow Angiogenesis by Osteoblasts during Bone Development and Homeostasis. *Front. Endocrinol.* **2013**, *4*, 85. [CrossRef]
17. Kusumbe, A.P.; Ramasamy, S.K.; Adams, R.H. Coupling of angiogenesis and osteogenesis by a specific vessel subtype in bone. *Nature* **2014**, *507*, 323–328. [CrossRef]
18. Carmeliet, P.; Ferreira, V.; Breier, G.; Pollefeyt, S.; Kieckens, L.; Gertsenstein, M.; Fahrig, M.; Vandenhoeck, A.; Harpal, K.; Eerhardt, C.; et al. Abnormal blood vessel development and lethality in embryos lacking a single VEGF allele. *Nature* **1996**, *380*, 435–439. [CrossRef]
19. Carulli, C.; Innocenti, M.; Brandi, M.L. Bone vascularization in normal and disease conditions. *Front. Endocrinol.* **2013**, *4*, 1–10. [CrossRef]
20. Ding, W.-G.; Yan, W.; Wei, Z.-X.; Liu, J.-B. Difference in intraosseous blood vessel volume and number in osteoporotic model mice induced by spinal cord injury and sciatic nerve resection. *J. Bone Miner. Metab.* **2012**, *30*, 400–407. [CrossRef]
21. Wang, L.; Zhou, F.; Zhang, P.; Wang, H.; Qu, Z.; Jia, P.; Yao, Z.; Shen, G.; Li, G.; Zhao, G.; et al. Human type H vessels are a sensitive biomarker of bone mass. *Cell Death Dis.* **2017**, *8*, e2760. [CrossRef] [PubMed]
22. Stegen, S.; Van Gastel, N.; Carmeliet, G. Bringing new life to damaged bone: The importance of angiogenesis in bone repair and regeneration. *Bone* **2014**, *70*, 19–27. [CrossRef] [PubMed]
23. Hu, K.; Olsen, B.R. Osteoblast-derived VEGF regulates osteoblast differentiation and bone formation during bone repair. *J. Clin. Invest.* **2016**, *126*, 509–526. [CrossRef]
24. Hu, K.; Olsen, B.R. The roles of vascular endothelial growth factor in bone repair and regeneration. *Bone* **2016**, *91*, 30–38. [CrossRef] [PubMed]
25. Maes, C.; Kobayashi, T.; Selig, M.; Torrekens, S.; Roth, S.; Mackem, S.; Carmeliet, G.; Kronenberg, H. Osteoblast precursors, but not mature osteoblasts, move into developing and fractured bones along with invading blood vessels. *Dev. Cell* **2010**, *19*, 329–344. [CrossRef] [PubMed]
26. Maes, C.; Carmeliet, G.; Schipani, E. Hypoxia-driven pathways in bone development, regeneration and disease. *Nat. Rev. Rheumatol.* **2012**, *8*, 358–366. [CrossRef] [PubMed]
27. Wang, Y.; Wan, C.; Deng, L.; Liu, X.; Cao, X.; Gilbert, S.R.; Bouxsein, M.L.; Faugere, M.; Guldberg, R.E.; Gerstenfeld, L.C.; et al. The hypoxia-inducible factor α pathway couples angiogenesis to osteogenesis during skeletal development. *J. Clin. Invest.* **2007**, *117*, 1616–1626. [CrossRef]
28. Pugh, C.W.; Ratcliffe, P.J. Regulation of angiogenesis by hypoxia: Role of the HIF system. *Nat. Med.* **2003**, *9*, 677–684. [CrossRef]
29. Ferrara, N.; Gerber, H.; Lecouter, J. The biology of VEGF and its receptors. *Nat. Med.* **2003**, *9*, 669–676. [CrossRef]
30. Ornitz, D.; Marie, P. Fibroblast growth factor signaling in skeletal development and disease. *Genes Dev.* **2015**, *29*, 1463–1468. [CrossRef]
31. Ramasamy, S.K.; Kusumbe, A.P.; Wang, L.; Adams, R.H. Endothelial Notch activity promotes angiogenesis and osteogenesis in bone. *Nature* **2014**, *507*, 376–380. [CrossRef] [PubMed]
32. Ortega, N.; Behonick, D.; Werb, Z. Matrix remodeling during endochondral ossification. *Trends Cell Biol.* **2004**, *14*, 86–93. [CrossRef] [PubMed]
33. Kleinheinz, J.; Stratmann, U.; Joos, U.; Wiesmann, H.-P. VEGF-Activated Angiogenesis During Bone Regeneration. *J. Oral Maxillofac. Surg.* **2005**, *63*, 1310–1316. [CrossRef] [PubMed]
34. Maes, C.; Coenegrachts, L.; Stockmans, I.; Daci, E.; Luttun, A.; Petryk, A.; Gopalakrishnan, R.; Moermans, K.; Smets, N.; Verfaillie, C.M.; et al. Placental growth factor mediates mesenchymal cell development, cartilage turnover, and bone remodeling during fracture repair. *J. Clin. Invest.* **2006**, *116*, 16–18. [CrossRef]

35. Kingsley, D. What do BMPs do in mammals? Clues from the mouse short-ear mutation. *Trends Genet.* **1994**, *10*, 16–21. [CrossRef]

36. Hassan, M.Q.; Tye, C.E.; Stein, G.S.; Lian, J.B. Non-coding RNAs: Epigenetic regulators of bone development and homeostasis. *Bone* **2015**, *81*, 746–756. [CrossRef] [PubMed]

37. Papaioannou, G.; Mirzamohammadi, F.; Kobayashi, T. MicroRNAs involved in bone formation. *Cell Mol. Life Sci.* **2014**, *71*, 4747–4761. [CrossRef] [PubMed]

38. Papaioannou, G. miRNAs in Bone Development. *Curr. Genom.* **2015**, *16*, 427–434. [CrossRef] [PubMed]

39. Ambros, V. MicroRNAs: Tiny regulators with great potential. *Cell* **2001**, *107*, 823–826. [CrossRef]

40. Lau, N.C.; Lim, L.P.; Weinstein, E.G.; Bartel, D.P. An abundant class of tiny RNAs with probable regulatory roles in Caenorhabditis elegans. *Science* **2001**, *294*, 858–862. [CrossRef]

41. Bartel, D.P.; Lee, R.; Feinbaum, R. MicroRNAs: Genomics, Biogenesis, Mechanism, and Function Genomics: The miRNA Genes. *Cell* **2004**, *116*, 281–297. [CrossRef]

42. Bartel, D.P. MicroRNAs: Target Recognition and Regulatory Functions. *Cell* **2009**, *136*, 215–233. [CrossRef] [PubMed]

43. Lee, R.C.; Ambros, V. An extensive class of small RNAs in Caenorhabditis elegans. *Science* **2001**, *294*, 858–862. [CrossRef] [PubMed]

44. Macfarlane, L.A.; Murphy, P.R. MicroRNA: Biogenesis, Function and Role in Cancer. *Curr. Genom.* **2010**, *11*, 537–561. [CrossRef] [PubMed]

45. Hutvagner, G.; McLachlan, J.; Pasquinelli, A.E.; Balint, E.; Tuschl, T.; Zamore, P.D. A cellular function for the RNA-interference enzyme Dicer in the maturation of the let-7 small temporal RNA. *Science* **2001**, *293*, 834–838. [CrossRef] [PubMed]

46. Grishok, A.; Pasquinelli, A.E.; Conte, D.; Li, N.; Parrish, S.; Ha, I.; Baillie, D.L.; Fire, A.; Ruvkun, G.; Mello, C. Genes and mechanisms related to RNA interference regulate expression of the small temporal RNAs that control C. elegans developmental timing. *Cell* **2001**, *106*, 23–34. [CrossRef]

47. Knight, S.W.; Bass, B.L. A role for the RNase III enzyme DCR-1 in RNA interference and germ line development in Caenorhabditis elegans. *Science* **2001**, *293*, 2269–2271. [CrossRef]

48. Yang, J.S.; Lai, E.C. Alternative miRNA biogenesis pathways and the interpretation of core miRNA pathway mutants. *Mol. Cell* **2011**, *43*, 892–903. [CrossRef]

49. Lai, E.C. Micro RNAs are complementary to 3′UTR sequence motifs that mediate negative post-transcriptional regulation. *Nat. Genet.* **2002**, *30*, 363–364. [CrossRef]

50. Bernstein, E.; Caudy, A.A.; Hammond, S.M.; Hannon, G.J. Role for a bidentate ribonuclease in the initiation step of RNA interference. *Nature* **2001**, *409*, 363–366. [CrossRef]

51. Wang, Y.; Medvid, C.; Melton, R.; Jaenisch, R.; Blelloch, R. DGCR8 is essential for microRNA biogenesis and silencing of embryonic stem cell self-renewal. *Nat. Genet.* **2007**, *39*, 380–385. [CrossRef] [PubMed]

52. Förstemann, K.; Tomari, Y.; Du, T.; Vagin, V.V.; Denli, A.M.; Bratu, D.P.; Klattenhoff, C.; Theurkauf, W.E.; Zamore, P.D. Normal microRNA maturation and germ-line stem cell maintenance requires Loquacious, a double stranded RNA-binding domain protein. *PLoS Biol.* **2005**, *3*, e236. [CrossRef] [PubMed]

53. Huang, Y.; Shen, X.J.; Zou, Q.; Wang, S.P.; Tang, S.M.; Zhang, G.Z. Biological functions of microRNAs: A review. *J. Physiol. Biochem.* **2011**, *67*, 129–139. [CrossRef] [PubMed]

54. Melton, C.; Judson, R.L.; Blelloch, R. Opposing micro-RNA families regulate self-renewal in mouse embryonic stem cells. *Nature* **2010**, *463*, 621–626. [CrossRef] [PubMed]

55. Brennecke, J.; Hipfner, D.R.; Stark, A.; Russell, R.B.; Cohen, S.M. bantam encodes a developmentally regulated microRNA that controls cell proliferation and regulates the proapoptotic gene hid in Drosophila. *Cell* **2003**, *113*, 25–36. [CrossRef]

56. Hipfner, D.R.; Weigmann, K.; Cohen, S.M. The bantam gene regulates Drosophila growth. *Genetics* **2002**, *161*, 1527–1537. [PubMed]

57. Esau, C.; Davis, S.; Murray, S.F.; Yu, X.X.; Pandey, S.K.; Pear, M.; Watts, L.; Booten, S.L.; Graham, M.; McKay, R.; et al. miR-122 regulation of lipid metabolism revealed by in vivo antisense targeting. *Cell Metab.* **2006**, *3*, 87–98. [CrossRef]

58. Krützfeldt, J.; Rajewsky, N.; Braich, R.; Rajeev, K.G.; Tuschl, T.; Manoharan, M.; Stoffel, M. Silencing of microRNAs in vivo with ′antagomirs′. *Nature* **2005**, *438*, 685–689. [CrossRef]

59. Miska, E.A. How microRNAs control cell division, differentiation and death. *Curr. Opin. Genet. Dev.* **2005**, *15*, 563–568. [CrossRef]

60. Chen, C.Z.; Li, L.; Lodish, L.F.; Bartel, D. MicroRNAs modulate hematopoietic lineage differentiation. *Science* **2004**, *303*, 83–86. [CrossRef]

61. Tay, Y.; Rinn, J.; Pandolfi, P.P. The multilayered complexity of ceRNA crosstalk and competition. *Nature* **2014**, *505*, 344–352. [CrossRef] [PubMed]

62. Alvarez-Garcia, I.; Miska, E.A. MicroRNA functions in animal development and human disease. *Development* **2005**, *132*, 4653–4662. [CrossRef] [PubMed]

63. Gennari, L.; Bianciardi, S.; Merlotti, D. MicroRNAs in bone diseases. *Osteoporos. Int.* **2017**, *28*, 1191–1213. [CrossRef] [PubMed]

64. Clark, E.; Kalomoiris, S.; Nolta, J.; Fierro, F. Concise Review: MicroRNA Function in Multipotent Mesenchymal Stromal Cells. *Stem Cells* **2014**, *32*, 1074–1082. [CrossRef]

65. Peng, S.; Gao, D.; Gao, C.; Wei, P.; Niu, M.; Shuai, C. MicroRNAs regulate signaling pathways in osteogenic differentiation of mesenchymal stem cells (Review). *Mol. Med. Rep.* **2016**, *14*, 623–629. [CrossRef] [PubMed]

66. Fang, S.; Deng, Y.; Gu, P.; Fan, X. MicroRNAs Regulate Bone Development and Regeneration. *Int. J. Mol. Sci.* **2015**, *16*, 8227–8253. [CrossRef]

67. Ji, X.; Chen, X.; Yu, X. MicroRNAs in Osteoclastogenesis and Function: Potential Therapeutic Targets for Osteoporosis. *Int. J. Mol. Sci.* **2016**, *17*, 349. [CrossRef]

68. Dong, S.; Yang, B.; Guo, H.; Kang, F. MicroRNAs regulate osteogenesis and chondrogenesis. *Biochem. Biophys. Res. Commun.* **2012**, *418*, 587–591. [CrossRef]

69. Kiga, K.; Mimuro, H.; Suzuki, M.; Shinozaki-Ushiku, A.; Kobayashi, T.; Sanada, T.; Kim, M.; Ogawa, M.; Iwasaki, Y.W.; Kayo, H.; et al. Epigenetic silencing of miR-210 increases the proliferation of gastric epithelium during chronic Helicobacter pylori infection. *Nat. Commun.* **2014**, *5*, 4497. [CrossRef]

70. Chen, J.; Qiu, M.; Dou, C.; Cao, Z.; Dong, S. MicroRNAs in Bone Balance and Osteoporosis. *Drug Dev. Res.* **2015**, *76*, 235–245. [CrossRef]

71. Nugent, M. MicroRNAs and Fracture Healing. *Calcif. Tissue Int.* **2017**, *101*, 355–361. [CrossRef] [PubMed]

72. Wu, C.; Tian, B.O.; Qu, X.; Liu, F.; Tang, T.; Qin, A.N.; Zhu, Z.; Dai, K. MicroRNAs play a role in chondrogenesis and osteoarthritis (Review). *Int. J. Mol. Med.* **2014**, *34*, 13–23. [CrossRef] [PubMed]

73. Min, Z.; Zhang, R.; Yao, J.; Jiang, C.; Guo, Y.; Cong, F.; Wang, W.; Tian, J.; Zhong, N.; Sun, J.; et al. MicroRNAs associated with osteoarthritis differently expressed in bone matrix gelatin (BMG) rat model. *Int. J. Clin. Exp. Med.* **2015**, *8*, 1009–1017. [PubMed]

74. Seeliger, C.; Balmayor, E.; van Griensven, M. miRNAs Related to Skeletal Diseases. *Stem Cells Dev.* **2016**, *25*, 1261–1281. [CrossRef] [PubMed]

75. Anand, S.; Cheresh, D.A. Emerging Role of Micro-RNAs in the Regulation of Angiogenesis. *Genes Cancer* **2011**, *2*, 1134–1138. [CrossRef] [PubMed]

76. Wang, S.; Olson, E.N. AngiomiRs—Key Regulators of Angiogenesis. *Curr. Opin. Genet. Dev.* **2009**, *19*, 205–211. [CrossRef] [PubMed]

77. Small, E.M.; Olson, E.N. Pervasive roles of microRNAs in cardiovascular biology. *Nature* **2011**, *469*, 336–342. [CrossRef]

78. Weis, S.M.; Caheresh, D.A. Tumor angiogenesis: Molecular pathways and therapeutic targets. *Nat. Med.* **2011**, *17*, 1359–1370. [CrossRef]

79. Salinas-Vera, Y.; Marchat, L.; Gallardo-Rincon, D.; Ruiz-Garcia, E.; Astudillo- De La Vega, H.; Echavarria-Zepeda, R.; Lopez-Camarillo, C. AngiomiRs: MicroRNAs driving angiogenesis in cancer (Review). *Int. J. Mol. Med.* **2018**, *2018*. [CrossRef]

80. Suarez, Y.; Sessa, W.C. MicroRNAs As Novel Regulators of Angiogenesis. *Circ. Res.* **2009**, *104*, 442–454. [CrossRef]

81. Yang, W.; Yang, D.; Na, S.; Sandusky, G.; Zhang, Q.; Zhao, G. Dicer is required for embryonic angiogenesis during mouse development. *J. Biol. Chem.* **2005**, *280*, 9330–9335. [CrossRef] [PubMed]

82. Albinsson, S.; Suarez, Y.; Skoura, A.; Offermann, S.; Miano, J.M.; Sessa, W.C. MicroRNAs are necessary for vascular smooth muscle growth, differentiation, and function. *Arterioscler. Thromb. Vasc. Biol.* **2010**, *30*, 1118–1126. [CrossRef] [PubMed]

83. Wang, S.; Aurora, A.B.; Johnson, B.A.; Qi, X.; McAnnaly, J.; Hill, J.A.; Richardson, J.A.; Bassel-Duby, R.; Olson, E.N. The endothelial-specific microRNA miR-126 governs vascular integrity and angiogenesis. *Dev. Cell* **2008**, *15*, 261–271. [CrossRef] [PubMed]

84. Suarez, Y.; Wang, C.; Manes, T.D.; Pober, J.S. Cutting edge: TNF-induced microRNAs regulate TNF-induced expression of E-selectin and intercellular adhesion molecule-1 on human endothelial cells: Feedback control of inflammation. *J. Immunol.* **2010**, *184*, 21–25. [CrossRef] [PubMed]

85. Suarez, Y.; Fernandez-Hernando, C.; Pober, J.; Sessa, W. Dicer dependent microRNAs regulate gene expression and functions in human endothelial cells. *Circ. Res.* **2007**, *100*, 1164–1173. [CrossRef] [PubMed]

86. Greco, S.; Gaetano, C.; Martelli, F. HypoxamiR Regulation and Function in Ischemic Cardiovascular Diseases. *Antioxid. Redox Signal.* **2014**, *21*, 1202–1219. [CrossRef] [PubMed]

87. Samanta, S.; Balasubramanian, S.; Rajasingh, S.; Patel, U.; Dhanasekaran, A.; Dawn, B.; Rajasingh, J. MicroRNA: A new therapeutic strategy for cardiovascular diseases. *Trends Cardiovasc. Med.* **2016**, *26*, 407–419. [CrossRef]

88. Suarez, Y.; Fernandez-Hernando, C.; Yu, J.; Gerber, S.A.; Harrison, K.D.; Pober, J.S.; Iruela-Arispe, M.L.; Merkenschlager, M.; Sessa, W.C. Dicer-dependent endothelial microRNAs are necessary for postnatal angiogenesis. *Proc. Natl. Acad. Sci. USA* **2008**, *105*, 14082–14087. [CrossRef]

89. Würdinger, T.; Tannous, B.A.; Saydam, O.; Skog, J.; Grau, S.; Soutschek, J.; Weissleder, R.; Breakefield, X.O.; Krichevsky, A.M. miR-296 regulates growth factor receptor overexpression in angiogenic endothelial cells. *Cancer Cell* **2008**, *14*, 382–393. [CrossRef]

90. Anand, S.; Cheresh, D.A. MicroRNA-mediated Regulation of the Angiogenic Switch. *Curr. Opin. Hematol.* **2011**, *18*, 171–176. [CrossRef]

91. Nallamshetty, S.; Chan, S.Y.; Loscalzo, J. Hypoxia: A master regulator of microRNA biogenesis and activity. *Free Radic. Biol. Med.* **2013**, *64*, 20–30. [CrossRef] [PubMed]

92. Madanecki, P.; Kapoor, N.; Bebok, Z.; Ochocka, R.; Collawn, J.F.; Bartoszewski, R. Regulation of angiogenesis by hypoxia: The role of microRNA. *Cell. Mol. Biol. Lett.* **2013**, *18*, 47–57. [CrossRef] [PubMed]

93. el Azzouzi, H.; Leptidis, S.; Doevendans, P.A.; De Windt, L.J. HypoxamiRs: Regulators of cardiac hypoxia and energy metabolism. *Trends Endocrinol. Metab.* **2015**, *26*, 502–508. [CrossRef] [PubMed]

94. Greco, S.; Martelli, F. MicroRNAs in Hypoxia Response. *Antioxid. Redox Signal.* **2014**, *21*, 1164–1166. [CrossRef] [PubMed]

95. Collet, G.; Skrzypek, K.; Grillon, C.; Matejuk, A.; El Hafni-Rahbi, B.; Fayel, N.L.; Kieda, C. Hypoxia control to normalize pathologic angiogenesis: Potential role for endothelial precursor cells and miRNAs regulation. *Vascul. Pharmacol.* **2012**, *56*, 252–261. [CrossRef] [PubMed]

96. Bertero, T.; Rezzonico, R.; Pottier, N.; Mari, B. Impact of MicroRNAs in the Cellular Response to Hypoxia. *Int. Rev. Cell Mol. Biol.* **2017**, *333*, 91–158. [CrossRef] [PubMed]

97. Hua, Z.; Lv, Q.; Ye, W.; Wong, A.C.-K.; Cai, G.; Gu, D.; Ji, Y.; Zhao, C.; Wang, J.; Yang, B.B.; et al. MiRNA-Directed Regulation of VEGF and Other Angiogenic Factors under Hypoxia. *PloS ONE* **2006**, *1*, e116. [CrossRef]

98. Loscalzo, J. The cellular response to hypoxia: Tuning the system with microRNAs. *J. Clin. Invest.* **2010**, *120*, 3815–3817. [CrossRef]

99. Devlin, C.; Greco, S.; Martelli, F.; Ivan, M. MiR-210: More than a silent player in hypoxia. *IUBMB Life* **2011**, *63*, 94–100. [CrossRef]

100. Chan, S.; Loscalzo, J. MicroRNA-210: A unique and pleiotropic hypoxamir. *Cell Cycle* **2010**, *9*, 1072–1083. [CrossRef]

101. Ghosh, G.; Subramanian, I.V.; Adhikari, N.; Zhang, X.; Joshi, H.P.; Basi, D.; Chandrashekhar, Y.S.; Hall, J.L.; Roy, S.; Zeng, Y.; et al. Hypoxia-induced microRNA-424 expression in human endothelial cells regulates HIF-α isoforms and promotes angiogenesis. *J. Clin. Invest.* **2010**, *120*, 4141–4154. [CrossRef] [PubMed]

102. Taguchi, A.; Yanagisawa, K.; Tanaka, M.; Cao, K.; Matsuyama, Y.; Goto, H.; Takahashi, T. Identification of hypoxia-inducible factor-1alpha as a novel target for miR-17-92 microRNA cluster. *Cancer Res.* **2008**, *68*, 5540–5545. [CrossRef] [PubMed]

103. Liu, C.; Tsai, M.; Hung, P.; Kao, S.; Liu, T.; Wu, K.; Chiou, S.; Lin, S.; Chang, K. miR31 ablates expression of the HIF regulatory factor FIH to activate the HIF pathway in head and neck carcinoma. *Cancer Res.* **2010**, *70*, 1635–1644. [CrossRef] [PubMed]

104. Yuva-Aydemir, Y.; Simkin, A.; Gascon, E.; Gao, F.-B. MicroRNA-9: Functional evolution of a conserved small regulatory RNA. *RNA Biol.* **2011**, *8*, 557–564. [CrossRef] [PubMed]

105. Han, R.; Kan, Q.; Sun, Y.; Wang, S.; Zhang, G.; Peng, T.; Jia, Y. MiR-9 promotes the neural differentiation of mouse bone marrow mesenchymal stem cells via targeting zinc finger protein 521. *Neurosci. Lett.* **2012**, *515*, 147–152. [CrossRef] [PubMed]

106. Qu, J.; Lu, D.; Guo, H.; Miao, W.; Wu, G. MicroRNA-9 regulates osteoblast differentiation and angiogenesis via the AMPK signaling pathway. *Mol. Cell Biochem.* **2016**, *411*, 23–33. [CrossRef] [PubMed]

107. Liu, X.; Xu, H.; Kou, J.; Wang, Q.; Zheng, X.; Yu, T. MiR-9 promotes osteoblast differentiation of mesenchymal stem cells by inhibiting DKK1 gene expression. *Mol. Biol. Rep.* **2016**, *43*, 939–946. [CrossRef]

108. Zhuang, G.; Wu, X.; Jiang, Z.; Kasman, I.; Yao, J.; Guan, Y.; Oeh, J.; Modrusan, Z.; Bais, C.; Sampath, D.; et al. Tumour-secreted miR-9 promotes endothelial cell migration and angiogenesis by activating the JAK-STAT pathway. *EMBO J.* **2012**, *31*, 3513–3523. [CrossRef]

109. Jones, S.; Watkins, G.; Le Good, N.; Roberts, S.; Murphy, C.; Brockbank, S.; Needham, M.; Read, S.; Newham, P. The identification of differentially expressed microRNA in osteoarthritic tissue that modulate the production of TNF-alpha and MMP13. *Osteoarthr. Cartil.* **2009**, *17*, 464–472. [CrossRef]

110. Wang, S.; Tang, C.; Zhang, Q.; Chen, W. Reduced miR-9 and miR-181a expression down-regulates Bim concentration and promote osteoclasts survival. *Int. J. Clin. Exp. Pathol.* **2014**, *7*, 2209–2218.

111. Li, J.; Zhang, Y.; Zhao, Q.; Wang, J.; He, X. MicroRNA-10a Influences Osteoblast Differentiation and Angiogenesis by Regulating ß-Catenin Expression. *Cell. Physiol. Biochem.* **2015**, *37*, 2194–2208. [CrossRef]

112. Day, T.F.; Guo, X.; Garrett-Beal, L.; Yang, Y. Wnt/beta-catenin signaling in mesenchymal progenitors controls osteoblast and chondrocyte differentiation during vertebrate skeletogenesis. *Dev. Cell* **2005**, *8*, 739–750. [CrossRef] [PubMed]

113. Zhu, S.; Deng, S.; Ma, Q.; Zhang, T.; Jia, C.; Zhuo, D.; Yang, F.; Wei, J.; Wang, L.; Dykxhoorn, D.M.; et al. MicroRNA-10A* and MicroRNA-21 Modulate Endothelial Progenitor Cell Senescence Via Suppressing High-Mobility Group A2. *Circ. Res.* **2013**, *112*, 152–164. [CrossRef]

114. Hassel, D.; Cheng, P.; White, M.P.; Ivey, K.N.; Kroll, J.; Augustin, H.G.; Katus, H.A.; Stainier, D.Y.R.; Srivastava, D. MicroRNA-10 Regulates the Angiogenic Behavior of Zebrafish and Human Endothelial Cells by Promoting Vascular Endothelial Growth Factor Signaling. *Circ. Res.* **2012**, *111*, 1421–1433. [CrossRef]

115. Wang, X.; Ling, C.C.; Li, L.; Qin, Y.; Qi, J.; Liu, X.; You, B.; Shi, Y.; Zhang, J.; Xu, Q.J.H.; et al. MicroRNA-10a/10b represses a novel target gene mib1 to regulate angiogenesis. *Cardiovasc. Res.* **2016**, *110*, 140–150. [CrossRef]

116. Zhang, J.; Fu, W.; He, M.; Xie, W.; Lv, Q.; Li, G.; Wang, H.; Lu, G.; Hu, X.; Jiang, S.; et al. MiRNA-20a promotes osteogenic differentiation of human mesenchymal stem cells by co-regulating BMP signaling. *RNA Biol.* **2011**, *8*, 829–838. [CrossRef] [PubMed]

117. Doebele, C.; Bonauer, A.; Fischer, A.; Scholz, A.; Ress, Y.; Urbich, C.; Hofmann, W.-K.; Zeiher, A.M.; Dimmeler, S. Members of the microRNA-17-92 cluster exhibit a cell-intrinsic antiangiogenic function in endothelial cells. *Blood* **2010**, *115*, 4944–4950. [CrossRef]

118. Deng, H.-T.; Liu, H.-L.; Zhai, B.-B.; Zhang, K.; Xu, G.-C.; Peng, X.-M. Vascular endothelial growth factor suppresses TNFSF15 production in endothelial cells by stimulating miR-31 and miR-20a expression via activation of Akt and Erk signals. *FEBS Open Bio.* **2017**, *7*, 108–117. [CrossRef] [PubMed]

119. Luzi, E.; Marini, F.; Sala, S.C.; Tognarini, I.; Galli, G.; Brandi, M.L. Osteogenic Differentiation of Human Adipose Tissue–Derived Stem Cells Is Modulated by the miR-26a Targeting of the SMAD1 Transcription Factor. *J. Bone Miner. Res.* **2008**, *23*, 287–295. [CrossRef] [PubMed]

120. Luzi, E.; Marini, F.; Tognarini, I.; Galli, G.; Falchetti, A.; Brandi, M.L. The regulatory network menin-microRNA 26a as a possible target for RNA-based therapy of bone diseases. *Nucleic. Acid Ther.* **2012**, *22*, 103–108. [CrossRef]

121. Su, X.; Liao, L.; Shuai, Y.; Jing, H.; Liu, S.; Zhou, H.; Liu, Y.; Jin, Y. MiR-26a functions oppositely in osteogenic differentiation of BMSCs and ADSCs depending on distinct activation and roles of Wnt and BMP signaling pathway. *Cell Death Dis.* **2015**, *6*, e1851. [CrossRef] [PubMed]

122. Trompeter, H.-I.; Dreesen, J.; Hermann, E.; Iwaniuk, K.M.; Hafner, M.; Renwick, N.; Tuschl, T.; Wernet, P. MicroRNAs miR-26a, miR-26b, and miR-29b accelerate osteogenic differentiation of unrestricted somatic stem cells from human cord blood. *BMC Genom.* **2013**, *14*, 1–13. [CrossRef] [PubMed]

123. Icli, B.; Wara, A.K.M.; Moslehi, J.; Sun, X.; Plovie, E.; Cahill, M.; Marchini, J.F.; Schissler, A.; Padera, R.F.; Shi, J.; et al. MicroRNA-26a regulates pathological and physiological angiogenesis by targeting BMP/SMAD1 signaling. *Circ. Res.* **2013**, *113*, 1231–1241. [CrossRef] [PubMed]

124. Li, Z.; Hassan, M.Q.; Jafferji, M.; Aqeilan, R.I.; Garzon, R.; Croce, C.M.; Van Wijnen, A.J.; Stein, J.L.; Stein, G.S.; Lian, J.B. Biological Functions of miR-29b Contribute to Positive Regulation of Osteoblast Differentiation. *J. Biol. Chem.* **2009**, *284*, 15676–15684. [CrossRef] [PubMed]

125. Rossi, M.; Pitari, M.R.; Amodio, N.; Di Martino, T.M.; Conforti, F.; Leone, E.; Botta, C.; Paolino, F.M.; Giudice, T.D.E.L.; Iuliano, E.; et al. miR-29b Negatively Regulates Human Osteoclastic Cell Differentiation and Function: Implications for the Treatment of Multiple Myeloma-Related Bone Disease. *J. Cell. Physiol.* **2013**, *228*, 1506–1515. [CrossRef] [PubMed]

126. Zhang, K.; Cai, H.-X.; Gao, S.; Yang, G.-L.; Deng, H.-T.; Xu, G.-C.; Han, J.; Zhang, Q.-Z.; Li, L.-Y. TNSF15 suppresses VEGF production in endothelial cells by stimulating miR-29b expression via activation of JNK-GATA3 Signals. *Oncotarget* **2016**, *7*, 69436–69449. [CrossRef] [PubMed]

127. Li, Y.; Cai, B.; Shen, L.; Dong, Y.; Lu, Q.; Sun, S.; Liu, S.; Ma, S.; Ma, P.X.; Chen, J. MiRNA-29b suppresses tumor growth through simultaneously inhibiting angiogenesis and tumorigenesis by targeting Akt3. *Cancer Lett.* **2017**, *397*, 111–119. [CrossRef]

128. Granchi, D.; Ochoa, G.; Leonardi, E.; Devescovi, V.; Baglìo, S.R.; Osaba, L.; Baldini, N.; Ciapetti, G. Gene expression patterns related to osteogenic differentiation of bone marrow-derived mesenchymal stem cells during ex vivo expansion. *Tissue Eng. Part. C Methods* **2010**, *16*, 511–523. [CrossRef]

129. Baglìo, S.R.; Devescovi, V.; Granchi, D.; Baldini, N. MicroRNA expression profiling of human bone marrow mesenchymal stem cells during osteogenic differentiation reveals Osterix regulation by miR-31. *Gene* **2013**, *527*, 321–331. [CrossRef]

130. Deng, Y.; Wu, S.; Zhou, H.; Bi, X.; Wang, Y.; Hu, Y.; Gu, P.; Fan, X. Effects of a miR-31, Runx2, and Satb2 regulatory loop on the osteogenic differentiation of bone mesenchymal stem cells. *Stem Cells Dev.* **2013**, *22*, 2278–2286. [CrossRef]

131. Luo, J.; Lin, J.; Paranya, G.; Bischoff, J. Angiostatin Upregulates E-Selectin in Proliferating Endothelial Cells. *Biochem. Biophys. Res. Commun.* **1998**, *911*, 906–911. [CrossRef] [PubMed]

132. Chen, L.; Holmstrom, K.; Qiu, W.; Ditzel, N.; Shi, K.; Hokland, L.; Kassem, M. MicroRNA-34a Inhibits Osteoblast Differentiation and In Vivo Bone Formation of Human Stromal Stem Cells. *Stem Cells* **2014**, *32*, 902–912. [CrossRef] [PubMed]

133. Kang, H.; Chen, H.; Huang, P.; Qi, J.; Qian, N.; Deng, L.; Guo, L. Glucocorticoids impair bone formation of bone marrow stromal stem cells by reciprocally regulating microRNA-34a-5p. *Osteoporos. Int.* **2016**, *27*, 1493–1505. [CrossRef] [PubMed]

134. Fan, C.; Jia, L.; Zheng, Y.; Jin, C.; Liu, Y.; Liu, H.; Zhou, Y. Mir-34a Promotes Osteogenic Differentiation of Human Adipose-Derived Stem Cells via the RBP2/NOTCH I/CYCLIN DI Coregulatory Network. *Stem Cell Rep.* **2016**, *7*, 236–248. [CrossRef] [PubMed]

135. Engin, F.; Yao, Z.; Yang, T.; Zhou, G.; Bertin, T.; Jiang, M.M.; Chen, Y.; Wang, L.; Zheng, H.; Sutton, R.E.; et al. Dimorphic effects of Notch signaling in bone homeostasis. *Nat. Med.* **2008**, *14*, 299–305. [CrossRef] [PubMed]

136. Galindo, M.; Pratap, J.; Young, D.W.; Hovhannisyan, H.; Im, H.J.; Choi, J.Y.; Lian, J.B.; Stein, J.L.; Stein, G.S.; van Wijnen, A.J. The bone-specific expression of Runx2 oscillates during the cell cycle to support a G1-related antiproliferative function in osteoblasts. *J. Biol. Chem.* **2005**, *280*, 20274–20285. [CrossRef] [PubMed]

137. Zha, X.; Sun, B.; Zhang, R.; Li, C.; Yan, Z.; Chen, J. Regulatory Effect of MicroRNA-34a on Osteogenesis and Angiogenesis in Glucocorticoid-Induced Osteonecrosis of the Femoral Head. *J. Orthop. Res.* **2018**, *36*, 417–424. [CrossRef]

138. Zhao, T.; Li, J.; Chen, A.F. MicroRNA-34a induces endothelial progenitor cell senescence and impedes its angiogenesis via suppressing silent information regulator 1. *Am. J. Endocrinol. Metab.* **2010**, *299*, E110–E116. [CrossRef]

139. Mattagajasingh, I.; Kim, C.; Naqvi, A.; Yamamori, T.; Hoffman, T.; Jung, S.; DeRicco, J.; Kasuno, K.; Irani, K. SIRT1 promotes endothelium-dependent vascular relaxation by activating endothelial nitric oxide synthase. *Proc. Natl. Acad. Sci. USA* **2007**, *104*, 14855–14860. [CrossRef]

140. Kumar, B.; Yadav, A.; Lang, J.; Teknos, T.N.; Kumar, P. Dysregulation of MicroRNA-34a Expression in Head and Neck Squamous Cell Carcinoma Promotes Tumor Growth and Tumor Angiogenesis. *PLoS ONE* **2012**, *7*, e37601. [CrossRef]

141. Chai, Z.T.; Kong, J.; Zhu, X.D.; Zhang, Y.Y.; Lu, L.; Zhou, J.M.; Wang, L.R.; Zhang, K.Z.; Zhang, Q.B.; Ao, J.Y.; et al. MicroRNA-26a Inhibits Angiogenesis by Down-Regulating VEGFA through the PIK3C2α/Akt/HIF-1α Pathway in Hepatocellular Carcinoma. *PLoS ONE* **2013**, *8*, 1–12. [CrossRef] [PubMed]

142. Murata, K.; Ito, H.; Yoshitomi, H.; Yamamoto, K.; Fukuda, A.; Yoshikawa, J.; Furu, M.; Ishikawa, M.; Shibuya, H.; Matsuda, S. Inhibition of miR-92a enhances fracture healing via promoting angiogenesis in a model of stabilized fracture in young mice. *J. Bone Miner. Res.* **2014**, *29*, 316–326. [CrossRef] [PubMed]

143. Kalinina, N.; Klink, G.; Glukhanyuk, E.; Lopatina, T.; Anastassia, E.; Akopyan, Z.; Tkachuk, V. miR-92a regulates angiogenic activity of adipose-derived mesenchymal stromal cells. *Exp. Cell Res.* **2015**, *339*, 61–66. [CrossRef] [PubMed]

144. Daniel, J.-M.; Penzkofer, D.; Teske, R.; Dutzmann, J.; Koch, A.; Bielenberg, W.; Bonauer, A.; Boon, R.A.; Fischer, A.; Bauersachs, J.; et al. Inhibition of miR-92a improves re-endothelialization and prevents neointima formation following vascular injury. *Cardiovasc. Res.* **2014**, *103*, 564–572. [CrossRef] [PubMed]

145. Goettsch, C.; Rauner, M.; Pacyna, N.; Hempel, U.; Bornstein, S.R.; Hofbauer, L.C. MiR-125b regulates calcification of vascular smooth muscle cells. *Am. J. Pathol.* **2011**, *179*, 1594–1600. [CrossRef] [PubMed]

146. Mizuno, Y.; Yagi, K.; Tokuzawa, Y.; Kanesaki-Yatsuka, Y.; Suda, T.; Katagiri, T.; Fukuda, T.; Maruyama, M.; Okuda, A.; Amemiya, T.; et al. miR-125b inhibits osteoblastic differentiation by down-regulation of cell proliferation. *Biochem. Biophys. Res. Commun.* **2008**, *368*, 267–272. [CrossRef] [PubMed]

147. Zhou, S.; Zhang, P.; Liang, P.; Huang, X. The expression of miR-125b regulates angiogenesis during the recovery of heat-denatured HUVECs. *Burns* **2015**, *41*, 803–811. [CrossRef]

148. Huang, K.; Fu, J.; Zhou, W.; Li, W.; Dong, S.; Yu, S.; Hu, Z.; Wang, H.; Xie, Z. MicroRNA-125b regulates osteogenic differentiation of mesenchymal stem cells by targeting Cbfb in vitro. *Biochimie* **2014**, *102*, 47–55. [CrossRef]

149. Xihong, L.U.; Min, D.; Honghui, H.E.; Dehui, Z.; Wei, Z. miR-125b regulates osteogenic differentiation of human bone marrow mesenchymal stem cells by targeting Smad4. *J. Cent. South. Univ. (Med. Sci.)* **2013**, *38*, 341–346. [CrossRef]

150. Muramatsu, F.; Kidoya, H.; Naito, H.; Sakimoto, S.; Takakura, N. microRNA-125b inhibits tube formation of blood vessels through translational suppression of VE-cadherin. *Oncogene* **2013**, *32*, 414–421. [CrossRef]

151. Schaap-Oziemlak, A.M.; Raymakers, R.A.; Bergevoet, S.M.; Gilissen, C.; Jansen, B.J.H.; Adema, G.J.; Kögler, G.; le Sage, C.; Agami, R.; van der Reijden, B.A.; et al. MicroRNA hsa-miR-135b Regulates Mineralization in Osteogenic Differentiation of Human Unrestricted Somatic Stem Cells. *Stem Cells Dev.* **2010**, *19*, 877–885. [CrossRef] [PubMed]

152. Umezu, T.; Tadokoro, H.; Azuma, K.; Yoshizawa, S.; Ohyashiki, K.; Ohyashiki, J.H. Exosomal miR-135b shed from hypoxic multiple myeloma cells enhances angiogenesis by targeting factor-inhibiting HIF-1. *Blood* **2014**, *124*, 3748–3757. [CrossRef] [PubMed]

153. Xu, S.; Santini, G.C.; De Veirman, K. ; Broek, I.V.; Leleu, X.; De, A.; Van Camp, B.; Vanderkerken, K.; Van Riet, I. Upregulation of miR-135b Is Involved in the Impaired Osteogenic Differentiation of Mesenchymal Stem Cells Derived from Multiple Myeloma Patients. *PLoS ONE* **2013**, *8*, e79752. [CrossRef]

154. Sumiyoshi, K.; Kubota, S.; Ohgawara, T.; Kawata, K.; Abd El Kader, T.; Nishida, T.; Ikeda, N.; Shimo, T.; Yamashiro, T.; Takigawa, M. Novel Role of miR-181a in Cartilage Metabolism. *J. Cell. Biochem.* **2013**, *114*, 2094–2100. [CrossRef] [PubMed]

155. Gabler, J.; Ruetze, M.; Kynast, K.L.; Grossner, T.; Diederichs, S.; Richter, W. Stage-Specific miRs in Chondrocyte Maturation: Differentiation-Dependent and Hypertrophy-Related miR Clusters and the miR-181 Family. *Tissue Eng. Part. A* **2015**, *21*, 2840–2851. [CrossRef] [PubMed]

156. Sun, X.; Charbonneau, C.; Wei, L.; Chen, Q.; Terek, R.M. miR-181a Targets RGS16 to Promote Chondrosarcoma Growth, Angiogenesis, and Metastasis. *Mol. Cancer Res.* **2015**, *13*, 1347–1357. [CrossRef] [PubMed]

157. Sun, X.; Wei, L.; Chen, Q.; Terek, R.M. MicroRNA Regulates Vascular Endothelial Growth Factor Expression in Chondrosarcoma Cells. *Clin. Orthop. Relat. Res.* **2015**, *473*, 907–913. [CrossRef] [PubMed]

158. Almeida, M.I.; Silva, A.M.; Vasconcelos, D.M.; Almeida, C.R.; Caires, H.; Pinto, M.T.; Calin, A.; Santos, S.G.; Barbosa, M.A. miR-195 in human primary mesenchymal stromal/stem cells regulates proliferation, osteogenesis and paracrine effect on angiogenesis. *Oncotarget* **2015**, *7*, 7–22. [CrossRef] [PubMed]

159. Wang, R.; Zhao, N.; Li, S.; Fang, J.; Chen, M.; Yang, J.; Jia, W.; Yuan, Y.; Zhuang, S. MicroRNA-195 Suppresses Angiogenesis and Metastasis of Hepatocellular Carcinoma by Inhibiting the Expression of VEGF, VAV2, and CDC42. *Hepatology* **2013**, *58*, 642–653. [CrossRef] [PubMed]

160. Li, Y.; Zeng, C.; Tu, M.; Jiang, W.; Dai, Z.; Hu, Y.; Deng, Z.; Xiao, W. MicroRNA-200b acts as a tumor suppressor in osteosarcoma via targeting ZEB1. *Onco Targets Ther.* **2016**, *9*, 3101–3111.

161. Fan, X.; Teng, Y.; Ye, Z.; Zhou, Y.; Tan, W.-S. The effect of gap junction-mediated transfer of miR-200b on osteogenesis and angiogenesis in a co-culture of MSCs and HUVECs. *J. Cell Sci.* **2018**, *131*, jcs216135. [CrossRef] [PubMed]

162. Choi, Y.; Yoon, S.; Jeong, Y.; Yoon, J.; Baek, K. Regulation of Vascular Endothelial Growth Factor Signaling by miR-200b. *Mol. Cells* **2011**, *32*, 77–82. [CrossRef] [PubMed]

163. Mizuno, Y.; Tokuzawa, Y.; Ninomiya, Y.; Yagi, K.; Yatsuka-Kanesaki, Y.; Suda, T.; Fukuda, T.; Katagiri, T.; Kondoh, Y.; Amemiya, T.; et al. miR-210 promotes osteoblastic differentiation through inhibition of AcvR1b. *FEBS Lett.* **2009**, *583*, 2263–2268. [CrossRef] [PubMed]

164. Liu, X.-D.; Cai, F.; Liu, L.; Zhang, Y.; Yang, A.-L. microRNA-210 is involved in the regulation of postmenopausal osteoporosis through promotion of VEGF expression and osteoblast differentiation. *Biol. Chem.* **2015**, *396*, 339–347. [CrossRef] [PubMed]

165. Fasanaro, P.; D'Alessandra, Y.; Di Stefano, V.; Melchionna, R.; Romani, S.; Pompilio, G.; Capogrossi, M.C.; Martelli, F. MicroRNA-210 modulates endothelial cell response to hypoxia and inhibits the receptor tyrosine kinase ligand ephrin-A3. *J. Biol. Chem.* **2008**, *283*, 15878–15883. [CrossRef] [PubMed]

166. Yan, J.; Guo, D.; Yang, S.; Sun, H.; Wu, B.; Zhou, D. Inhibition of miR-222-3p activity promoted osteogenic differentiation of hBMSCs by regulating Smad5-RUNX2 signal axis. *Biochem. Biophys. Res. Commun.* **2016**, *470*, 498–503. [CrossRef] [PubMed]

167. Takigawa, S.; Chen, A.; Wan, Q.; Na, S.; Sudo, A.; Yokota, H.; Hamamura, K. Role of miR-222-3p in c-Src-Mediated Regulation of Osteoclastogenesis. *Int. J. Mol. Sci.* **2016**, *17*, 240. [CrossRef] [PubMed]

168. Poliseno, L.; Tuccoli, A.; Mariani, L.; Evangelista, M.; Citti, L.; Woods, K.; Mercatanti, A.; Hammond, S.; Rainaldi, G. MicroRNAs modulate the angiogenic properties of HUVECs. *Blood* **2006**, *108*, 3068–3071. [CrossRef]

169. Gao, J.; Yang, T.; Han, J.; Yan, K.; Qiu, X.; Zhou, Y.; Fan, Q.; Ma, B. MicroRNA Expression During Osteogenic Differentiation of Human Multipotent Mesenchymal Stromal Cells From Bone Marrow. *J. Cell. Biochem.* **2011**, *112*, 1844–1856. [CrossRef]

170. Vimalraj, S.; Selvamurugan, N. MicroRNAs expression and their regulatory networks during mesenchymal stem cells differentiation toward osteoblasts. *Int. J. Biol. Macromol.* **2014**, *66*, 194–202. [CrossRef]

171. Li, L.; Qi, Q.; Luo, J.; Huang, S.; Ling, Z.; Gao, M.; Zhou, Z.; Stiehler, M.; Zou, X. FOXO1-suppressed miR-424 regulates the proliferation and osteogenic differentiation of MSCs by targeting FGF2 under oxidative stress. *Sci. Rep.* **2017**, *7*, 1–12. [CrossRef]

172. de Pontual, L.; Yao, E.; Callier, P.; Faivre, L.; Drouin, V.; Cariou, S.; Van Haeringen, A.; Geneviève, D.; Goldenberg, A.; Oufadem, M.; Manouvrier, S.; Munnich, A.; et al. Germline deletion of the miR-17 ~ 92 cluster causes skeletal and growth defects in humans. *Nat. Genet.* **2011**, *43*, 1026–1030. [CrossRef] [PubMed]

173. Penzkofer, D.; Bonauer, A.; Fischer, A.; Tups, A.; Brandes, R.P.; Zeiher, A.M.; Dimmeler, S. Phenotypic Characterization of miR-92a - /- Mice Reveals an Important Function of miR-92a in Skeletal Development. *PLoS ONE* **2014**, *9*, e101153. [CrossRef] [PubMed]

174. Zhou, M.; Ma, J.; Chen, S.; Chen, X.; Yu, X. MicroRNA-17-92 cluster regulates osteoblast proliferation and differentiation. *Endocrine* **2014**, *45*, 302–310. [CrossRef] [PubMed]

175. Mao, G.; Wu, P.; Zhang, Z.; Zhang, Z.; Liao, W.; Li, Y.; Kang, Y. MicroRNA-92a-3p Regulates Aggrecanase-1 and Aggrecanase-2 Expression in Chondrogenesis and IL-1β- Induced Catabolism in Human Articular Chondrocytes. *Cell. Physiol. Biochem.* **2017**, *44*, 38–52. [CrossRef]

176. Zhang, Z.; Kang, Y.; Zhang, Z.; Zhang, H.; Duan, X.; Liu, J.; Li, X.; Liao, W. Expression of microRNAs during chondrogenesis of human adipose-derived stem cells. *Osteoarthr. Cartil.* **2012**, *20*, 1638–1646. [CrossRef] [PubMed]

177. Mao, G.; Zhang, Z.; Huang, Z.; Chen, W.; Huang, G.; Meng, F.; Zhang, Z.; Kang, Y. MicroRNA-92a-3p regulates the expression of cartilage-specific genes by directly targeting histone deacetylase 2 in chondrogenesis and degradation. *Osteoarthr. Cartil.* **2017**, *25*, 521–532. [CrossRef] [PubMed]

178. Bonauer, A.; Carmona, G.; Iwasaki, M.; Mione, M.; Koyanagi, M.; Fischer, A.; Burchfield, J.; Fox, H.; Doebele, C.; Ohtani, K.; et al. MicroRNA-92a Controls Angiogenesis and Functional Recovery of Ischemic Tissues in Mice. *Science* **2009**, *324*, 1710–1713. [CrossRef] [PubMed]

179. Elmén, J.; Lindow, M.; Schütz, S.; Lawrence, M.; Petri, A.; Obad, S.; Lindholm, M.; Hedtjärn, M.; Hansen, H.; Berger, U.; et al. LNA-mediated microRNA silencing in non-human primates. *Nature* **2008**, *452*, 896–899. [CrossRef] [PubMed]

180. Zhang, L.; Zhou, M.; Qin, G.; Weintraub, N.L.; Tang, Y. MiR-92a regulates viability and angiogenesis of endothelial cells under oxidative stress. *Biochem. Biophys. Res. Commun.* **2015**, *446*, 952–958. [CrossRef] [PubMed]

181. Rippe, C.; Blimline, M.; Magerko, K.A.; Lawson, B.R.; LaRocca, T.; Donato, A.; Seals, D.R. MicroRNA Changes in Human Arterial Endothelial Cells with Senescence: Relation to Apoptosis, eNOS and Inflammation Catarina. *Exp. Gerontol.* **2012**, *47*, 45–51. [CrossRef] [PubMed]

182. Lee, Y.S.; Kim, H.K.; Chung, S.; Kim, K.S.; Dutta, A. Depletion of human micro-RNA miR-125b reveals that it is critical for the proliferation of differentiated cells but not for the down-regulation of putative targets during differentiation. *J. Biol. Chem.* **2005**, *280*, 16635–16641. [CrossRef] [PubMed]

183. Xu, J.-F.; Yang, G.-H.; Pan, X.-H.; Zhang, S.-J.; Zhao, C.; Qiu, B.-S.; Gu, H.-F.; Hong, J.-F.; Cao, L.; Chen, Y.; et al. Altered MicroRNA Expression Profile in Exosomes during Osteogenic Differentiation of Human Bone Marrow- Derived Mesenchymal Stem Cells. *PLoS ONE* **2014**, *9*, e114627. [CrossRef] [PubMed]

184. Fan, G. Hypoxic exosomes promote angiogenesis Platelets: Balancing the septic triad. *Blood* **2014**, *124*, 3669–3670. [CrossRef] [PubMed]

185. Zhang, L.; Sun, Z.-J.; Bian, Y.; Kulkarni, A.B. MicroRNA-135b acts as a tumor promoter by targeting the hypoxia-inducible factor pathway in genetically defined mouse model of head and neck squamous cell carcinoma. *Cancer Lett.* **2013**, *331*, 230–238. [CrossRef] [PubMed]

186. Kazenwadel, J.; Michael, M.Z.; Harvey, N.L. Prox1 expression is negatively regulated by miR-181 in endothelial cells. *Blood* **2010**, *116*, 2395–2401. [CrossRef]

187. Naguibneva, I.; Ameyar-Zazoua, M.; Polesskaya, A.; Ait-Si-Ali, S.; Groisman, R.; Souidi, M.; Cuvellier, S.; Harel-Bellan, A. The microRNA miR-181 targets the homeobox protein Hox-A11 during mammalian myoblast differentiation. *Nat. Cell Biol.* **2006**, *8*, 278–284. [CrossRef]

188. Akiyama, T.; Bouillet, P.; Miyazaki, T.; Kadono, Y.; Chikuda, H.; Chung, U.; Fukuda, A.; Hikita, A.; Seto, H.; Okada, T.; et al. Regulation of osteoclast apoptosis by ubiquitination of proapoptotic BH3-only Bcl-2 family member Bim. *EMBO J.* **2003**, *22*, 6653–6664. [CrossRef]

189. Palmieri, A.; Pezzetti, F.; Brunelli, G.; Zollino, I.; Scapoli, L.; Martinelli, M.; Arlotti, M.; Carinci, F. Differences in osteoblast miRNA induced by cell binding domain of collagen and silicate-based synthetic bone. *J. Biomed. Sci.* **2007**, *14*, 777–782. [CrossRef]

190. Christoffersen, N.R.; Silahtaroglu, A.; Ørom, U.L.F.A.; Kauppinen, S.; Lund, A.H. miR-200b mediates post-transcriptional repression of ZFHX1B. *RNA* **2007**, *13*, 1172–1178. [CrossRef]

191. Baglio, S.R.; Rooijers, K.; Koppers-lalic, D.; Verweij, F.J.; Lanzón, M.P.; Zini, N.; Naaijkens, B.; Perut, F.; Niessen, H.W.M.; Baldini, N.; et al. Human bone marrow- and adipose- mesenchymal stem cells secrete exosomes enriched in distinctive miRNA and tRNA species. *Stem Cell Res. Ther.* **2015**, *6*, 1–20. [CrossRef] [PubMed]

192. Zong, L.; Zhu, Y.; Liang, R.; Zhao, H.-B. Gap junction mediated miRNA intercellular transfer and gene regulation: A novel mechanism for intercellular genetic communication. *Sci. Rep.* **2016**, *6*, 19884. [CrossRef] [PubMed]

193. Liu, G.-T.; Chen, H.-T.; Tsou, H.-K.; Tan, T.-W.; Fong, Y.-C.; Chen, P.-C.; Yang, W.-H.; Wang, S.-W.; Chen, J.-C.; Tang, C.-H. CCL5 promotes VEGF-dependent angiogenesis by downregulating miR-200b through PI3K/Akt signaling pathway in human chondrosarcoma cells. *Oncotarget* **2014**, *5*, 10718–10731. [CrossRef] [PubMed]

194. Chan, Y.C.; Khanna, S.; Roy, S.; Sen, C.K. miR-200b targets Ets-1 and is down-regulated by hypoxia to induce angiogenic response of endothelial cells. *J. Biol. Chem.* **2011**, *286*, 2047–2056. [CrossRef] [PubMed]

195. Lou, Y.-L.; Guo, F.; Liu, F.; Gao, F.-L.; Zhang, P.-Q.; Niu, X.; Guo, S.-C.; Yin, J.-H.; Wang, Y.; Deng, Z.-F. miR-210 activates notch signaling pathway in angiogenesis induced by cerebral ischemia. *Mol. Cell Biochem.* **2012**, *370*, 45–51. [CrossRef] [PubMed]

196. Fasanaro, P.; Greco, S.; Lorenzi, M.; Pescatori, M.; Brioschi, M.; Kulshreshta, R.; Banfi, C.; Stubbs, A.; Calin, G.A.; Ivan, M.; et al. An integrated approach for experimental target identification of hypoxia-induced miR-210. *J. Biol. Chem.* **2009**, *284*, 35134–35143. [CrossRef] [PubMed]

197. Ivan, M.; Huang, X. miR-210: Fine-Tuning the Hypoxic Response. *Adv. Exp. Med. Biol.* **2014**, *772*, 205–227. [CrossRef]

198. Ivan, M.; Harris, A.L.; Martelli, F.; Kulshreshtha, R. Hypoxia response and microRNAs: No longer two separate worlds. *J. Cell Mol. Med.* **2008**, *12*, 1426–1431. [CrossRef]

199. Kuijper, S.; Turner, C.J.; Adams, R.H. Regulation of angiogenesis by Eph-ephrin interactions. *Trends Cardiovasc. Med.* **2007**, *17*, 145–151. [CrossRef]

200. Pandey, A.; Shao, H.; Marks, R.M.; Polverini, P.J.; Dixit, V.M. Role of B61, the ligand for the Eck receptor tyrosine kinase, in TNF-alpha-induced angiogenesis. *Science* **1995**, *268*, 567–569. [CrossRef]

201. Matsui, J.; Wakabayashi, T.; Asada, M.; Yoshimatsu, K. Stem cell factor/c-kit signaling promotes the survival, migration, and capillary tube formation of human umbilical vein endothelial cells. *J. Biol. Chem.* **2004**, *279*, 18600–18607. [CrossRef] [PubMed]

202. Jung, S.; Kleinheinz, J. Angiogenesis—The Key to Regeneration. In *Tissue Engineering and Regenerative Medicine*; Andrades, J.A., Ed.; InTechOpen: London, UK, 2013; pp. 453–473.

203. Kanczler, J.M.; Oreffo, R.O.C. Osteogenesis and angiogenesis: The potential for engineering bone. *Eur. Cells Mater.* **2008**, *15*, 100–114. [CrossRef]

204. Hou, H.; Zhang, X.; Tang, T.; Dai, K.; Ge, R. Enhancement of bone formation by genetically-engineered bone marrow stromal cells expressing BMP-2, VEGF and angiopoietin-1. *Biotechnol. Lett.* **2009**, *31*, 1183–1189. [CrossRef] [PubMed]

205. Zhang, F.; Qiu, T.; Wu, X.; Wan, C.; Shi, W.; Wang, Y.; Chen, J.; Wan, M.; Clemens, T.L.; Cao, X. Sustained BMP Signaling in Osteoblasts Stimulates Bone Formation by Promoting Angiogenesis and Osteoblast Differentiation. *J. Bone Miner. Res.* **2009**, *24*, 1224–1233. [CrossRef] [PubMed]

206. Shi, Z.; Wang, K. Effects of recombinant adeno-associated viral vectors on angiopoiesis and osteogenesis in cultured rabbit bone marrow stem cells via co-expressing hVEGF and hBMP genes: A preliminary study in vitro. *Tissue Cell* **2010**, *42*, 314–321. [CrossRef] [PubMed]

207. Hon, L.S.; Zhang, Z. The roles of binding site arrangement and combinatorial targeting in microRNA repression of gene expression. *Genome Biol.* **2007**, *8*, R166. [CrossRef]

208. Grimson, A.; Farh, K.K.; Johnston, W.K.; Garrett-Engele, P.; Lim, L.P.; Bartel, D.P. MicroRNA targeting specificity in mammals: Determinants beyond seed pairing. *Mol. Cell* **2007**, *27*, 91–105. [CrossRef] [PubMed]

209. Betel, D.; Wilson, M.; Gabow, A.; Marks, D.S.; Sander, C. The microRNA.org resource: Targets and expression. *Nucleic Acids Res.* **2008**, *36*, D149–D153. [CrossRef]

210. Krek, A.; Grün, D.; Poy, M.N.; Wolf, R.; Rosenberg, L.; Epstein, E.J.; MacMenamin, P.; da Piedade, I.; Gunsalus, K.C.; Stoffel, M.; et al. Combinatorial microRNA target predictions. *Nat. Genet.* **2005**, *37*, 495–500. [CrossRef]

211. Brennecke, J.; Stark, A.; Russell, R.; Cohen, S. Principles of microRNA-target recognition. *PLoS Biol.* **2005**, *3*, e85. [CrossRef]

212. Tsang, J.; Zhu, J.; van Oudenaarden, A. MicroRNAmediated feedback and feedforward loops are recurrent network motifs in mammals. *Mol. Cell* **2007**, *26*, 753–767. [CrossRef] [PubMed]

213. Lewis, B.; Burge, C.; Bartel, D. Conserved seed pairing, often flanked by adenosines, indicates that thousands of human genes are microRNA targets. *Cell* **2005**, *120*, 15–20. [CrossRef] [PubMed]

214. Li, H.; Xie, H.; Liu, W.; Hu, R.; Huang, B.; Tan, Y.; Xu, K.; Sheng, Z.; Zhou, H.; Wu, X.; et al. A novel microRNA targeting HDAC5 regulates osteoblast differentiation in mice and contributes to primary osteoporosis in humans. *J. Clin. Invest.* **2009**, *119*, 3666–3677. [CrossRef] [PubMed]

215. Mariner, P.; Johannesen, E.; Anseth, K. Manipulation of miRNA activity accelerates osteogenic differentiation of hMSCs in engineered 3D scaffolds. *J. Tissue Eng. Regen. Med.* **2012**, *6*, 314–324. [CrossRef] [PubMed]

216. Lee, W.Y.; Li, N.; Lin, S.; Wang, B.; Lan, H.Y.; Li, G. miRNA-29b improves bone healing in mouse fracture model. *Mol. Cell. Endocrinol.* **2016**, *430*, 97–107. [CrossRef] [PubMed]

217. Li, Y.; Fan, L.; Liu, S.; Liu, W.; Zhang, H.; Zhou, T.; Wu, D.; Yang, P.; Shen, L.; Chen, J.; et al. The promotion of bone regeneration through positive regulation of angiogenic-osteogenic coupling using microRNA-26a. *Biomaterials* **2013**, *34*, 5048–5058. [CrossRef] [PubMed]

218. Yoshizuka, M.; Nakasa, T.; Kawanishi, Y.; Hachisuka, S.; Furuta, T.; Miyaki, S.; Adachi, N.; Ochi, M. Inhibition of microRNA-222 expression accelerates bone healing with enhancement of osteogenesis, chondrogenesis, and angiogenesis in a rat refractory fracture model. *J. Orthop. Sci.* **2016**, *21*, 852–858. [CrossRef] [PubMed]

219. Ørom, U.A.; Kauppinen, S.; Lund, A.H. LNA-modified oligonucleotides mediate specific inhibition of microRNA function. *Gene* **2006**, *372*, 137–141. [CrossRef] [PubMed]

220. Adams, B.D.; Parsons, C.; Walker, L.; Zhang, W.C.; Slack, F.J. Targeting noncoding RNAs in disease. *J. Clin. Investig.* **2017**, *127*, 761–771. [CrossRef]

221. Simonson, B.; Das, S. MicroRNA Therapeutics: The Next Magic Bullet? *Mini Rev. Med. Chem.* **2016**, *15*, 467–474. [CrossRef]

222. Esquela-Kerscher, A.; Slack, F.J. Oncomirs-microRNAs with a role in cancer. *Nat. Rev. Cancer* **2006**, *6*, 259–269. [CrossRef] [PubMed]

223. López-Camarillo, C.; Marchat, L.A.; Aréchaga-Ocampo, E.; Azuara-Liceaga, E.; Pérez-Plasencia, C.; Fuentes-Mera, L.; Fonseca-Sánchez, M.A.; Flores-Pérez, A. *Functional Roles of microRNAs in Cancer: microRNomes and oncomiRs Connection*; Oncogenomi.; In Tech Open Science: London, UK, 2013.

224. Senanayake, U.; Das, S.; Vesely, P.; Alzoughbi, W.; Fröhlich, L.F.; Chowdhury, P.; Leuschner, I.; Hoefler, G.; Guertl, B. miR-192, miR-194, miR-215, miR-200c and miR-141 are downregulated and their common target ACVR2B is strongly expressed in renal childhood neoplasms. *Carcinogenesis* **2012**, *33*, 1014–1021. [CrossRef] [PubMed]

Review

MicroRNA Expression is Associated with Sepsis Disorders in Critically Ill Polytrauma Patients

Alexandru Florin Rogobete [1,2], Dorel Sandesc [1,2], Ovidiu Horea Bedreag [1,2], Marius Papurica [1,2], Sonia Elena Popovici [1], Tiberiu Bratu [1], Calin Marius Popoiu [1,*], Razvan Nitu [1], Tiberiu Dragomir [1], Hazzaa I. M. AAbed [1] and Mihaela Viviana Ivan [1]

[1] Faculty of Medicine, "Victor Babes" University of Medicine and Pharmacy, 300041 Timisoara, Romania; alexandru.rogobete@umft.ro (A.F.R.); dsandesc@yahoo.com (D.S.); bedreag.ovidiu@umft.ro (O.H.B.); marius.papurica@umft.ro (M.P.); popovici.sonia@yahoo.com (S.E.P.); tiberiu.bratu@umft.ro (T.B.); razvan.nitu@umft.ro (R.N.); tiberiu.dragomir@umft.ro (T.D.); hazzaa.aabed@gmail.com (H.I.M.A.); viviana.ivan@umft.ro (M.V.I.)
[2] Clinic of Anesthesia and Intensive Care, Emergency County Hospital "Pius Brinzeu", 300723 Timisoara, Romania
*   Correspondence: mcpopoiu@yahoo.com; Tel.: +40-0729-101-221

Received: 2 November 2018; Accepted: 6 December 2018; Published: 13 December 2018

Abstract: A critically ill polytrauma patient is one of the most complex cases to be admitted to the intensive care unit, due to both the primary traumatic complications and the secondary post-traumatic interactions. From a molecular, genetic, and epigenetic point of view, numerous biochemical interactions are responsible for the deterioration of the clinical status of a patient, and increased mortality rates. From a molecular viewpoint, microRNAs are one of the most complex macromolecular systems due to the numerous modular reactions and interactions that they are involved in. Regarding the expression and activity of microRNAs in sepsis, their usefulness has reached new levels of significance. MicroRNAs can be used both as an early biomarker for sepsis, and as a therapeutic target because of their ability to block the complex reactions involved in the initiation, maintenance, and augmentation of the clinical status.

Keywords: microRNAs; epigenetic biomarker; sepsis; inflammation

---

1. Introduction

Critically ill polytrauma patients present one of the most complex clinical pictures that the intensivist and trauma team will encounter in their careers [1–5]. The complexity of these cases is due both to the initial traumatic injury, and to the secondary post-traumatic responses to injury [2–16]. Moreover, through the interactions of molecular mechanisms with other, initially functional systems, and through molecular denaturation reactions, the critically ill polytrauma patient becomes a complex medical case from a clinical and molecular point of view. A series of complex mechanisms involved in the pathophysiology and biochemistry of sepsis have been studied for the past several years. However, the critically ill polytrauma patient is so complex biochemically and molecularly that no specific biochemical pathways have been found in which intervention could increase survival rates, or decrease the incidence of sepsis and multiple organ dysfunction syndrome (MODS) [3–25]. However, in the last few years, a series of microRNA epigenetic species have been identified. These species are responsible for the modulation of certain complex molecular reactions. Furthermore, numerous studies have shown the importance of microRNAs in early diagnosis and possible future epigenetic therapies [6–33]. By examining microRNAs with respect to critically ill polytrauma patients, we can see important links between the development and modulation of the systemic inflammatory response, the immune system, coagulation status, and response to infections [7–22]. The paper aims to systematize the microRNA expressions that are closely related to the pathophysiological events involved in a critically ill polytrauma patient with sepsis. Moreover, we wished to highlight the most important microRNA expression studies conducted to date that could be used as biomarkers for the early diagnosis of sepsis.

2. Biochemical and Biosynthesis Aspects of MicroRNAs

From a molecular point of view, microRNAs are synthesized in the cell nucleus through the action of RNA polymerase II on certain specific genes. Hence, the initial species, the pri-microRNAs, are formed following complex reactions [13–35]. In the next step, RNAse III endonuclease, also called Drosha, activates the pri-microRNAs. This reaction is catalyzed by the DiGeorge Syndrome Critical Region 8 (DGCR8) complex, which leads to the formation of pre-microRNAs [15,34–74].

Once these almost final species are formed, the pre-microRNAs bind with the Exportin-5 transporter protein, which shifts them from the nucleus into the cytoplasm. Inside the cytoplasm, a new reaction, initiated by an RNAse III endonuclease called Dicer and by the RNA binding protein (TRBP), takes place, which leads to the formation of the final microRNA species [17,20,66–88]. The last step involves coupling the RNA-induced silencing complex (RISC) [18,78–91]. The final molecular species is then transported outside the cell through different mechanisms and in various forms, such as ribonucleoprotein complexes, microvesicles, exosomes, and high-density lipoproteins (Figure 1) [14].

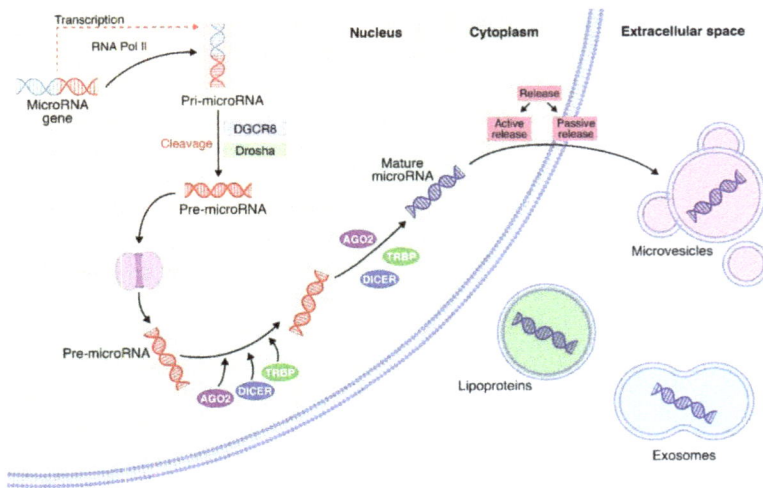

Figure 1. MicroRNA biosynthesis mechanisms. For further explanation, please see the details in the text. RNA pol II—RNA polymerase II; pri-microRNA—primitive microRNA; DGCR8—DiGeorge Syndrome Critical Region 8; Drosha—RNAse III endonuclease; pre-microRNA—precursor of microRNA; AGO2—endonuclease Argonaute 2; Dicer—Rnase III endonuclease; TRBP—transactivation response element RNA-binding protein.

3. MicroRNA Identification from Different Body Fluids

MicroRNAs have been proposed as possible biomarkers because of the research evidence that shows that changes in a range of cellular microRNAs correlate with various pathophysiological conditions, including inflammation, oxidative stress, sepsis diabetes and different types of cancer [33–45,73–93]. These molecules have also been known for their low complexity, simple detection and amplification, tissue-restricted expression profiles, and sequence conservation between human and model organisms [94–98]. However, they have not been incorporated into clinical practice due to several factors such as the lack of a universal and comprehensive measurement technique that would be convenient enough in terms of handling, the rate of analysis, and reliability [96–101]. Apart from the measurement technique, another factor that has been holding back the use of microRNAs is that their concentration in the body is relatively low. However, there exist measurement methods that have been routinely used, although they have their advantages and disadvantages. These techniques include small RNA sequencing, quantitative reverse transcription polymerase chain reaction (qPCR), and microarray hybridization. All of these are applied according to the respective propose of analysis (Figure 2). When it comes to the successful identification of these microRNAs, the factors that are critical, such as the choice of the measurement sample and the appropriate normalization strategy come into play. The profiles of these important biomarkers are also influenced considerably by exogenous factors such as medication, nutrition, and certain environmental conditions [98–114].

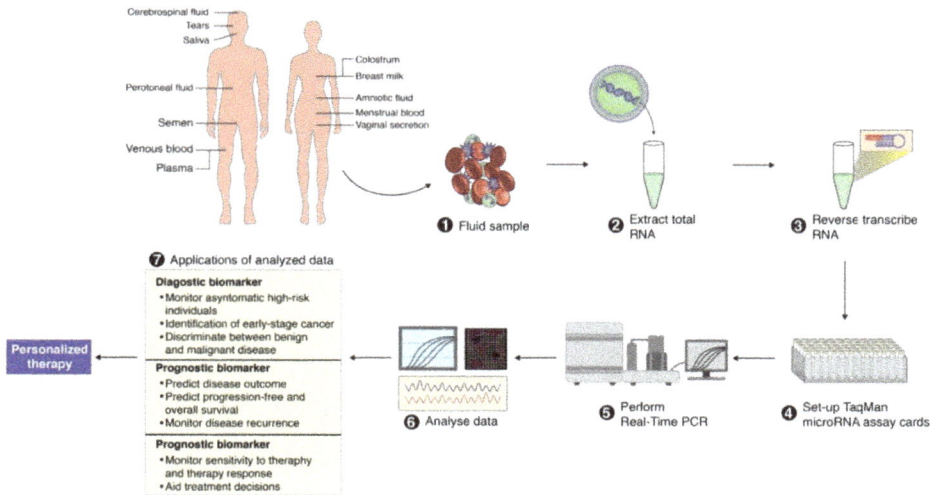

Figure 2. MicroRNA identification workflow from different body fluids.

4. Importance of MicroRNAs for Clinical Use

Due to their unique features, such as disease specificity, relative stability, and easy accessibility, microRNAs are considered the future biomarkers for the diagnosis and prognosis of specific diseases as well as monitoring therapeutic responses in clinical settings [75–79]. MicroRNAs have been identified in different clinical settings, and their importance as biomarkers is still under investigation [80–117]. For example, a number of these molecules have been associated with sepsis, acute lung injury and acute organ dysfunction diagnoses. MicroRNAs are also being considered for therapeutic purposes where up-regulatory or down-regulatory molecules, targeting specific microRNAs, can be administered with the aim of managing specific pathological conditions. A study of this is currently at the clinical trial stage. Studies have shown that the anesthetics and medications used in post-operative patient care affect the expression of microRNA, which in turn affects the functioning or survival of certain types of cells in the body, such as neurocytes. The expression of microRNAs in their various cells are highly specific, and therefore, they have a distinct display pattern in different tissues, which contributes to their characteristic features and functions. With this in mind, these molecules have been used to detect the presence of disease or tissue malfunction due to their recognizable pattern of appearance [97–105]. For instance, rough relations have been created linking specific microRNA expressions to the manifestation of certain pathological conditions, including microR-21 being shown as a proto-oncogene in adenocarcinoma, and microR-146a acting as an inhibitory factor to inflammatory processes by dampening the nuclear factor-kB (NF-kB) signaling [114–116]. MicroRNA has also been considered in forensic investigations due to their initially-named properties. It helps address the challenge of sensitivity and specificity when it comes to criminal identification. However, this is entirely dependent on the method applied in such endeavors. In criminal investigations, the disadvantages of microRNA profiling have not yet been studied [116,117].

5. Roles of MicroRNAs in the Pathophysiology of Oxidative Stress Associated with Sepsis

Under normal conditions, the human body synthesizes numerous biochemical species with increased reactivity compared to the existent macromolecules. These species are called free radicals and are divided depending on their origin into reactive oxygen species, reactive lipid species, reactive nitrogen species, and other more complex redox systems [20–74]. Biochemically speaking, the most aggressive free radicals are oxygen radicals such as hydroxyl radicals (HO^-), superoxide anions (O_2^-), or hydrogen peroxide (H_2O_2). Moreover, nitrogen radicals such as peroxynitrite ($ONOO^-$) and nitric oxide (NO), or the lipid radicals, especially the lipid peroxyl (LOO^-), also present similarly high reactivity [3,19,20]. The oxidative stress appears once the free radicals accumulate over the endogenous antioxidants. Under circumstances of traumatic stress, a series of endogenous systems are responsible for generating an excessive amount of free radicals. Among these, the most studied are mitochondrial respiration, the xanthine reduction mechanisms, and the NADPH oxidase enzymatic system [15,21–24]. Admittedly, under physiological conditions, the human body has a series of biomacromolecules with antioxidant capacities such as catalase (CAT) [25], superoxide dismutase (SOD) [26], peroxiredoxins (PRXs) [27], glutathione (GSH) [28], and glutaredoxins (GRXs). However, in the case of critically ill polytrauma patients, the production of free radicals overcomes the endogenous antioxidant capacity of the body, and therefore oxidative stress associated with the systemic inflammatory response appears very quickly. In the case of critically ill polytrauma patients, the pro-oxidative phenomenon appears at the moment of trauma because of the associated organic injuries. A short time after the traumatic impact, the molecular injury will be transmitted, augmented, and multiplied in the cell, especially inside the cellular organelles [29–31]. From a clinical point of view, the molecular lesions have important implications in the clinical evolution of a patient due to the significant increase in the morbidity and mortality rates through their association with the systemic inflammatory response, and also because of their association with generalized infections [32–36]. With regard to the association with infections, oxidative stress has significant implications in increasing the incidence of sepsis due to the release of free radicals, cytokines, and adhesion molecules. Hence, immunosuppression, the increased concentration of pro-inflammatory factors, and the aggressive attack of free radicals all lead to MODS in the critically ill polytrauma patient, despite complex treatment options (Figure 3) [37–85].

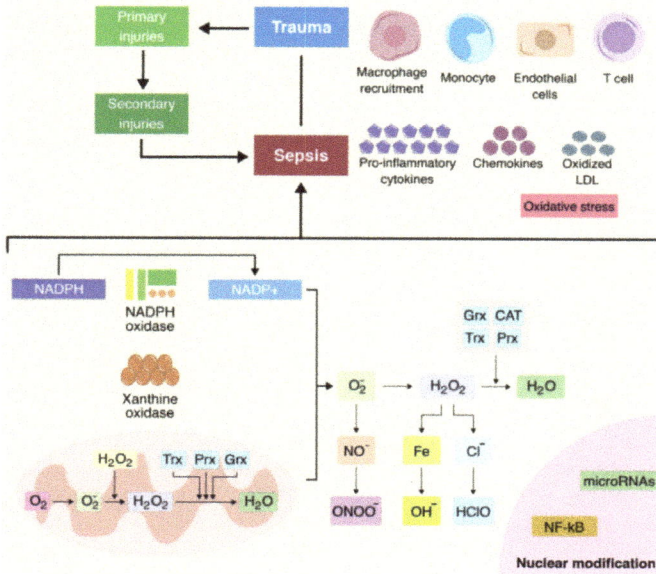

Figure 3. The critically ill polytrauma patient is characterized by a series of secondary, post-traumatic injuries, represented especially by cellular and molecular damage. Oxidative stress is an important molecular phenomenon, and it has important links with a series of bio-macromolecular systems. An important source of free radicals is the mitochondria, where huge amounts of free oxygen radicals are produced that will further lead to the augmentation of the pro-oxidative phenomena. Moreover, the molecular disaster will continue as other systems are affected such as the endovascular system, lipid molecules, proteins, and cellular organelles.

The microRNA species play an important role in the propagation of pro-oxidative signals by modifying the reactivity of the molecular receptors. Numerous studies have identified important implications of microRNAs for the cis-acting DNA sequences [41–43]. Practically, the cellular proliferation is influenced by microRNA-9 by modulating the activity of orphan nuclear receptor TLX, located in the neuroepithelium. Another example is the estrogen and androgen receptors that have microRNA-21, microRNA-222, microRNA-221, microRNA-101, microRNA-206, microRNA-433, microRNA-34a, microRNA-125b, and microRNA-127 as genetic substrates [44,45].

The nuclear transcription factor kB (NF-kB) represents another interesting aspect of a molecular attack. From a biochemical viewpoint, NF-kB is involved in modifying the reactions of certain genes and is influenced in most cases by a series of external or internal factors such as the IKB and IKK proteins [9,46,47]. If we were to discuss the links between NF-kB and oxidative stress, and the implications of NF-kB in the clinical outcome of these patients, one could highlight the cellular adaptability induced by the pathophysiological changes arising from inflammation, infections, and the immune response. This can be explained through the implications that NF-kB has in the production of pro- and anti-inflammatory cytokines such as interleukin-1 (IL-1) and tumor necrosis factor-alpha (TNF-alpha) [48–51]. Moreover, in this complex series of events that make up a molecular disaster, there are numerous important links caused by the reciprocal activation of certain factors that are decisive in the augmentation of the molecular disaster. In the case of critically ill patients, a series of specific secondary phenomena occur, such as tissue hypoxia, generalized inflammation, and infections [52]. With regard to this, researchers have identified the microRNAs that play a decisive role in the modification of the biochemical pathways. An important study carried out by Scott et al. reported significant changes in the expression of microRNA-17-92, microRNA-221, microRNA-126, and microRNA-222 [53]. In the literature, other microRNA species that have important implications for endothelial damage have also been reported, such as microRNA-278 and microRNA-146 [54]. Another study carried out by Kung et al. reported reduced activity for microRNA-26a, microRNA-126, and microRNA-24 [55]. The same study showed increased expression of microRNA-346, microRNA-30b, microRNA-999, and microRNA-30a.

With regard to epigenetic expression in tissue hypoxia, microRNAs have been shown to have multiple implications, both augmenting cellular destruction and increasing the pro-inflammatory and pro-oxidative status [80–84]. Numerous microRNAs are responsible for dictating the biosynthesis for adhesion molecules, free oxygen, nitrogen, or lipid radicals, and affecting cell and mitochondria energy. Among these, the most microRNAs that have been most studied in-depth are microRNA-213, microRNA-210, microRNA-24, microRNA-27, microRNA-23, microRNA-26, microRNA-210-3p, microRNA 23b-3p, microRNA-1275, microRNA-210-3p, microRNA-145-5p, microRNA-92b-3p, microRNA-181a-2-3p, microRNA-185-5p, microrRNA-20a-5p, and microRNA-92b-3p [84–87]. Another associated phenomenon is ischemia–reperfusion syndrome. From a clinical and molecular point of view, ischemia–reperfusion is an important generator of free radicals and inflammatory molecules that are responsible for aggravating the clinical status of these patients, especially in the context of inflammation and infection. Important changes in epigenetic expression have also been identified in the case of ischemia–reperfusion syndrome. Among these, the most commonly studied are microRNA-290, microRNA-26, microRNA-192, microRNA-805, microRNA-194, microRNA-187, microRNA-145, and microRNA-21 [14,88,89].

A high proportion of critically ill polytrauma patients develop acute respiratory distress syndrome (ARDS). From a cellular and molecular viewpoint, in ARDS, the neutrophils invade the pulmonary tissue leading to the initiation of aggressive pro-inflammatory mechanisms [48,56,70]. The molecular cascade in this case is activated and augmented by the excess production of interleukin 6 (IL-6), interleukin 1 beta (IL-1), and tumor necrosis factor alpha (TNF-alpha). Furthermore, this molecular cascade leads to increased vascular permeability in the pulmonary tissue. The molecular reactions are extremely complex, with the inhibition of apoptosis in the alveolar capillaries through the action of vascular endothelial growth factor (VEGF) [60,71,72]. The VEGF receptors, including vascular endothelial growth factor receptor 1 (VEGFR1) and vascular endothelial growth factor receptor 2 (VEGFR2) are further activated, leading to increased vascular permeability. In this case, the expression of microRNAs also plays an important role, modulating a series of complex molecular reactions [73,74]. Yehya et al. reported important changes for microRNA-466c-5p, microRNA-466d-5p, microRNA-15b, microRNA-154, microRNA-466c, microRNA-466b, microRNA-466f-3p, microRNA-375, microRNA-378, microRNA-347, and microRNA-32* [75]. A similar study carried out on the same group of patients by Kulshreshtha et al. reported changes in the expressions of miRNA-27, miRNA-103, miRNA-107, miRNA-26, miRNA-181, miRNA-210, miRNA-23, miRNA-24, and miRNA-213 [76]. Likewise, important changes have been noted in these situations for miRNA-194, miRNA-214, miRNA-223, miRNA-100, miRNA-140, miRNA-142-3p, miRNA-25, miRNA-27b, miRNA-181c, miRNA-21, and microRNA-224 activity [77]. Tacke et al. reported an increased expression for microRNA-133a in patients with sepsis [78]. Wang et al. also reported the decreased expression of microRNA-223, microRNA-181b, and microRNA-146a [79] (Figure 4).

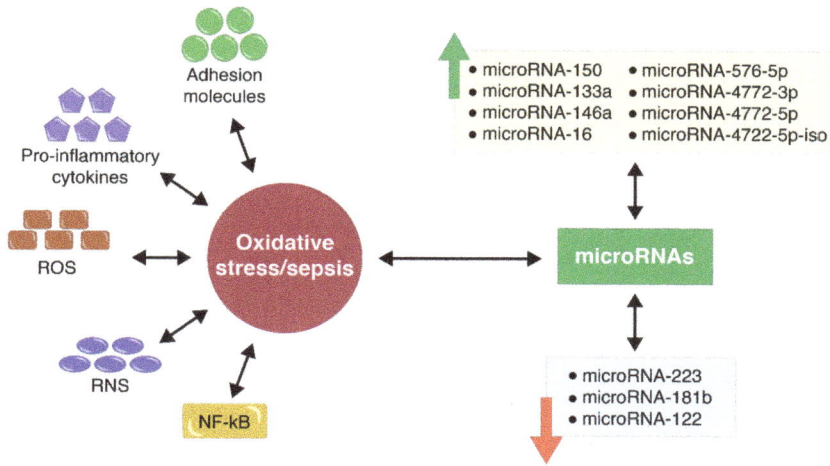

Figure 4. MicroRNA expression in a critically ill polytrauma patient with sepsis. A short time after the primary traumatic injury, the critically ill polytrauma patient develops a series of secondary post-traumatic injuries, especially molecular and cellular injuries. Among these, the most studied are the excess biosynthesis of free radicals reactive oxygen species (ROS) and reactive nitrogen species (RNS), and the augmentation of the pro-oxidative chain. Moreover, together with the involvement of the immune system, the activation of nuclear transcription factor kappa B (NF-kB), the emergence of adhesion molecules, the release of excess pro-inflammatory factors, and infections will determine a series of microRNA species that will intervene in the modulation of this complex molecular cycle. Numerous studies have shown both an increase in the activity of certain microRNA species, and a decrease in the expression of other species in certain selected cases [14,80–83].

Hyperoxia is closely related to ARDS. This phenomenon is mostly induced by the intensive care physician because of difficult ventilation and inadequate oxygen concentrations in the circulatory system. In these situations, the intensive care unit (ICU) Fi-inspired oxygen fraction (FiO_2) is usually increased to 1.0 (100% O_2). On the other hand, increasing FiO_2 to 1.0 directly affects the mitochondria and the microvascular system. Together with the impairment of the microvascular system, the vascular perfusion in the pulmonary tissue will drop significantly, leading to a decreased gas-exchange capacity and the progressive deterioration of the patient's clinical status.

Vascular endothelial growth factor (VEGF) has been widely discussed in the literature in relation to microvascular injury. Specifically, a series of reactions involved in the inhibition of apoptosis in the alveolar capillaries has been mentioned in the literature. Moreover, an important increase in the expression of vascular endothelial growth factor receptor 1 (VEGFR1) in patients with ARDS has been demonstrated [58–60].

There are complex mechanisms that are closely related to the biofunctionality of the endothelial surface. An important system in this category is the KL-6 glycoprotein that can be found on the surface of type II alveolar cells [61]. From an immunological point of view, T-cell expression is widely influenced by all of these mechanisms. Recent studies have shown changes in the Foxp3+ regulatory T-cell (Tregs), CD4+, CD3+, CD25hi, CD127lo, and CD25+ expression [62].

6. MicroRNA Expression in the Case of Polytrauma Patients with Sepsis

From a pathophysiological and molecular viewpoint, in the case of polytrauma patients with sepsis, an important phenomenon appears due to excess cytokine synthesis. This is determined by the complex reactions between the lipopolysaccharide macromolecules (LPS) and lipopolysaccharide binding protein (LPB) [63,64]. The molecular bond required for these reactions to take place exists due to the CD14 receptor found on the surface of macrophages. Apart from these complex links, there are a series of other reactions represented especially by the synthesis of certain mediators, such as histamines, chemokines, or different hormones. Moreover, the coagulation cascade plays an important role in the augmentation and self-propagation of the molecular phenomena involved in sepsis. The most important pro-inflammatory and anti-inflammatory mediators are interleukin 4 (IL-4), interleukin 10 (IL-10), interleukin 17 (IL-17), interleukin 1 (IL-1), interleukin 2 (IL-2), interleukin 6 (IL-6), interleukin 12 (IL-12), interleukin 8 (IL-8), Procalcitonin (PCT), N-terminal C natriuretic peptide (NT-CNP), C-reactive proteins (CRP), tumor necrosis factor alpha (TNF-alpha), interferon gamma (INF-gamma), and transforming growth factor beta (TGF-beta) [14,49,65,66].

Recent studies have shown a series of implications for microRNA species in the pathophysiology of sepsis, pro-inflammatory and pro-oxidative phenomena (Table 1). MicroRNA-146a, microRNA-150, and microRNA-233 have complex implications in the molecular damage in sepsis [67]. Moreover, Puskarich et al. reported strong statistical correlations between the changes in microRNA-150 expression and the increase in mortality rates. Another important study, carried out by Vasilescu et al. reported decreased activity for microRNA-150 and microRNA-342-5p in the case of septic patients. On the other hand, there was an increased expression of microRNA-486 and microRNA-182 in these patients [67]. A similar study carried out by Benz et al., identified important changes in microRNA-233 in patients with sepsis [68]. Other species involved in the molecular and genetic sepsis reactions include microRNA-340, microRNA-324-3p, microRNA-16, microRNA-210, let-7b, microRNA-15b, microRNA-484, microRNA-486-5p, and microRNA-324-3p [69].

Table 1. MicroRNA expression in sepsis.

Involved MicroRNAs	Body Fluid of Identification	Expression	References
microRNA-4772-5p	Serum	Up-regulated	[90–92]
microRNA-4772-5p Iso	Serum	Up-regulated	[92]
microRNA-15a	Serum	Up-regulated	[90,91]
microRNA-16	Serum	Up-regulated	[90]
microRNA-574-5p	Serum	Up-regulated	[91]
microRNA-4772-3p	Serum	Up-regulated	[92]
microRNA-4516	Serum	Up-regulated	[93]
microRNA-454-3p	Serum	Up-regulated	[93]
miR-155-3p	Serum	Up-regulated	[93]
microRNA-219b	Serum	Up-regulated	[94]
microRNA-1889	Serum	Up-regulated	[94]
microRNA-106a	Serum	Up-regulated	[94]
microRNA-106b	Serum	Up-regulated	[94]
microRNA-205	Serum	Up-regulated	[94]
microRNA-20a	Serum	Up-regulated	[94]
miR-150	Serum	Up-regulated	[91]
microRNA-27a	Serum	Up-regulated	[95]
microRNA-122	Serum	Up-regulated	[96]
microRNA-146a	Serum	Up-regulated	[91]
microRNA-422	Serum	Up-regulated	[91]

Table 1. *Cont.*

Involved MicroRNAs	Body Fluid of Identification	Expression	References
microRNA-133a	Serum	Up-regulated	[90–95]
microRNA-4532	Serum	Up-regulated	[95]
microRNA-576-5p	Serum	Up-regulated	[80–83]
microRNA-483-5p	Serum	Down-regulated	[91]
microRNA-499-5p	Serum	Down-regulated	[91]
microRNA-193b*	Serum	Down-regulated	[91]
miR-146a-5p	Serum	Down-regulated	[93]
Let-7g-5p	Serum	Down-regulated	[93]
microRNA-30	Serum	Down-regulated	[94]
microRNA-199a-3p	Serum	Down-regulated	[93]
microRNA-29	Serum	Down-regulated	[95,96]
microRNA-297	Serum	Down-regulated	[96]
microRNA-125	Serum	Down-regulated	[96]
microRNA-25	Serum	Down-regulated	[96]
microRNA-19	Serum	Down-regulated	[95]
microRNA-182	Blood	Up-regulated	[95,96]
microRNA-15b	Blood	Up-regulated	[95]
microRNA-486	Blood	Up-regulated	[94–96]
microRNA-25	Blood	Down-regulated	[90]
microRNA-223	Blood	Down-regulated	[92–95]
microRNA-181b	Blood	Down-regulated	[90]
microRNA-342-5p	Blood	Down-regulated	[94,95]
microRNA-126	Blood	Down-regulated	[90]
microRNA-499-5p	Blood	Down-regulated	[90]

Moreover, numerous studies have reported a series of microRNAs that did not show significant changes regarding their expression in sepsis. Such microRNA species have been identified both in the patients' serum (microRNA-451, [95], microRNA-494 [90]), as well as in the plasma (let-7i [90]) and blood (microRNA-21, microRNA-503, microRNA-155, microRNA-486-5p, microRNA-132, microRNA-203, and microRNA-1249 [90–95]) (Table 2).

Table 2. MicroRNA expression unchanged in sepsis.

Involved MicroRNAs	Body Fluid of Identification	Expression	References
microRNA-451	Serum	Unchanged	[95]
microRNA-494	Serum	Unchanged	[90]
Let-7i	Plasma	Unchanged	[90]
microRNA-21	Blood	Unchanged	[90]
microRNA-503	Blood	Unchanged	[90]
microRNA-155	Blood	Unchanged	[91]
microRNA-486-5p	Blood	Unchanged	[95]
microRNA-132	Blood	Unchanged	[95]
microRNA-203	Blood	Unchanged	[90]
microRNA-1249	Blood	Unchanged	[90]

7. Conclusions

The complexity of the pathophysiological and molecular mechanisms in critically ill polytrauma patients is very high. They are responsible for the worsening of the patient's clinical status under certain circumstances. Understanding and preventing certain biomolecular and epigenetic mechanisms could lead to decreased molecular and cellular injury, as well as a lower overall risk for these patients. MicroRNA expression is a strong candidate for the future of intensive care because of the early diagnosis opportunity and because of its capacity to interact with certain key points of the biochemical pathways. Among these, the most widely-studied species are represented by microRNA-150, microRNA-133a, microRNA-146a, microRNA-576-5p, microRNA-4772-3p, microRNA4772-5p, and microRNA4722-5p-iso—the expression of which is highly augmented—as well as microRNA-223, microRNA-181b, and microRNA-122, which have lower levels in sepsis patients. Moreover, these microRNA species can be determined in different body fluids, such as serum, plasma, and blood, widening the range of options for the epigenetic determination of sepsis in critically ill polytrauma patients. However, until now, the epigenetic interactions in a clinical context have not been clearly reported, and further studies are necessary to identify the correct context for microRNA expression.

Author Contributions: Conceptualization, A.F.R. and D.S.; methodology, O.H.B. and M.P.; software, S.E.P.; validation, D.S., T.B., M.V.I. and A.F.R.; formal analysis, R.N.; investigation, C.M.P.; resources, T.D.; data curation, A.F.R.; writing-original draft preparation, A.F.R.; writing-review and editing, S.E.P.; visualization, D.S.; supervision, D.S.; project administration, T.B.; funding acquisition, O.H.B. and H.I.M.A.

Funding: This research received no external funding.

Acknowledgments: This article was published with the help from the "Aurel Mogoseanu" Association for Anesthesia and Intensive Care, Timisoara, Romania. The authors wish to thank the Clinical Research Network Company for their support during the completion of this paper.

Conflicts of Interest: The authors declare no conflict of interest.

References

1. Rogobete, A.F.; Sandesc, D.; Papurica, M.; Stoicescu, E.R.; Popovici, S.E.; Bratu, L.M.; Vernic, C.; Sas, A.M.; Stan, A.T.; Bedreag, O.H. The influence of metabolic imbalances and oxidative stress on the outcome of critically ill polytrauma patients: A review. *Burn Trauma* **2017**, *5*, 8. [CrossRef] [PubMed]

2. Papurica, M.; Rogobete, A.F.; Sandesc, D.; Cradigati, C.A.; Sarandan, M.; Dumache, R.; Horhat, F.G.; Bratu, L.M.; Nitu, R.; Crisan, D.C.; et al. Using the expression of damage-associated molecular pattern (DAMP) for the evaluation and monitoring of the critically Ill polytrauma patient. *Clin. Lab.* **2016**, *62*, 1829–1840. [CrossRef] [PubMed]

3. Horhat, F.G.; Rogobete, A.F.; Papurica, M.; Sandesc, D.; Tanasescu, S.; Dumitrascu, V.; Licker, M.; Nitu, R.; Cradigati, C.A.; Sarandan, M.; et al. The use of lipid peroxidation expression as a biomarker for the molecular damage in the critically Ill polytrauma patient. *Clin. Lab.* **2016**, *62*, 1601–1607. [CrossRef] [PubMed]

4. Berger, M.M.; Soguel, L.; Shenkin, A.; Revelly, J.-P.; Pinget, C.; Baines, M.; Chioléro, R.L. Influence of early antioxidant supplements on clinical evolution and organ function in critically ill cardiac surgery, major trauma, and subarachnoid hemorrhage patients. *Crit. Care* **2008**, *12*, R101. [CrossRef] [PubMed]

5. Papurica, M.; Rogobete, A.F.; Sandesc, D.; Dumache, R.; Nartita, R.; Sarandan, M.; Cradigati, A.C.; Luca, L.; Vernic, C.; Bedreag, O.H. Redox changes induced by general anesthesia in critically Ill patients with multiple traumas. *Mol. Biol. Int.* **2015**, *2015*. [CrossRef]

6. Ning, B.; Gao, L.; Liu, R.; Liu, Y.; Zhang, N.; Chen, Z. MicroRNAs in spinal cord injury: Potential roles and therapeutic implications. *Int. J. Biol. Sci.* **2014**, *10*, 997–1006. [CrossRef]

7. Dumache, R.; Ciocan, V.; Muresan, C.; Rogobete, A.F.; Enache, A. Circulating microRNAs as promising biomarkers in forensic body fluids identification. *Clin. Lab.* **2015**, *61*, 1129–1135. [CrossRef] [PubMed]

8. Saugstad, J.A. MicroRNAs as effectors of brain function with roles in ischemia and injury, neuroprotection, and neurodegeneration. *J. Cereb. Blood Flow Metab.* **2010**, *30*, 1564–1576. [CrossRef] [PubMed]

9. Papurica, M.; Rogobete, A.F.; Sandesc, D.; Cradigati, C.A.; Sarandan, M.; Crisan, D.C.; Horhat, F.G.; Boruga, O.; Dumache, R.; Nilima, K.R.; et al. The expression of nuclear transcription factor kappa B (NF-κB) in the case

of critically Ill polytrauma patients with sepsis and its interactions with microRNAs. *Biochem. Genet.* **2016**, *54*, 337–347. [CrossRef]

10. Bedreag, O.H.; Rogobete, A.F.; Sandesc, D.; Cradigati, C.A.; Sarandan, M.; Popovici, S.E.; Dumache, R.; Horhat, F.G.; Vernic, C.; Sima, L.V.; et al. Modulation of the redox expression and inflammation response in the crtically Ill polytrauma patient with thoracic injury. Statistical correlations between antioxidant therapy and clinical aspects. A retrospective single center study. *Clin. Lab.* **2016**, *62*, 1747–1759. [CrossRef]

11. Chandrasekaran, S.; Bonchev, D. Network topology analysis of post-mortem brain microarrays identifies more Alzheimer's related genes and microRNAs and points to novel routes for fighting with the disease. *PLoS ONE* **2016**, *11*, e0144052. [CrossRef] [PubMed]

12. Halvorsen, A.R.; Bjaanæs, M.; Leblanc, M.; Holm, A.M.; Rubio, L.; Peñalver, J.C.; Cervera, J.; Mojarrieta, J.C.; Bolstad, N.; Lopez-Guerrero, J.A.; et al. A unique set of 6 circulating microRNAs for early detection of non-small cell lung cancer. *Oncotarget* **2016**, *7*, 37250–37259. [CrossRef] [PubMed]

13. David, L.V.; Ercisli, M.F.; Rogobete, A.F.; Boia, E.S.; Horhat, R.; Nitu, R.; Diaconu, M.M.; Pirtea, L.; Ciuca, I.; Horhat, D.; et al. Early prediction of sepsis incidence in critically Ill patients using specific genetic polymorphisms. *Biochem. Genet.* **2017**, *55*, 193–203. [CrossRef] [PubMed]

14. Essandoh, K.; Fan, G.C. Role of extracellular and intracellular microRNAs in sepsis. *Biochim. Biophys. Acta* **2014**, *1842*, 2155–2162. [CrossRef] [PubMed]

15. MacFarlane, L.A.; Murphy, P.R. MicroRNA: Biogenesis, function and role in cancer. *Curr. Genom.* **2010**, *11*, 537–561. [CrossRef] [PubMed]

16. Bratu, L.M.; Rogobete, A.F.; Papurica, M.; Sandesc, D.; Cradigati, C.A.; Sarandan, M.; Dumache, R.; Popovici, S.E.; Crisan, D.C.; Stanca, H.; et al. Literature research regarding miRNAs' expression in the assessment and evaluation of the critically Ill polytrauma patient with traumatic brain and spinal cord injury. *Clin. Lab.* **2016**, *62*, 2019–2024. [CrossRef] [PubMed]

17. Macias, S.; Michlewski, G.; Cáceres, J.F. Hormonal regulation of microRNA biogenesis. *Mol. Cell* **2009**, *36*, 172–173. [CrossRef]

18. Bedreag, O.H.; Papurica, M.; Rogobete, A.F.; Sandesc, D.; Dumache, R.; Cradigati, C.A.; Sarandan, M.; Bratu, L.M.; Popovici, S.E.; Sima, L.V. Using circulating miRNAs as biomarkers for the evaluation and monitoring of the mitochondrial damage in the critically Ill polytrauma patients. *Clin. Lab.* **2016**, *62*, 1397–1403. [CrossRef]

19. Bedreag, O.H.; Sandesc, D.; Chiriac, S.D.; Rogobete, A.F.; Cradigati, A.C.; Sarandan, M.; Dumache, R.; Nartita, R.; Papurica, M. The use of circulating miRNAs as biomarkers for oxidative stress in critically Ill polytrauma patients. *Clin. Lab.* **2016**, *62*, 263–274. [CrossRef]

20. Bedreag, O.H.; Rogobete, A.F.; Sarandan, M.; Cradigati, A.C.; Papurica, M.; Dumbuleu, M.C.; Chira, A.M.; Rosu, O.M.; Sandesc, D. Oxidative stress in severe pulmonary trauma in critical ill patients. Antioxidant therapy in patients with multiple trauma—A review. *Anaesthesiol. Intens. Ther.* **2015**, *47*, 351–359. [CrossRef]

21. Ticlea, M.; Bratu, L.M.; Bodog, F.; Bedreag, O.H.; Rogobete, A.F.; Crainiceanu, Z.P. The use of exosomes as biomarkers for evaluating and monitoring critically Ill polytrauma patients with sepsis. *Biochem. Genet.* **2017**, *55*, 1–9. [CrossRef] [PubMed]

22. Sandesc, M.; Dinu, A.; Rogobete, A.F.; Bedreag, O.H.; Sandesc, D.; Papurica, M.; Bratu, L.M.; Negoita, S.; Vernic, C.; Popovici, S.E.; et al. Circulating microRNAs expressions as genetic biomarkers in pancreatic cancer patients continuous non-invasive monitoring. *Clin. Lab.* **2017**, *63*, 1561–1566. [CrossRef] [PubMed]

23. Negoita, S.I.; Sandesc, D.; Rogobete, A.F.; Dutu, M.; Bedreag, O.H.; Papurica, M.; Ercisli, M.F.; Popovici, S.E.; Dumache, R.; Sandesc, M.; et al. MiRNAs expressions and interaction with biological systems in patients with Alzheimer's disease. Using miRNAs as a diagnosis and prognosis biomarker. *Clin. Lab.* **2017**, *63*, 1315–1321. [CrossRef] [PubMed]

24. Etheridge, A.; Lee, I.; Hood, L.; Galas, D.; Wang, K. Extracellular microRNA: A new source of biomarkers. *Mutat. Res. Fundam. Mol. Mech. Mutagen.* **2011**, *717*, 85–90. [CrossRef] [PubMed]

25. Guerra, R.C.; Zuñiga-muñoz, A.; Lans, V.G.; Díaz-díaz, E.; Alberto, C.; Betancourt, T.; Pérez-torres, I. Modulation of the activities of catalase, Cu-Zn, Mn superoxide dismutase, and glutathione peroxidase in adipocyte from ovariectomised female rats with metabolic syndrome. *Int. J. Endocrinol.* **2014**, *2014*. [CrossRef] [PubMed]

26. Arcaroli, J.J.; Hokanson, J.E.; Abraham, E.; Geraci, M.; Murphy, J.R.; Bowler, R.P.; Dinarello, C.A.; Silveira, L.; Sankoff, J.; Heyland, D.; et al. Extracellular superoxide dismutase haplotypes are associated with acute lung injury and mortality. *Am. J. Respir. Crit. Care Med.* **2009**, *179*, 105–112. [CrossRef] [PubMed]

27. Horhat, F.G.; Gundogdu, F.; David, L.V.; Boia, E.S.; Pirtea, L.; Horhat, R.; Cucui-Cozma, A.; Ciuca, I.; Diaconu, M.; Nitu, R.; et al. Early evaluation and monitoring of critical patients with acute respiratory distress syndrome (ARDS) using specific genetic polymorphisms. *Biochem. Genet.* **2017**, *55*, 204–211. [CrossRef] [PubMed]

28. Schmitt, B.; Vicenzi, M.; Garrel, C.; Denis, F.M. Redox biology effects of N-acetylcysteine, oral glutathione (GSH) and a novel sublingual form of GSH on oxidative stress markers: A comparative crossover study. *Redox Biol.* **2015**, *6*, 198–205. [CrossRef]

29. Hwang, J.H.; Ryu, J.; An, J.N.; Kim, C.T.; Kim, H.; Yang, J.; Ha, J.; Chae, D.W.; Ahn, C.; Jung, I.M.; et al. Pretransplant malnutrition, inflammation, and atherosclerosis affect cardiovascular outcomes after kidney transplantation. *BMC Nephrol.* **2015**, *16*, 109. [CrossRef]

30. Yin, G.; Wang, Y.; Cen, X.; Yang, M.; Liang, Y.; Xie, Q. Lipid peroxidation-mediated inflammation promotes cell apoptosis through activation of NF-kB pathway in rheumatoid arthritis synovial cells. *Mediat. Inflamm.* **2015**, *2015*, 460310. [CrossRef]

31. Sies, H. Oxidative stress: A concept in redox biology and medicine. *Redox Biol.* **2015**, *4*, 180–183. [CrossRef] [PubMed]

32. Tavladaki, T.; Spanaki, A.; Dimitriou, H.; Kozlov, A.; Duvigneau, J.; Weidinger, A.; Kondili, E.; Georgopoulos, D.; Briassoulis, G. Bioenergetics and metabolic patterns in early onset severe sepsis or trauma. *Intensiv. Care Med. Exp.* **2015**, *3*, A43. [CrossRef]

33. Denk, S.; Perl, M.; Huber-Lang, M. Damage- and pathogen-associated molecular patterns and alarmins: Keys to sepsis? *Eur. Surg. Res.* **2012**, *48*, 171–179. [CrossRef] [PubMed]

34. Soud, D.E.M.; Amin, O.A.I.; Amin, A.A.I. New era "soluble triggering receptor expressed on myeloid cells-I" as a marker for early detection of infection in trauma patients. *Egypt. J. Anaesth.* **2011**, *27*, 267–272. [CrossRef]

35. Horton, J.W. Free radicals and lipid peroxidation mediated injury in burn trauma: The role of antioxidant therapy. *Toxicology* **2003**, *189*, 75–88. [CrossRef]

36. Van der Kuip, M.; De Meer, K.; Oosterveld, M.J.S.; Lafeber, H.N.; Gemke, R.J.B.J. Simple and accurate assessment of energy expenditure in ventilated paediatric intensive care patients. *Clin. Nutr.* **2004**, *23*, 657–663. [CrossRef]

37. Andruszkow, H.; Fischer, J.; Sasse, M.; Brunnemer, U.; Andruszkow, J.H.K.; Gänsslen, A.; Hildebrand, F.; Frink, M. Interleukin-6 as inflammatory marker referring to multiple organ dysfunction syndrome in severely injured children. *Scand. J. Trauma Resusc. Emerg. Med.* **2014**, *22*, 16. [CrossRef]

38. Gu, W.; Jiang, J. Genetic polymorphisms and posttraumatic complications. *Comp. Funct. Genom.* **2010**, *2010*. [CrossRef]

39. Burkhardt, M.; Nienaber, U.; Pizanis, A.; Maegele, M.; Culemann, U.; Bouillon, B.; Flohé, S.; Pohlemann, T.; Paffrath, T. Acute management and outcome of multiple trauma patients with pelvic disruptions. *Crit. Care* **2012**, *16*, R163. [CrossRef]

40. Gao, J.; Zeng, L.; Zhang, A.; Wang, X.; Pan, W.; Du, D. Identification of haplotype tag single-nucleotide polymorphisms within the PPAR family genes and their clinical relevance in patients with major trauma. *Int. J. Environ. Res. Public Health* **2016**, *13*, 374. [CrossRef]

41. Zhao, C.; Sun, G.Q.; Ye, P.; Li, S.; Shia, Y. MicroRNA let-7d regulates the TLX/microRNA-9 cascade to control neural cell fate and neurogenesis. *Sci. Rep.* **2013**, *3*, 1329. [CrossRef] [PubMed]

42. Ramachandran, S.; Coffin, S.L.; Tang, T.Y.; Jobaliya, C.D.; Spengler, R.M.; Davidson, B.L. Cis-acting single nucleotide polymorphisms alter microRNA-mediated regulation of human brain-expressed transcripts. *Hum. Mol. Genet.* **2016**, *25*, 4939–4950. [CrossRef]

43. Yuva-Aydemir, Y.; Simkin, A.; Gascon, E.; Gao, F.B. MicroRNA-9—Functional evolution of a conserved small regulatory RNA. *RNA Biol.* **2011**, *8*, 557–564. [CrossRef] [PubMed]

44. Klinge, C.M. Estrogen regulation of microRNA expression. *Curr. Genom.* **2009**, *10*, 169–183. [CrossRef] [PubMed]

45. Fibach, E.; Dana, M. Oxidative stress in paroxysmal nocturnal hemoglobinuria and other conditions of complement-mediated hemolysis. *Free Radic. Biol. Med.* **2015**, *88*, 63–69. [CrossRef]

46. Abraham, E. Nuclear factor—kB and its role in sepsis-associated organ failure. *J. Infect. Dis.* **2003**, *187*, S364–S369. [CrossRef]

47. Bernardes, M.; Vieira, T.S.; Martins, M.J.; Lucas, R.; Costa, L.; Pereira, J.G.; Ventura, F.; Martins, E. Myocardial perfusion in rheumatoid arthritis patients: Associations with traditional risk factors and novel biomarkers. *Biomed. Res. Int.* **2017**, *2017*. [CrossRef]

48. Salsano, E.; Rizzo, A.; Bedini, G.; Bernard, L.; Olio, V.D.; Volorio, S.; Lazzaroni, M.; Ceccherini, I.; Lazarevic, D.; Cittaro, D.; et al. An autoinflammatory neurological disease due to interleukin 6 hypersecretion. *J. Neuroinflamm.* **2013**, *10*, 802. [CrossRef]

49. Treszl, A.; Kocsis, I.; Szathmári, M.; Schuler, Á.; Héninger, E.; Tulassay, T.; Vásárhelyi, B. Genetic variants of TNF-α, IL-1β, IL-4 receptor α-Chain, IL-6 and IL-10 genes are not risk factors for sepsis in low-birth-weight infants. *Neonatology* **2003**, *83*, 241–245. [CrossRef]

50. Harding, D.; Dhamrait, S.; Millar, A.; Humphries, S.; Marlow, N.; Whitelaw, A.; Montgomery, H. Is interleukin-6–174 genotype associated with the development of septicemia in preterm infants? *Pediatrics* **2003**, *112*, 800–803. [CrossRef]

51. Chuang, T.-Y.; Chang, H.-T.; Chung, K.-P.; Cheng, H.-S.; Liu, C.-Y.; Liu, Y.-C.; Huang, H.-H.; Chou, T.-C.; Chang, B.-L.; Lee, M.-R.; et al. High levels of serum macrophage migration inhibitory factor and interleukin 10 are associated with a rapidly fatal outcome in patients with severe sepsis. *Int. J. Infect. Dis.* **2014**, *20*, 13–17. [CrossRef] [PubMed]

52. Bedreag, O.H.; Rogobete, A.F.; Cradigati, C.A.; Sarandan, M.; Nartita, R.; Horhat, F.G.; Popovici, S.E.; Sandesc, D.; Papurica, M. A novel evaluation of microvascular damage in critically ill polytrauma patients by using circulating micrornas. *Rev. Rom. Med. Lab.* **2016**, *24*, 21–30. [CrossRef]

53. Scott, E.; Loya, K.; Mountford, J.; Milligan, G.; Baker, A.H. MicroRNA regulation of endothelial homeostasis and commitment—Implications for vascular regeneration strategies using stem cell therapies. *Free Radic. Biol. Med.* **2013**, *64*, 52–60. [CrossRef] [PubMed]

54. Hulsmans, M.; Holvoet, P. MicroRNA-containing microvesicles regulating inflammation in association with atherosclerotic disease. *Cardiovasc. Res.* **2013**, *100*, 7–18. [CrossRef] [PubMed]

55. Kung, C.; Hsiao, S.; Tsai, T.; Su, C.; Chang, W.; Huang, C.; Wang, H.; Lin, W.; Chang, H.; Lin, Y.; et al. Plasma nuclear and mitochondrial DNA levels as predictors of outcome in severe sepsis patients in the emergency room. *J. Transl. Med.* **2012**, *10*, 130. [CrossRef] [PubMed]

56. Haas, C. Lung protective mechanical ventilation in acute respiratory distress syndrome. *Respir. Care Clin.* **2003**, *9*, 363–396. [CrossRef]

57. Haas, S.; Trepte, C. Prediction of volume responsiveness using pleth variability index in patients undergoing cardiac surgery after cardiopulmonary bypass. *J. Anesth.* **2012**, *26*, 696–701. [CrossRef] [PubMed]

58. Wada, T.; Jesmin, S.; Gando, S.; Yanagida, Y.; Mizugaki, A.; Sultana, S.N.; Zaedi, S. The role of angiogenic factors and their soluble receptors in acute lung injury (ALI)/acute respiratory distress syndrome (ARDS) associated with critical illness. *J. Inflamm.* **2013**, *10*, 6. [CrossRef]

59. Medford, A.R.L.; Godinho, S.I.H.; Keen, L.J. Fluid vascular endothelial growth factor protein levels in patients with and at risk for ARDS. *Chest* **2009**, *136*, 457–464. [CrossRef]

60. Abadie, Y.; Bregeon, F.; Papazian, L.; Lange, F.; Thomas, P.; Duvaldestin, P.; Adnot, S. Decreased VEGF concentration in lung tissue and vascular injury during ARDS. *Eur. Respir. J.* **2005**, *25*, 139–146. [CrossRef]

61. Kondo, T.; Hattori, N.; Ishikawa, N.; Murai, H.; Haruta, Y.; Hirohashi, N.; Tanigawa, K.; Kohno, N. KL-6 concentration in pulmonary epithelial lining fluid is a useful prognostic indicator in patients with acute respiratory distress syndrome. *Respir. Res.* **2011**, *12*, 32. [CrossRef] [PubMed]

62. Aggarwal, N.R.; Alessio, F.R.D.; Tsushima, K.; Sidhaye, V.K.; Cheadle, C.; Grigoryev, D.N.; Barnes, K.C.; King, L.S. Regulatory T cell-mediated resolution of lung injury: Identification of potential target genes via expression profiling. *Physiol. Genom.* **2010**, *41*, 109–119. [CrossRef] [PubMed]

63. Pena, O.M.; Hancock, D.G.; Lyle, N.H.; Linder, A.; Russell, J.A.; Xia, J.; Fjell, C.D.; Boyd, J.H.; Hancock, R.E.W. An endotoxin tolerance signature predicts sepsis and organ dysfunction at initial clinical presentation. *EBioMedicine* **2014**, *1*, 64–71. [CrossRef] [PubMed]

64. McClure, C.; Brudecki, L.; Ferguson, D.A.; Yao, Z.Q.; Moorman, J.P.; McCall, C.E.; Gazzar, M. El microRNA 21 (miR-21) and miR-181b couple with NFI-A to generate myeloid-derived suppressor cells and promote immunosuppression in late sepsis. *Infect. Immun.* **2014**, *82*, 3816–3825. [CrossRef] [PubMed]

65. Belopolskaya, O.B.; Smelaya, T.V. Clinical associations of host genetic variations in the genes of cytokines in critically ill patients. *Clin. Exp. Immunol.* **2015**, *180*, 531–541. [CrossRef] [PubMed]

66. Jacobi, J. Pathophysiology of sepsis. *Am. J. Heal Pharm.* **2002**, *59*, 1435–1444. [CrossRef]

67. Puskarich, M.; Shapiro, N.; Trzeciak, S. Plasma levels of mitochondrial DNA in patients presenting to the emergency department with sepsis. *Shock* **2012**, *38*, 337–340. [CrossRef] [PubMed]

68. Fejes, Z.; Szilágyi, B.; Kappelmayer, J.; Ifj Nagy, B. Alteration in the expression of platelet microRNAs in diseases with abnormal platelet activation. *Orv. Hetil.* **2018**, *159*, 1962–1970. [CrossRef]

69. Hassan, F.I.; Didari, T.; Khan, F.; Mojtahedzadeh, M.; Abdollahi, M. The role of epigenetic alterations involved in sepsis: An overview. *Curr. Pharm. Des.* **2018**, *24*, 2862–2869. [CrossRef]

70. Ehrnthaller, C.; Flierl, M.; Perl, M.; Denk, S.; Unnewehr, H.; Ward, P.A.; Radermacher, P.; Ignatius, A.; Gebhard, F.; Chinnaiyan, A.; et al. The molecular fingerprint of lung inflammation after blunt chest trauma. *Eur. J. Med. Res.* **2015**, *20*, 70. [CrossRef]

71. Zhai, R.; Gong, M.N.; Zhou, W.; Thompson, T.B.; Kraft, P.; Su, L.; Christiani, D.C. Genotypes and haplotypes of the VEGF gene are associated with higher mortality and lower VEGF plasma levels in patients with ARDS. *Thorax* **2007**, *62*, 718–722. [CrossRef] [PubMed]

72. Medford, A.R.; Millar, A.B. Vascular endothelial growth factor (VEGF) in acute lung injury (ALI) and acute respiratory distress syndrome (ARDS): Paradox or paradigm? *Thorax* **2006**, *61*, 621–626. [CrossRef]

73. Duan, Q.; Chen, C.; Yang, L.; Li, N.; Gong, W.; Li, S.; Wang, D.W. MicroRNA regulation of unfolded protein response transcription factor XBP1 in the progression of cardiac hypertrophy and heart failure in vivo. *J. Transl. Med.* **2015**, *13*, 363. [CrossRef] [PubMed]

74. Blass, S.C.; Goost, H.; Tolba, R.H.; Stoffel-Wagner, B.; Kabir, K.; Burger, C.; Stehle, P.; Ellinger, S. Time to wound closure in trauma patients with disorders in wound healing is shortened by supplements containing antioxidant micronutrients and glutamine: A. PRCT. *Clin. Nutr.* **2012**, *31*, 469–475. [CrossRef] [PubMed]

75. Yehya, N.; Yerrapureddy, A.; Tobias, J.; Margulies, S.S. MicroRNA modulate alveolar epithelial response to cyclic stretch. *BMC Genom.* **2012**, *13*, 154. [CrossRef]

76. Kulshreshtha, R.; Ferracin, M.; Wojcik, S.E.; Garzon, R.; Alder, H.; Agosto-Perez, F.J.; Davuluri, R.; Liu, C.-G.; Croce, C.M.; Negrini, M.; et al. A microRNA signature of hypoxia. *Mol. Cell Biol.* **2007**, *27*, 1859–1867. [CrossRef]

77. Moschos, S.A.; Williams, A.E.; Perry, M.M.; Birrell, M.A.; Belvisi, M.G.; Lindsay, M.A. Expression profiling in vivo demonstrates rapid changes in lung microRNA levels following lipopolysaccharide-induced inflammation but not in the anti-inflammatory action of glucocorticoids. *BMC Genom.* **2007**, *8*, 240. [CrossRef]

78. Tacke, F.; Roderburg, C.; Benz, F.; Cardenas, D.V.; Luedde, M.; Hippe, H.-J.; Frey, N.; Vucur, M.; Gautheron, J.; Koch, A.; et al. Levels of circulating miR-133a are elevated in sepsis and predict mortality in critically Ill patients. *Crit. Care Med.* **2014**, *42*, 1096–1104. [CrossRef]

79. Wang, X.; Huang, W.; Yang, Y.; Wang, Y.; Peng, T.; Chang, J.; Caldwell, C.C.; Zingarelli, B.; Fan, G. Biochimica et biophysica acta loss of duplexmiR-223 (5p and 3p) aggravates myocardial depression and mortality in polymicrobial sepsis. *BBA Mol. Basis Dis.* **2014**, *1842*, 701–711. [CrossRef]

80. Shao, Y.; Li, J.; Cai, Y.; Xie, Y.; Ma, G.; Li, Y.; Chen, Y.; Liu, G.; Zhao, B.; Cui, L.; et al. The functional polymorphisms of miR-146a are associated with susceptibility to severe sepsis in the Chinese population. *Mediat. Inflamm.* **2014**, *2014*. [CrossRef]

81. Søndergaard, E.S.; Alamili, M.; Coskun, M.; Gögenur, I. MicroRNA's are novel biomarkers in sepsis—A systematic review. *Trends Anaesth. Crit. Care* **2015**, *5*, 151–156. [CrossRef]

82. Roderburg, C.; Luedde, M.; Vargas Cardenas, D.; Vucur, M.; Scholten, D.; Frey, N.; Koch, A.; Trautwein, C.; Tacke, F.; Luedde, T. Circulating microRNA-150 serum levels predict survival in patients with critical illness and sepsis. *PLoS ONE* **2013**, *8*, e54612. [CrossRef]

83. La Rosa, A.H.; Acker, M.; Swain, S.; Manoharan, M. The role of epigenetics in kidney malignancies. *Cent European. J. Urol.* **2015**, *68*, 157–164. [CrossRef]

84. Xu, S.; Zhang, R.; Niu, J.; Cui, D.; Xie, B.; Zhang, B. Oxidative stress mediated-alterations of the microRNA expression profile in mouse hippocampal neurons. *Int. J. Mol. Sci.* **2012**, *13*, 16945–16960. [CrossRef] [PubMed]

85. Yao, L.; Liu, Z.; Zhu, J.; Li, B.; Chai, C.; Tian, Y. Clinical evaluation of circulating microRNA-25 level change in sepsis and its potential relationship with oxidative stress. *Int. J. Clin. Exp. Pathol.* **2015**, *8*, 7675–7684. [PubMed]

86. Rodrigues, C.E.; Capcha, J.M.C.; De Bragança, A.C.; Sanches, T.R.; Gouveia, P.Q.; De Oliveira, P.A.F.; Malheiros, D.M.A.C.; Volpini, R.A.; Santinho, M.A.R.; Santana, B.A.A.; et al. Human umbilical cord-derived mesenchymal stromal cells protect against premature renal senescence resulting from oxidative stress in rats with acute kidney injury. *Stem Cell Res. Ther.* **2017**, *8*, 19. [CrossRef] [PubMed]

87. Fredriksson, K.; Tjäder, I.; Keller, P.; Petrovic, N.; Ahlman, B.; Schéele, C.; Wernerman, J.; Timmons, J.A.; Rooyackers, O. Dysregulation of mitochondrial dynamics and the muscle transcriptome in ICU patients suffering from sepsis induced multiple organ failure. *PLoS ONE* **2008**, *11*, e3686. [CrossRef]

88. Huang, C.; Xiao, X.; Chintagari, N.R.; Breshears, M.; Wang, Y.; Liu, L. MicroRNA and mRNA expression profiling in rat acute respiratory distress syndrome. *BMC Med. Genom.* **2014**, *7*, 46. [CrossRef]

89. Bedreag, O.H.; Rogobete, A.F.; Dumache, R.; Sarandan, M.; Cradigati, A.C.; Papurica, M.; Craciunescu, M.C.; Popa, D.M.; Luca, L.; Nartita, R.; et al. Use of circulating microRNAs as biomarkers in critically ill polytrauma patients. *Biomarkers Genom. Med.* **2015**, *7*, 131–138. [CrossRef]

90. Benz, F.; Roy, S.; Traukwein, C.; Roderburg, C.; Luedde, T. Circulating microRNAs as biomarkers for sepsis. *Int. J. Mol. Sci.* **2016**, *17*, 78. [CrossRef]

91. Huang, J.; Sun, Z.; Yan, W.; Zhu, Y.; Lin, Y.; Chen, J.; Shen, B.; Wang, J. Identfifcation of microRNAs as sepsis biomarkers based on MiRNA regulatory network analysis. *Biomed. Res. Int.* **2014**, *2014*, 594350. [PubMed]

92. Dumache, R.; Rogobete, A.F.; Bedreag, O.H.; Sarandan, M.; Cradigati, A.C.; Papurica, M.; Dumbuleu, C.M.; Nartita, R.; Sandesc, D. Use of miRNAs as biomarkers in sepsis. *Anal. Cell. Pathol.* **2015**, *2015*, 186716. [CrossRef]

93. Sun, Z.; Zhang, Q.; Cui, X.; Yang, J.; Zhang, B.; Song, G. Differential expression of miRNA and its role in sepsis. *Pediatrics* **2018**, *142*. [CrossRef]

94. Ansari, M.; Gupta, P. Nex-Gen biomarkers—A genetic model of sepsis. *Biomark. J.* **2016**, *2*, 8. [CrossRef]

95. Vasilescu, C.; Dragomir, M.; Tanase, M.; Giza, D.; Purnichescu-Purtan, R.; Chen, M.; Yeung, S.C.-J.; Calin, G.A. Circulating miRNAs in sepsis—A network under attack: An in silico prediction of the potential existance of microRNAs sponges in sepsis. *PLoS ONE* **2017**, *12*, e01833334. [CrossRef] [PubMed]

96. Puskarich, M.A.; Nandi, U.; Shapiro, N.I.; Trzeciak, S.; Kline, J.A.; Jones, A.E. Detection of microRNAs in patients with sepsis. *J. Acute Dis.* **2015**, *4*, 101–106. [CrossRef]

97. Weber, J.A.; Baxter, D.H.; Zhang, S.; Huang, D.Y.; Huang, K.H.; Lee, M.J.; Galas, D.J.; Wang, K. The microRNA spectrum in 12 body fluids. *Clin. Chem.* **2010**, *56*, 1733–1741. [CrossRef]

98. Díaz-Prado, S.; Cicione, C.; Muiños-López, E.; Hermida-Gómez, T.; Oreiro, N.; Fernández-López, C.; Blanco, F.J. Characterization of microRNA expression profiles in normal and osteoarthritic human chondrocytes. *BMC Musculoskelet. Disord.* **2012**, *13*, 144. [CrossRef]

99. Campomenosi, P.; Gini, E.; Noonan, D.M.; Poli, A.; D'Antona, P.; Rotolo, N.; Dominioni, L.; Imperatori, A. A comparison between quantitative PCR and droplet digital PCR technologies for circulating microRNA quantification in human lung cancer. *BMC Biotechnol.* **2016**, *16*, 60. [CrossRef]

100. Warburton, A.; Savage, A.L.; Myers, P.; Peeney, D.; Bubb, V.J.; Quinn, J.P. Molecular signatures of mood stabilisers highlight the role of the transcription factor REST/NRSF. *J. Affect. Disord.* **2014**, *172*, 63–73. [CrossRef]

101. Inchley, C.S.; Sonerud, T.; Fjærli, H.O.; Nakstad, B. Nasal mucosal microRNA expression in children with respiratory syncytial virus infection. *BMC Infect. Dis.* **2015**, *15*, 150. [CrossRef] [PubMed]

102. Dong, J.; Liu, Y.; Liao, W.; Liu, R.; Shi, P.; Wang, L. MiRNA-223 is a potential diagnostic and prognostic marker for osteosarcoma. *J. Bone Oncol.* **2016**, *5*, 74–79. [CrossRef] [PubMed]

103. Hu, Z.; Chen, X.; Zhao, Y.; Tian, T.; Jin, G.; Shu, Y.; Chen, Y.; Xu, L.; Zen, K.; Zhang, C.; et al. Serum microRNA signatures identified in a genome-wide serum microRNA expression profiling predict survival of non-small-cell lung cancer. *J. Clin. Oncol.* **2010**, *28*, 1721–1726. [CrossRef] [PubMed]

104. Pandey, A.C.; Semon, J.A.; Kaushal, D.; O'Sullivan, R.P.; Glowacki, J.; Gimble, J.M.; Bunnell, B.A. MicroRNA profiling reveals age-dependent differential expression of nuclear factor kappaB and mitogen-activated protein kinase in adipose and bone marrow-derived human mesenchymal stem cells 928. *Stem Cell Res. Ther.* **2011**, *2*, 49. [CrossRef] [PubMed]

105. Haider, B.A.; Baras, A.S.; McCall, M.N.; Hertel, J.A.; Cornish, T.C.; Halushka, M.K. A critical evaluation of microRNA biomarkers in non-neoplastic disease. *PLoS ONE* **2014**, *9*. [CrossRef] [PubMed]

106. Zheng, D.; Yu, Y.; Li, M.; Wang, G.; Chen, R.; Fan, G.C.; Martin, C.; Xiong, S.; Peng, T. Inhibition of microRNA 195 prevents apoptosis and multiple-organ injury in mouse models of sepsis. *J. Infect. Dis.* **2016**, *213*, 1661–1670. [CrossRef] [PubMed]

107. Fang, C.; Zhao, J.; Li, J.; Qian, J.; Liu, X.; Sun, Q.; Liu, W.; Tian, Y.; Ji, A.; Wu, H.; et al. Massively parallel sequencing of microRNA in bloodstains and evaluation of environmental influences on miRNA candidates using realtime polymerase chain reaction. *Forensic. Sci. Int. Genet.* **2018**, *38*, 32–38. [CrossRef]

108. Tian, F.; Yuan, C.; Hu, L.; Shan, S. MicroRNA-93 inhibits inflammatory responses and cell apoptosis after cerebral ischemia reperfusion by targeting interleukin-1 receptor-associated kinase 4. *Exp. Ther. Med.* **2017**, *14*, 2903–2910. [CrossRef]

109. Qiu, J.; Zhou, X.; Zhou, X.; Cheng, R.; Liu, H.; Li, Y. Neuroprotective effects of microRNA-210 on hypoxic-ischemic encephalopathy. *BioMed Res. Int.* **2013**, *2013*, 350419. [CrossRef]

110. Hara, N.; Kikuchi, M.; Miyashita, A.; Hatsuta, H.; Saito, Y.; Kasuga, K.; Murayama, S.; Ikeuchi, T.; Kuwano, R. Serum microRNA miR-501-3p as a potential biomarker related to the progression of Alzheimer's disease. *Acta Neuropathol. Commun.* **2017**, *5*, 10. [CrossRef]

111. Xiao, B.; Liu, H.; Gu, Z. Expression of microRNA-133 inhibits epithelial—Mesenchymal transition in lung cancer cells by directly targeting FOXQ1. *Arch. Bronconeumol.* **2016**, *52*, 505–511. [CrossRef] [PubMed]

112. Liz, J.; Esteller, M. lncRNAs and microRNAs with a role in cancer development. *Biochim. Biophys. Acta Gene Regul. Mech.* **2016**, *1859*, 169–176. [CrossRef] [PubMed]

113. Bera, A.; Ghosh-Choudhury, N.; Dey, N.; Das, F.; Kasinath, B.S.; Abboud, H.E.; Ghosh, G. NF-κB-mediated cyclin D1 expression by microRNA-21 in fluences renal cancer cell proliferation. *Cell Signal.* **2013**, *25*, 2575–2586. [CrossRef] [PubMed]

114. Huang, J.; Liu, J.; Chen-Xiao, K.; Zhang, X.; Lee, W.N.P.; Go, V.L.W.; Xiao, G.G. Advance in microRNA as a potential biomarker for early detection of pancreatic cancer. *Biomark. Res.* **2016**, *4*, 20. [CrossRef] [PubMed]

115. Li, S.; Liang, Z.; Xu, L.; Zou, F. MicroRNA-21: A ubiquitously expressed pro-survival factor in cancer and other diseases. *Mol. Cell. Biochem.* **2012**, *360*, 147–158. [CrossRef] [PubMed]

116. Zhou, X.; Su, S.; Li, S.; Pang, X.; Chen, C.; Li, J. MicroRNA-146a down-regulation correlates with neuroprotection and targets pro-apoptotic genes in cerebral ischemic injury in vitro. *Brain Res.* **2016**, *1648*, 136–143. [CrossRef] [PubMed]

117. Dumache, R.; Ciocan, V.; Muresan, C.; Enache, A. Molecular DNA analysis in forensic identification. *Clin. Lab.* **2016**, *62*, 245–248. [CrossRef] [PubMed]

cells

MDPI

Brief Report

Substantial Dysregulation of miRNA Passenger Strands Underlies the Vascular Response to Injury

Karine Pinel [†,*], Louise A. Diver, Katie White, Robert A. McDonald and Andrew H. Baker [‡,*]

Institute of Cardiovascular and Medical Sciences, BHF Glasgow Cardiovascular Research Centre, University of Glasgow, Glasgow G12 8TA, UK; drlouisediver@gmail.com (L.A.D.); kt_m_white@hotmail.com (K.W.); robertamcdonald8@gmail.com (R.A.M.)

* Correspondence: pinel.karine.m@gmail.com (K.P.); Andy.Baker@ed.ac.uk (A.H.B.), Tel.: +33-559515974 (K.P.); +44-0131-242-6728 (A.H.B.)

† Current address: INRA, UMR 1419 NuMeA (Nutrition Metabolism Aquaculture), Aquapôle, 64310 St-Pee-sur-Nivelle, France.

‡ Current address: BHF/University Centre for Cardiovascular Science, University of Edinburgh, Edinburgh EH16 4UU, Scotland, UK.

Received: 19 December 2018; Accepted: 21 January 2019; Published: 23 January 2019

Abstract: Vascular smooth muscle cell (VSMC) dedifferentiation is a common feature of vascular disorders leading to pro-migratory and proliferative phenotypes, a process induced through growth factor and cytokine signaling cascades. Recently, many studies have demonstrated that small non-coding RNAs (miRNAs) can induce phenotypic effects on VSMCs in response to vessel injury. However, most studies have focused on the contribution of individual miRNAs. Our study aimed to conduct a detailed and unbiased analysis of both guide and passenger miRNA expression in vascular cells in vitro and disease models in vivo. We analyzed 100 miRNA stem loops by TaqMan Low Density Array (TLDA) from primary VSMCs in vitro. Intriguingly, we found that a larger proportion of the passenger strands was significantly dysregulated compared to the guide strands after exposure to pathological stimuli, such as platelet-derived growth factor (PDGF) and IL-1α. Similar findings were observed in response to injury in porcine vein grafts and stent models in vivo. In these studies, we reveal that the miRNA passenger strands are predominantly dysregulated in response to vascular injury.

Keywords: miRNA expression and regulation; passenger miRNA; biomarker; vascular injury; smooth muscle cells; porcine vein graft and stent models

1. Introduction

MicroRNAs (miRNAs) are a class of non-coding RNAs known to play a prominent role in gene regulation at a post-transcriptional level [1–3]. Pri-miRNAs are transcribed by RNA polymerase II and are processed successively by two RNase III enzymes, Drosha and Dicer, to a ~22 nt miRNA:miRNA * duplex [4], which, on processing, generates two single-stranded RNA molecules. miRNAs are loaded onto Argonaute to produce a RNA-induced silencing complex (RISC), which exerts translational repression via an imperfect binding to the mRNA target, generally localized in the 3′UTR. An equal amount of the two strands from the miRNA:miRNA * duplex is produced by the transcription but their accumulation becomes asymmetric [5]. Current miRNA nomenclature provides information on the direction of the mature miRNA strand using the -3p or -5p suffixes and, normally, the miRNA guide strand is thought to be the strand that preferentially accumulates in the RISC and the miRNA * or passenger strand (its partner) is degraded [6]. However, recent studies suggest that these passenger strands can accumulate in a number of disease pathologies and mediate strand-specific roles based on their distinct seed sequences and targets. Indeed, the mature sequences of the two strand forms can regulate distinct sets of mRNA transcripts or common targets through different hybridization sites on the mRNA targeted.

miRNAs have been shown to induce phenotypic effects on smooth muscle cells (SMCs) in vascular disorders. Vascular SMC (VSMCs) within the vessel wall perform both contractile and proliferative functions in response to diverse cellular stimuli [7]. Under normal physiological conditions, VSMCs are relatively quiescent, harboring a contractile phenotype responding to changes in vascular flow to mediate relaxation and contraction. However, vascular injury or insult results in the initiation of phenotypic switching to a dedifferentiated phenotype. Indeed, the transition of VSMCs from a differentiated to a dedifferentiated phenotype is a common feature of vascular disorders where VSMCs harbor pro-proliferative and pro-migratory status. The VSMC phenotype can be modulated by many environmental cues and triggers, including a number of cytokines and growth factors released in response to injury. Indeed, platelet-derived growth factor (PDGF) promotes VSMC proliferation and migration resulting in neointimal formation after artery injury [8,9].

Since there is some recent evidence of the importance of the miRNA passenger strands [10–12], their individual functions have been studied in settings such as vascular cells with miR-126-5p and miR-10A*, [13,14] and cancer [15,16]. However, a more detailed or global evaluation of the importance of the miRNA passenger strands in vascular remodeling is still lacking. As miRNAs emerge as potential prognostic biomarkers and as attractive targets for disease intervention, we utilized array technology to evaluate the global expression of miRNA guide and passenger strands from hairpin precursors in pre-clinical porcine models of vascular injury and human VSMCs.

2. Materials and Methods

2.1. Animals and Tissues

All the procedures for surgery, for the pig vein graft model and the stent model, were performed in reference [17] and [18], respectively, and were performed in accordance with the UK Home Office Guidance on the Operation of the Animals (Scientific Procedures) Act 1986 and with the institutional ethical approval (PPL60/4114 and PPL 60/4429). Male white Landrace pigs (SAC Commercial Ltd, Edinburgh, UK) were maintained on a 12 h light/dark cycle with free access to food and water at a designated Biological Procedures Unit. Briefly, the animals (weighing 19–26 kg) were pre-dosed orally with aspirin and clopidogrel 24 h prior to surgery. The animals were sedated by intramuscular (IM) injection of Tiletamine/Zolazepam (Zoletil; Virbac, France); general anesthesia was conducted using intravenous (IV) Propofol (Rapinovet; Schering-Plough, Welwyn Garden City, UK) and maintained with isoflurane (Abbott Laboratories Ltd, Berkshire, UK). Animals received 100 IU/kg of IV heparin (Leo Laboratories, London, UK) before surgery. After vascular access, coronary angiography was performed prior to the deployment of either bare metal stents (BMS) (Gazelle™, Biosensors, Morges, Switzerland) or drug eluting stents (DES) (biolimus A9 eluting stents; Biomatrix Flex™, Biosensors, Hägglingen, Switzerland). The unstented animals were used as controls. The pigs received 0.15 mg of buprenorphine IM (Vetergesic: Alstore Ltd, Sheriff Hutton, York, UK) and 350 mg of ampicillin IM (Amfipen LA, Intervet, UK) immediately following the procedure and were euthanized after 7 or 28 days by an intravenous overdose of euthatal, and their coronary vessels carefully removed. For the porcine vein graft surgery, following the induction with Ketaset (100 mg/mL ketamine hydrochloride), the animals were intubated and anesthetized with halothane and allowed to spontaneously ventilate. The method for the saphenous vein-carotid interpositional graft model has been described previously [19]. The saphenous vein was surgically exposed and harvested from the hind leg by a 'no-touch' technique. The vein was cannulated and gently irrigated with heparinized iso-osmotic sodium chloride (NaCl) solution (0.9 g/L). Each carotid was exposed via a longitudinal neck incision. The animal was heparinized by the IV administration of 100 IU/kg of heparin (CP Parmaceuticals Ltd, Wrexham, UK). A 4–5 cm segment of the carotid artery was isolated between the vascular clamps and 10 mm was then removed. The residual ends were beveled at 45°

and a small longitudinal incision was made to lengthen the anastomotic area. The ends of the vein were similarly beveled and anastomosed as an interposition graft under optical magnification using continuous 7/0 Surgipro (auto Suture, Dagford, UK) sutures. The saphenous veins from the ungrafted animals were used as controls. The animals were recovered, returned to their pen and maintained on a normal chow diet for the duration of the experiment.

Surplus human saphenous vein tissue was obtained with informed consent from patients undergoing CABG (Coronary Artery Bypass Grafting). All procedures received local ethical approval (Research Ethical Committee number: 06/S0703/110) and the experimental procedures conformed to the principles outlined in the Declaration of Helsinki.

2.2. Vessel Storage and RNA Isolation from the Pig Vein Graft Model and Stent Model

All the procedures for RNA extraction from the pig vein graft model and stent model are described in detail in [17,18], respectively. Briefly, the harvested vessels were placed in RNAlater®-ice (Invitrogen, Paisley, UK) and stored at −80 °C until the day of isolation. RNAs from the veins or arteries were isolated following disruption of the vessels under liquid nitrogen using a pestle and mortar, and these vessel fragments were placed in Qiazol (Qiagen, Hilden, Germany) and homogenized using a tissue homogenizer (Polytron, Switzerland). The RNAs were processed through miRNEasy Mini Kit (Qiagen) following the manufacturer's instructions, treated with DNAse 1 (amplification grade; Sigma, St. Louis, MO, USA) in order to eliminate genomic DNA contamination and quantified using a NanoDrop ND-1000 Spectrophotometer (Nano-Drop Technologies, Wilmington, DE, USA). RNA integrity was assessed using the RNA 6000 Nano LabChip kit (Agilent Technologies, Santa Clara, CA, USA). Only The RNAs with an RNA integrity number value >7 were used for the TLDA experiments.

2.3. Cell Culture

VSMCs were isolated from human saphenous vein segments within 24 h of surgery using the explant technique as previously described by Southgate et al. [20]. Briefly, following removal of the adventitial layer using sterile forceps and scissors, the vein was longitudinally opened. The lumenal surface of the vein was scraped gently with a rubber policeman to remove the endothelium. Using fine forceps, thin segments of the medial layer were stripped from the vein. The stripped medial segments were then cut into approximately 1 mm^2 pieces using a tissue chopper and then transferred into a sterile tube containing wash media (DMEM medium supplemented with 2 mmol/L of L-glutamine (Invitrogen), 50 µg/mL of penicillin (Life Technologies, Paisley, UK) and 50 µg/L of streptomycin (Life Technologies)). Following two washes in the wash media, the segments were resuspended in a small volume of SMC complete medium (SMC growth medium 2 (PromoCell, Heidelberg, Germany) supplemented with 15% foetal calf serum (FCS) (PAA laboratories, Yeovil, UK), 2 mmol/L of L-glutamine (Invitrogen), 50 µg/mL of penicillin (Life Technologies) and 50 µg/L of streptomycin (Life Technologies)) and were then spread onto the base of 25 cm^2 tissue culture flasks. Following overnight culture at 37 °C in a humidified atmosphere containing 5% CO$_2$ (*v/v*), 5 mL of SMC complete medium was carefully added to each flask. The media were changed every 3–5 days and after 2–3 weeks, the outgrowth of VSMCs was passaged into 75 cm^2 flasks and grown to confluence. All experiments were performed using VSMCs between passages 3 and 6. The VSMCs were plated in SMC complete medium, quiesced in medium (DMEM medium supplemented with 2 mmol/L of L-glutamine (Invitrogen), 50 µg/mL of penicillin (Life Technologies) and 50 µg/L of streptomycin (Life Technologies)) containing 0.2% FCS for 48 h. The plated cells were either placed in fresh 0.2% FCS medium (control) or stimulated with 0.2% medium supplemented with PDGF (PDGF-BB, 20 ng/mL) or with IL-1α (10 ng/mL) or a combination of PDGF and IL-1α for 48 h before RNA isolation.

2.4. RNA Extraction and Purification

For the extraction of RNAs from VSMCs, the lysis of the plated cells was performed with 700 μL of QIAzol lysis reagent (Qiagen). The purification of total RNA was achieved using miRNEasy Mini Kit (Qiagen) according to the manufacturer's instructions. The RNA integrity and concentration were determined using the NanoDrop ND-1000 spectrophotometer.

2.5. TaqMan Low Density Array (TLDA) Experiment

miRNA profiling was performed with the Human MicroRNA Array v2.0 Card A and B which is enabled to quantify 667 human miRNAs catalogued in the miRBase v10. For three independent patients (Pt 1, Pt 2 and Pt 3), 90 ng of total RNA was reverse transcribed using the Megaplex RT set pool A and B and then preamplified following the manufacturer's instructions (Applied Biosystems, Carlsbad, CA, USA). The cDNA that was produced was loaded on the Human MicroRNA Array cards and run on the 7900HT Fast Real Time PCR system (Applied Biosystems) using SDS2.3 software. Data analysis was performed using DataAssist v2.0 Software (Applied Biosystems). The miRNA expression data were normalized using the RNU48 housekeeper expression because of their low Ct variance.

The TLDA experiments for the pig vein graft model and the pig stent model are described in detail in ref [17] and [18], respectively.

2.6. Global Analysis

The literature and databases such as the miRBase Sequence Database (Release 21) permit the determination of the guide and the passenger strands for a thousand stem loops from the data obtained by the TLDA experiments. The level of expression (relative expression, %) was classified into three groups: high (Ct < 23), medium (Ct between 23 and 28) and low (Ct > 28) expressions. The dysregulated miRNAs are represented by a relative quotient (RQ) higher than two or less than 0.5 in expression level.

2.7. Gene Expression Quantitative RealTime-PCR (qRT-PCR)

The cDNA for gene expression analysis was synthesized from 400 ng of RNA using TaqMan Reverse Transcription Reagents (Applied Biosystems). The reverse transcription (RT) reactions were carried out in a volume of 20 μL following the manufacturer's instructions and run in a thermocycler under the conditions: 25 °C for 10 min, 48 °C for 30 min and 95 °C for 5 min, then held at 4 °C.

The TaqMan real-time PCR reactions were performed in triplicate comprising 5 μL of TaqMan Universal Master Mix II, no UNG (Applied Biosystems), 0.5 μL of 20× gene-specific primer and a probe mix (TaqMan gene expression assays, Cat number # 4331182 (CDKN1B, Assay ID: Hs01597588_m1; UBC, Assay ID: Hs00824723_m1)), 1.2 μL of the reverse transcription product and 3.3 μL of nuclease free water. RT-PCR was carried out by using a 7900HT Fast Real Time PCR system (Applied Biosystems) under the conditions: 50 °C for 2 min, 95 °C for 10 min, and then 40 cycles of 95 °C for 15 s, 60 °C for 60 s, then held at 4 °C. The gene expression level was normalized to UBC (Ubiquitin C) and the fold-changes were calculated using the ΔΔCT method.

2.8. miRNA Expression Quantitative RealTime-PCR (qRT-PCR)

All the miRNA specific probes (TaqMan microRNA assays) are commercial and available from Life Technologies: Cat number # 4427975 (hsa-miR-204, Assay ID: 000508; hsa-miR-218, Assay ID: 000521; hsa-miR-1275, Assay ID: 002840; hsa-miR-625*, Assay ID: 002432; hsa-miR-222*, Assay ID: 002097).

The RT reactions were performed using the TaqMan miRNA Reverse Transcription Kit (Applied Biosystems) and miRNA-specific RT stem-loop primers (Applied Biosystems). The RT reactions were carried out in a volume of 7.5 μL following the manufacturer's instructions and run in a thermocycler under the conditions: 16 °C for 30 min, 42 °C for 45 min and 85 °C for 5 min, then held at 4 °C.

The TaqMan real-time PCR reactions were performed in triplicate comprising 5 μL TaqMan Universal Master Mix II, no UNG (Applied Biosystems), 0.5 μL of 20 × miRNA-specific primer and a probe mix of the TaqMan miRNA Assay Kit (Applied Biosystems), 0.7 μL of the RT product and 3.8 μL of nuclease free water. RT-PCR was carried out by using a 7900HT Fast Real Time PCR system (Applied Biosystems) under the conditions: 95 °C for 10 min, then 40 cycles of 95 °C for 15 s, 60 °C for 60 s, then held at 4 °C. The expression level of miRNA was normalized to RNU48 and the fold-changes were calculated using the ΔΔCT method. Only the cells from patients harboring a low variation concerning the baseline expression of miRNAs have been used for the miRNA expression analysis (Figure S1).

2.9. Statistical Analysis

The statistical analysis was performed according to figure legends. The statement N = 3 means that three experiments were performed, three independent times, all in triplicate. Data are expressed as mean ± SEM. Comparisons between the two groups were analyzed using the 2-tailed Student's *t*-test. The one-way ANOVA with Tukey's post hoc multiple comparison test, via Graph Pad Prism version 5.0 (GraphPad Software, San Diego, CA, USA), was used for comparisons among three or more groups. A statistical difference was considered as $p < 0.05$. Heat maps were generated using the R program (version 3.2.4) (R Foundation for Statistical Computing, Vienna, Austria).

3. Results

3.1. Global Analysis of miRNA in VSMCs in Response to Cytokine and Growth Factor Stimulation and in Vascular Injury In Vivo

Human VSMCs were treated with PDGF, interleukin 1α (IL-1α) or a combination of both to mimic a number of the pathologic signals induced by vascular injury. As expected, PDGF induced a significant increase in VSMC migration (Figure S2A), and the pro-proliferative phenotype was confirmed by a reduction in the expression of the cyclin-dependent kinase inhibitor 1B (CDKN1B) (Figure S2B). Following confirmation of the phenotypic alternation in VSMCs, we performed a detailed and unbiased analysis of both the guide and passenger miRNAs expression in VSMCs (Table S1). Indeed, using TaqMan Low Density Arrays (TLDA) we were able to quantify a large number of miRNAs simultaneously in each individual sample. The relative expression in each condition was compared to the unstimulated quiested VSMCs and expressed in fold change. Globally, our data highlighted a low variation between the samples. Bioinformatics tools including the miRBase (Realease 21) (http://www.mirbase.org/) have been used to discriminate "guide" versus "passenger" strands. We validated five miRNAs from our array by subsequent assays (Figure S3) and found miRNA markers in our global analysis such as miR-146a, an inflammation-associated miRNA [21], which showed overexpression following exposure to IL-1α. From our array, the level of expression was classified into three groups following pre-amplification according to cycle threshold (Ct) values (low [Ct > 28], medium [Ct between 23 and 28] and high [Ct < 23] expression) for each miRNA. The global analysis revealed that the miRNA guide strands are more abundantly expressed than the passenger strands (Figure 1A) in all the conditions tested (Figure S4A). A similar pattern was identified following the computational analysis of the array data from the porcine models of vein graft failure [17] and in stent restenosis [18] (Figure S4B and Figure S4C, respectively, Figure 1B and Figure S5). A total of one hundred hairpins were analyzed from the data generated by TLDA. Consistently across the three models, we found that the hairpin of miR-10b, miR-193b, miR-199a, miR-214, miR-30a, miR-335 and miR-99b contained medium or highly expressed passenger strands. Furthermore, a direct comparison of the relative expression of miRNA strands obtained from their corresponding hairpins confirmed that the guide strand was more abundantly expressed than its corresponding passenger strand in more than 60% of cases, a pattern that was consistent across all three data sets (Figure 1C). In human VSMCs, the guide strand appeared to be the most predominantly expressed variant of the hairpin, although there are exceptions where the passenger strands were more abundantly expressed than their corresponding guide strand such as miR-625 and miR-30a (heat map, Figure 1A). The same type of exception has been found across the three models for miR-199a, miR-30a, and miR-335.

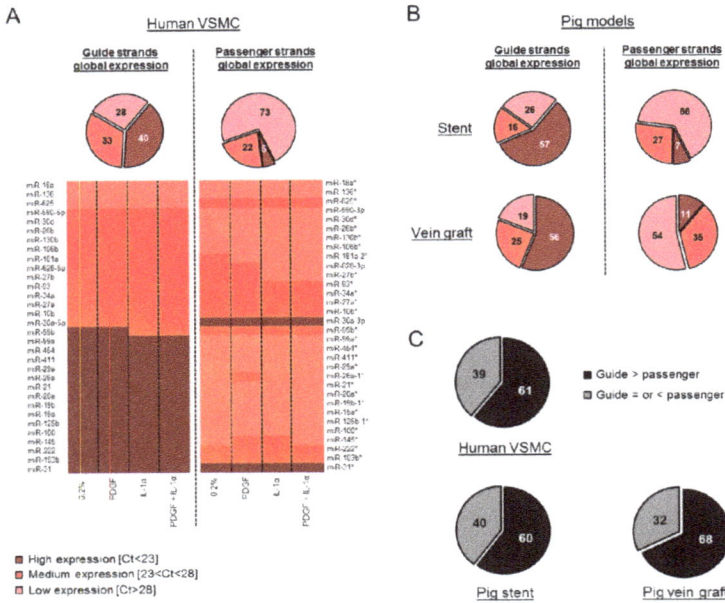

Figure 1. Low passenger strands expression compared to the guide strands in VSMCs and in vivo. One hundred hairpins were analyzed by TaqMan Low Density Array (TLDA) in human VSMCs (average of *n* = 3), the porcine models of vein graft failure and in stent restenosis (average of *n* = 6 for both the porcine models, except for the 7 days DES (drug eluting stent) condition, where *n* = 5). The level of expression (relative expression, %) was classified into three groups: high (Ct < 23), medium (Ct between 23 and 28) and low (Ct > 28) expressions. These groups are represented as a % in each pie chart. (**A**) Each pie chart represents the global expression of the guide and passenger strands in human VSMCs. The heat map illustrates the relative abundance of the number of miRNA hairpins that were consistently expressed across three independent patients and represents the level of expression of the guide versus passenger strands. (**B**) Each pie chart represents the global expression of the guide and passenger strands in the porcine model of in stent restenosis and vein graft failure. (**C**) Each pie chart represents the expression of the guide versus passenger strands within each independent hairpin.

3.2. Passenger Strands Are Dysregulated More Frequently than Guide Strands in Vascular Cells and Vascular Injury In Vivo

Despite their generally lower levels of expression compared to the guide strands, we then quantified the dysregulation of the passenger strands in vascular cells in vitro and in injury models in vivo. Interestingly, in all three settings, we found a substantial and predominant dysregulation of the passenger strands in response to pathological stimuli (Figure 2). The relative number of dysregulated miRNAs included 39% and 32% of the passenger strands and only 15% and 10% of the guide strands in human VSMCs treated with PDGF and IL-1α, respectively (Figure 2A). Furthermore, all together, 20% of the dysregulation concerned the two strands in a hairpin. In vivo, 36% and 43% of the passenger strands were dysregulated compared to 25% and 29% for the guide strands at 7 days following stenting by BMS (bare-metal stent) and DES (drug eluting stent), respectively (Figure 2B). This dysregulation was still at 32% and 39% for the passenger strands compared to 19% and 23% for the guide strands at

28 days following stenting by BMS and DES, respectively (Figure S6A). In our vein graft failure model, we found 59% of the passenger strands dysregulated compared to 40% the guide strands at 7 days following engraftment (Figure 2B). This dysregulation reached 68% for the passenger strands and 33% for the guide strands at 28 days (Figure S6B). Additionally, all together, we found 47% and 31% of the dysregulation concerned the two strands in a hairpin in the pig vein graft model and in the pig stent model, respectively.

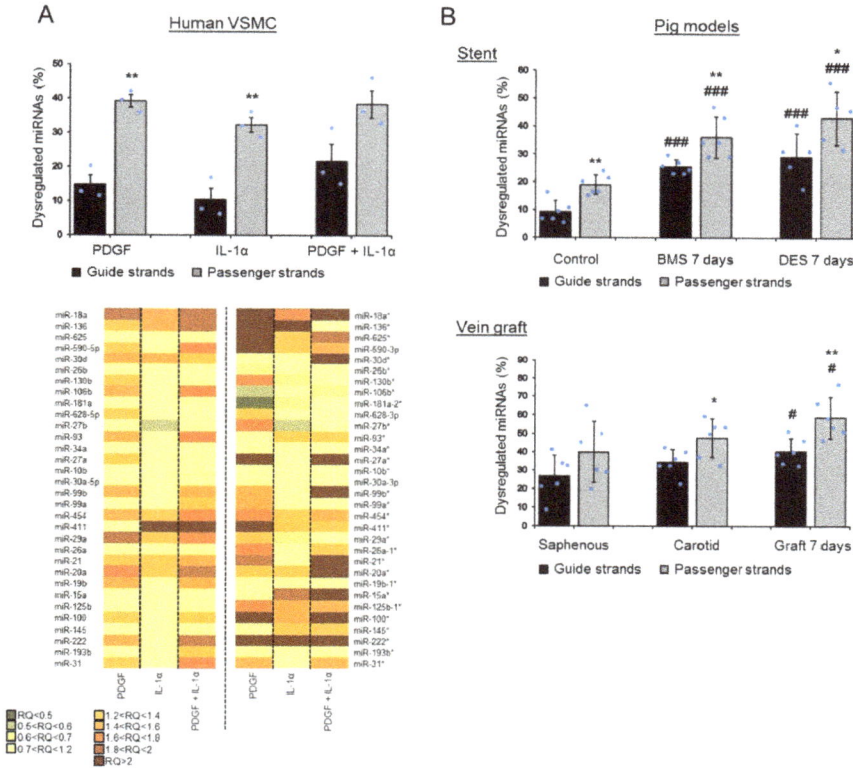

Figure 2. High passenger strand dysregulation compared to the guide strand dysregulation in VSMCs and in vivo. The graphs show the percentage of dysregulated miRNAs for each form in the three independent models. (**A**) The relative dysregulation of each strand of the miRNA hairpin was quantified after stimulation by platelet-derived growth factor (PDGF) and IL-1α in human VSMCs (the statistics have been made between the passenger and guide strands within each condition; unpaired t-test; ** $p < 0.01$ for passenger vs guide strands). The heat map illustrates the relative dysregulation (in relative quotient (RQ)) of a number of miRNA hairpins after stimulation by PDGF and IL-1α across three independent patient samples (average of $n = 3$). (**B**) The dysregulation of the guide and passenger strands was quantified at 7 days following stenting (BMS (bare-metal stent) or DES (drug eluting stent)) or at 7 days following vein grafting in vivo ($n = 6$ for both the porcine models, except for the 7 days in the DES condition where $n = 5$) (the statistics have been made between the passenger and guide strands within each condition; unpaired t-test; * $p < 0.05$ and ** $p < 0.01$; and vs. the corresponding control (unstented control animals or saphenous), # $p < 0.05$ and ### $p < 0.001$).

4. Discussion

VSMCs are typically quiescent and contractile under normal physiological conditions. However, following vascular injury, the release of cytokines and growth factors initiates a transcriptional cascade resulting in the initiation of a process called "phenotypic switching" and this leads the VSMCs to harbor a synthetic, pro-proliferative and pro-migratory state. Since miRNAs regulate mRNA expression and thereby many fundamental biological processes, they are intensely studied as candidates for diagnostic and prognostic biomarkers in a number of diseases. Indeed, miRNAs have been shown to play important roles in many cellular processes such as cell differentiation, proliferation and migration. We decided to conduct a detailed analysis in order to understand the microRNAome through TLDA in the pre-clinical porcine models of vascular injury and human VSMCs. It is important to note that we utilized human arrays in the porcine samples to try and identify a miRNA which would extrapolate to the pathology in the clinical setting, so it is possible that a proportion of miRNAs may be under-represented due to the sequence variation or chromosomal locations between pig and human. Our global analysis shows the dysregulation of a number of miRNAs already described in the literature as phenotypic regulators of VSMCs. For example, we found that miR-222, previously reported to be involved in vascular pathologies associated with excessive rates of SMC proliferation and migration [22], was substantially up-regulated in response to pathological stimuli. Indeed, in our arrays, miR-222 was overexpressed upon PDGF stimulation in VSMCs (Figure 2 and Figure S3) as well as following DES implantation and vein grafting in the porcine models (Figure S6). Furthermore, we found that miR-18a was upregulated following vein grafting and stenting in pig models (Figure S6). This finding is in accordance with the literature, in which miR-18a has been found to be upregulated after rat carotid balloon injury and to promote VSMC differentiation [23]. Furthermore, miR-145 was downregulated following vein grafting and during the early stage after stenting in our porcine models (Figure S6). MiR-145 is extensively studied in vascular biology for its ability to control vascular neointimal lesion formation. Indeed, the restoration of miR-145 in balloon-injured arteries inhibits neointimal growth in rat [24]. Moreover, our global analysis gave a global landscape of the miRNA guide strand as well as passenger strand expression. Indeed, recent studies suggest that these passenger strands can mediate strand-specific roles based on their distinct seed sequences and targets, highlighting their importance [13,14]. We decided to study the behavior of miRNA depending on the strand form after exposure to pathological stimuli via our detailed analysis of the miRNA transcriptome through TLDA. Our results confirmed a higher global expression of the guide strands compared to passenger strands. Few exceptions with a higher expression of the passenger strands of the miRNA:miRNA * duplex have been noted, and this suggests the possibility that they can mediate functional effects on mRNA targets. However, our study is the first to show that a greater proportion of passenger strand miRNAs was significantly altered after exposure to the pathological mediators of vascular remodeling, PDGF and/or IL-1α in vitro and post-stenting or grafting in vivo. Some studies reported a preferred arm switched between samples and tissues [25,26]. In our study, after vascular injury, the selection of the mature miRNA strand via an "arm switching" mechanism could be responsible for the greater proportion of the dysregulated passenger strands that was obtained. It is widely accepted that vascular injury is responsible for the "phenotypic switching" induction in VSMCs but our study suggests that vascular injury could also be implicated in the aforementioned miRNA "arm switching" phenomenon.

In conclusion, we demonstrated via a global analysis that the miRNA passenger strands are frequently dysregulated in acute vascular injury. Our detailed and unbiased analysis emphasized an interesting landscape of miRNAs in vascular biology where the passenger strands contribute more broadly than previously thought and may be attractive targets for disease intervention.

Supplementary Materials: The following are available online, Figure S1: Low variation of the baseline miRNAs expression, Figure S2: Validation of VSMC stimulation, Figure S3: Validation of TLDA arrays expression profiles by qRT-PCR, Figure S4: Higher expression of the guide strands compared to the passenger strands in VSMCs and in vivo, Figure S5: Low passenger strands expression compared to the guide strands in stent and vein graft porcine models, Figure S6: High passenger strands dysregulation compared to the guide strands 28 days following stenting or grafting in porcine models, Table S1: miRNAs expression analyzed by TLDA in human VSMCs.

Author Contributions: Conceptualization, K.P., R.A.M. and A.H.B.; methodology, K.P., L.A.D., K.W. and R.A.M.; investigation and formal analysis, K.P.; writing, K.P. and A.H.B.; funding acquisition, A.H.B.

Funding: This work was supported by British Heart Foundation Programme Grants [RG/09/005/27915, RG/14/3/30706]. Andrew H. Baker is supported by the British Heart Foundation Chair of Translational Cardiovascular Sciences.

Acknowledgments: We thank Gregor Aitchison, Elaine Friel and Nicola Britton for technical assistance.

Conflicts of Interest: The authors declare no conflict of interest.

References

1. Wu, L.; Belasco, J.G. Let me count the ways: Mechanisms of gene regulation by miRNAs and siRNAs. *Mol. Cell* **2008**, *29*, 1–7. [CrossRef] [PubMed]
2. Eulalio, A.; Huntzinger, E.; Izaurralde, E. Getting to the root of miRNA-mediated gene silencing. *Cell* **2008**, *132*, 9–14. [CrossRef]
3. Filipowicz, W.; Bhattacharyya, S.N.; Sonenberg, N. Mechanisms of post-transcriptional regulation by microRNAs: Are the answers in sight? *Nat. Rev. Genet.* **2008**, *9*, 102–114. [CrossRef] [PubMed]
4. Bartel, D.P. MicroRNAs: Genomics, biogenesis, mechanism, and function. *Cell* **2004**, *116*, 281–297. [CrossRef]
5. Guo, L.; Zhang, H.; Zhao, Y.; Yang, S.; Chen, F. Selected isomiR expression profiles via arm switching? *Gene* **2014**, *533*, 149–155. [CrossRef] [PubMed]
6. Winter, J.; Diederichs, S. Argonaute-3 activates the let-7a passenger strand microRNA. *RNA Biol.* **2013**, *10*, 1631–1643. [CrossRef]
7. Owens, G.K.; Kumar, M.S.; Wamhoff, B.R. Molecular regulation of vascular smooth muscle cell differentiation in development and disease. *Physiol. Rev.* **2004**, *84*, 767–801. [CrossRef]
8. Dzau, V.J.; Braun-Dullaeus, R.C.; Sedding, D.G. Vascular proliferation and atherosclerosis: New perspectives and therapeutic strategies. *Nat. Med.* **2002**, *8*, 1249–1256. [CrossRef]
9. Ferns, G.A.; Raines, E.W.; Sprugel, K.H.; Motani, A.S.; Reidy, M.A.; Ross, R. Inhibition of neointimal smooth muscle accumulation after angioplasty by an antibody to PDGF. *Science* **1991**, *253*, 1129–1132. [CrossRef]
10. Yang, J.-S.; Phillips, M.D.; Betel, D.; Mu, P.; Ventura, A.; Siepel, A.C.; Chen, K.C.; Lai, E.C. Widespread regulatory activity of vertebrate microRNA* species. *RNA Soc.* **2011**, *17*, 312–326. [CrossRef]
11. Okamura, K.; Phillips, M.D.; Tyler, D.M.; Duan, H.; Chou, Y.; Lai, E.C. The regulatory activity of microRNA* species has substantial influence on microRNA and 3′ UTR evolution. *Nat. Struct. Mol. Biol.* **2008**, *15*, 354–363. [CrossRef]
12. Griffiths-Jones, S.; Hui, J.H.L.; Marco, A.; Ronshaugen, M. microRNA evolution by arm switching. *EMBO Rep.* **2011**, *12*, 172–177. [CrossRef]
13. Poissonnier, L.; Villain, G.; Soncin, F.; Mattot, V. miR126-5p repression of ALCAM and SetD5 in endothelial cells regulates leucocyte adhesion and transmigration. *Cardiovasc. Res.* **2014**, *102*, 436–447. [CrossRef] [PubMed]
14. Zhu, S.; Deng, S.; Ma, Q.; Zhang, T.; Jia, C.; Zhuo, D.; Yang, F.; Wei, J.; Wang, L.; Dykxhoorn, D.M.; et al. microRNA-10A* and microRNA-21 modulate endothelial progenitor cell senescence via suppressing high-mobility group A2. *Circ. Res.* **2013**, *112*, 152–164. [CrossRef] [PubMed]
15. Chang, K.-W.; Kao, S.-Y.; Wu, Y.-H.; Tsai, M.-M.; Tu, H.-F.; Liu, C.-J.; Lui, M.-T.; Lin, S.-C. Passenger strand miRNA miR-31* regulates the phenotypes of oral cancer cells by targeting RhoA. *Oral Oncol.* **2013**, *49*, 27–33. [CrossRef] [PubMed]
16. Yang, X.; Du, W.W.; Li, H.; Liu, F.; Khorshidi, A.; Rutnam, Z.J.; Yang, B.B. Both mature miR-17-5p and passenger strand miR-17-3p target TIMP3 and induce prostate tumor growth and invasion. *Nucleic Acids Res.* **2013**, *41*, 9688–9704. [CrossRef] [PubMed]
17. McDonald, R.A.; White, K.M.; Wu, J.; Cooley, B.C.; Robertson, K.E.; Halliday, C.A.; McClure, J.D.; Francis, S.; Lu, R.; Kennedy, S.; et al. miRNA-21 is dysregulated in response to vein grafting in multiple models and genetic ablation in mice attenuates neointima formation. *Eur. Heart J.* **2013**, *34*, 1636–1643. [CrossRef]

18. McDonald, R.A.; Halliday, C.A.; Miller, A.M.; Diver, L.A.; Dakin, R.S.; Montgomery, J.; McBride, M.W.; Kennedy, S.; McClure, J.D.; Robertson, K.E.; et al. Reducing In-Stent Restenosis: Therapeutic Manipulation of miRNA in Vascular Remodeling and Inflammation. *J. Am. Coll. Cardiol.* **2015**, *65*, 2314–2327. [CrossRef]

19. Angelini, G.D.; Bryan, A.J.; Williams, H.M.; Morgan, R.; Newby, A.C. Distention promotes platelet and leukocyte adhesion and reduces short-term patency in pig arteriovenous bypass grafts. *J. Thorac. Cardiovasc. Surg.* **1990**, *99*, 433–439. [PubMed]

20. Southgate, K.M.; Davies, M.; Booth, R.F.; Newby, A.C. Involvement of extracellular-matrix-degrading metalloproteinases in rabbit aortic smooth-muscle cell proliferation. *Biochem. J.* **1992**, *288*, 93–99. [CrossRef]

21. Raitoharju, E.; Lyytikäinen, L.-P.; Levula, M.; Oksala, N.; Mennander, A.; Tarkka, M.; Klopp, N.; Illig, T.; Kähönen, M.; Karhunen, P.J.; et al. miR-21, miR-210, miR-34a, and miR-146a/b are up-regulated in human atherosclerotic plaques in the Tampere Vascular Study. *Atherosclerosis* **2011**, *219*, 211–217. [CrossRef] [PubMed]

22. Liu, X.; Cheng, Y.; Yang, J.; Xu, L.; Zhang, C. Cell-specific effects of miR-221/222 in vessels: Molecular mechanism and therapeutic application. *J. Mol. Cell. Cardiol.* **2012**, *52*, 245–255. [CrossRef]

23. Kee, H.J.; Kim, G.R.; Cho, S.-N.; Kwon, J.-S.; Ahn, Y.; Kook, H.; Jeong, M.H. miR-18a-5p MicroRNA Increases Vascular Smooth Muscle Cell Differentiation by Downregulating Syndecan4. *Korean Circ. J.* **2014**, *44*, 255–263. [CrossRef]

24. Cheng, Y.; Liu, X.; Yang, J.; Lin, Y.; Xu, D.-Z.; Lu, Q.; Deitch, E.A.; Huo, Y.; Delphin, E.S.; Zhang, C. MicroRNA-145, a novel smooth muscle cell phenotypic marker and modulator, controls vascular neointimal lesion formation. *Circ. Res.* **2009**, *105*, 158–166. [CrossRef]

25. Chiang, H.R.; Schoenfeld, L.W.; Ruby, J.G.; Auyeung, V.C.; Spies, N.; Baek, D.; Johnston, W.K.; Russ, C.; Luo, S.; Babiarz, J.E.; et al. Mammalian microRNAs: Experimental evaluation of novel and previously annotated genes. *Genes Dev.* **2010**, *24*, 992–1009. [CrossRef] [PubMed]

26. Ro, S.; Park, C.; Young, D.; Sanders, K.M.; Yan, W. Tissue-dependent paired expression of miRNAs. *Nucleic Acids Res.* **2007**, *35*, 5944–5953. [CrossRef] [PubMed]

MDPI

St. Alban-Anlage 66

4052 Basel

Switzerland

Tel. +41 61 683 77 34

Fax +41 61 302 89 18

www.mdpi.com

Cells Editorial Office

E-mail: cells@mdpi.com

www.mdpi.com/journal/cells

www.ingramcontent.com/pod-product-compliance
Lightning Source LLC
Chambersburg PA
CBHW051712210326
41597CB00032B/5448